Fourier Series and Boundary Value Problems with Engineering Applications

This book is a comprehensive and time-tested guide to the mathematical theory of Fourier series and boundary value problems, with a strong emphasis on engineering applications. Over the past two decades, *Fourier Series and Boundary Value Problems with Engineering Applications* has been rigorously refined and tested in classroom settings, ensuring its effectiveness as a teaching and learning resource.

The journey begins with a thorough development of Fourier series, a cornerstone of modern mathematics and engineering. The Fourier series provides a powerful framework for analyzing periodic functions and decomposing complex signals into simpler sinusoidal components. This foundational knowledge is then extended to boundary value problems, which arise naturally in the study of physical phenomena such as heat flow, vibrations, and wave propagation.

A distinctive feature of this book is its focus on applications in both rectangular and spherical coordinates. These coordinate systems are essential for modeling problems in diverse engineering contexts. Additionally, the book addresses partial differential equations on unbounded domains and ordinary differential equations.

Whether you are a student encountering Fourier series and boundary value problems for the first time, an educator seeking a reliable and classroom-tested resource, or a professional looking to refresh your knowledge, this book offers a clear exposition, practical focus, and extensive problem sets, making it an indispensable companion for mastering the mathematical tools that underpin modern engineering.

Youssef N. Raffoul is a professor and Graduate Program Director at the University of Dayton. He holds a PhD from Southern Illinois University. He is the recipient of the University of Dayton College of Arts and Sciences' Award for Outstanding Scholarship (twice) and the University of Dayton Alumni Award in Scholarship. Prof. Raffoul has published over one hundred and sixty articles in prestigious journals in the areas of functional differential equations, difference equations, and dynamical systems on time scales. He was honored by the Lebanese government with the Career in Science Award. The Archbishop of Lebanon awarded him the Lifetime Achievement Award. Most notably, he is the recipient of the Order of Merit, Silver Medal with Distinction awarded by the president of Lebanon, General Aoun.

Textbooks in Mathematics
Series editors:
Al Boggess, Kenneth H. Rosen

Algebra, Second Edition
Groups, Rings, and Fields
Louis Halle Rowen and Uzi Vishne

Functional Analysis for the Applied Mathematician
Todd Arbogast and Jerry L. Bona

Exploring Linear Algebra, Second Edition
Labs and Projects with Mathematica®
Crista Arangala

Measure Theory and Fine Properties of Functions, Second Edition
Lawrence Craig Evans and Ronald F. Gariepy

Set Theory
An Introduction to Axiomatic Reasoning
Robert André

Introduction to Differential and Difference Equations Through Modeling
William P. Fox and Robert E. Burks

Abstract Algebra, Third Edition
An Interactive Approach
William Paulsen

Elements of Algebraic Topology, Second Edition
James R. Munkres, Steven G. Krantz, and Harold R. Parks

One Complex Variable from the Several Variable Point of View
Peter V. Dovbush and Steven G. Krantz

Math Anxiety How to Beat It
Brian Cafarella

Lectures on Differential Geometry with Maple
Mayer Humi

Numerical Analysis for Engineers
Methods and Applications
Bilal M. Ayyub and Richard H. McCuen

An Invitation to Real Analysis
Andrew D. Hwang

Fourier Series and Boundary Value Problems with Engineering Applications
Youssef N. Raffoul

https://www.routledge.com/Textbooks-in-Mathematics/book-series/CANDHTEX-BOOMTH

Fourier Series and Boundary Value Problems with Engineering Applications

Youssef N. Raffoul

CRC Press
Taylor & Francis Group
Boca Raton London New York

CRC Press is an imprint of the
Taylor & Francis Group, an **informa** business
A CHAPMAN & HALL BOOK

Designed cover image: Dr. Joseph Raffoul

First edition published 2026
by CRC Press
2385 NW Executive Center Drive, Suite 320, Boca Raton FL 33431

and by CRC Press
4 Park Square, Milton Park, Abingdon, Oxon, OX14 4RN

CRC Press is an imprint of Taylor & Francis Group, LLC

ISBN: 978-1-041-09925-3 (hbk)
ISBN: 978-1-041-09923-9 (pbk)
ISBN: 978-1-003-65254-0 (ebk)

DOI: 10.1201/9781003652540

Typeset in Nimbus font
by KnowledgeWorks Global Ltd.

To my grandchildren
Aurora, Everett, and Quinn

Contents

Preface xi

Author xiii

1 Fourier Series 1
1.1 Fourier Series 1
1.1.1 Exercises 11
1.2 Sine and Cosine Series 13
1.2.1 Exercises 18
1.3 Fourier Series on Arbitrary Periods 19
1.3.1 Exercises 27
1.4 Complex Form of Fourier Series 29
1.4.1 Exercises 33

2 Theory of Fourier Series 35
2.1 Bessel's Inequality and its Consequences 35
2.1.1 Exercises 42
2.2 Pointwise Convergence of Fourier Series 45
2.2.1 Exercises 53
2.3 Gibb's Phenomenon 55
2.3.1 Exercises 59

3 Uniform Convergence and Operations on Fourier Series 61
3.1 Preliminaries 61
3.1.1 Exercises 65
3.2 Uniform and Absolute Convergence of Fourier Series 65
3.3 More on the Dirichlet Kernel 69
3.3.1 Exercises 71
3.4 Integrations and Differentiations of Fourier Series 72
3.4.1 Exercises 79
3.5 Generalized Fourier Series and Orthonormal Functions 81
3.5.1 Exercises 86
3.6 Best Approximation in the Mean 87
3.6.1 Exercises 91

4 Heat Equation and Boundary Value Problems **93**
4.1 Introduction to PDEs 93
4.2 Derivation of the Heat Equation 97
4.2.1 Exercises 99
4.3 Separable Equations 100
4.3.1 Exercises 101
4.4 BVP-Heat Equation 101
4.5 Insulated Homogeneous Heat Problem with Homogeneous Boundaries 102
4.6 Insulated Homogeneous Heat Problem with Non-Homogeneous Boundaries 106
4.6.1 Exercises 108
4.7 Non-Insulated Homogeneous Heat Problem with Homogeneous Boundaries 110
4.8 Non-Insulated Homogeneous Heat Problem with Non-Homogeneous Boundaries 114
4.8.1 Exercises 117
4.9 Heat Source 119
4.10 Heat Flow with Variable Sources and Time-Dependent Boundaries . 123
4.11 Method of Eigenfunction Expansion Using Green's Formula 127
4.11.1 Exercises 131
4.12 Laplace Equation on Rectangles and Unbounded Domains 133
4.12.1 Exercises 137
4.13 Laplacian in Polar, Cylindrical and Spherical Coordinates 139
4.13.1 Exercises 142
4.14 The Laplacian on an Annulus and a Disk: Analysis and Applications 142
4.14.1 Exercises 152
4.15 Poisson's Equation with a Source 154
4.15.1 Exercises 155

5 The Wave Equation and Boundary Value Problems **158**
5.1 Derivation of the Wave Equation 158
5.2 Vibrating String with Fixed Ends 160
5.2.1 Air Pressure in Thin Pipe 165
5.2.2 Exercises 165
5.3 Forced Vibrating String at Resonance 167
5.3.1 Exercises 173
5.4 Vibrating String and Double Fourier Series 174
5.4.1 Exercises 179
5.5 Vibration of a Beam 180
5.5.1 Exercises 184

6 Sturm–Liouville Systems and Applications **186**
6.1 Sturm–Liouville Problem 186
6.2 Expansion in Series of Eigenfunctions 192

6.2.1 Exercises 194
6.3 Rayleigh Quotient 196
6.4 Reformulation of the Zero Eigenvalue and Minimum Eigenvalue Concepts 198
6.4.1 Exercises 199
6.5 Applications to Nonuniform String 201
6.5.1 Exercises 204
6.6 Robin's Boundary Conditions in Heat Flow 205
6.6.1 Exercises 210
6.7 Robin's Boundary Conditions in Laplacian 213
6.7.1 Exercises 215
6.8 Bessel's Function 216
6.8.1 Exercises 227
6.9 Orthogonality of Bessel's Functions 227
6.10 Vibrations of Circular Membrane; Bessel's Equation 232
6.10.1 Exercises 237
6.11 Legendre Polynomials 240
6.11.1 Exercises 246
6.12 The Generating Function and Eigenfunctions Expansion 247
6.12.1 Exercises 254
6.13 Applications of Legendre's Equation 255
6.13.1 Exercises 264
6.14 Non-Homogeneous Sturm–Liouville Problem 265
6.14.1 Exercises 269

7 Fourier Integrals and Applications **271**
7.1 Transition from Fourier Series to Fourier Integral 272
7.1.1 Exercises 276
7.2 Fourier Integral Theorem 279
7.2.1 Exercises 284
7.3 Complex Form of Fourier Integral Representation 285
7.3.1 Exercises 287
7.4 Heat Flow in an Infinite Rod 288
7.4.1 Exercises 293
7.5 Wave Motion for Infinite String 297
7.5.1 Exercises 299
7.6 Fourier Transforms 300
7.6.1 Exercises 306
7.7 Properties of Fourier Transform 308
7.8 Parseval's Formula 319
7.9 Normalized Auto-Correlation Function 321
7.9.1 Exercises 323
7.10 Application of Fourier Transform 325
7.10.1 Heat Equation 325
7.10.2 Schrödinger Equation 328

7.10.3 Laplace Equation 329
7.10.4 Vibrating String 332
7.10.5 Exercises 333

8 Classic PDEs **337**
8.1 Linear Equations 337
8.1.1 Linear Equations with Constant Coefficients 337
8.1.2 Exercises 341
8.2 Equations with Variable Coefficients 342
8.2.1 Exercises 345
8.3 Quasi-Linear Equations 345
8.3.1 Exercises 354
8.4 Burger's Equation 356
8.4.1 Shock Path 362
8.4.2 Exercises 364
8.5 Second-Order PDEs 366
8.5.1 Exercises 371
8.6 Inequalities and Energy Estimates 373
8.6.1 Exercises 377

9 Heat Equation on Unbounded Domains **379**
9.1 Uniqueness of Solution 379
9.2 Solution of the Heat Equation 382
9.3 Heat Equation on Semi-Infinite Domain—Dirichlet Condition . . . 388
9.4 Heat Equation on Semi-Infinite Domain—Neumann Condition . . . 392
9.4.1 Exercises 395

10 Wave Equation on Unbounded Domains **398**
10.1 Wave equation and D'Alembert's Solution 398
10.1.1 Exercises 406
10.2 Non-Homogeneous Wave Equation 408
10.2.1 Exercises 412
10.3 Vibrating String with Fixed Ends 414
10.3.1 Exercises 420
10.4 Wave Equation on Semi-Infinite Domain 422
10.4.1 Exercises 426
10.5 Duhamel's Principle for the Wave Equation 427
10.5.1 Exercises 432

Bibliography **435**

Index **445**

Preface

This book is a comprehensive and time-tested guide to the mathematical theory of Fourier series and boundary value problems, with a strong emphasis on their engineering applications. Over the past two decades, it has been rigorously refined and tested in classroom settings, ensuring its effectiveness as a teaching and learning resource. The text is designed to bridge the gap between abstract mathematical theory and practical engineering applications, making it an invaluable tool for students, educators, and professionals alike. The journey begins with a thorough development of Fourier series, a cornerstone of modern mathematics and engineering. Fourier series provide a powerful framework for analyzing periodic functions and decomposing complex signals into simpler sinusoidal components. This foundational knowledge is then extended to boundary value problems, which arise naturally in the study of physical phenomena such as heat flow, vibrations, and wave propagation. Through clear explanations and illustrative examples, the book demonstrates how these mathematical tools can be applied to solve real-world engineering problems.

A distinctive feature of this book is its focus on applications in both rectangular and spherical coordinates. These coordinate systems are essential for modeling problems in diverse engineering contexts, from heat conduction in solid materials to the analysis of vibrating membranes and acoustic waves. By exploring these applications, readers gain a deeper understanding of how mathematical theory translates into practical solutions. Additionally, the book addresses partial differential equations on unbounded domains, providing readers with the tools to tackle problems that extend beyond finite regions, such as those encountered in wave propagation and diffusion processes.

To reinforce learning, the book includes a wealth of exercises carefully designed to solidify conceptual understanding and encourage advanced exploration. These exercises range from foundational problems to more challenging applications, catering to learners at various stages of their academic and professional journeys. Additionally, the text is structured to facilitate further study, providing a springboard for those interested in delving deeper into the subject.

The first three chapters of this work lay the foundational principles of Fourier series, a crucial mathematical tool used across science and engineering for representing periodic functions as infinite sums of sines and cosines. Chapter 1 introduces the Fourier series and its applications in solving problems related to waveforms, heat transfer, signal processing, and vibrations. A key focus is on the conditions under which the Fourier series of a function $f(x)$ converges to $f(x)$, particularly addressing pointwise convergence and the challenges posed by discontinuities. This naturally leads to Chapter 2, which examines the limitations of Fourier series convergence, including cases where the series fails to match the function across its entire domain, especially at points of jump discontinuities. This discussion brings forth the well-known Gibb's phenomenon, an overshoot effect occurring near discontinuities. Finally, Chapter 3 builds upon these concepts by investigating the conditions for absolute and uniform convergence of Fourier series, providing a deeper understanding of how the properties of a function influence different modes of convergence. Through this progression, these chapters establish a rigorous framework for analyzing Fourier series and their practical implications in mathematical modeling and engineering applications.

Chapters 4 and 5 extend these ideas to partial differential equations PDEs, focusing on the heat equation and the wave equation. The heat equation, a fundamental PDE, describes temperature distribution over time and is analyzed using separation of variables and Fourier series expansions. Similarly, the wave equation models oscillatory behaviors in vibrating strings, sound waves, and electromagnetic waves, emphasizing boundary value problems and solution techniques.

Chapter 6 delves into Sturm–Liouville problems, a critical area in mathematical physics, where eigenvalues and eigenfunctions play a vital role in understanding stability and resonance in physical systems. Special functions such as Bessel functions and Legendre polynomials, which arise naturally in cylindrical and spherical coordinate systems, are explored for their practical significance in solving PDEs.

Chapters 7 and 8 extend Fourier analysis to infinite domains through Fourier integrals, offering solutions to problems in heat conduction and wave propagation over unbounded regions. These topics are essential for modeling real-world phenomena where natural boundaries are absent. Chapter 9 revisits the heat equation, focusing on uniqueness theorems, while Chapter 10 explores solutions to homogeneous and non-homogeneous wave equations on unbounded domains using the method of characteristics.

Designed for students, researchers, and professionals, it aims to deepen their understanding of these fundamental concepts and their wide-ranging applications. The content is accessible to advanced undergraduate and graduate students and will appeal to mathematicians, scientists, and engineers alike. Additionally, the book is structured to accommodate a two-semester course, with flexibility depending on the instructor's objectives.

Youssef N. Raffoul
University of Dayton
Dayton, Ohio
July, 2025

Author

Youssef N. Raffoul
University of Dayton
Dayton, Ohio, USA

Youssef N. Raffoul is a professor and the graduate program director in the Department of Mathematics at the University of Dayton. He has been a faculty member since 1999. Dr. Raffoul earned both his BS (1987) and MS (1989) in mathematics from the University of Dayton. He later obtained a second MS in pure mathematics from Indiana University, Bloomington in 1991. In 1996, he completed his PhD in mathematics at Southern Illinois University at Carbondale. Before joining the University of Dayton, he served as a faculty member in the Mathematics Department at Tougaloo College in Mississippi, where he chaired the department for two years.

A prolific researcher, Prof. Raffoul has published over 160 articles in prestigious journals, primarily focusing on functional differential equations, difference equations, and dynamical systems on time scales. His contributions have earned him numerous accolades, including the University of Dayton College of Arts and Sciences' Award for Outstanding Scholarship (2005) and the University of Dayton Alumni Award in Scholarship (2009). Internationally, he was honored with the Career in Science Award by the Lebanese government in 2018. In December 2021, the Archbishop of Lebanon presented him with the Lifetime Achievement Award. His most distinguished honor came on April 16, 2022 when the President of Lebanon, General Michel Aoun, awarded him the Order of Merit, Silver Medal with Distinction.

In addition to his research, Prof. Raffoul is the sole author of four books and a co-author of one, further cementing his impact on the field of mathematics.

1

Fourier Series

This chapter is dedicated to the comprehensive development of Fourier series, a powerful mathematical tool widely used in various fields of science and engineering. The Fourier series serves as a method for representing periodic functions as an infinite sum of sines and cosines, each multiplied by specific coefficients. This representation is fundamental in solving problems involving waveforms, heat transfer, signal processing, and vibrations, among other areas. By breaking complex periodic phenomena into simpler trigonometric components, the Fourier series provides an essential framework for analyzing and modeling engineering systems and other physical phenomena with precision and clarity.

1.1 Fourier Series

This section begins with the formal definition of Fourier series, followed by the derivation of its coefficients. Broadly speaking, suppose that we have a function $f(x)$ that we wish to express as an infinite sum of sine and cosine terms. Specifically, we aim to represent $f(x)$ in the form

$$f(x) = \frac{a_0}{2} + \sum_{n=1}^{\infty} (a_n \cos nx + b_n \sin nx) \tag{1.1}$$

where a_0, a_n, and b_n are coefficients that depend on the function f. Loosely speaking, throughout this chapter, we treat $f(x)$ as equal to its corresponding right-hand Fourier series. A rigorous justification for this will be provided in Chapters 2–3. This representation, known as a Fourier series, allows us to decompose $f(x)$ into its constituent trigonometric components, enabling detailed analysis of its periodic behavior. Since each term in the infinite sum of (1.1) is periodic of period 2π, that is, $\cos n(x+2\pi) = \cos nx$ and $\sin n(x+2\pi) = \sin nx$, we expect the function f to be periodic of period 2π in the sense that $f(x+2\pi) = f(x)$. Henceforth, given a function $f(x)$ defined on $-\pi < x \leq \pi$, we shall require that $f(x)$ is defined outside of this basic interval in such a way $f(x+2\pi) = f(x)$, for $-\infty < x < \infty$. This process is referred to as the *periodic extension* of f on the whole real line (see Fig. 1.1). Throughout this book we assume the period T to be the smallest positive number such that $f(x+T) = f(x)$ for all x in the domain of f. For instance, the period of the

DOI: 10.1201/9781003652540-1

functions

$$f(x) = \cos nx \quad \text{and} \quad h(x) = \sin nx,$$

where n is a positive integer, is $\frac{2\pi}{n}$ because

$$\cos\left(n(x + \frac{2\pi}{n})\right) = \cos(nx + 2\pi) = \cos nx,$$

and

$$\sin\left(n(x + \frac{2\pi}{n})\right) = \sin(nx + 2\pi) = \sin nx.$$

Furthermore, 2π is also a period of the functions $f(x)$ and $h(x)$. In most cases, it is unnecessary to specify the smallest period of a function $f(x)$. Instead, we simply state that $f(x)$ has period T if T is any valid period of $f(x)$.

Example 1.1 Fig. 1.1 shows the periodic extension $f(x) = x, \ -\pi < x \leq \pi$. □

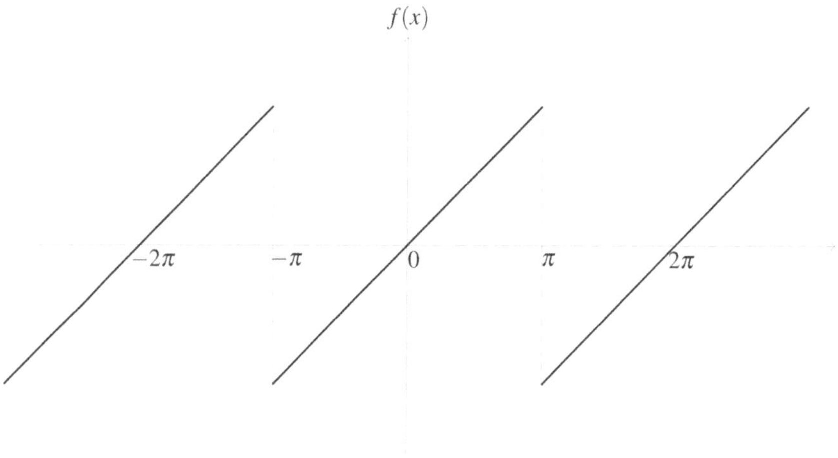

FIGURE 1.1: Periodic extension of $f(x) = x$.

Since the right-side of (1.1) is an infinite sum, the following questions arise naturally:

(1) How can we determine the coefficients a_0, a_n, and b_n?

(2) Under what conditions on f does the right-hand side of (1.1) converge uniformly to f?

(3) What conditions must f satisfy to ensure that the right-hand side of (1.1) is differentiable or integrable?

One of the major requirements for f to have a Fourier representation is that it must be *piecewise continuous*. The requirement of piecewise continuity for a function f

is important in Fourier series representation because it ensures that the Fourier series converges to $f(x)$ in a well-defined manner, and the mathematical operations involved are valid. We have the following definition.

Definition 1.1 *Let x_0 be a point in the domain of a function f. Then,*
(a) the right-hand limit of f at x_0, denoted by $f(x_0^+)$ is defined by

$$\lim_{\substack{x\to x_0\\ x>x_0}} f(x) = f(x_0^+),$$

(b) the left-hand limit of f at x_0, denoted by $f(x_0^-)$ is defined by

$$\lim_{\substack{x\to x_0\\ x<x_0}} f(x) = f(x_0^-),$$

(c) the right-hand derivative of f at x_0, denoted by $f'(x_0^+)$ is defined by

$$f'(x_0^+) = \lim_{\substack{x\to x_0\\ x>x_0}} \frac{f(x)-f(x_0^+)}{x-x_0},$$

(d) the left-hand derivative of f at x_0, denoted by $f'(x_0^-)$ is defined by

$$f'(x_0^-) = \lim_{\substack{x\to x_0\\ x<x_0}} \frac{f(x)-f(x_0^-)}{x-x_0}.$$

Definition 1.2 *Let f be a function defined on (a_*, b_*). If both limits in conditions (a) and (b) of Definition 1.1 exist as finite numbers for every $x \in (a_*, b_*)$, then we say that f is* ***piecewise continuous*** *on (a_*, b_*) and denote it as $f \in C_p(a_*, b_*)$.*

Furthermore, if, in addition to (a) and (b), the limits in conditions (c) and (d) of Definition 1.1 also exist as finite numbers for every $x \in (a_*, b_*)$, and ***if the derivative*** *$f'(x)$* ***exists and is continuous between consecutive jump discontinuity*** *x_j and x_{j+1}, then we say that f is* ***piecewise smooth*** *on (a_*, b_*) and denote it as $f \in C_p'(a_*, b_*)$.*

In other words, if a function f is piecewise smooth on the open interval $a < x < b$, then at every point x_0 within the closed interval $a \le x \le b$, the one-sided derivatives of f, including the limits from the interior at the endpoints, exist and match their corresponding one-sided limits. Specifically,

$$f'_+(x_0) = f'(x_0^+), \quad f'_-(x_0) = f'(x_0^-), \tag{1.2}$$

where

$$\lim_{\substack{x\to x_0\\ x>x_0}} f'(x) = f'_+(x_0) \quad \text{and} \quad \lim_{\substack{x\to x_0\\ x<x_0}} f'(x) = f'_-(x_0).$$

Remark 1 *If a function $f(x)$ is piecewise continuous on a finite interval, it must satisfy the following conditions:*

- *$f(x)$ is continuous on each subinterval within the finite interval, except possibly at a finite number of points where $f(x)$ may have jump discontinuities.*

- *At each point of discontinuity, the left-hand and right-hand limits of $f(x)$ exist and are finite.*

These properties imply from the Extreme Value Theorem that $f(x)$ is bounded on the finite interval and consequently, its integral exists.

Let the function $f(x)$ be defined for $x \in (-\infty, \infty)$.

- If $f \in C_p(a_*, b_*)$ for every finite interval (a_*, b_*), then we say $f \in C_p(-\infty, \infty)$.
- If $f \in C_p(-\infty, \infty)$ and $f \in C_p'(a_*, b_*)$ for every finite interval (a_*, b_*), then we say $f \in C_p'(-\infty, \infty)$.

Note that in Definition 1.1 we do not require $f(x_0^+) = f(x_0^-)$. However, when $f(x_0^+) \neq f(x_0^-)$, then f has a **jump discontinuity** at x_0. Recall, f is continuous at x_0 if and only if

$$f(x_0^+) = f(x_0^-) = f(x_0).$$

For example, the periodic extension of the function in Example 1.1 as shown in Fig. 1.1 has jump discontinuities at $x_k = k\pi,\ k = \pm 1, \pm 3, \ldots$

Example 1.2 Consider

$$f(x) = \begin{cases} -x, & x < 0 \\ x+1, & x > 0 \end{cases}$$

Then

$$\lim_{\substack{x \to 0 \\ x > 0}} f(x) = f(0^+) = 1 \quad \text{and} \quad \lim_{\substack{x \to 0 \\ x < 0}} f(x) = f(0^-) = 0.$$

Hence $f \in C_p(-\infty, \infty)$. Moreover,

$$f'(0^+) = \lim_{\substack{x \to 0 \\ x > 0}} \frac{f(x) - 1}{x} = \lim_{\substack{x \to 0 \\ x > 0}} \frac{(x+1) - 1}{x} = 1,$$

and

$$f'(0^-) = \lim_{\substack{x \to 0 \\ x < 0}} \frac{f(x) - 0}{x} = \lim_{\substack{x \to 0 \\ x < 0}} \frac{-x}{x} = -1.$$

Finally,

$$f'(x) = \begin{cases} -1, & x < 0 \\ 1, & x > 0 \end{cases}$$

and hence $f_+'(0) = 1$ and $f_-'(0) = -1$. We see that $f \in C_p'(-\infty, \infty)$. □

Be aware that not every piecewise continuous function belongs to the space $C_p'(-\pi, \pi)$. For example, the function $f(x) = x^{\frac{1}{3}}$ is not smooth since its derivative

$$f'(x) = \frac{1}{3x^{2/3}}$$

does not exist at $x = 0$. Consequently, $f'(x)$ is not piecewise continuous on any open interval containing $x = 0$. Nonetheless, it turns out that its Fourier series exists. In the same spirit, not every continuous function is piecewise continuous. For example, the function

$$f(x) = \frac{1}{x-2}, \quad x > 2$$

is continuous on its domain of definition. However,

$$\lim_{x \to 2^+} \frac{1}{x-2} = \infty,$$

which is not a finite number. In the following, we will address (1) by deriving formulas for the Fourier coefficients. We begin with the following definition regarding the orthogonality of functions.

Definition 1.3 *If for any two functions h and g we have*

$$\int_{-l}^{l} h(x)g(x)p(x)dx = 0, \tag{1.3}$$

then we say h and g are orthogonal on $[-l, l]$ with respect to the weight function $p(x)$.

Finding the Fourier coefficients depends on the concept of orthogonality of the sine and cosine functions. Recall the following trigonometric identities

$$\cos nx \cos mx = \frac{1}{2}\Big(\cos(n+m)x + \cos(n-m)x\Big)$$

$$\sin nx \sin mx = \frac{1}{2}\Big(\cos(n-m)x - \cos(n+m)x\Big),$$

and

$$\sin nx \cos mx = \frac{1}{2}\Big(\sin(n+m)x + \sin(n-m)x\Big).$$

Thus, by utilizing these trigonometric identities along with (1.3) for $p(x) = 1$, we conclude that, for any integers $m \neq n$

$$\begin{aligned} \int_{-\pi}^{\pi} \sin nx \cos mx dx &= \frac{1}{2}\int_{-\pi}^{\pi} \Big(\sin(n+m)x + \sin(n-m)x\Big)\, dx \\ &= -\frac{1}{2}\left(\frac{\cos(n+m)x}{n+m} + \frac{\cos(n-m)x}{n-m}\right)\Bigg|_{-\pi}^{\pi} = 0. \end{aligned}$$

On the other hand, if $n = m$, then

$$\sin nx \cos nx = \frac{1}{2} \sin 2nx$$

and therefore,

$$\int_{-\pi}^{\pi} \sin nx \cos nx dx = \frac{1}{2}\int_{-\pi}^{\pi} \sin 2nx dx$$

$$= -\frac{1}{2}\frac{\cos 2nx}{2n}\bigg|_{-\pi}^{\pi} = 0.$$

Thus, we have shown that $\sin nx$ and $\cos mx$ are orthogonal on the interval $[-\pi, \pi]$ for all integers n and m. Likewise,

$$\begin{aligned}\int_{-\pi}^{\pi} \cos nx \cos mx\, dx &= \frac{1}{2}\int_{-\pi}^{\pi}\Big(\cos(n+m)x + \cos(n-m)x\Big)\,dx \\ &= \frac{1}{2}\left(\frac{\sin(n+m)x}{n+m} + \frac{\sin(n-m)x}{n-m}\right)\bigg|_{-\pi}^{\pi} = 0.\end{aligned}$$

Whereas if $n = m$, then

$$\cos nx \cos nx = \frac{1}{2}(1 + \cos 2nx)$$

and

$$\sin nx \sin nx = \frac{1}{2}(1 - \cos 2nx).$$

Therefore,

$$\begin{aligned}\int_{-\pi}^{\pi} \cos nx \cos nx\, dx &= \frac{1}{2}\int_{-\pi}^{\pi}(1 + \cos 2nx)\,dx \\ &= \frac{1}{2}\left(x + \frac{\sin 2nx}{2n}\right)\bigg|_{-\pi}^{\pi} = \pi.\end{aligned}$$

hence,

$$\int_{-\pi}^{\pi} \cos nx \cos mx\, dx = \begin{cases} 0, & n \neq m \\ \pi, & n = m \end{cases}$$

Moreover,

$$\begin{aligned}\int_{-\pi}^{\pi} \sin nx \sin nx\, dx &= \frac{1}{2}\int_{-\pi}^{\pi}(1 - \cos 2nx) \\ &= \frac{1}{2}\left(x - \frac{\sin 2nx}{2n}\right)\bigg|_{-\pi}^{\pi} = \pi.\end{aligned}$$

and

$$\int_{-\pi}^{\pi} \sin nx \sin mx\, dx = 0, \quad n \neq m.$$

Thus,

$$\int_{-\pi}^{\pi} \sin nx \sin mx\, dx = \begin{cases} 0, & n \neq m \\ \pi, & n = m \end{cases}$$

Before we embark on finding the Fourier coefficients, we note that it is not generally true that the integral of an infinite sum of functions is necessarily equal to the sum of the integrals of those functions. However, if the function f is sufficiently well-behaved, then the trigonometric series for the function f will converge sufficiently rapidly for this interchange of integration and summation to be valid, so that

$$\int_{-\pi}^{\pi}\left(\sum_{n=1}^{\infty} a_n \cos nx\right) dx = \sum_{n=1}^{\infty} a_n \int_{-\pi}^{\pi} \cos nx\, dx \quad \text{etc.}$$

Hereafter, we assume that $f \in C_p(-\pi,\pi)$ so that f is sufficiently well-behaved. To find the coefficient a_0, we integrate both sides of (1.1) from $x=-\pi$ to $x=\pi$ and obtain

$$\begin{aligned}\int_{-\pi}^{\pi} f(x)\,dx &= \int_{-\pi}^{\pi} \frac{a_0}{2}\,dx + \sum_{n=1}^{\infty}\left(a_n \int_{-\pi}^{\pi} \cos nx\,dx + b_n \int_{-\pi}^{\pi} \sin nx\,dx\right) \\ &= \frac{a_0}{2}x\Big|_{-\pi}^{\pi} + \sum_{n=1}^{\infty}\left(a_n \frac{\sin nx}{n} - b_n \frac{\cos nx}{n}\right)\Big|_{-\pi}^{\pi} = 2\pi\frac{a_0}{2} + 0.\end{aligned}$$

Solving for a_0 gives

$$a_0 = \frac{1}{\pi}\int_{-\pi}^{\pi} f(x)\,dx. \tag{1.4}$$

Note that the term

$$\frac{a_0}{2} = \frac{1}{2\pi}\int_{-\pi}^{\pi} f(x)\,dx,$$

is the *mean,* or *average,* value of $f(x)$ over the interval $-\pi < x < \pi$.

To find the coefficient a_n, where $n = 1,2,\ldots$ we begin by multiplying both sides of (1.1) by $\cos mx$ and integrating term by term. Using the orthogonality property, we obtain

$$\begin{aligned}\int_{-\pi}^{\pi} f(x)\cos mx\,dx &= \int_{-\pi}^{\pi} \frac{a_0}{2}\cos mx\,dx \\ &+ \sum_{n=1}^{\infty}\left(a_n \int_{-\pi}^{\pi} \cos nx \cos mx\,dx + b_n \int_{-\pi}^{\pi} \sin nx \cos mx\,dx\right) \\ &= 0 + a_m\pi + 0.\end{aligned}$$

Hence by replacing m with n, we get

$$a_n = \frac{1}{\pi}\int_{-\pi}^{\pi} f(x)\cos nx\,dx, \quad n = 1,2\ldots \tag{1.5}$$

Finally, multiplying both sides of (1.1) by $\sin mx$, integrating term by term, and applying the orthogonality property yield

$$\begin{aligned}\int_{-\pi}^{\pi} f(x)\sin mx\,dx &= \int_{-\pi}^{\pi} \frac{a_0}{2}\sin mx\,dx \\ &+ \sum_{n=1}^{\infty}\left(a_n \int_{-\pi}^{\pi} \cos nx \sin mx\,dx + b_n \int_{-\pi}^{\pi} \sin nx \sin mx\,dx\right)\end{aligned}$$

$$= \quad 0 + a_n(0) + b_m\pi.$$

Hence by replacing m with n, we get

$$b_n = \frac{1}{\pi}\int_{-\pi}^{\pi} f(x)\sin nx\,dx, \quad n = 1,2\ldots \tag{1.6}$$

The coefficient a_n given by (1.5) is known as the Fourier cosine coefficient of f, while b_n, as defined in (1.6), is referred to as the Fourier sine coefficient of f.

Remark 2 *The formula* (1.4) *for* a_0 *can be combined with equation* (1.5) *when the integer* n *is allowed to run from* $n = 0$, *rather than* $n = 1$. *However, this approach may not always be applicable in certain cases.*

In Chapter 2, it will be necessary to write (1.1) in terms of the values of the coefficients a_0, a_n, and b_n. Thus, if we replace x with s in (1.4)–(1.6) and then substitute into (1.1), we obtain the alternative formula

$$\begin{aligned} f(x) &= \frac{1}{2\pi}\int_{-\pi}^{\pi} f(s)\,ds \\ &\quad + \sum_{n=1}^{\infty}\left(\frac{1}{\pi}\int_{-\pi}^{\pi} f(s)\cos ns\,ds\cos nx + \frac{1}{\pi}\int_{-\pi}^{\pi} f(s)\sin ns\,ds\sin nx\right) \\ &= \frac{1}{2\pi}\int_{-\pi}^{\pi} f(s)\,ds + \frac{1}{\pi}\sum_{n=1}^{\infty}\int_{-\pi}^{\pi} f(s)\left(\cos ns\cos nx + \sin ns\sin nx\right)ds \\ &= \frac{1}{2\pi}\int_{-\pi}^{\pi} f(s)\,ds + \frac{1}{\pi}\sum_{n=1}^{\infty}\int_{-\pi}^{\pi} f(s)\cos n(s-x)\,ds. \end{aligned} \tag{1.7}$$

Example 1.3 Let

$$f(x) = \begin{cases} 2, & 0 < x < \pi \\ -1, & -\pi < x < 0 \end{cases}$$

The function $f(x)$ has period 2π. By periodically extending the function, we observe that it is clearly piecewise continuous, exhibiting jump discontinuities at $x = k\pi$ for $k = 0, \pm1, \pm2, \ldots$ as shown in Fig. 1.2. In the interval $(x_k, x_{k+1}) = (0,\pi)$, $f'(x) = 0$ is continuous. Now we worry about the endpoints 0 and π.

$$f'(0^+) = \lim_{\substack{x\to 0\\ x>0}} \frac{f(x) - f(0^+)}{x - 0} = \lim_{\substack{x\to 0\\ x>0}} \frac{2-2}{x-0} = 0.$$

Similarly,

$$f'(\pi^-) = \lim_{\substack{x\to \pi\\ x<\pi}} \frac{f(x) - f(\pi^-)}{x - \pi} = \lim_{\substack{x\to \pi\\ x<\pi}} \frac{2-2}{x-\pi} = 0.$$

This shows that $f \in C'_p(0,\pi)$. By a similar argument, one can easily show that $f \in C'_p(-\pi,0)$. Thus, this work can be easily generalized on any interval

$(k\pi,(k+1)\pi)$, $k=0,\pm1,\pm2,\ldots$ and with that we have shown $f\in C_p'(-\infty,\infty)$. The resulting function is graphed in Fig. 1.2. Now we compute the Fourier coefficients.

$$a_0=\frac{1}{\pi}\int_{-\pi}^{\pi}f(x)dx \;=\; \frac{1}{\pi}\left(\int_{-\pi}^{0}-1dx+\int_{0}^{\pi}2dx\right)=\frac{1}{\pi}(-\pi+2\pi)=1,$$

$$\begin{aligned}a_n=\frac{1}{\pi}\int_{-\pi}^{\pi}f(x)\cos nx\,dx &= \frac{1}{\pi}\left(\int_{-\pi}^{0}-\cos nx\,dx+\int_{0}^{\pi}2\cos nx\,dx\right)\\ &= \frac{1}{\pi}\left(-\frac{\sin nx}{n}\bigg|_{-\pi}^{0}+2\frac{\sin nx}{n}\bigg|_{0}^{\pi}\right)=0,\quad n=1,2,\ldots.\end{aligned}$$

Finally,

$$\begin{aligned}b_n &= \frac{1}{\pi}\int_{-\pi}^{\pi}f(x)\sin nx\,dx=\frac{1}{\pi}\left(\int_{-\pi}^{0}-\sin nx\,dx+\int_{0}^{\pi}2\sin nx\,dx\right)\\ &= \frac{1}{\pi}\left(\frac{\cos nx}{n}\bigg|_{-\pi}^{0}-2\frac{\cos nx}{n}\bigg|_{0}^{\pi}\right)\\ &= \frac{1}{\pi}\left(\frac{1}{n}-\frac{\cos n\pi}{n}-2\frac{\cos n\pi}{n}+\frac{2}{n}\right)=\frac{3}{n\pi}(1-\cos n\pi),\quad n=1,2,\ldots\end{aligned}$$

Note that when n is even then $\cos n\pi=1$ and when n is odd then $\cos n\pi=-1$. Hence

$$b_n=\frac{3}{n\pi}(1-\cos n\pi)=\frac{6}{n\pi},$$

if n is odd and $b_n=0$ if n is even. This means that n can be replaced by $2n-1$ in the sum, resulting in

$$f(x)=\begin{cases}2, & 0\le x\le\pi\\ -1, & -\pi<x<0\end{cases}=\frac{1}{2}+\sum_{n=1}^{\infty}\frac{6}{(2n-1)\pi}\sin(2n-1)x.$$

In Chapter 2, we develop the theory to show that the infinite sum given by the above expression converges to the function

$$g(x)=\begin{cases}\frac{1}{2}, & \text{if } x=k\pi\\ 2, & \text{if } 2k\pi<x<(2k+1)\pi\\ -1, & \text{if } (2k-1)\pi<x<2k\pi\end{cases}$$

for $k=0,\pm1,\pm2,\ldots$. Furthermore, Fig. 1.2 provides additional insight into the behavior of the function $g(x)$. □

We end this section with the following basic but useful result.

Theorem 1.1 *Suppose that f is piecewise continuous and T-periodic. Then, for any real number c, we have*

$$\int_0^T f(x)dx=\int_c^{c+T}f(x)dx.$$

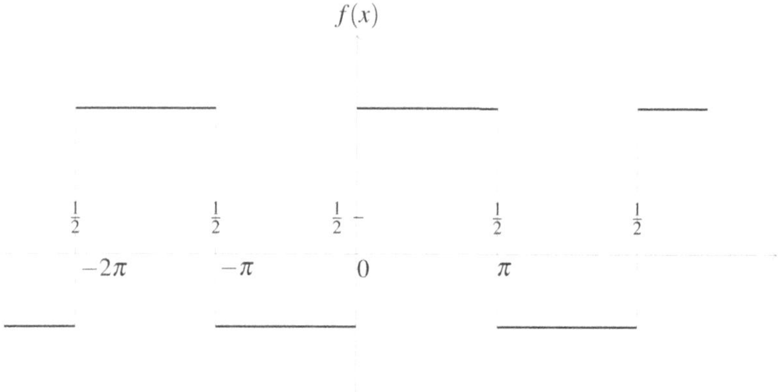

FIGURE 1.2: The periodic extension of the function $f(x)$ in Example 1.3 is shown in the graph. The graph represents a function belonging to $C_p'(-\infty,\infty)$ and exhibits jump discontinuities of magnitude $\frac{1}{2}$ at $x = k\pi$ for $k = 0, \pm 1, \pm 2, \ldots$.

Proof *Let*

$$F(c) = \int_c^{c+T} f(x)\,dx.$$

Then $F'(c) = f(c+T) - f(c) = 0$, *since* f *is periodic with period* T. *Therefore,* $F(c)$ *is constant for all* c *and hence* $F(0) = F(c)$, *which implies the result.*

We end this section with the following example, in which the function is extended over the real line.

Example 1.4 For $k = 0, \pm 1, \pm 2, \ldots$ consider the function $f : \mathbb{R} \to \mathbb{R}$ defined by

$$f(x) = \begin{cases} \frac{1}{2}, & \text{if } x = k\pi \\ 1, & \text{if } 2k\pi < x < (2k+1)\pi \\ 0, & \text{if } (2k-1)\pi < x < 2k\pi \end{cases}$$

This function f has the property that $f(x) = f(x + 2k\pi)$ for all real numbers x and integers k, and $f(x) \in C_p(-\infty,\infty)$. Hence, it can be represented by a Fourier series. Let's consider the function over one period, namely we let $k = 0$. Then $f(x) = 0$ if $-\pi < x < 0$, and $f(x) = 1$ if $0 < x < \pi$. Therefore, $a_0 = 1$, and

$$a_n = \frac{1}{\pi}\int_0^{\pi} \cos nx\,dx = \frac{1}{n\pi} \sin nx\Big|_0^{\pi} = 0,$$

$$b_n = \frac{1}{\pi}\int_0^{\pi} \sin nx\,dx = \frac{1}{n\pi}[-\cos nx]_0^{\pi} = \frac{1}{n\pi}(1 - \cos n\pi).$$

Thus,

$$b_n = \frac{1}{n\pi}(1 - (-1)^n) = \begin{cases} \frac{2}{n\pi}, & \text{if } n \text{ is odd and } n > 0, \\ 0, & \text{if } n \text{ is even and } n > 0. \end{cases}$$

Finally,

$$f(x) = \frac{1}{2} + \frac{2}{\pi} \sum_{n=1}^{\infty} \frac{1}{2n-1} \sin(2n-1)x, \quad x \in \mathbb{R}.$$

□

1.1.1 Exercises

Exercise 1.1 *Determine the constant c so that the functions $h(x) = x + 1$ and $g(x) = x^2 + c$ are orthogonal on the interval $[-1, 1]$ with respect to the weight function $p(x) = 1$.*

Exercise 1.2 *Show that the set of function $\varphi_n(x) = \sin(2n-1)x$, $n = 1, 2, \ldots$ is orthogonal on the set $(0, \frac{\pi}{2})$ with weight function $p(x) = 1$.*

Exercise 1.3 *Recall that $e^{inx} = \cos nx + i \sin nx$, and hence*

$$\cos nx = \frac{e^{inx} + e^{-inx}}{2}, \quad \sin nx = \frac{e^{inx} - e^{-inx}}{2}.$$

Use the right-hand side of these relations to show that

$$\int_{-\pi}^{\pi} \cos nx \cos mx dx = \begin{cases} 0, & n \neq m \\ \pi, & n = m \end{cases}$$

Exercise 1.4 *Find the Fourier series of $f(x) = x$, $-\pi < x \le \pi$ where $f(x + 2\pi) = f(x)$.*

Ans: $\sum_{n=1}^{\infty} \frac{2}{n} (-1)^{n+1} \sin nx.$

Exercise 1.5 *Find the Fourier series of*

$$f(x) = \begin{cases} 1, & 0 \le x \le \pi \\ -1, & -\pi < x \le 0 \end{cases}, \quad f(x + 2\pi) = f(x)$$

and draw its periodic extension.

Ans: $\frac{4}{\pi} \sum_{n=1}^{\infty} \frac{\sin(2n-1)x}{2n-1}.$

Exercise 1.6 *For $k = 0, \pm 1, \pm 2, \ldots$, find the Fourier series of*

$$f(x) = \begin{cases} \frac{\pi}{2}, & \text{if } x = k\pi,\ k = \pm 1, \pm 2, \ldots \\ x - 2k\pi, & \text{if } 2k\pi < x < (2k+1)\pi \\ 0, & \text{if } (2k-1)\pi < x < 2k\pi \end{cases}, \quad f(x + 2k\pi) = f(x).$$

Ans: $\frac{\pi}{4} - 2 \sum_{n=1}^{\infty} \frac{\cos(2n-1)x}{(2n-1)^2 \pi} + \sum_{n=1}^{\infty} \frac{(-1)^{n+1}}{n} \sin nx,\ x \in \mathbb{R}.$

Exercise 1.7 *Show that $f(x) \in C_p(-2,6)$,*

$$f(x) = \begin{cases} x^2, & -2 \le x < 2 \\ 6, & 2 \le x < 4 \\ 10 - x, & 2 < x \le 6 \end{cases}$$

Is $f(x) \in C_p'(-2,6)$?

Exercise 1.8 *Find the Fourier series of*

$$f(x) = \begin{cases} 2 + \frac{2x}{\pi}, & -\pi \le x \le -\frac{\pi}{2} \\ 1, & -\frac{\pi}{2} \le x \le \frac{\pi}{2} \\ 2 - \frac{2x}{\pi}, & \frac{\pi}{2} \le x \le \pi \end{cases}$$

where $f(x+2\pi) = f(x)$.

Exercise 1.9 *Let*

$$f(x) = \begin{cases} x^2 \sin(\frac{1}{x}), & x \neq 0 \\ 0, & x = 0 \end{cases}$$

Show that $f(x) \in C_p(-\infty,\infty)$, but $f(x) \notin C_p'(-\infty,\infty)$.

Exercise 1.10 *Consider the step function*

$$f(x) = \begin{cases} 0, & x < 0 \\ x, & x \ge 0 \end{cases}$$

Show that $f(x) \in C_p(-\infty,\infty)$, $f(x) \in C_p'(-\infty,\infty)$, but $f'(0)$ does not exist.

Exercise 1.11 *Find the Fourier series of the following functions and sketch several periods of its graphs.*

(a) $f(t) = |t|, \quad -\pi \le t \le \pi,$

(b) $f(t) = \begin{cases} \pi + t, & -\pi \le t \le 0 \\ \pi - t, & 0 < t \le \pi \end{cases}$

(c) $f(t) = \begin{cases} \pi + t, & -\pi \le t < 0 \\ 0, & 0 \le t \le \pi \end{cases}$

(d) $f(t) = \begin{cases} 0, & -\pi \le t \le 0 \\ t^2, & 0 \le t \le \pi \end{cases}$

(e) $f(t) = |\sin t|, \quad -\pi < t < \pi,$

(f) $f(t) = \cos^2(2t), \quad -\pi \le t \le \pi$

(g) $f(t) = \begin{cases} 0, & -\pi \le t \le 0 \\ \sin t, & 0 < t \le \pi \end{cases}$

Exercise 1.12 *Let F be T-periodic and piecewise continuous. Set*

$$F(c) = \int_0^c f(x)\,dx.$$

Show that F is continuous and $F'(c) = f(c)$ at all points c where f is continuous.

Exercise 1.13 *For $h \in (0, \pi)$ define* $f_h(x) = \begin{cases} \frac{1}{h}, & |x| < h, \\ 0, & h < |x| < \pi \end{cases}$

Let $a_n(h)$ and $b_n(h)$ denote the Fourier coefficients of $f_h(x)$. Find

$$\lim_{h \to 0} a_n(h) \quad \text{and} \quad \lim_{h \to 0} b_n(h).$$

1.2 Sine and Cosine Series

The Fourier sine series and Fourier cosine series are fundamentally tied to the concepts of odd and even extensions of a function. For a given function defined on a finite interval, the Fourier sine series represents its odd extension, where the function is reflected across the origin with a change in sign, ensuring odd symmetry. Similarly, the Fourier cosine series corresponds to the even extension, where the function is mirrored symmetrically across the y-axis, preserving its values. These extensions allow us to express the function over an extended domain, making it compatible with the orthogonality properties of sine and cosine functions, which are defined over symmetric intervals like $[-L, L]$.

Definition 1.4 *A function $f : \mathbb{R} \to \mathbb{R}$ is said to be even if $f(-x) = f(x)$ for all real numbers x. A function $f : \mathbb{R} \to \mathbb{R}$ is said to be odd if $f(x) = -f(-x)$ for all real numbers x.*

Even functions are symmetric about the y-axis and odd functions are symmetric about the origin. For example, $f(x) = \cos(x)$ is an even function since $\cos(-x) = \cos(x)$ and $f(x) = \sin(x)$ is an odd function since $\sin(-x) = -\sin(x)$. The product of two even functions is even, the product of two odd functions is even, and the product of an even and an odd function is odd.

Lemma 1 *Suppose that $h : \mathbb{R} \to \mathbb{R}$.*
(a) If h is even, then

$$\int_{-c}^{c} h(x)\, dx = 2 \int_{0}^{c} h(x)\, dx.$$

(b) If h is odd, then

$$\int_{-c}^{c} h(x)\, dx = 0.$$

Proof *We only prove part (a) and leave part (b) as an exercise. Let the function h be even. Then $h(-x) = h(x)$. Hence*

$$\int_{-c}^{c} h(x)\, dx = \int_{-c}^{0} h(x)\, dx + \int_{0}^{c} h(x)\, dx.$$

By letting $x = -s$, we have that

$$\int_{-c}^{0} h(x)\, dx = \int_{c}^{0} h(-s)\,(-ds) = \int_{0}^{c} h(s)\, ds.$$

Substituting into the preceding equation gives the desired result. □

If a function is defined on the interval $[0, \pi]$, then it is possible to expand the function periodically onto the interval $[-\pi, 0]$ by either using an even extension (reflection about the y-axis), or an odd extension (reflection about the origin). Both expansions are called half-range extensions. The Fourier series of an even half-range extension is the Fourier cosine series, and the Fourier series of an odd half-range extension is the Fourier sine series. We have the following. Let f_e and f_o denote the even and odd extensions of 2π-periodic function f, respectively. Then,

$$f_e(x) = \begin{cases} f(x), & 0 < x < \pi \\ f(-x), & -\pi < x < 0 \end{cases}, \quad f_o(x) = \begin{cases} f(x), & 0 < x < \pi \\ -f(-x), & -\pi < x < 0 \end{cases}$$

This type of extension enables us to represent even or odd functions defined on $[0, \pi]$ using their respective Fourier series. Note that the definitions of odd and even extensions of a function can be applied to any interval $[0, L]$ by replacing π with L. The same holds for odd and even extensions of functions on any interval $[0, L]$. In Figs. 1.3 and 1.4, we display the periodic even extension and odd extension of

$$f(x) = \frac{x}{L} + 1, \quad 0 < x < L,$$

respectively. Notice that the periodic even extension does not introduce new jumps. Similarly, the periodic odd extension of $f(x) = \frac{x}{L} + 1, \ 0 < x < L$, does not introduce

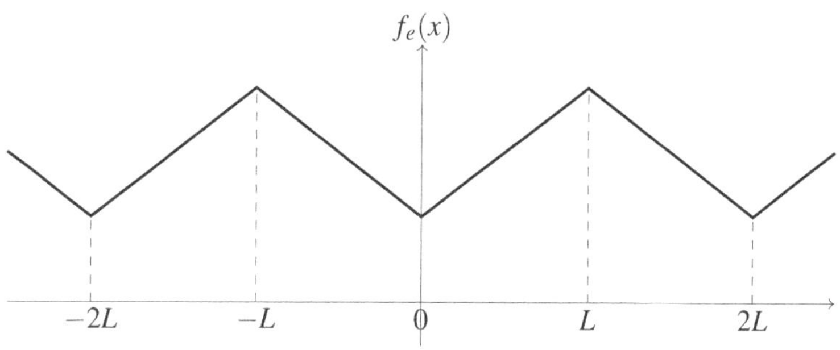

FIGURE 1.3: Periodic even extension.

new jumps if and only if $f(0) = f(L) = 0$, (see Fig. 1.5).

Comparing Figs. 1.4 and 1.5, we see that ensuring continuity of the odd extension involves addressing any discontinuities at the boundaries of the original interval. For example, in Fig. 1.4, the function has a jump discontinuity at $x = 0$, the odd extension will reflect this discontinuity across the origin, creating a continuous but possibly non-smooth transition.

Remember, odd symmetry requires that $f(-x) = -f(x)$. At $x = 0$, this implies $f(0) = -f(0)$, which means $f(0) = 0$. Therefore, if the original function is not zero at $x = 0$, as the case in Fig. 1.4, the odd extension must be adjusted so that $f(0)$ is set to zero to maintain both odd symmetry and continuity. Periodicity is preserved

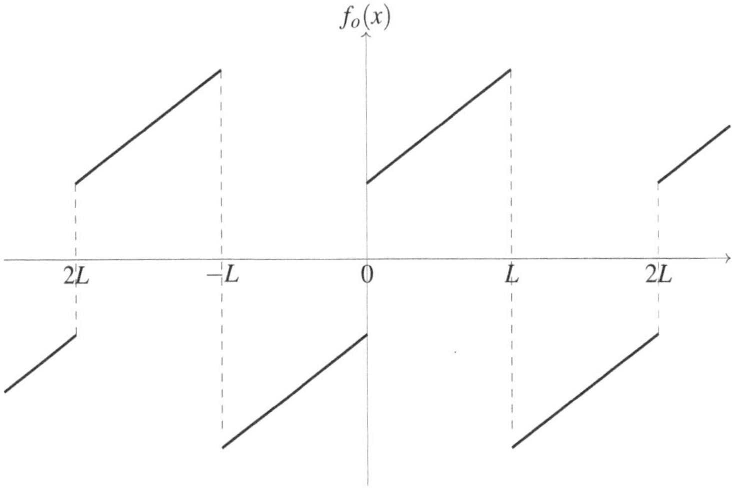

FIGURE 1.4: Jumps in the periodic odd extension.

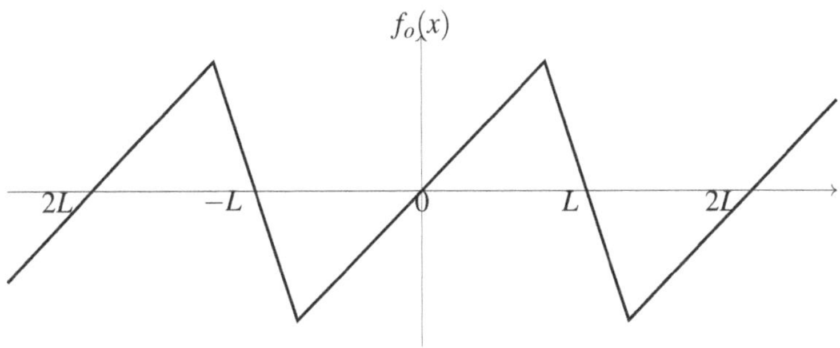

FIGURE 1.5: The continuous periodic odd extension under the condition that $f(0) = f(L) = 0$.

because the Fourier sine series inherently assumes periodic boundary conditions for the extended function. By constructing an odd extension, you're essentially creating a periodic odd function that repeats over symmetric intervals.

Thus, when you force an odd extension to be continuous, the result is still periodic but may involve adjustments to the function values at specific points like $x = 0$ to satisfy both odd symmetry and continuity.

We have the following theorem.

Theorem 1.2 *Suppose that $f(x) \in C_p(-\pi, \pi)$ and is periodic with period 2π.*

- *If f is even, then its Fourier series has only cosine terms. The Fourier coefficients* *are given by*

$$b_n = 0, \; n = 1, 2 \ldots \; and \; a_n = \frac{2}{\pi} \int_0^\pi f(x) \cos nx dx, \; n = 0, 1, \ldots \tag{1.8}$$

and the corresponding Fourier series is called a Fourier cosine series.

- *If f is odd, then its Fourier series has only sine terms. The Fourier coefficients are given by*

$$a_n = 0, \; n = 0, 1 \ldots \; and \; b_n = \frac{2}{\pi} \int_0^\pi f(x) \sin nx dx, \; n = 0, 1, \ldots \tag{1.9}$$

and the corresponding Fourier series is called a Fourier sine series.

Proof *We only prove the first part of the Theorem and leave the other part as an exercise. By (1.5) and part (a) of Lemma 1 we have that*

$$a_n = \frac{1}{\pi} \int_{-\pi}^\pi f(x) \cos nx dx = \frac{1}{\pi}\left(2 \int_0^\pi f(x) \cos nx dx\right),$$

for $n = 0, 1 \ldots$ Similarly, By (1.6) and part (b) of Lemma 1 we have that

$$b_n = \frac{1}{\pi} \int_{-\pi}^\pi f(x) \sin nx dx = \frac{1}{\pi}(0),$$

for $n = 1, 2 \ldots$ This completes the proof.

The literature often remains ambiguous regarding the origins of the Fourier sine and cosine series. In many cases, a function $f(x)$ is defined only on the half-interval $[0, \pi]$. To construct a periodic extension of f with period 2π, we have two primary options: an even extension $f_e(x)$ or an odd extension $f_o(x)$. The even extension $f_e(x)$ leads to the Fourier cosine series, while the odd extension $f_o(x)$ gives rise to the Fourier sine series. In summary, the *periodic odd extension* of a function f that is piecewise continuous on the interval $(0, \pi)$, is the *Fourier sine series* of f given by

$$f(x) = \sum_{n=1}^\infty b_n \sin nx, \quad 0 < x < \pi, \tag{1.10}$$

where

$$b_n = \frac{2}{\pi} \int_0^\pi f(x) \sin nx dx, \quad n = 1, 2, \ldots. \tag{1.11}$$

Similarly, the *periodic even extension* of a function f that is piecewise continuous on the interval $(0, \pi)$, is the *Fourier cosine series* of f given by

$$f(x) = \frac{a_0}{2} + \sum_{n=1}^\infty a_n \cos nx, \quad 0 < x < \pi, \tag{1.12}$$

where

$$a_0 = \frac{2}{\pi}\int_0^{\pi} f(x)dx, \quad a_n = \frac{2}{\pi}\int_0^{\pi} f(x)\cos nx\,dx, \quad n = 1,2,\ldots.$$

If the function f is defined over the interval $[-\pi,\pi]$ and periodic with period 2π, we test to determine whether f is even, odd, or neither. If f is an even function, it gives rise to a Fourier cosine series. If f is an odd function, it gives rise to a Fourier sine series. If f is neither even nor odd, it gives rise to the general Fourier series, given by (1.1). Before presenting an example, we review the following concepts. Finding the Fourier cosine series of the function $f(x) = x$ for $0 < x < \pi$ is equivalent to finding the Fourier series of $f(x) = |x|$ on the interval $-\pi < x < \pi$. This equivalence holds because $f(x) = |x|$ is an even function. Moreover, the periodic even extension of $f(x)$ is

$$f_e(x) = \begin{cases} x, & 0 < x < \pi, \\ -x, & -\pi < x < 0, \end{cases} = |x|$$

Similarly, finding the Fourier sine series of $f(x) = x$ for $0 < x < \pi$ is equivalent to finding the Fourier series of $f(x) = x$ on the interval $-\pi < x < \pi$, which corresponds to the periodic odd extension of $f(x) = x$ for $0 < x < \pi$. This holds true because

$$f_o(x) = \begin{cases} x, & 0 < x < \pi, \\ x, & -\pi < x < 0, \end{cases} = x, \quad -\pi < x < \pi$$

We provide the following example.

Example 1.5 Let us find the Fourier cosine series of $f(x) = x,\ 0 < x < \pi$. Since we are asked to find the Fourier cosine of f on $0 < x < \pi$, the function must be extended evenly over $-\pi < x < 0$. Making use of (1.12), we have that

$$a_0 = \frac{2}{\pi}\int_0^{\pi} x\,dx = \pi.$$

Using integration by parts, we find that

$$a_n = \frac{2}{\pi}\int_0^{\pi} x\cos nx\,dx = \frac{2}{\pi}\frac{(-1)^n - 1}{n^2}, \quad n = 1,2,\ldots.$$

By noticing that $a_n = 0$ for n is even and $a_n = -2$ for n is odd, we may replace $(-1)^n - 1$ with -2 and use $2n-1$ for n in the summation. Thus,

$$x = \frac{\pi}{2} - \frac{4}{\pi}\sum_{n=1}^{\infty}\frac{\cos(2n-1)x}{(2n-1)^2}, \quad 0 < x < \pi,$$

which is the periodic even extension of $f(x) = x$. Note that $f_e(x) = |x|, \quad -\pi \le x \le \pi$. In conclusion we may safely write

$$|x| = \frac{\pi}{2} - \frac{4}{\pi}\sum_{n=1}^{\infty}\frac{\cos(2n-1)x}{(2n-1)^2}, \quad -\pi \le x \le \pi.$$

□

1.2.1 Exercises

Exercise 1.14 *Prove part (b) of Lemma 1.*

Exercise 1.15 *Prove the second part of Theorem 1.2.*

In Exercises 1.16–1.19 we assume all functions are 2π periodic.

Exercise 1.16 *Find the Fourier sine series of* $f(x) = x, \quad 0 < x < \pi$, *and graph its* $f_o(x)$.

Exercise 1.17 *Find the Fourier series of* $f(x) = x^2, \quad -\pi < x < \pi$.
Ans: $\dfrac{\pi^2}{3} + 4\sum_{n=1}^{\infty} \dfrac{(-1)^n}{n^2}\cos(nx)$.

Exercise 1.18 *Find the Fourier series of* $f(x) = x^3, \quad -\pi < x < \pi$.

Exercise 1.19 *Find the Fourier sine series of* $f(x) = 1, \quad 0 < x < \pi$.

Exercise 1.20 *Let f be 2π- periodic function given by*

$$f(x) = \begin{cases} 0, & 0 < x < \frac{\pi}{2} \\ 1, & \frac{\pi}{2} < x < \frac{3\pi}{2} \\ 0, & \frac{3\pi}{2} < x < 2\pi \end{cases}, \; f(0) = 0, \; f(\frac{\pi}{2}) = f(\frac{3\pi}{2}) = \frac{1}{2}.$$

(a) Show that f is even.
(b) Find the Fourier series of f.

Exercise 1.21 *Let f be 2π−periodic function given by*

$$f(x) = \begin{cases} 2, & 0 < x < \frac{\pi}{2} \\ 0, & \frac{\pi}{2} < x < \frac{3\pi}{2} \\ -2, & \frac{3\pi}{2} < x < 2\pi \end{cases}, \; f(0) = 0, \; f(\frac{\pi}{2}) = 1, \; f(\frac{3\pi}{2}) = -1.$$

(a) Show that f is odd.
(b) Find the Fourier series of f.

Exercise 1.22 *Compute the Fourier series for f where*

$$f(x) = \begin{cases} x^2, & 0 \le x < \pi, \\ -(x-\pi)^2, & \pi \le x < 2\pi, \end{cases} \quad f(x+2\pi) = f(x).$$

Graph $f(x)$ on the interval $-3\pi < x < 5\pi$.

Exercise 1.23 *Let f be 2π-periodic function with $f(x+\pi) = -f(x)$.*
(a) Show that its Fourier coefficients a_n and b_n are zero whenever n is even.
(b) Show that for odd n, we have

$$a_n = \frac{2}{\pi}\int_0^{\pi} f(x)\cos nx\,dx, \quad b_n = \frac{2}{\pi}\int_0^{\pi} f(x)\sin nx\,dx.$$

(c) Show that the square wave given by

$$f(x) = \begin{cases} 1, & 0 < x < \pi \\ -1, & -\pi < x < 0 \end{cases}$$

satisfies the property that $f(x+\pi) = -f(x)$ and find its Fourier series.

Exercise 1.24 *Verify that the Fourier series of the even periodic extension of*

$$f(x) = \begin{cases} x^2, & 0 < x < \frac{\pi}{2} \\ 0, & \frac{\pi}{2} < x < \pi \end{cases}$$

can be written in the form

$$\frac{\pi^2}{24} + \sum_{n=1}^{\infty} \left[\frac{2(-1)^n}{4n^2} \cos 2nx + \frac{2}{\pi(2n-1)^3} \left(\frac{(2n-1)^2\pi^2}{4} - 2 \right) (-1)^{n-1} \cos(2n-1)x \right].$$

1.3 Fourier Series on Arbitrary Periods

In this section, we extend the concepts introduced in Section 1.1 to functions $f : \mathbb{R} \to \mathbb{R}$ that are periodic with period $2L$. Specifically, we focus on the behavior of $f(x)$ over the interval $-L < x < L$. There will be no need to redevelop all the results from scratch. Let

$$g(x) = f\left(\frac{Lx}{\pi}\right) \quad \text{so that} \quad f(x) = g\left(\frac{\pi x}{L}\right).$$

If we require that $f \in C_p(-L, L)$ then the function g has the Fourier series

$$g(x) = \frac{a_0}{2} + \sum_{n=1}^{\infty} (a_n \cos nx + b_n \sin nx).$$

The coefficients of this Fourier series are then given by the formulae

$$a_0 = \frac{1}{\pi} \int_{-\pi}^{\pi} g(s) ds,$$

$$a_n = \frac{1}{\pi} \int_{-\pi}^{\pi} g(s) \cos ns\, ds, \quad n = 1, 2 \ldots$$

$$b_n = \frac{1}{\pi} \int_{-\pi}^{\pi} g(s) \sin ns\, ds, \quad n = 1, 2 \ldots$$

If we make the substitution

$$s = \frac{\pi x}{L}$$

in the above four equations, we arrive at the Fourier series of $f(x)$ over $(-L, L)$ given by

$$f(x) = g\left(\frac{\pi x}{L}\right) = \frac{a_0}{2} + \sum_{n=1}^{\infty} \left(a_n \cos\frac{n\pi x}{L} + b_n \sin\frac{n\pi x}{L}\right), \tag{1.13}$$

where the Fourier coefficients of $f(x)$ are given by the formulas

$$a_0 = \frac{1}{\pi}\int_{-\pi}^{\pi} g(s)ds = \frac{1}{\pi}\int_{-L}^{L} g\left(\frac{\pi x}{L}\right)\frac{\pi}{L}dx = \frac{1}{L}\int_{-L}^{L} f(x)\,dx.$$

Thus,

$$a_0 = \frac{1}{L}\int_{-L}^{L} f(x)\,dx. \tag{1.14}$$

Similarly,

$$\begin{aligned} a_n &= \frac{1}{\pi}\int_{-\pi}^{\pi} g(s)\cos ns\,ds = \frac{1}{\pi}\int_{-L}^{L} g\left(\frac{\pi x}{L}\right)\frac{\pi}{L}\cos\frac{n\pi x}{L}dx \\ &= \frac{1}{L}\int_{-L}^{L} f(x)\cos\frac{n\pi x}{L}dx. \end{aligned}$$

Consequently, we have

$$a_n = \frac{1}{L}\int_{-L}^{L} f(x)\cos\frac{n\pi x}{L}dx, \quad n = 1, 2\ldots \tag{1.15}$$

Along the same line, we can show that

$$b_n = \frac{1}{L}\int_{-L}^{L} f(x)\sin\frac{n\pi x}{L}dx, \quad n = 1, 2\ldots \tag{1.16}$$

Example 1.6 Consider the function with period 8

$$f(x) = \begin{cases} 0 & -4 < x < -2 \\ 1 & -2 < x < 2 \\ 0 & 2 < x < 4 \end{cases}$$

This function is 1 for $-2 < x < 2$, $6 < x < 10$, etc. It is a regular pulse which is on for 4 units of time, and then off for four units of time. Since the period is not 2π, but instead $2L = 8$, we have $L = 4$. The Fourier coefficients are

$$a_0 = \frac{1}{4}\int_{-4}^{4} f(x)dx = \frac{1}{4}\int_{-2}^{2} dx = 1,$$

$$a_n = \frac{1}{L}\int_{-4}^{4} f(x)\cos\frac{n\pi x}{4}dx = \frac{1}{4}\int_{-2}^{2}\cos\frac{n\pi x}{4}dx$$

$$\begin{aligned}
&= \frac{1}{4}\frac{4}{n\pi}\sin\frac{n\pi x}{4}\bigg|_{-2}^{2} = \frac{1}{n\pi}\sin\frac{n\pi x}{4}\bigg|_{-2}^{2} \\
&= \frac{1}{n\pi}\left(\sin\frac{2n\pi}{4} - \sin\frac{-2n\pi}{4}\right) = \frac{1}{n\pi}(2\sin\frac{n\pi}{2}),
\end{aligned}$$

$$\begin{aligned}
b_n &= \frac{1}{L}\int_{-4}^{4} f(x)\sin\frac{n\pi x}{4}dx = \frac{1}{4}\int_{-2}^{2}\sin\frac{n\pi x}{4}dx \\
&= -\frac{1}{4}\frac{4}{n\pi}\cos\frac{n\pi x}{4}\bigg|_{-2}^{2} = -\frac{1}{n\pi}\cos\frac{n\pi x}{4}\bigg|_{-2}^{2} \\
&= -\frac{1}{n\pi}\left(\cos\frac{n\pi}{2} - \cos\frac{-n\pi}{2}\right) = \frac{1}{n\pi}(0) = 0.
\end{aligned}$$

If n is even, then $a_n = 0$ as sine is 0 at integer values. Thus, a_n contribute nonzero values for odd n, and so we may replace n by $2n-1$. With this in mind, the Fourier series can be written as

$$f(x) = \begin{cases} 0 & -4 < x < -2 \\ 1 & -2 < x < 2 \\ 0 & 2 < x < 4 \end{cases} = \frac{1}{2} + \frac{2}{\pi}\sum_{n=1}^{\infty}\frac{\sin\left(\frac{(2n-1)\pi}{2}\right)}{2n-1}\cos\left(\frac{(2n-1)\pi x}{4}\right).$$

□

Example 1.7 Use the result of Exercise 1.4 to obtain the Fourier series of

$$g(x) = x, \quad -3 < x \le 3, \quad \text{where} \quad g(x+6) = g(x).$$

The Fourier series obtained in Exercise 1.4 for

$$f(x) = x, \quad -\pi < x \le \pi,$$

is given by

$$x = \sum_{n=1}^{\infty}\frac{2}{n}(-1)^{n+1}\sin nx.$$

To adapt this result for $g(x)$, we rescale x in the Fourier series of $f(x)$ by replacing x with $\frac{\pi}{3}x$. This yields

$$\frac{\pi}{3}x = \sum_{n=1}^{\infty}\frac{2}{n}(-1)^{n+1}\sin\frac{n\pi}{3}x.$$

Solving for x, we obtain

$$x = \frac{3}{\pi}\sum_{n=1}^{\infty}\frac{2}{n}(-1)^{n+1}\sin\frac{n\pi}{3}x.$$

In conclusion, the Fourier series of $g(x)$ is

$$g(x) = \frac{6}{\pi}\sum_{n=1}^{\infty}\frac{(-1)^{n+1}}{n}\sin\frac{n\pi}{3}x.$$

□

The next theorem is an extension of Theorem 1.2 to arbitrary periods.

Theorem 1.3 *Suppose that $f(x) \in C_p(-L, L)$ and is periodic with period $2L$.*

- *If f is even, then its Fourier series has only cosine terms. The Fourier coefficients are given by*

$$b_n = 0, \ \ n = 1, 2 \ldots \ \ \textit{and} \ \ a_n = \frac{2}{L}\int_0^L f(x)\cos\frac{nx}{L}\,dx, \ \ n = 0, 1, \ldots \tag{1.17}$$

and the corresponding Fourier series is called a Fourier cosine series.

- *If f is odd, then its Fourier series has only sine terms. The Fourier coefficients are given by*

$$a_n = 0, \ \ n = 0, 1 \ldots \ \ \textit{and} \ \ b_n = \frac{2}{L}\int_0^L f(x)\sin\frac{nx}{L}\,dx, \ \ n = 0, 1, \ldots \tag{1.18}$$

and the corresponding Fourier series is called a Fourier sine series.

Now we define f_e and f_o that denote the even and odd extensions of $2L$-periodic function f, respectively.

$$f_e(x) = \begin{cases} f(x), & 0 < x < L \\ f(-x), & -L < x < 0 \end{cases}, \quad f_o(x) = \begin{cases} f(x), & 0 < x < L \\ -f(-x), & -L < x < 0 \end{cases}$$

For clarifications on such extensions we refer to Figs. 1.3–1.5. This type of expansion enables us to represent even or odd functions defined on $[0, L]$ in terms of their respective Fourier series. In summary, the *periodic odd extension* of a function f that is piecewise continuous on the interval $(0, L)$, is the *Fourier sine series* of f given by

$$f(x) = \sum_{n=1}^{\infty} b_n \sin\frac{nx}{L}, \quad 0 < x < L, \tag{1.19}$$

where

$$b_n = \frac{2}{\pi}\int_0^L f(x)\sin\frac{nx}{L}\,dx, \quad n = 1, 2, \ldots. \tag{1.20}$$

Similarly, the *periodic even extension* of a function f that is piecewise continuous on the interval $(0, L)$, is the *Fourier cosine series* of f given by

$$f(x) = \frac{a_0}{2} + \sum_{n=1}^{\infty} a_n \cos\frac{nx}{L}, \quad 0 < x < L, \tag{1.21}$$

where

$$a_0 = \frac{2}{L}\int_0^L f(x)dx \quad \text{and} \quad a_n = \frac{2}{L}\int_0^L f(x)\cos\frac{nx}{L}\,dx, \quad n = 1, 2, \ldots.$$

If the function f is defined over the interval $[-L, L]$ and periodic with period $2L$, we test to determine whether f is even, odd, or neither. If f is an even function, it gives rise to a Fourier cosine series. If f is an odd function, it gives rise to a Fourier sine series. If f is neither even nor odd, it gives rise to the general Fourier series, given by (1.13).

Example 1.8 Consider the L periodic triangle function

$$f(x) = \begin{cases} x, & 0 \leq x \leq \frac{L}{2} \\ L - x, & \frac{L}{2} \leq x \leq L \end{cases}$$

The Fourier cosine series has coefficients

$$a_0 = \frac{2}{L}\int_0^{\frac{L}{2}} x\,dx + \frac{2}{L}\int_{\frac{L}{2}}^{L} (L-x)\,dx = \frac{1}{2},$$

and

$$a_n = \frac{2}{L}\int_0^{\frac{L}{2}} x\cos\frac{n\pi x}{L}\,dx + \frac{2}{L}\int_{\frac{L}{2}}^{L} (L-x)\cos\frac{n\pi x}{L}\,dx.$$

We note that in this example we have to adjust our period since it is L and not $2L$. Integration by parts gives

$$\begin{aligned} \int_0^{\frac{L}{2}} x\cos\frac{n\pi x}{L}\,dx &= \left(x\frac{L}{n\pi}\sin\frac{n\pi x}{L} + \frac{L^2}{n^2\pi^2}\cos\frac{n\pi x}{L} \right)\Bigg|_0^{\frac{L}{2}} \\ &= \left(\frac{L^2}{2n\pi}\sin\frac{n\pi}{2} + \frac{L^2}{n^2\pi^2}\cos\frac{n\pi}{2} \right) - \frac{L^2}{n^2\pi^2}. \end{aligned}$$

Similarly, we compute

$$\begin{aligned} \int_{\frac{L}{2}}^{L} (L-x)\cos\frac{n\pi x}{L}\,dx &= \left((L-x)\frac{L}{n\pi}\sin\frac{n\pi x}{L} - \frac{L^2}{n^2\pi^2}\cos\frac{n\pi x}{L} \right)\Bigg|_{\frac{L}{2}}^{L} \\ &= \left(0 - \frac{L^2}{n^2\pi^2}\cos n\pi \right) - \left(\frac{L^2}{2n\pi}\sin\frac{n\pi}{2} - \frac{L^2}{n^2\pi^2}\cos\frac{n\pi}{2} \right). \end{aligned}$$

Combining these results, we arrive at

$$\begin{aligned} a_n &= \frac{2}{L}\left(\frac{L^2}{2n\pi}\sin\frac{n\pi}{2} + \frac{L^2}{n^2\pi^2}\cos\frac{n\pi}{2} - \frac{L^2}{n^2\pi^2} \right. \\ &\quad \left. - \frac{L^2}{n^2\pi^2}\cos n\pi - \frac{L^2}{2n\pi}\sin\frac{n\pi}{2} + \frac{L^2}{n^2\pi^2}\cos\frac{n\pi}{2} \right) \\ &= \frac{2L}{n^2\pi^2}\left(2\cos\frac{n\pi}{2} - 1 - \cos n\pi \right), \quad n = 1, 2\ldots \end{aligned}$$

In conclusion

$$a_0 = \frac{1}{2}, \quad a_2 = -\frac{8L}{2^2\pi^2}, \quad a_6 = -\frac{8L}{6^2\pi^2}, \quad a_{10} = -\frac{8L}{10^2\pi^2}, \ldots$$

and that $a_n = 0$ for all odd n and multiples of 4. Hence, the even expansion of f has the Fourier series

$$f(x) = \frac{1}{2} - \frac{8L}{\pi^2}\left(\frac{1}{2^2}\cos\frac{2\pi x}{L} + \frac{1}{6^2}\cos\frac{6\pi x}{L} + \frac{1}{10^2}\cos\frac{10\pi x}{L} + \cdots\right).$$

Similar computations show that

$$b_n = \frac{4L}{n^2\pi^2}\sin\frac{n\pi}{2}, \quad n = 1, 2, \ldots$$

which means $b_n = 0$ for all even n. Therefore, the half-range sine series for f is

$$f(x) = \frac{4L}{\pi^2}\left(\frac{1}{1^2}\sin\frac{\pi x}{L} - \frac{1}{3^2}\sin\frac{3\pi x}{L} + \frac{1}{5^2}\sin\frac{5\pi x}{L} - \cdots\right).$$

□

Example 1.9 For a positive constant L, let $f : [0, L] \to \mathbb{R}$ be the function defined by

$$f(x) = x(L - x), \quad 0 \le x \le L.$$

This function can be expanded in a sine series of the form

$$f(x) = \sum_{n=1}^{\infty} b_n \sin\frac{n\pi x}{L},$$

where

$$b_n = \frac{2}{L}\int_0^L f(x)\sin\frac{n\pi x}{L}dx.$$

In other words, we are tasked with creating a periodic odd extension of $f(x)$, where

$$f_o(x) = \begin{cases} x(L-x), & 0 < x < L \\ x(L+x), & -L < x < 0 \end{cases}$$

Using the method of integration by parts, and the result that $\sin(n\pi) = 0$ *and* $\cos(n\pi) = (-1)^n$ for all integers n, we find

$$b_n = \frac{2}{L}\int_0^L x(L-x)\sin\frac{n\pi x}{L}dx = -\frac{2}{n\pi}\int_0^L x(L-x)\frac{d}{dx}\left[\cos\frac{n\pi x}{L}\right]dx.$$

Expanding further, we obtain

$$b_n = -\frac{2}{n\pi}\left[x(L-x)\cos\frac{n\pi x}{L}\right]_0^L + \frac{2}{n\pi}\int_0^L (L-2x)\cos\frac{n\pi x}{L}dx.$$

The boundary terms vanish, so

$$b_n = \frac{2}{n\pi}\int_0^L (L-2x)\cos\frac{n\pi x}{L}dx = \frac{2L}{n^2\pi^2}\int_0^L (L-2x)\frac{d}{dx}\left[\sin\frac{n\pi x}{L}\right]dx.$$

An integration by parts yields,

$$b_n = \frac{2L}{n^2\pi^2}\left[(L-2x)\sin\frac{n\pi x}{L}\right]_0^L - \frac{2L}{n^2\pi^2}\int_0^L (-2)\sin\frac{n\pi x}{L}dx.$$

Simplifying further

$$b_n = \frac{4l}{n^2\pi^2}\int_0^L \sin\frac{n\pi x}{L}dx = -\frac{4L^2}{n^3\pi^3}\left[\cos\frac{n\pi x}{L}\right]_0^L.$$

Using $\cos(n\pi) = (-1)^n$ and $\cos(0) = 1$, we have

$$b_n = \frac{4L^2}{n^3\pi^3}(1-(-1)^n), \quad n = 1, 2, \ldots.$$

Thus

$$b_n = \begin{cases} \frac{8L^2}{n^3\pi^3}, & \textit{if n is odd} \\ 0, & \textit{if n is even} \end{cases}$$

and consequently,

$$f(x) = \sum_{n=1}^{\infty} \frac{8L^2}{(2n-1)^3\pi^3}\sin\frac{(2n-1)\pi x}{L}.$$

□

Example 1.10 Suppose that we want to find the Fourier series for

$$f(x) = \begin{cases} 2-x, & 0<x<4 \\ x-6, & 4<x<8 \end{cases}, \quad f(x+8) = f(x)$$

To save time and effort, we analyze the given function to determine whether it is even, odd, or neither. Hence, we need to verify this by examining how the function is defined for negative values of x. In the interval $-4 < x < 0$ we can see from the periodicity that $f(x) = f(x+8)$ so for $x \in (-4, 0)$

$$f(x) = f(x+8) = (x+8) - 6 = x+2.$$

Consequently, for, $x \in (-4, 0)$ we have $f(-x) = -x+2 = f(x)$. Hence, the function is even. To be surer, let's consider the other interval $(-8, -4)$. Since $f(x) = f(x+8)$, we have for $x \in (-8, -4)$

$$f(x) = f(x+8) = 2-(x+8) = -x-6.$$

As a result, we have $f(-x) = x-6 = f(x)$ for $x \in (-8, 4)$. Hence, the function is even and luckily $b_n = 0$ for all n. Now,

$$a_n = \frac{1}{4}\int_0^8 f(x)\cos\frac{\pi n x}{4}dx, \quad n = 1, 2\ldots$$

and, using the definition of $f(x)$, we get

$$4a_n = \int_0^4 (2-x)\cos\frac{\pi nx}{4}\,dx + \int_4^8 (x-6)\cos\frac{\pi nx}{4}\,dx.$$

We will attempt to combine both integrals by using an appropriate substitution. By performing a change of variables in the second integral, we hope to make it equal to the first. So if we let $x-6=2-s$, then $x=8-s$, and consequently,

$$\cos\frac{\pi nx}{4} = \cos\frac{\pi n(8-s)}{4} = \cos\left(-\frac{\pi ns}{4}+2\pi\right) = \cos\left(-\frac{\pi ns}{4}\right) = \cos\frac{\pi ns}{4}$$

where we have used the periodicity $\cos(\theta+2\pi) = \cos\theta$ and the evenness $\cos(-\theta) = \cos\theta$ of cosine.
Since $ds = -dx$, the second integral is the same as the first and so

$$4a_n = 2\int_0^4 (2-x)\cos\frac{\pi nx}{4}\,dx.$$

Lets integrate that by parts, with $u = 2-x$, so

$$2a_n = \left[(2-x)\frac{4}{n\pi}\sin\frac{\pi nx}{4}\right]_0^4 + \frac{4}{n\pi}\int_0^4 \sin\frac{\pi nx}{4}\,dx.$$

Now the first term has a zero from the sine at both ends, the second term needs to be integrated again to give

$$2a_n = \left[-\frac{16}{n^2\pi^2}\cos\frac{\pi nx}{4}\right]_0^4$$

and using $\cos\pi n = (-1)^n$, we get $a_n = \frac{16}{n^2\pi^2}$ for n odd, and $a_n = 0$ for n even. Finally

$$a_n = \begin{cases} \frac{16}{n^2\pi^2}, & \textit{if n is odd} \\ 0, & \textit{if n is even} \end{cases}$$

It is easy to show that

$$a_0 = \frac{1}{4}\int_0^4 (2-x)\,dx + \int_4^8 (x-6)\,dx = 0.$$

Hence,

$$f(x) = \sum_{n=1}^{\infty} \frac{16}{(2n-1)^2\pi^2}\cos\frac{(2n-1)\pi x}{8}.$$

□

1.3.1 Exercises

Exercise 1.25 *Provide all details for the Fourier coefficient b_n in Example 1.8.*

Exercise 1.26 *For nonzero constant k, find the Fourier series of the function f, where $f(x+2) = f(x)$ and*

$$f(x) = \begin{cases} e^{kx}, & 0 \le x \le 1 \\ 0, & -1 \le x < 0 \end{cases}$$

Ans: $\frac{1}{2k}(e^k - 1) + \sum_{n=1}^{\infty} \frac{k}{k^2 + n^2\pi^2}(e^k - 1)\cos n\pi x - \sum_{n=1}^{\infty} \frac{2n\pi}{k^2 + n^2\pi^2}(e^k - 1)\sin n\pi x.$

Exercise 1.27 *Let $f(x) = e^{kx}$, $0 < x < 1$, where k is a nonzero constant and $f(x+1) = f(x)$. Use the result of Exercise 1.26 and show that the Fourier series of $f(x)$ is given by*

$$\frac{1}{k}(e^k - 1) + \sum_{n=1}^{\infty} \frac{2k}{k^2 + 4n^2\pi^2}(e^k - 1)\cos 2n\pi x - \sum_{n=1}^{\infty} \frac{4n\pi}{k^2 + 4n^2\pi^2}(e^k - 1)\sin 2n\pi x.$$

Exercise 1.28 *Find $f_e(x)$ of the function of Example 1.9 and find its Fourier cosine series.*

Exercise 1.29 *Find the Fourier series of*

$$f(x) = |x|, \quad -\sqrt{3} < x < \sqrt{3}, \quad f(x+2\sqrt{3}) = f(x).$$

Ans: $\frac{\sqrt{3}}{2} - \frac{4\sqrt{3}}{\pi^2} \sum_{n=1}^{\infty} \frac{\cos\left((2n-1)\frac{\pi x}{\sqrt{3}}\right)}{(2n-1)^2}.$

Exercise 1.30 *Find the Fourier series of*

$$f(x) = \cos x, \quad -\frac{\pi}{2} < x < \frac{\pi}{2}, \quad f(x+\pi) = f(x).$$

Ans: $\frac{2}{\pi} + \frac{4}{\pi} \sum_{n=1}^{\infty} \frac{(-1)^{n+1}}{4n^2 - 1} \cos 2nx.$

Exercise 1.31 *Use the results Find the Fourier series of $f(x) = x^2, \; -\pi < x < \pi$.*

Exercise 1.32 *Find the Fourier series of*

$$f(x) = \sin^2\left(\frac{\pi x}{\sqrt{3}}\right), \quad -\sqrt{3} < x < \sqrt{3}, \quad f(x+2\sqrt{3}) = f(x).$$

Exercise 1.33 *Let f be $2l$ periodic function with $f(x+l) = -f(x)$.*
(a) Show that its Fourier coefficients a_n and b_n are zero whenever n is even.
(b) Show that for odd n, we have

$$a_n = \frac{2}{l} \int_0^l f(x) \cos \frac{n\pi x}{l} dx, \quad b_n = \frac{2}{l} \int_0^l f(x) \sin \frac{n\pi x}{l} dx.$$

Exercise 1.34 *Show the function*

$$f(x)=\begin{cases}1-x, & 0<x<2\\ x-3, & 2<x<4\end{cases}, \quad f(x+4)=f(x)$$

is even and find its Fourier series.

Exercise 1.35 *Find the Fourier series of*

$$f(x)=2-2\left(\frac{x}{c}\right)^2, \quad -c\le x\le c.$$

Ans: $\dfrac{4}{3}+8\sum_{n=1}^{\infty}\dfrac{(-1)^{n+1}}{n^2\pi^2}\cos\dfrac{n\pi x}{c}.$

Exercise 1.36 *For $a\neq 0$, find the Fourier series of*

$$f(x)=\begin{cases}\frac{-2a}{d}\left(x-\frac{d}{2}\right), & 0\le x\le d\\ \frac{2a}{d}\left(x+\frac{d}{2}\right), & -d\le x\le 0\end{cases}$$

Ans: $\dfrac{8a}{\pi^2}\sum_{n=1}^{\infty}\dfrac{1}{(2n+1)^2}\cos\dfrac{(2n+1)\pi x}{d}.$

Exercise 1.37 *Let f be a 2ℓ-periodic function with $f(x+2\ell)=f(x)$.*
(a) Show that for any real constant c,

$$\int_c^{c+2\ell}f(x)\,dx=\int_0^{2\ell}f(x)\,dx=\int_{-\ell}^{\ell}f(x)\,dx.$$

(b) Show that the Fourier coefficients of f satisfy

$$a_n=\frac{1}{\ell}\int_c^{2\ell+c}f(x)\cos\frac{n\pi x}{\ell}\,dx, \quad n=0,1,\dots$$

and

$$b_n=\frac{1}{\ell}\int_c^{2\ell+c}f(x)\sin\frac{n\pi x}{\ell}\,dx, \quad n=1,2,\dots$$

Exercise 1.38 ***(Odd half-multiple sine series)*** *Assume $f(x)$ for $0<x<L$, and define $F(x)$ for $0<x<2L$ by*

$$F(x)=\begin{cases}f(x), & 0<x<L\\ f(2L-x), & L<x<2L\end{cases}$$

In other words, the graph of $F(x)$ is symmetric around the line $x=L$.

(a) Show that the period-4L Fourier sine series of F is

$$F(x)=\sum_{n=1}^{\infty}b_n\sin\frac{n\pi x}{2L},$$

where

$$b_n = \frac{1}{L}\int_0^L f(x)\sin\frac{n\pi x}{2L}\,dx + \frac{1}{L}\int_L^{2L} f(2L-x)\sin\frac{n\pi x}{2L}\,dx.$$

(b) Substitute $s = 2L - x$ in the second integral to derive the series

$$f(x) = \sum_{n=1} b_{2n-1}\sin\frac{(2n-1)\pi x}{2L}, \quad 0 < x < L,$$

where

$$b_{2n-1} = \frac{2}{L}\int_0^L f(x)\sin\frac{(2n-1)\pi x}{2L}dx, \quad n = 1,2,\ldots$$

Exercise 1.39 ***(Odd half-multiple cosine series)*** *Assume $f(x)$ for $0 < x < L$, and define $F(x)$ for $0 < x < 2L$ by*

$$F(x) = \begin{cases} f(x), & 0 < x < L \\ -f(2L-x), & L < x < 2L \end{cases}$$

Show by using the period-4L Fourier cosine series of H that

$$f(x) = \sum_{n=1} a_{2n-1}\cos\frac{(2n-1)\pi x}{2L}, \quad 0 < x < L,$$

where

$$a_{2n-1} = \frac{2}{L}\int_0^L f(x)\cos\frac{(2n-1)\pi x}{2L}dx, \quad n = 1,2,\ldots$$

Exercise 1.40 *With the aid of Exercise 1.38, show that if :*

$$f(x) = x, \quad 0 < x < \pi$$

then its odd half-range sine series is

$$f(x) = 8\sum_{n=1}^{\infty}\frac{(-1)^{n-1}}{(2n-1)^2}\sin(2n-1)x.$$

1.4 Complex Form of Fourier Series

In this section we derive the complex form of Fourier series. The complex form is particularly useful in physics and electrical engineering. It can also help identify the function corresponding to a Fourier series more easily. Let $f(x)$ be a function that is

$2L$-periodic with $f(x) \in C'_p(-L,L)$. As before, we assume the function $f(x)$ equals its Fourier series

$$f(x) = \frac{a_0}{2} + \sum_{n=1}^{\infty} \left(a_n \cos \frac{n\pi x}{L} + b_n \sin \frac{n\pi x}{L} \right).$$

Using the famous Euler formula

$$e^{i\theta} = \cos(\theta) + i\sin(\theta), \quad \text{where} \quad i^2 = -1,$$

along with

$$\cos(\theta) = \frac{1}{2}\left(e^{i\theta} + e^{-i\theta}\right), \quad \sin(\theta) = \frac{1}{2i}\left(e^{i\theta} - e^{-i\theta}\right),$$

and then substituting the exponential forms into the above Fourier series of $f(x)$, we arrive at the equivalent form

$$\begin{aligned} f(x) &= \frac{a_0}{2} + \frac{1}{2}\sum_{n=1}^{\infty} \left[a_n \left(e^{i\frac{n\pi x}{L}} + e^{-i\frac{n\pi x}{L}} \right) - ib_n \left(e^{i\frac{n\pi x}{L}} - e^{-i\frac{n\pi x}{L}} \right) \right] \\ &= \frac{a_0}{2} + \frac{1}{2}\sum_{n=1}^{\infty} \left[(a_n - ib_n)e^{i\frac{n\pi x}{L}} + (a_n + ib_n)e^{-i\frac{n\pi x}{L}} \right] \\ &= c_0 + \sum_{n=1}^{\infty} \left[c_n e^{i\frac{n\pi x}{L}} + c_{-n} e^{-i\frac{n\pi x}{L}} \right], \end{aligned}$$

where

$$c_0 = \frac{a_0}{2}, \quad c_n = \frac{1}{2}(a_n - ib_n), \quad c_{-n} = \frac{1}{2}(a_n + ib_n), \quad n = 1, 2, 3, \ldots$$

Notice that c_n is the conjugate of c_{-n}. That is

$$c_n = \bar{c}_{-n}, \quad \text{or} \quad c_n \cdot c_{-n} = |c_n|^2.$$

Next, we compute the Fourier coefficients c_n, $n = 0, \pm 1 \ldots$

$$\begin{aligned} c_n &= \frac{1}{2}(a_n - ib_n) = \frac{1}{2}\left[\frac{1}{L}\int_{-L}^{L} f(x)\cos\frac{n\pi x}{L}\,dx - \frac{i}{L}\int_{-L}^{L} f(x)\sin\frac{n\pi x}{L}\,dx \right] \\ &= \frac{1}{2L}\left[\int_{-L}^{L} f(x)\left(\cos\frac{n\pi x}{L} - i\sin\frac{n\pi x}{L}\,dx \right) \right] = \frac{1}{2L}\int_{-L}^{L} f(x)e^{-i\frac{n\pi x}{L}}\,dx. \end{aligned}$$

Along the same lines, it can be shown that

$$c_{-n} = \frac{1}{2L}\int_{-L}^{L} f(x)e^{i\frac{n\pi x}{L}}\,dx, \quad n = 1, 2, \ldots \tag{1.22}$$

Thus, if $f(x) \in C'_p(-L,L)$ and $2L$-periodic, then the complex form of its Fourier series is given by

$$f(x) = \sum_{n=-\infty}^{\infty} c_n e^{i\frac{n\pi x}{L}}, \quad -L < x < L \tag{1.23}$$

where

$$c_n = \frac{1}{2L}\int_{-L}^{L} f(x)e^{-i\frac{n\pi x}{L}}\,dx, \quad n = 0, \pm 1\ldots \tag{1.24}$$

Remark 3 *If $f(x) \in C_p'(-\pi,\pi)$ and 2π-periodic, then replace L with π.*

We provide the following example.

Example 1.11 For a nonzero constant a, we want to use the complex form to find the Fourier series of $f(x) = e^{ax}$, where $f(x+2l) = f(x)$. We compute the coefficients c_n using (1.24).

$$c_n = \frac{1}{2l}\int_{-l}^{l} f(x)e^{-i\frac{n\pi x}{l}}\,dx = \frac{1}{2l}\int_{-l}^{l} e^{ax}e^{-\frac{n\pi x}{l}}\,dx = \frac{1}{2l}\int_{-l}^{l} e^{\left(a-i\frac{n\pi}{l}\right)x}\,dx.$$

After evaluating the integral and then simplifying, we get

$$c_n = \frac{1}{2l}\frac{1}{a-i\frac{n\pi}{l}}\left[e^{al-in\pi} - e^{-al+in\pi}\right].$$

Using Euler's formula $e^{in\pi} = (-1)^n + i\cdot 0 = (-1)^n$ and $e^{-in\pi} = (-1)^n$, the above expression becomes

$$c_n = \frac{(-1)^n}{2l}\frac{e^{al} - e^{-al}}{a-i\frac{n\pi}{l}}.$$

Since $2\sinh(al) = e^{al} - e^{-al} = 2\sinh(al)$, we find, after multiplying and dividing c_n by the conjugate of its denominator, that

$$c_n = \frac{(-1)^n\sinh(al)}{l\left(a-i\frac{n\pi}{l}\right)} = \sinh(al)\frac{(-1)^n(al+in\pi)}{a^2l^2+n^2\pi^2}. \tag{1.25}$$

Substitute into (1.23), we obtain the complex form of the Fourier series

$$f(x) = \sinh(al)\sum_{n=-\infty}^{\infty}\frac{(-1)^n(al+in\pi)}{a^2l^2+n^2\pi^2}e^{i\frac{n\pi x}{l}}. \tag{1.26}$$

To find the coefficients a_n and b_n we make use of (1.25). That is

$$a_n - ib_n = 2c_n = 2\sinh(al)\left(\frac{(-1)^n al}{a^2l^2+n^2\pi^2} + i\frac{(-1)^n n\pi}{a^2l^2+n^2\pi^2}\right).$$

By equating both sides of the above expression we arrive at

$$a_n = 2\sinh(al)\frac{(-1)^n al}{a^2l^2+n^2\pi^2}, \quad n = 0, 1\ldots$$

$$b_n = 2\sinh(al)\frac{(-1)^{n+1} n\pi}{a^2l^2+n^2\pi^2}, \quad n = 1, 2\ldots$$

Hence, in terms of sine and cosine, the Fourier series of

$$f(x) = e^{ax}, \quad -l < x < l,$$

is

$$\begin{aligned} f(x) &= \frac{a_0}{2} + \sum_{n=1}^{\infty}\left(a_n \cos\frac{n\pi x}{l} + b_n \sin\frac{n\pi x}{l}\right) \\ &= \frac{\sinh(al)}{al} + 2\sinh(al)\sum_{n=1}^{\infty}\left(\frac{(-1)^n al}{a^2l^2+n^2\pi^2}\cos\frac{n\pi x}{l} + \frac{(-1)^{n+1}n\pi}{a^2l^2+n^2\pi^2}\sin\frac{n\pi x}{l}\right). \end{aligned}$$

It can be deduced from Chapter 3 that the Fourier sine and cosine series obtained above converges to $\frac{e^{-al}+e^{al}}{2}$ at the jump discontinuity $x = jl,\ j = \pm 1, \pm 2 \ldots$ and to the periodic extension of $f(x)$ otherwise. □

Example 1.12 The following Fourier cosine series

$$\sum_{n=1}^{\infty}\frac{(-1)^{n+1}}{n}\cos nx, \quad -\pi < x < \pi,$$

may be considered the real part of

$$\sum_{n=1}^{\infty}\frac{(-1)^{n+1}}{n}e^{inx} = \sum_{n=1}^{\infty}\frac{(-1)^{n+1}}{n}(e^{ix})^n$$

since the real part of $e^{i\theta}$ is $\cos\theta$. Moreover, the right-hand series is recognized as a Taylor series,

$$\sum_{n=1}^{\infty}\frac{(-1)^{n+1}}{n}e^{ixn} = \ln(1+e^{ix}).$$

By rewriting

$$1+e^{ix} = e^{ix/2}\left(e^{ix/2}+e^{-ix/2}\right) = 2e^{ix/2}\cos\frac{x}{2},$$

we obtain

$$\ln(1+e^{ix}) = i\frac{x}{2} + \ln\left(2\cos\frac{x}{2}\right).$$

The real part of

$$\ln(1+e^{ix}) \text{ is } \ln(2\cos(x/2)), \quad -\pi < x < \pi.$$

Consequently, we arrive at the relation

$$\ln\left(2\cos\frac{x}{2}\right) \sim \sum_{n=1}^{\infty}\frac{(-1)^{n+1}}{n}\cos nx, \quad -\pi < x < \pi.$$

Notice that the series actually converges except at $x = \pm\pi, \pm 3\pi, \ldots$.

We end this section with the following theorem regarding the **frequency** of function. Its proof is left as an exercise.

Theorem 1.4 *(Frequency form) Suppose that the function f is T-periodic. Let $\omega = \frac{2\pi}{T}$, where ω is the frequency of f. The complex form of its Fourier series can then be expressed as*

$$f(x) = \sum_{n=-\infty}^{\infty} d_n e^{in\omega x},$$

where

$$d_n = \frac{\omega}{2\pi} \int_0^{\frac{2\pi}{\omega}} f(x) e^{-in\omega x}\, dx, \quad n = 0, \pm 1, \ldots$$

1.4.1 Exercises

Exercise 1.41 Prove (1.22).

Exercise 1.42 Show that

$$\int_{-\pi}^{\pi} e^{inx} e^{-imx}\, dx = \begin{cases} 0, & n \neq m \\ 2\pi, & n = m. \end{cases}$$

The orthogonality over the interval $-\pi \leq x \leq \pi$ is a fundamental property of the exponential functions in the Fourier series. Consequently, show that the coefficients c_n of the complex Fourier of a function $f(x)$ with period 2π is

$$c_n = \frac{1}{2\pi} \int_{-\pi}^{\pi} f(x) e^{-inx}\, dx, \quad \text{where} \quad n = 0, \pm 1, \pm 2, \ldots$$

Exercise 1.43 For a nonzero constant a, find the complex form of the Fourier series of

$$f(x) = \begin{cases} ax, & 0 < x < l \\ 0, & l < x < 2l. \end{cases}$$

Ans: $c_n = \dfrac{al\left[(n\pi i + 1)(-1)^n - 1\right]}{2(n\pi)^2}, \quad n = \pm 1, \pm 2 \ldots$

Exercise 1.44 Let the function f be 2π-periodic. Use the identity

$$\sin(\theta) = \frac{1}{2i}\left(e^{i\theta} - e^{-i\theta}\right),$$

to obtain the complex form of the Fourier sine series

$$f(x) = \sum_{n=1}^{\infty} b_n \sin nx.$$

Exercise 1.45 For a nonzero constant $a \neq n$, find the complex form of the Fourier series of $f(x) = \cos ax, \ -\pi < x < \pi$.

Ans: $c_n = \dfrac{a \sin(a\pi)}{\pi(n^2 - a^2)}(-1)^{n+1}$.

Exercise 1.46 Prove Theorem 1.4.

Exercise 1.47 Suppose that $f(x)$ is periodic with period $2c$ such that

$$f(x+c) = -f(x).$$

(a) Show that the complex Fourier coefficient, c_n satisfies $c_{2n} = 0$, $n = 1, 2\ldots$
(b) With the aid of part (a), find the complex form of the Fourier series of $f(x) = \cos(x)$, $-\pi < x < \pi$.

Exercise 1.48 *(Parseval formula) Multiply* (1.23) *by* $\frac{1}{2L}f(x)$ *and then integrate the resulting equation from* $-L$ *to* L *to obtain the Parseval formula*

$$\frac{1}{2L}\int_{-L}^{L} f^2(x)\,dx = \sum_{n=-\infty}^{\infty} |c_n|^2.$$

Exercise 1.49 *In Example 1.11, let* $a = 1$, *and* $L = \pi$. *With the aid of Parseval formula, show that*

$$\sum_{n=-\infty}^{\infty} \frac{1}{1+n^2} = \pi \coth \pi.$$

Exercise 1.50 *Relate the functions and series below by using the complex form and Taylor series.*

(a)

$$1 + \sum_{n=1}^{\infty} a^n \cos nx = \frac{1 - a\cos x}{1 - 2a\cos x + a^2}, \quad 0 \le a < 1.$$

(b)

$$\sum_{n=1}^{\infty} \frac{\sin nx}{n!} = e^{\cos x} \sin(\sin x).$$

2

Theory of Fourier Series

In Chapter 1, we posed several questions about the conditions under which the equality holds between $f(x)$ and its Fourier series. In this chapter, we aim to address these questions by exploring the necessary and sufficient conditions that the function f must satisfy.

2.1 Bessel's Inequality and its Consequences

Square-integrable functions are typically associated with physical systems that have finite energy. For a finite Fourier series with N terms, we derive the Bessel Inequality, where N can be extended to infinity, given that the function f is square-integrable. The Bessel Inequality becomes an equality if and only if the partial sum of Fourier series $S_N(x)$ converges to f in the mean.

Definition 2.1 *We say a function $f(x)$ is square-integrable on $[-L,L]$, if*

$$\int_{-L}^{L} [f(x)]^2\,dx < \infty.$$

In which case we write $f \in L^2[-L,L]$.

In the next lemma we state and prove Bessel's inequality.

Lemma 2 *(Bessel's Inequality) Suppose that the function $f(x)$ is defined on $[-\pi,\pi]$ and $f \in L^2[-\pi,\pi]$. Then,*

$$\frac{a_0^2}{2} + \sum_{n=1}^{\infty}\left(a_n^2 + b_n^2\right) \le \frac{1}{\pi}\int_{-\pi}^{\pi} f^2(x)\,dx, \tag{2.1}$$

where a_n and b_n are the Fourier coefficients for f.

Proof *We begin by mentioning that the orthogonality of sine and cosine functions is crucial in this proof, as it allows us to eliminate all cross-terms when integrating the expanded square of the Fourier series.*

DOI: 10.1201/9781003652540-2

To establish the equality between f and its Fourier series, we start by writing

$$f(x) \sim \frac{a_0}{2} + \sum_{n=1}^{\infty} (a_n \cos nx + b_n \sin nx).$$

Then the partial sum of the above Fourier series, denoted by $S_N(x)$, is given by

$$S_N(x) = \frac{a_0}{2} + \sum_{n=1}^{N} (a_n \cos nx + b_n \sin nx),$$

where a_0, a_n, and b_n represent the Fourier coefficients of the function f.

Now,

$$[f(x) - S_N(x)]^2 = f^2(x) - 2f(x)S_N(x) + S_N^2(x).$$

An integration from $-\pi$ to π of the above expression yields

$$\begin{aligned}\int_{-\pi}^{\pi} [f(x) - S_N(x)]^2 \, dx &= \int_{-\pi}^{\pi} f^2(x)\, dx - 2\int_{-\pi}^{\pi} f(x)S_N(x)\, dx \\ &+ \int_{-\pi}^{\pi} S_N^2(x)\, dx. \end{aligned} \tag{2.2}$$

Next, we evaluate each of the last two terms of the right-hand side of (2.2). In doing so, we repeatedly make use of (1.4) – (1.6). We begin by multiplying $S_N(x)$ with $f(x)$ and then integrate from $-\pi$ to π and obtain

$$\begin{aligned}\int_{-\pi}^{\pi} f(x)S_N(x)\, dx &= \frac{a_0}{2}\int_{-\pi}^{\pi} f(x)\, dx + \sum_{n=1}^{N} a_n \int_{-\pi}^{\pi} f(x) \cos n\pi x\, dx + b_n \int_{-\pi}^{\pi} f(x) \sin nx\, dx \\ &= \left[\frac{a_0}{2}(\pi a_0) + \sum_{n=1}^{N} \Big(a_n(\pi a_n) + b_n(\pi b_n)\Big)\right]. \\ &= \pi\left[\frac{a_0^2}{2} + \sum_{n=1}^{N} \left(a_n^2 + b_n^2\right)\right].\end{aligned}$$

We square $S_N(x)$ then integrate from $-\pi$ to π.

$$\int_{-\pi}^{\pi} S_N^2(x)\, dx = \int_{-\pi}^{\pi} \left[\frac{a_0}{2} + \sum_{n=1}^{N} (a_n \cos nx + b_n \sin nx)\right]^2 dx.$$

By applying the orthogonality properties of cosine and sine on the interval $[-\pi, \pi]$, and considering the finite nature of the summation, we arrive at

$$\begin{aligned}\int_{-\pi}^{\pi} S_N^2(x)\, dx &= \int_{-\pi}^{\pi} \frac{a_0^2}{4}\, dx + \sum_{n=1}^{N} a_n^2 \int_{-\pi}^{\pi} \cos^2 nx\, dx + b_n^2 \int_{-\pi}^{\pi} \sin^2 nx\, dx \\ &= \pi\left[\frac{a_0^2}{2} + \sum_{n=1}^{N} (a_n^2 + b_n^2)\right].\end{aligned}$$

Therefore, (2.2) *reduces to*

$$\begin{aligned}\int_{-\pi}^{\pi}[f(x)-S_N(x)]^2\,dx &= \int_{-\pi}^{\pi} f^2(x)\,dx - 2\pi\Big[\frac{a_0^2}{2}+\sum_{n=1}^{N}(a_n^2+b_n^2)\Big] \\ &+ \pi\Big[\frac{a_0^2}{2}+\sum_{n=1}^{N}(a_n^2+b_n^2)\Big] \\ &= \int_{-\pi}^{\pi} f^2(x)\,dx - \pi\Big[\frac{a_0^2}{2}+\sum_{n=1}^{N}(a_n^2+b_n^2)\Big]. \qquad (2.3)\end{aligned}$$

Since the left-hand side of (2.3) *is nonnegative, we see that*

$$\frac{a_0^2}{2}+\sum_{n=1}^{N}(a_n^2+b_n^2) \le \frac{1}{\pi}\int_{-\pi}^{\pi} f^2(x)\,dx. \qquad (2.4)$$

Since $f \in L^2[-\pi,\pi]$, *the right-hand side of* (2.4) *is uniformly bounded, then by letting* $N \to \infty$ *in* (2.4) *we arrive at the Bessel's inequality*

$$\frac{a_0^2}{2}+\sum_{n=1}^{\infty}(a_n^2+b_n^2) \le \frac{1}{\pi}\int_{-\pi}^{\pi} f^2(x)\,dx.$$

This completes the proof. □

Bessel's inequality is easily extended to arbitrary intervals of the form $[-L,L]$, and as a result we have

$$\frac{a_0^2}{2}+\sum_{n=1}^{\infty}(a_n^2+b_n^2) \le \frac{1}{L}\int_{-L}^{L} f^2(x)\,dx. \qquad (2.5)$$

We have the immediate corollary.

Corollary 1 *Let* a_n, b_n *be the Fourier coefficients of* f *given by* (1.5) *and* (1.6). *Then, under the assumptions of Lemma 2 we have*

$$\lim_{n\to\infty} a_n = \lim_{n\to\infty} b_n = 0.$$

Proof *From Bessel's inequality, or* (2.1) *we have*

$$\sum_{n=1}^{\infty}(a_n^2+b_n^2) < \infty,$$

or converges. Hence

$$\lim_{n\to\infty}(a_n^2+b_n^2)=0 \quad \textit{if and only if} \quad \lim_{n\to\infty} a_n = \lim_{n\to\infty} b_n = 0.$$

This completes the proof. □

Remark 4 *Under the assumptions of Lemma 2 and the results of Corollary 1, we derive the following immediate results, which state*

$$\lim_{n\to\infty}\int_{-\pi}^{\pi} f(x)\cos n\pi x\,dx = \lim_{n\to\infty}\int_{-\pi}^{\pi} f(x)\sin n\pi x\,dx = 0.$$

The following definition will be examined in greater detail later in this chapter.

Definition 2.2 *Let f_n be a sequence of functions in $L^2[-\ell,\ell]$. The sequence is said to converge in the mean, or in L^2 to a function $f \in L^2[-\ell,\ell]$, if the mean square error*

$$\int_{-\ell}^{\ell} |f_n(x) - f(x)|^2\,dx \to 0 \quad \textit{as} \quad n \to \infty.$$

The sequence converges pointwise to f if

$$\lim_{n\to\infty} f_n(x) = f(x) \quad \textit{for all} \quad x \in [-\ell,\ell].$$

The sequence converges uniformly to f if

$$\max_{x\in[-\ell,\ell]} |f_n(x) - f(x)| \to 0 \quad \textit{as} \quad n \to \infty.$$

Remark 5 *If we replace $f_n(x)$ with $S_N(x)$ in the above definition, then L^2 convergence implies that the mean-square error approaches zero, meaning the weighted average of the squared difference between the partial sums and the function vanishes. This type of convergence does not require the partial sums to converge at each individual point.*

Pointwise convergence, on the other hand, ensures that for each point x, the value of the partial sums $S_N(x)$ approaches the value of $f(x)$. However, it does not guarantee that the rate of convergence is uniform across all points. At some points, $S_N(x) \to f(x)$ quickly, while at others, convergence may occur arbitrarily slowly.

Uniform convergence is a stronger form of convergence. It requires that the maximum error between $S_N(x)$ and $f(x)$ over the entire interval decreases to zero as $N \to \infty$.

In general, pointwise convergence and L^2 (norm) convergence do not imply each other. Uniform convergence, however, is stronger and implies both pointwise and L^2 convergence.

For clarification of such concepts, we provide the following example.

Example 2.1 *(*Pointwise but not Uniform Convergence*)*

Consider the sequence of functions

$$f_n(x) = x^n \quad \text{on} \quad [0,1], \quad n = 1,2,\ldots$$

which converges pointwise to the function

$$f(x) = \begin{cases} 0 & \text{if } 0 \le x < 1 \\ 1 & \text{if } x = 1 \end{cases}$$

To see this, note that $f_n(1) = 1$ for all n (so $f_n(1) \to 1$), and

$$\lim_{n \to \infty} x^n = 0 \quad \text{if} \quad 0 < x < 1.$$

However, for $x < 1$,

$$|f_n(x) - f(x)| = x^n,$$

and x^n can be made arbitrarily close to 1 by taking x close enough to 1 (for any n). Thus, the maximum error is always 1. The interval where the error is near 1 shrinks in size as n increases, but the maximum error never decreases. □

We conclude from Remark 5 that $S_N(x)$ converges to $f(x)$ in the mean if

$$\lim_{N \to \infty} \int_{-\pi}^{\pi} |S_N(x) - f(x)|^2 \, dx = 0.$$

Consequently, (2.3) implies that

$$\int_{-\pi}^{\pi} f^2(x)\, dx - \pi \left[\frac{a_0^2}{2} + \sum_{n=1}^{\infty} \left(a_n^2 + b_n^2\right) \right] = 0,$$

from which we conclude the **Parseval's equation**

$$\int_{-\pi}^{\pi} f^2(x)\, dx = \pi \left[\frac{a_0^2}{2} + \sum_{n=1}^{\infty} \left(a_n^2 + b_n^2\right) \right]. \tag{2.6}$$

Parseval's equation holds for any function $f \in L^2[-L, L]$. That is

$$\int_{-L}^{L} f^2(x)\, dx = L \left[\frac{a_0^2}{2} + \sum_{n=1}^{\infty} \left(a_n^2 + b_n^2\right) \right]. \tag{2.7}$$

Parseval's equation can be used to compute infinite sums as the next examples show.

Example 2.2 Use Parseval's equation to compute

$$\sum_{n=1}^{\infty} \frac{1}{(2n-1)^2}.$$

From Example 1.3 we obtained

$$f(x) = \begin{cases} 2, & 0 < x < \pi \\ -1, & -\pi < x < 0 \end{cases} = \frac{1}{2} + \sum_{n=1}^{\infty} \frac{6}{(2n-1)\pi} \sin(2n-1)x.$$

Thus,

$$\frac{a_0}{2} = \frac{1}{2}, \quad b_n = \frac{6}{(2n-1)\pi} \quad \text{and} \quad a_n = 0.$$

Next, we evaluate both sides of (2.6). We begin with the left-hand side and obtain

$$\int_{-\pi}^{\pi} f^2(x)\,dx = \int_{-\pi}^{0} dx + \int_{0}^{\pi} 4\,dx = 5\pi.$$

As for the right-hand side of (2.6), we have that

$$\pi\left[\frac{a_0^2}{2} + \sum_{n=1}^{\infty}(a_n^2 + b_n^2)\right] = \pi\left[\frac{1}{2} + \sum_{n=1}^{\infty}\left(\frac{36}{(2n-1)^2\pi^2} + 0\right)\right].$$

By simplifying and setting both sides equal, we obtain

$$5\pi = \frac{\pi}{2} + \frac{36}{\pi}\sum_{n=1}^{\infty}\frac{1}{(2n-1)^2}.$$

Therefore,

$$\sum_{n=1}^{\infty}\frac{1}{(2n-1)^2} = \frac{\pi^2}{8}.$$

□

Another interesting example.

Example 2.3 The Fourier sine series of $f(x) = x, \quad 0 \le x \le 2$, is given by

$$f(x) = \frac{4}{\pi}\sum_{n=1}^{\infty}\frac{(-1)^{n+1}}{n}\sin\frac{n\pi x}{2}.$$

Therefore, equating both sides of (2.7) yields

$$\int_{-2}^{2} f^2(x)\,dx = 2\int_{0}^{2} x^2 dx = 2\left(\frac{4}{\pi}\right)^2\sum_{n=1}^{\infty}\frac{1}{n^2}.$$

Thus,

$$\left.\frac{x^3}{3}\right|_0^2 = \left(\frac{4}{\pi}\right)^2\sum_{n=1}^{\infty}\frac{1}{n^2},$$

and from which we obtain

$$\frac{\pi^2}{6} = \sum_{n=1}^{\infty}\frac{1}{n^2}.$$

We observe that

$$\sum_{n=1}^{\infty}\frac{1}{(2n)^2} = \frac{1}{2^2}\sum_{n=1}^{\infty}\frac{1}{n^2} = \frac{\pi^2}{24}.$$

Also note that

$$\frac{\pi^2}{6} = \sum_{n=1}^{\infty} \frac{1}{n^2} = \sum_{m=1}^{\infty} \frac{1}{(2m)^2} + \sum_{m=0}^{\infty} \frac{1}{(2m+1)^2} = \frac{\pi^2}{24} + \sum_{m=0}^{\infty} \frac{1}{(2m+1)^2}.$$

Therefore,

$$\sum_{m=0}^{\infty} \frac{1}{(2m+1)^2} = \frac{\pi^2}{6} - \frac{\pi^2}{24} = \frac{\pi^2}{8}. \tag{2.8}$$

It is important to observe that (2.8) does not contradict

$$\sum_{n=1}^{\infty} \frac{1}{(2n-1)^2} = \frac{\pi^2}{8}$$

of Example 2.2. They actually match. To see this, let $n = m+1$. Then (2.8) is transformed into

$$\sum_{m=0}^{\infty} \frac{1}{(2m+1)^2} = \sum_{n=1}^{\infty} \frac{1}{(2n-1)^2} = \frac{\pi^2}{8}.$$

□

In the next two examples, we draw the Fourier series against few terms of its partial sum to observe the rate of convergence.

Example 2.4 In Example 1.3 we showed that

$$f(x) = \begin{cases} 2, & 0 < x < \pi \\ -1, & -\pi < x < 0 \end{cases} = \frac{1}{2} + \sum_{n=1}^{\infty} \frac{6}{(2n-1)\pi} \sin(2n-1)x.$$

Set the partial sum

$$S_N(x) = \frac{1}{2} + \sum_{n=1}^{N} \frac{6}{(2n-1)\pi} \sin(2n-1)x.$$

In Fig. 2.1, we illustrate the 2-periodic cycles of $f(x)$ along with $S_1(x)$, $S_2(x)$, and $S_3(x)$.

□

Example 2.5 In Example 1.5 we showed that the Fourier series for $f(x) = |x|,\ -\pi \le x \le \pi$ is given by

$$f(x) = \frac{\pi}{2} - \frac{4}{\pi} \sum_{n=1}^{\infty} \frac{\cos(2n-1)x}{(2n-1)^2}, \quad -\pi \le x \le \pi.$$

In Figs. 2.1–2.4, we draw 2-periodic cycles of $f(x)$ along with $S_1(x), S_3(x)$, and $S_5(x)$.

□

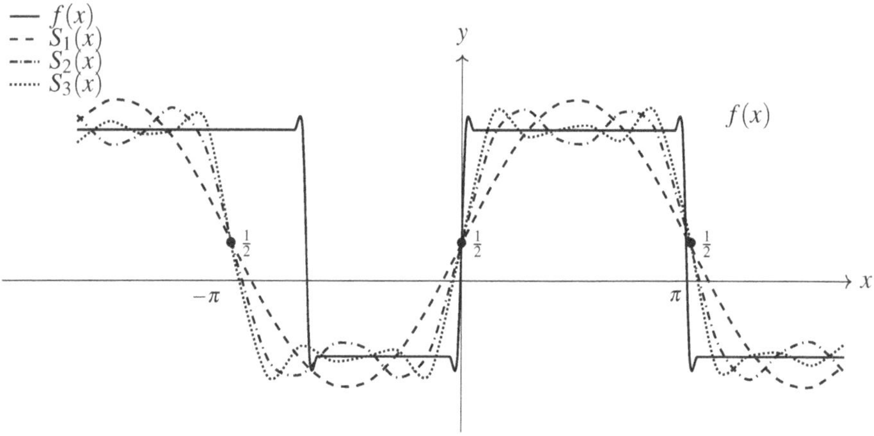

FIGURE 2.1: $f(x)$ plotted against $S_1(x), S_2(x)$ and $S_3(x)$. (Pointwise convergence.)

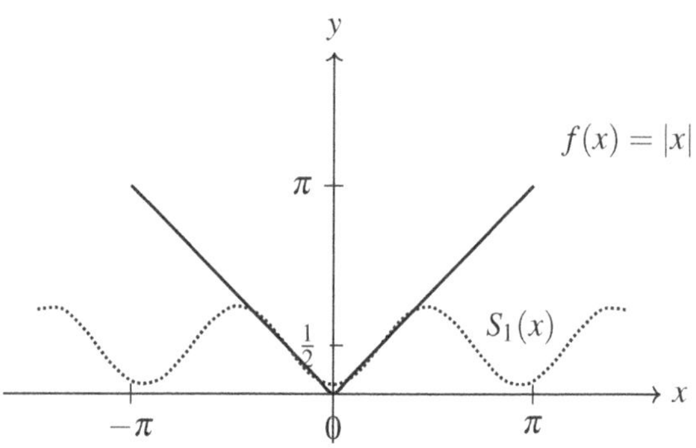

FIGURE 2.2: $S_1(x) = \frac{1}{2} - \frac{4}{\pi^2}\cos(\pi x)$.

2.1.1 Exercises

Exercise 2.1 *Suppose that the function $f(x)$ is defined on $[-L, L]$ and $f \in L^2[-L, L]$. Show that*

$$\frac{a_0^2}{2} + \sum_{n=1}^{\infty} \left(a_n^2 + b_n^2\right) \leq \frac{1}{L} \int_{-L}^{L} f^2(x)\, dx,$$

where a_n and b_n are the Fourier coefficients for f.

Exercise 2.2 *Use the results of Exercise 1.17 to compute*

$$\sum_{n=1}^{\infty} \frac{1}{n^4}.$$

Exercise 2.3 *Use the results of Exercise 1.5 to compute*

$$\sum_{n=1}^{\infty} \frac{1}{(2n-1)^2}.$$

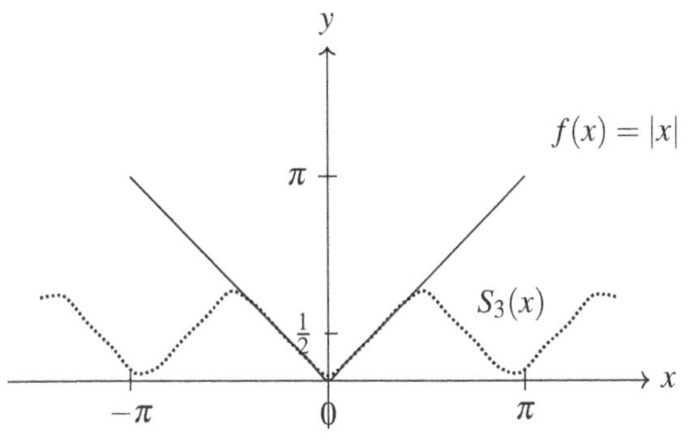

FIGURE 2.3: $S_3(x) = \frac{1}{2} - \frac{4}{\pi^2}\left(\cos(\pi x) + \frac{1}{9}\cos(3\pi x)\right)$.

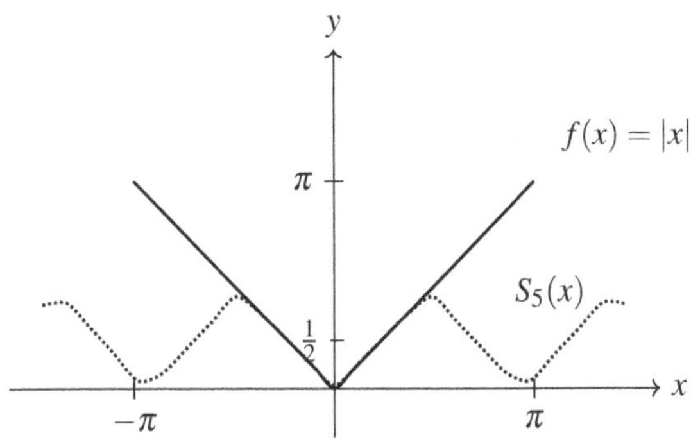

FIGURE 2.4: $S_5(x) = \frac{1}{2} - \frac{4}{\pi^2}\left(\cos(\pi x) + \frac{1}{9}\cos(3\pi x) + \frac{1}{25}\cos(5\pi x)\right)$.

Exercise 2.4 *Use the results of Exercise 1.6 to compute*

$$\sum_{n=1}^{\infty} \frac{1}{(2n-1)^4}.$$

Exercise 2.5 *Use the results of Example 1.9 to compute*

$$\sum_{n=1}^{\infty} \frac{1}{(2n-1)^6}.$$

Exercise 2.6 *Use the results of Exercise 1.26 to compute*

$$\sum_{n=1}^{\infty} \frac{1+4n^2\pi^2}{(1+n^2\pi^2)^2}.$$

Exercise 2.7 *For a fixed positive number α with $\alpha \neq 1,2\ldots$ consider the function $f(x) = \cos\alpha x, \ -\pi < x < \pi$.*
(a) Find the Fourier series of $f(x)$.

(b) Use part (a) to evaluate

$$\sum_{n=1}^{\infty} \frac{1}{(\alpha^2 - n^2)^2}.$$

Ans (a):

$$\frac{\alpha \sin \pi\alpha}{\pi}\left(\frac{1}{\alpha^2} + \sum_{n=1}^{\infty} (-1)^n \frac{2}{\alpha^2 - n^2} \cos nx\right)$$

Exercise 2.8 *Let $f(x)$ be a piecewise continuous function in $[-\pi, \pi]$ given by*

$$f(x) = \begin{cases} 1, & 0 < x \leq \pi \\ -1, & -\pi \leq x < 0 \end{cases}$$

Given $\alpha > 0$, define $f_\alpha(x)$ in $[-\pi, \pi]$ by

$$f_\alpha(x) = \begin{cases} \frac{x}{\alpha}, & |x| \leq \alpha \\ f(x), & otherwise \end{cases}$$

(i) Sketch a graph of $f(x)$ and $f_\alpha(x)$, verifying that $f_\alpha(x)$ is continuous.

(ii) Show (directly) that $\|f - f_\alpha\| \to 0$ as $\alpha \to 0$, where

$$\|f - f_\alpha\| = \sqrt{\int_{-\pi}^{\pi} |f(x) - f_\alpha(x)|^2 \, dx}.$$

Exercise 2.9 *Consider the sequence of functions $f_N(x)$ $(N = 1, 2, \ldots)$ in $C_p(0,1)$ that are defined by*

$$f_N(x) = \begin{cases} 0 & when\ 0 \leq x \leq \frac{1}{N} \\ \sqrt{N} & when\ \frac{1}{N} < x < \frac{2}{N} \\ 0 & when\ \frac{2}{N} \leq x \leq 1 \end{cases}$$

(a) Argue that $f_N(x)$ converges to the function $f(x) = 0$, pointwise. That is $\lim_{N\to\infty} f_N(x) = f(x) = 0$.

(b) Show that $f_N(x)$ does not converge in L^2 to $f(x) = 0$ on $[0,1]$. In other words, show that

$$\lim_{N\to\infty} \int_0^1 \left|f_N(x) - f(x)\right|^2 dx \neq 0.$$

Exercise 2.10 *Let $f(x)$ be 2π-periodic function with continuous second derivative. Let a_n and b_n be its Fourier cosine and sine coefficients, respectively. Show that*

$$\lim_{n\to\infty} n^2 a_n = 0 \quad and \quad \lim_{n\to\infty} n^2 b_n = 0.$$

Exercise 2.11 *Suppose that the function $f(x)$ has a continuous first derivative on the interval $[0, \pi]$ with $f(0) = f(\pi) = 0$. Show that*

$$\int_0^{\pi} f^2(x)\, dx \leq \int_0^{\pi} (f'(x))^2 \, dx.$$

For what functions equality holds?

Exercise 2.12 *Given a function $f(x)$ defined on the interval $-L<x<L$ with Fourier coefficients*

$$a_n = 0, \quad b_n = \frac{1}{\sqrt{n}}.$$

Show that

$$\int_{-L}^{L} f^2(x)\,dx$$

diverges.

2.2 Pointwise Convergence of Fourier Series

The convergence of Fourier series, or in particular, the *Fourier Convergence Theorem,* is a cornerstone of Fourier analysis and holds significant importance in both theoretical and practical applications. The Fourier Convergence Theorem establishes that, under certain conditions, a function can be approximated by a series of sines and cosines (a Fourier series). It also describes how well the Fourier series converges to the original function. The theorem ensures pointwise convergence, meaning the Fourier series converges to the original function at all points where the function is continuous. On the other hand, at points of discontinuity, the theorem asserts that the Fourier series converges to the average of the left- and right-hand limits. We begin with the following definition.

Definition 2.3 *The Dirichlet kernel $K_N(u)$ is defined by the relation*

$$K_N(u) = \frac{1}{2} + \sum_{n=1}^{N} \cos nu. \tag{2.9}$$

It is evident that the Dirichlet kernel is even and 2π-periodic. Additionally,

$$\int_{-\pi}^{\pi} K_N(u)\,du = \pi.$$

The proof of the Fourier Convergence Theorem is both technical and somewhat lengthy. To prepare for it, we will first state and prove three supporting lemmas.

Lemma 3 *Let $K_N(u)$ be given by (2.9). Then*

$$K_N(u) = \frac{\sin\left[\left(N+\frac{1}{2}\right)u\right]}{2\sin\frac{u}{2}}, \quad \textit{for} \quad u \neq 0, \pm 2\pi \ldots \tag{2.10}$$

Proof *By letting $A = \frac{u}{2}$ and $B = nu$ in*

$$2\sin A\cos B = \sin(A+B) + \sin(A-B),$$

we arrive at

$$\cos nu = \frac{\sin(\frac{1}{2}+n)u + \sin(\frac{1}{2}-n)u}{2\sin\frac{u}{2}}.$$

Consequently,

$$\frac{1}{2}+\sum_{n=1}^{N}\cos nu = \frac{1}{2}+\sum_{n=1}^{N}\frac{\sin(\frac{1}{2}+n)u+\sin(\frac{1}{2}-n)u}{2\sin\frac{u}{2}}. \tag{2.11}$$

Let $k = n-1$. Then

$$\sum_{n=1}^{N}\sin(\frac{1}{2}-n)u = -\sum_{k=0}^{N-1}\sin(\frac{1}{2}+k)u = -\sum_{n=0}^{N-1}\sin(\frac{1}{2}+n)u.$$

Substituting into the right-hand side of (2.11) *yields*

$$\begin{aligned}\frac{1}{2} &+ \sum_{n=1}^{N}\frac{\sin(\frac{1}{2}+n)u+\sin(\frac{1}{2}-n)u}{2\sin\frac{u}{2}} = \frac{1}{2}+\frac{1}{2\sin\frac{u}{2}}\left[\sum_{n=1}^{N}\sin(\frac{1}{2}+n)u-\sum_{n=0}^{N-1}\sin(\frac{1}{2}+n)u\right]\\ &= \frac{1}{2}+\frac{1}{2\sin\frac{u}{2}}\left[\sin(\frac{1}{2}+N)u+\sum_{n=1}^{N-1}\sin(\frac{1}{2}+n)u-\sum_{n=1}^{N-1}\sin(\frac{1}{2}+n)u-\sin\frac{u}{2}\right]\\ &= \frac{\sin(\frac{1}{2}+N)u}{2\sin\frac{u}{2}}.\end{aligned}$$

□

Lemma 4 *Let $K_N(u)$ be defined by* (2.9)*, and let $S_N(x)$ be the partial sum of the Fourier series of $f(x) \in C_p'(-\pi,\pi)$. Then*

$$S_N(x) = \frac{1}{\pi}\int_{-\pi}^{\pi} f(s)K_N(s-x)\,ds. \tag{2.12}$$

Proof *The Fourier series of $f(x)$ given by* (1.7) *can be written as*

$$\frac{1}{2\pi}\int_{-\pi}^{\pi} f(s)\,ds + \frac{1}{\pi}\sum_{n=1}^{\infty}\int_{-\pi}^{\pi} f(s)\cos n(s-x)\,ds$$

from which we have

$$S_N(x) = \frac{1}{2\pi}\int_{-\pi}^{\pi} f(s)\,ds + \frac{1}{\pi}\sum_{n=1}^{N}\int_{-\pi}^{\pi} f(s)\cos n(s-x)\,ds.$$

After rearranging the terms, we obtain

$$\begin{aligned} S_N(x) &= \frac{1}{\pi}\int_{-\pi}^{\pi} f(s)\left[\frac{1}{2}+\sum_{n=1}^{N}\cos n(s-x)\right]ds\\ &= \frac{1}{\pi}\int_{-\pi}^{\pi} f(s)K_N(s-x)\,ds.\end{aligned}$$

□

Lemma 5 *Suppose that the function $h(u)$ is piecewise continuous on the interval $0 < u < \pi$, and let N denotes positive integers. Then*

$$\lim_{N\to\infty} \int_0^{\pi} h(u) \sin\left(\frac{u}{2} + Nu\right) du = 0.$$

Proof *By letting $A = \frac{u}{2}$ and $B = Nu$ in*

$$\sin(A+B) = \sin A \cos B + \cos A \sin B,$$

we obtain

$$\sin\left(\frac{u}{2} + Nu\right) = \sin \frac{u}{2} \cos Nu + \cos \frac{u}{2} \sin Nu.$$

Consequently,

$$\begin{aligned} \int_0^{\pi} h(u) \sin\left(\frac{u}{2} + Nu\right) du &= \int_0^{\pi} h(u) \sin \frac{u}{2} \cos Nu du \\ &+ \int_0^{\pi} h(u) \cos \frac{u}{2} \sin Nu du. \end{aligned} \tag{2.13}$$

Since $h(u) \sin\left(\frac{u}{2}\right)$ and $h(u) \cos\left(\frac{u}{2}\right)$ are piecewise continuous on the interval $0 < u < \pi$, we have from (2.13) that

$$\int_0^{\pi} h(u) \sin\left(\frac{u}{2} + Nu\right) du = \frac{\pi}{2}\left(a_N + b_N\right),$$

where a_N and b_N are the Fourier coefficients for the cosine and sine of the functions $h(u) \sin\left(\frac{u}{2}\right)$ and $h(u) \cos\left(\frac{u}{2}\right)$, respectively. It follows from Corollary 1 that

$$\lim_{N\to\infty} \int_0^{\pi} h(u) \sin\left(\frac{u}{2} + Nu\right) du = \lim_{N\to\infty} \frac{\pi}{2}\left(a_N + b_N\right) = 0.$$

□

Theorem 2.1 (Fourier Pointwise Convergence Theorem) *Suppose that $f(x) \in C_p'(-\pi, \pi)$, meaning that $f(x)$ is piecewise smooth on the interval $-\pi \le x \le \pi$. The Fourier series of $f(x)$ is given by*

$$\frac{a_0}{2} + \sum_{n=1}^{\infty} \left(a_n \cos nx + b_n \sin nx\right),$$

where a_0, a_n, and b_n are defined by (1.4), (1.5), and (1.6), respectively. This Fourier series will converge to:

1. *The periodic extension of $f(x)$ if the periodic extension is continuous.*
2. *The average of the two one-sided limits,*

$$\frac{1}{2}\left[f(x^+) + f(x^-)\right]$$

   if the periodic extension has a jump discontinuity at x.

Proof *From Lemma 4, we have*

$$S_N(x) = \frac{1}{\pi}\int_{-\pi}^{\pi} f(s)K_N(s-x)\,ds.$$

Making the substitution $u = s - x$, the above expression becomes

$$S_N(x) = \frac{1}{\pi}\int_{-\pi-x}^{\pi-x} f(x+u)K_N(u)\,du.$$

Since f and K_N are 2π-periodic, we have that

$$S_N(x) = \frac{1}{\pi}\int_{-\pi}^{\pi} f(x+u)K_N(u)\,du. \tag{2.14}$$

By substituting u with $-u$ in (2.14) *and noting that $K_N(u)$ is an even function, we can rewrite* (2.14) *as*

$$S_N(x) = \frac{1}{\pi}\int_{-\pi}^{\pi} f(x-u)K_N(u)\,du. \tag{2.15}$$

By averaging (2.14) *and* (2.15)*, we obtain*

$$\begin{aligned} S_N(x) &= \frac{1}{\pi}\int_{-\pi}^{\pi} \frac{f(x+u)+f(x-u)}{2}K_N(u)\,du \\ &= \frac{2}{\pi}\int_{0}^{\pi} \frac{f(x+u)+f(x-u)}{2}K_N(u)\,du, \end{aligned} \tag{2.16}$$

because the integrand

$$\frac{f(x+u)+f(x-u)}{2}K_N(u)$$

is even. As $\int_{-\pi}^{\pi} K_N(u)\,du = \pi$,

$$\frac{f(x^+)+f(x^-)}{2} = \frac{2}{\pi}\int_{0}^{\pi} \frac{f(x^+)+f(x^-)}{2}K_N(u)\,du. \tag{2.17}$$

Combining (2.16) *and* (2.17) *yields,*

$$\begin{aligned} \frac{f(x^+)+f(x^-)}{2} - S_N(x) &= \frac{1}{\pi}\int_{0}^{\pi} \left[f(x^+) - f(x+u)\right]K_N(u)\,du \\ &+ \frac{1}{\pi}\int_{0}^{\pi} \left[f(x^-) - f(x-u)\right]K_N(u)\,du. \end{aligned} \tag{2.18}$$

We will be done once we show the right-hand side of (2.18) *approaches zero as N approaches infinity.*

Using the identity

$$\sin(\frac{u}{2} + Nu) = \sin\frac{u}{2}\cos Nu + \cos\frac{u}{2}\sin Nu,$$

and utilizing Lemma 3, the first integral on the right-hand side of (2.18) *can be written as*

$$\begin{aligned}\frac{1}{\pi}\int_0^{\pi}\left[f(x^+)-f(x+u)\right]K_N(u) &= \frac{1}{\pi}\int_0^{\pi}\left[\frac{f(x^+)-f(x+u)}{2\sin(u/2)}\right]\sin\frac{u}{2}\cos Nu\,du\\ &+\frac{1}{\pi}\int_0^{\pi}\left[\frac{f(x^+)-f(x+u)}{2\sin(u/2)}\right]\cos\frac{u}{2}\sin Nu\,du.\end{aligned} \tag{2.19}$$

We must ensure that the hypothesis of Lemma 5 is satisfied in order to apply it. As $f(x)\in C_p'(-\pi,\pi)$ *we have*

$$\lim_{u\to 0^+}\frac{f(x^+)-f(x+u)}{2\sin(u/2)} = \lim_{u\to 0^+}\frac{f(x^+)-f(x+u)}{u}\frac{u}{2\sin(u/2)}$$

exists as a finite number. We conclude that the two functions

$$\left[\frac{f(x^+)-f(x+u)}{2\sin(u/2)}\right]\sin\frac{u}{2},\ \left[\frac{f(x^+)-f(x+u)}{2\sin(u/2)}\right]\cos\frac{u}{2},$$

are bounded and square integrable on $[-\pi,\pi]$. *Along the same lines, the second integral on the right-hand side of* (2.18) *can be written as*

$$\begin{aligned}\frac{1}{\pi}\int_0^{\pi}\left[f(x^-)-f(x-u)\right]K_N(u) = &\ \frac{1}{\pi}\int_0^{\pi}\left[\frac{f(x^-)-f(x-u)}{2\sin(u/2)}\right]\sin\frac{u}{2}\cos Nu\,du\\ + &\ \frac{1}{\pi}\int_0^{\pi}\left[\frac{f(x^-)-f(x-u)}{2\sin(u/2)}\right]\cos\frac{u}{2}\sin Nu\,du.\end{aligned} \tag{2.20}$$

By similar argument as before, the two functions

$$\left[\frac{f(x^-)-f(x-u)}{2\sin(u/2)}\right]\sin\frac{u}{2},\quad \left[\frac{f(x^-)-f(x-u)}{2\sin(u/2)}\right]\cos\frac{u}{2}$$

are bounded and square integrable on $[-\pi,\pi]$. *Applying Lemma 5 to* (2.20) *in combination with* (2.18) *we arrive at*

$$\begin{aligned}\lim_{N\to\infty}\left[\frac{f(x^+)+f(x^-)}{2}-S_N(x)\right] &= \lim_{N\to\infty}\left[\frac{1}{\pi}\int_0^{\pi}\left[f(x^+)-f(x+u)\right]K_N(u)\,du\right.\\ &+\left.\frac{1}{\pi}\int_0^{\pi}\left[f(x^-)-f(x-u)\right]K_N(u)\,du\right]=0.\end{aligned}$$

That is,

$$\frac{a_0}{2}+\sum_{n=1}^{\infty}(a_n\cos nx+b_n\sin nx)=\lim_{N\to\infty}S_N(x)=\frac{f(x^+)+f(x^-)}{2}.$$

This proves item 2. of the theorem. If f *is continuous at a point* $x\in(-\pi,\pi)$, *then* $f(x^-)=f(x^+)$ *and as a result, we have*

$$\frac{f(x^+)+f(x^-)}{2}=\frac{f(x)+f(x)}{2}=f(x).$$

This proves item 1. of the theorem. Next we argue that this result holds outside $[-\pi,\pi]$. The first thing to note is that on the interval $-\pi \le x \le \pi$, both the function and its periodic extension are equal. Therefore, where the function is continuous on $-\pi \le x \le \pi$, the periodic extension will also be continuous. Consequently, at these points, the Fourier series will converge to the function. The only points in the interval $-\pi \le x \le \pi$ where the Fourier series will not converge to the function are those where the function exhibits a jump discontinuity.

In Fig. 2.5, we have a function $f(x)$ on $-\pi < x < \pi$ that is 2π-periodic. We are interested in evaluating $\frac{f(x^+)+f(x^-)}{2}$ at the jump points $x = \pm\pi, \pm 2\pi \ldots$ Let $f^e(x)$ be the periodic extension outside $(-\pi,\pi)$. Then, from the graph we have that at $x = \pi$,

$$f^e(\pi^+) = f(-\pi^+) \quad \text{and} \quad f^e(\pi^-) = f(\pi^-).$$

Therefore, at $x = \pi$, we have the Fourier series converges to

$$\frac{f(x^+)+f(x^-)}{2} = \frac{f(-\pi^+)+f(\pi^-)}{2}. \tag{2.21}$$

Due to the periodicity of the series, the same conclusion holds when $x = -\pi$.

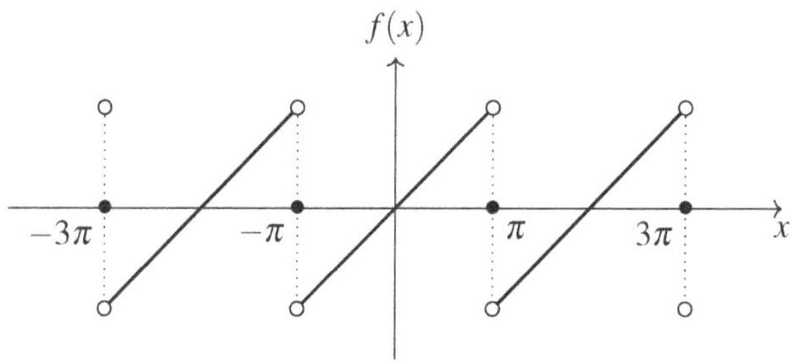

FIGURE 2.5: Periodic extension of a function $f(x)$ on $-\pi < x < \pi$ over the whole real line.

Example 2.6 Consider the function

$$f(x) = \begin{cases} 1, & 0 < x < \pi \\ 0, & -\pi < x < 0 \end{cases}, \quad f(x+2\pi) = f(x)$$

According to Theorem 2.1, the Fourier series will converge to the 2π-periodic extension $g(x)$ of the step function f, where

$$g(x) = \begin{cases} 0, & (2k-1)\pi < x < 2k\pi, \\ 1, & 2k\pi < x < (2k+1)\pi, \\ \frac{1}{2}, & x = k\pi,\ k = 0, \pm 1, \pm 2, \ldots \end{cases}$$

where $k = 0, \pm 1, \pm 2, \ldots$ Observe that, in accordance with Theorem 2.1, $g(x)$ takes the midpoint value $\frac{1}{2}$ at the jump discontinuities $0, \pm\pi, \pm 2\pi, \ldots$. □

Example 2.7 Determine the function $g(x)$ that the Fourier series of the function $f(x) = x, \quad -\pi < x < \pi$ converges to. The graph of $f(x)$ is given in Fig. 2.5. Since $f(x)$ is continuous on $-\pi < x < \pi$, Theorem 2.1 tells us that the Fourier series of f will converge to the function $f(x) = x, \quad -\pi < x < \pi$. The graph of the periodic extension of f has jump discontinuities at $x = \pm\pi, \pm 3\pi \ldots$. However, at $x = \pm\pi$, the above discussion tells us that the Fourier series of f will converge to (see Fig. 2.5)

$$\frac{f(-\pi^+) + f(\pi^-)}{2} = \frac{-\pi + \pi}{2} = 0.$$

In conclusion the Fourier series of $f(x)$ converges pointwise to

$$g(x) = \begin{cases} 0, & x = \pm\pi, \pm 3\pi \ldots \\ x - 2k\pi, & (2k-1)\pi < x < (2k+1)\pi, \quad k = 0, \pm 1, \pm 2, \ldots \end{cases}$$

□

Example 2.8 In Exercise 1.6 we found the Fourier series of

$$f(x) = \begin{cases} x, & 0 \leq x \leq \pi \\ 0, & -\pi < x < 0 \end{cases}, \quad f(x + 2\pi) = f(x)$$

to be

$$\frac{\pi}{4} - 2\sum_{n=1}^{\infty} \frac{\cos(2n-1)x}{(2n-1)^2\pi} + \sum_{n=1}^{\infty} \frac{(-1)^{n+1}}{n} \sin nx.$$

We ask you the student, to verify that $f(x) \in C_p'(-\pi, \pi)$. The periodic extension of f has jump discontinuities at $x = \pm\pi, \pm 3\pi, \ldots$. The Fourier series of $f(x)$ converges to the periodic extension of $f(x)$, except at these discontinuities. Thus, at $x = \pm\pi, \pm 3\pi \ldots$ the Fourier series will converge to

$$\frac{f(-\pi^+) + f(\pi^-)}{2} = \frac{0 + \pi}{2} = \frac{\pi}{2}.$$

In particular, since the series converges to $\pi/2$ when $x = \pi$, we have

$$\frac{\pi}{4} - 2\sum_{n=1}^{\infty} \frac{-1}{(2n-1)^2} = \frac{\pi}{2},$$

which can be written as

$$\sum_{n=1}^{\infty} \frac{1}{(2n-1)^2} = \frac{\pi^2}{8}.$$

We ask that you compare this result with Example 2.2. This demonstrates how Fourier series can be employed to determine the sums of convergent series that arise in calculus. Substituting $x = 0$ into the series similarly produces this specific summation formula. □

Now we address the convergence of Fourier sine and Fourier cosine series. Consider the Fourier sine series of a function $g(x)$ defined on the interval $0 \leq x \leq \pi$. This series corresponds to the Fourier series of the odd extension of $g(x)$, restricted to the interval $0 \leq x \leq \pi$. As a result, the Fourier sine series will converge to the odd extension on the interval $-\pi \leq x \leq \pi$ wherever the extension is continuous. At points where the odd extension exhibits a jump discontinuity, the series will converge to the average of the left and right limits at those points.

On the interval $0 \leq x \leq \pi$, the function $g(x)$ coincides with its odd extension. Therefore, the Fourier sine series exhibits the same convergence behavior as the full Fourier series on this interval.

A similar reasoning applies to the Fourier cosine series, which is derived from the even extension of a function defined on $0 \leq x \leq \pi$. The Fourier cosine series for such a function will also share the same convergence properties as the full Fourier series.

The next theorem extends Theorem 2.1 to arbitrary intervals.

Theorem 2.2 (Fourier Pointwise Convergence Theorem on Arbitrary Intervals) *Suppose that $f(x) \in C'_p(-L,L)$. That is, $f(x)$ is piecewise smooth on the interval $-L \leq x \leq L$. Then the Fourier series of $f(x)$ given by*

$$\frac{a_0}{2} + \sum_{n=1}^{\infty} \left(a_n \cos \frac{n\pi x}{L} + b_n \sin \frac{n\pi x}{L}\right),$$

where a_0, a_n and b_n are given by (1.14), (1.15) *and* (1.16), *respectively, will converge to:*

1. *The periodic extension of $f(x)$ if the periodic extension is continuous.*
2. *The average of the two one-sided limits,*

$$\frac{1}{2}\left[f(x^+) + f(x^-)\right]$$

if the periodic extension has a jump discontinuity at x.

□

In Section 1.1 we mentioned that the function $f(x) = x^{\frac{1}{3}}$ is not smooth since its derivative does not exist at $x = 0$. Consequently, $f'(x)$ is not piecewise continuous on any open interval containing $x = 0$. Nonetheless, it turns out that it admits a Fourier series. Since f is not smooth, Theorem 2.1 does not apply. However, the same theorem can be adjusted to the periodic extension $F(x)$ of $f(x)$. To see this, we let $f(x) = x^{1/3}$ on $(-\pi, \pi)$, and define its periodic extension $F(x)$ by

$$F(x) = f(x - 2\pi n), \quad \text{for } x \in (2n\pi - \pi, 2n\pi + \pi), n \in \mathbb{Z}.$$

Then $F(x) = f(x)$ on $(-\pi, \pi)$, and F is 2π-periodic. The function $f(x) = x^{1/3}$ is continuous on $(-\pi, \pi)$, including at $x = 0$, and differentiable everywhere except at

$x = 0$. Its derivative is

$$f'(x) = \frac{1}{3}x^{-2/3},$$

which is unbounded at $x = 0$. Thus, $f(x)$ is not differentiable at $x = 0$, but it is continuous and piecewise differentiable on $(-\pi, 0) \cup (0, \pi)$, so $F(x)$ is piecewise continuous on $[-\pi, \pi]$.

Since $f(x) = x^{1/3}$ is strictly increasing on $(-\pi, \pi)$, it has no local maxima or minima except possibly at the endpoints. Therefore, $F(x)$ has a finite number of extrema on $[-\pi, \pi]$.

The function $f(x)$ is continuous on $(-\pi, \pi)$, and the only possible discontinuities in the periodic extension $F(x)$ occur at the endpoints of the interval, i.e., at integer multiples of π.

$$\int_{-\pi}^{\pi} |x^{1/3}|\, dx = 2\int_{0}^{\pi} x^{1/3}\, dx = 2 \cdot \frac{3}{4}\pi^{4/3} = \frac{3}{2}\pi^{4/3} < \infty,$$

so $f(x)$ is absolutely integrable on $(-\pi, \pi)$. Since $f(x) = x^{1/3}$ is an odd function with $f(0) = 0$, its Fourier series will contain only sine terms. Therefore, the Fourier series takes the form

$$f(x) = \sum_{n=1}^{\infty} b_n \sin\frac{n\pi x}{L},$$

where b_n are the Fourier sine coefficients given by

$$b_n = \frac{2}{L}\int_{0}^{L} x^{1/3} \sin\frac{n\pi x}{L}\, dx.$$

On $-\pi < x < \pi$ the Fourier series for $F(x)$ converges to $F(x)$ when $-\pi < x < 0$ and when $0 < x < \pi$ and converges to the average of the left and right limits at the endpoints $\pm\pi$. In this case, since the periodic extension of $f(x)$ has a jump discontinuity at $\pm\pi$, the Fourier series converges to

$$\frac{1}{2}\left(f(\pi^-) + f(\pi^+)\right) = \frac{1}{2}\left(\pi^{1/3} + (-\pi)^{1/3}\right) = 0.$$

Similarly, its Fourier series converges to zero at $x = 0$. Thus, the function $f(x) = x^{1/3}$ admits a Fourier series on the interval $[-\pi, \pi]$.

2.2.1 Exercises

Exercise 2.13 *Consider*

$$f(x) = \begin{cases} \pi, & \text{if } -\pi \le x \le 0 \\ 2x, & \text{if } 0 < x \le \pi \end{cases}$$

(a) Find the Fourier series of $f(x)$.
(b) Indicate all the points at which f has jump discontinuities and find the function

$g(x)$ that the Fourier series of f converges to?
(c) Use part (b) to find $\sum_{n=1}^{\infty} \frac{1}{(2n-1)^2}$

Ans (a): $\pi + \sum_{n=1}^{\infty} \frac{2}{n^2\pi}\left((-1)^n - 1\right)\cos nx - \sum_{n=1}^{\infty} \frac{1}{n}\left(1+(-1)^n\right)\sin nx.$

Exercise 2.14 *Repeat Exercise 2.13 for the function*

$$f(x) = \begin{cases} L, & \text{if } -L \le x \le 0 \\ 2x, & \text{if } 0 < x \le L \end{cases}$$

Exercise 2.15 *Find the Fourier series of the 2L-periodic function*

$$f(x) = \begin{cases} -\frac{L}{2}\sin\frac{\pi x}{L}, & -L < x < 0 \\ x, & 0 < x < \frac{L}{2} \\ L - x, & \frac{L}{2} < x < L \end{cases}$$

Also, find the function $g(x)$ that the Fourier series of $f(x)$ converges to.

Exercise 2.16 *Make use of Exercise 1.26 to find*

$$\sum_{n=1}^{\infty} \frac{2}{4+n^2\pi^2}.$$

Exercise 2.17 *Let f be a 2π-periodic and integrable function defined as follows:*

$$f(x) = \begin{cases} 1 + \sin\frac{\pi^2}{x}, & \text{if } x \in [-\pi, \pi), x \neq 0, \\ 1, & \text{if } x = 0 \end{cases}$$

Does the Fourier series of f converge to $f(0)$?

Exercise 2.18 *Here is another proof for* (2.10).
(a) Use the identity

$$\sin(A+B) = \sin A \cos B + \cos A \sin B,$$

to obtain

$$\frac{1}{2} + \cos u = \frac{\sin\left(u + \frac{u}{2}\right)}{2}$$

(b) Extrapolate from part (a) that

$$\frac{1}{2} + \cos u + \cos 2u + \cdots + \cos Nu = \frac{\sin\left(\frac{(2N+1)u}{2}\right)}{2\sin\left(\frac{u}{2}\right)}$$

(c) Use an induction argument to prove that the formula in part (b) is valid for all $N = 1, 2 \ldots$. We have already established the base case of $N = 1$ since

$$\frac{1}{2} + \cos u = \frac{\sin\left(u + \frac{u}{2}\right)}{2\sin\left(\frac{u}{2}\right)} = \frac{\sin\left(\frac{3u}{2}\right)}{2\sin\left(\frac{u}{2}\right)}.$$

Assume the formula holds for any N, and then show it is true for $N+1$.

2.3 Gibb's Phenomenon

Gibbs' phenomenon refers to the overshoot that occurs when a Fourier series is used to approximate a function with a jump discontinuity. Specifically, when the Fourier series converges to a step function, the partial sums of the series exhibit an oscillation that peaks at around 9% above the jump's magnitude. This overshoot stabilizes at this level as more terms are added, even though the series eventually converges to the function almost everywhere. Despite the oscillations, the Fourier series will converge pointwise everywhere except at the discontinuities. For better illustration we consider the 2π- periodic function

$$f(x) = \frac{x}{2}, \quad -\pi < x \leq \pi,$$

that has the Fourier series

$$\sum_{n=1}^{N} \frac{(-1)^{n+1}}{n} \sin(nx).$$

Consequently, its partial sum is

$$S_N(x) = \sum_{n=1}^{N} \frac{(-1)^{n+1}}{n} \sin(nx). \tag{2.22}$$

In Fig. 2.6, we approximate the function $f(x) = \frac{x}{2}$ for $-\pi < x < \pi$ using its partial sums $S_4(x)$ and $S_{10}(x)$. We first note that the 2π-periodic extension of f has jump discontinuities at $x = j\pi$, $j = \pm 1, \pm 2 \ldots$ Fig. 2.6 shows the approximations of $f(x)$ for $N = 4$ and $N = 10$. It is evident from the graph that $S_{10}(x)$ is a better approximation than $S_4(x)$. However, neither of them is a good approximation near the points of discontinuities, specifically at $\pm\pi$. Near a discontinuity point, $S_N(x)$ spikes above $f(x)$ and quickly dips to a negative peak below $f(x)$. This type of behavior is typical of Fourier series and is known as *Gibb's phenomenon.* Gibbs' phenomenon is observed in Example 2.4 but not in Example 2.5, as the function $f(x) = |x|$, $-\pi \leq x \leq \pi$, is continuous in the latter case. It is known that the amount of the overshoots at the discontinuities is proportional to the height of the discontinuity. According to Gibbs, it is found to be around 9% of the height of the discontinuity, irrespective of the number of terms in the Fourier series. The exact proportion is given by the Wilbraham-Gibbs constant

$$G = \lim_{N\to\infty} (\text{max overshoot of partial sum}) = \frac{1}{\pi}\int_0^{\pi} \frac{\sin t}{t}\, dt - \frac{1}{2} \approx 0.0894898722\ldots$$

Note: Sometimes this is expressed slightly differently depending on the function being approximated, but the result is equivalent. This result highlights a fundamental property of Fourier series, explaining why the overshoot remains consistent near discontinuities, regardless of the number of terms in the approximations. The partial sum of the he function represented in Fig. 2.6 is challenging to analyze due to the

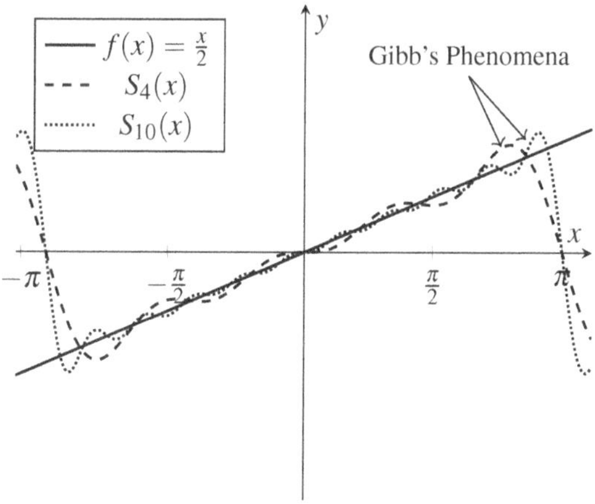

FIGURE 2.6: Gibb's phenomenon; (pointwise convergence), $S_N(x) - \sum_{n=1}^{N} \frac{(-1)^{n+1}}{n} \sin(nx)$.

presence of the term $(-1)^{n+1}$ in the Fourier series. Instead, we analyze the Fourier series of a simpler function to offer a clearer and more insightful explanation of Gibbs' phenomenon.

Example 2.9 In Example 1.3 we considered

$$f(x) = \begin{cases} 2, & 0 < x < \pi \\ -1, & -\pi < x < 0 \end{cases}$$

and obtained the Fourier series

$$\frac{1}{2} + \sum_{n=1}^{\infty} \frac{6}{(2n-1)\pi} \sin(2n-1)x.$$

The 2π-periodic extension of f has jump discontinuities at $x = j\pi$, $j = 0, \pm 1, \pm 2 \ldots$ Moreover, according to Fourier Convergence Theorem, (see also Fig. 2.1) the Fourier series converges to

$$g(x) = \begin{cases} \frac{1}{2} & \text{at } x = 0, \pm\pi \\ f(x) & \text{otherwise} \end{cases}$$

extended periodically. Then the partial sum of the Fourier series is

$$S_N(x) = \frac{1}{2} + \sum_{n=1}^{N} \frac{6}{(2n-1)\pi} \sin(2n-1)x.$$

Let us restrict ourselves to the interval $0 < x < \pi$. Evidently $S_N(x)$ converges to $f(x)$ inside $0 < x < \pi$ and $-\pi < x < 0$. We want to understand what happens near the jump discontinuity $x = 0$, or $x = \pi$. Differentiating the partial sum, we arrive at

$$S'_N(x) = \frac{6}{\pi} \sum_{n=1}^{N} \cos(2n-1)x.$$

The question is now how to make any sense out of $S'_N(x)$ in relation to finding the critical points. Recall,

$$2\sin A \cos B = \sin(A+B) + \sin(A-B).$$

First, we set $A = x$ and $B = (2n-1)x$ and sum both sides of the resulting equation from $n = 1$ to $n = N$ yields

$$2\sin x \sum_{n=1}^{N} \cos(2n-1)x = \Big(\sum_{n=1}^{N} \sin 2nx - \sum_{n=1}^{N} \sin 2(n-1)x \Big) = \sin 2Nx.$$

Solving for $\sum_{n=1}^{N} \cos(2n-1)x$, results into

$$\sum_{n=1}^{N} \cos(2n-1)x = \frac{\sin 2Nx}{2\sin x}.$$

Or,

$$S'_N(x) = \frac{3}{\pi} \frac{\sin 2Nx}{\sin x}, \quad \text{for} \quad 0 < x < \pi. \tag{2.23}$$

Setting $S'_N(x) = 0$ gives the critical points in the interval $0 < x < \pi$

$$x = \frac{k\pi}{2N}, \quad k = 1, 2, \ldots, 2N-1.$$

By differentiating (2.23) once more and substituting $x = \frac{\pi}{2N}$, we find

$$S''_N\left(\frac{\pi}{2N}\right) = -\frac{6N}{\pi \sin\left(\frac{\pi}{2N}\right)} < 0. \tag{2.24}$$

This indicates that a relative maximum occurs at $x = \frac{\pi}{2N}$, as shown in Fig. 2.6. For $k = 1$, the first jump occurs at $x = \frac{\pi}{2N}$, and so on. The values $N = 1$ and $k = 1$ correspond to the jump of $S_1(x)$ at $x = \frac{\pi}{2}$, as depicted in Fig. 2.6. The height of the jumps increases as we approach the jump discontinuities at $x = 0$ and $x = \pi$ within the interval $(0, \pi)$. To have an insight to what happened to the partial sum at $x = \frac{\pi}{2N}$, we integrate both sides of (2.23) from $x = 0$ to $x = \frac{\pi}{2N}$. This yield the relation

$$S_N\left(\frac{\pi}{2N}\right) - S_N(0) = \frac{3}{\pi} \int_0^{\frac{\pi}{2N}} \frac{\sin 2Nx}{\sin x}\, dx.$$

Since $S_N(0) = \frac{1}{2}$, we have that

$$S_N\left(\frac{\pi}{2N}\right) = \frac{1}{2} + \frac{3}{\pi} \int_0^{\frac{\pi}{2N}} \frac{\sin 2Nx}{\sin x}\, dx. \tag{2.25}$$

Next, we try to simplify the integral in (2.25). We rewrite

$$\frac{\sin 2Nx}{\sin x} = \frac{x-\sin x}{x\sin x}\sin 2Nx + \frac{\sin 2Nx}{x},$$

and consequently,

$$\int_0^{\frac{\pi}{2N}} \frac{\sin 2Nx}{\sin x}\,dx = \int_0^{\frac{\pi}{2N}} \frac{x-\sin x}{x\sin x}\sin 2Nx\,dx + \int_0^{\frac{\pi}{2N}} \frac{\sin 2Nx}{x}\,dx. \tag{2.26}$$

Using L'Hôpital's rule twice we arrive at

$$\begin{aligned}\lim_{\substack{x\to 0\\ x>0}} \frac{x-\sin x}{x\sin x} &= \lim_{\substack{x\to 0\\ x>0}} \frac{1-\cos x}{\sin x + x\cos x}\\ &= \lim_{\substack{x\to 0\\ x>0}} \frac{\sin x}{2\cos x - x\sin x} = 0.\end{aligned}$$

Along the same lines, we can easily show that

$$\lim_{\substack{x\to 0\\ x>0}} \frac{\sin 2Nx}{x} = 2N.$$

Thus, the integrands in both integrals are piecewise continuous, and consequently both integrals exist as finite numbers on the interval $(0, \frac{\pi}{2N})$. In addition, there is a positive constant M so that

$$\left|\frac{x-\sin x}{x\sin x}\sin 2Nx\right| \le M, \quad 0 < x < \frac{\pi}{2N}.$$

This implies that

$$\int_0^{\frac{\pi}{2N}} \left|\frac{x-\sin x}{x\sin x}\sin 2Nx\right| dx \le M(\frac{\pi}{2N}) \to 0, \quad \text{as} \quad N\to\infty.$$

As for the second integral in (2.26) we let $s = 2Nx$, and transform it into

$$\int_0^{\frac{\pi}{2N}} \frac{\sin 2Nx}{x}\,dx = \int_0^{\pi} \frac{\sin s}{s}\,ds \approx 1.85.$$

Taking the limit as N approaches infinity in (2.25), we obtain

$$\lim_{N\to\infty} S_N\left(\frac{\pi}{2N}\right) = \frac{1}{2} + \frac{3}{\pi}\int_0^{\pi} \frac{\sin s}{s}\,ds \approx 2.27.$$

The value $\sigma \approx 2.27$ is shown in Fig. 2.7. This demonstrates how the "spikes" observed in the graphs of the partial sums $y = S_N(x)$ shift toward the left as N grows, with their peaks approaching the value σ on the y-axis. A similar pattern is evident on the interval $-\pi < x < 0$. The corresponding Fourier series does not converge uniformly to $f(x)$ over $-\pi < x < \pi$ since $f(x)$ is not continuous. This is true even on the interval $-\pi < x < \pi$. Non-uniform convergence is the underlying cause of the Gibbs phenomenon, which persists regardless of how large N becomes. There will always be an overshoot of approximately 9%, where the overshoot is defined as the ratio of the absolute amount exceeding the target value to the size of the jump in the function. □

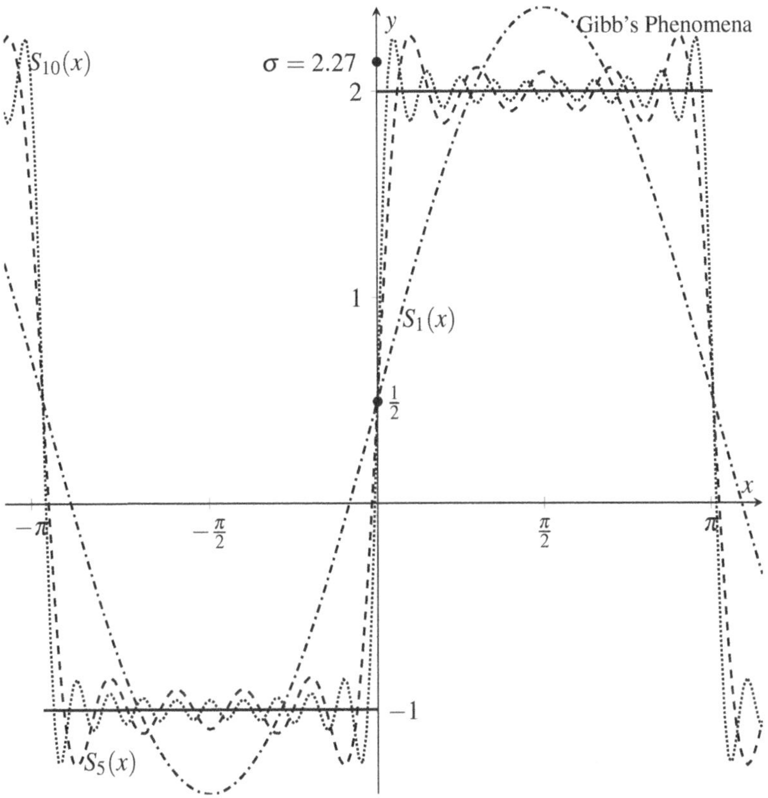

FIGURE 2.7: Gibb's phenomenon (pointwise convergence); $S_N(x) = \frac{1}{2} + \sum_{n=1}^{N} \frac{6}{(2n-1)\pi} \sin(2n-1)x$.

2.3.1 Exercises

Exercise 2.19 *Verify* (2.24).

Exercise 2.20 *Consider $S_N(x)$ given by 2.22. Find the critical points for $S_1(x)$. Do the same for $S_2(x)$.*

Exercise 2.21 *Let*

$$I(t) = \int_0^t \frac{\sin s}{s} ds.$$

Show that $\lim_{t\to\infty} I(t) = \frac{\pi}{2}$.

Exercise 2.22 *Consider the L-periodic function*

$$f(x) = \begin{cases} 1, & 0 \le x \le \frac{L}{2} \\ 0, & \frac{L}{2} \le x < L \end{cases}$$

Find the Fourier series of $f(x)$ *and show that there is a Gibb's phenomenon near* $x = \frac{L}{2}$.

Exercise 2.23 *In the same spirit as Example 2.9, discuss Gibb's phenomenon in the context of the square wave problem*

$$f(x) = \begin{cases} -1, & -1 \le x < 0 \\ 0, & 0 < x \le 1 \end{cases}$$

Exercise 2.24 *In the same spirit as Example 2.9, discuss Gibb's phenomenon in the context of*

$$f(x) = \begin{cases} -\frac{\pi}{2}, & -\pi < x < 0 \\ \frac{\pi}{2}, & 0 < x < \pi \end{cases}$$

3

Uniform Convergence and Operations on Fourier Series

In Chapter 1 , we encountered in many occasions that the Fourier series did not converge to the function $f(x)$ on its whole defined domain, in particular at the points where the function f had jump discontinuities. Consequently, this gave rise to Gibb's phenomenon. In this chapter, we will explore conditions that guarantee absolute and uniform convergence of Fourier series to the function f. Convergence of the Fourier series depends on several factors related to the properties of the target function and the type of convergence considered. We have seen that the Fourier series converges pointwise to the function $f(x)$ at a given point x if the partial sums $S_N(x)$ of the series approach $f(x)$ as $n \to \infty$.

3.1 Preliminaries

In this section we develop Cauchy's inequality for finite sums and state definitions regarding pointwise convergence, uniform convergence and Weierstrass M-test. For clarity, the **discriminant** of the quadratic equation in x,

$$ax^2 + bx + c = 0$$

is defined as

$$\Delta = b^2 - 4ac.$$

We have the following lemma.

Lemma 6 *If A_n $(n = 1, 2, \ldots, N)$ and B_n $(n = 1, 2, \ldots, N)$ are real, where at least one of the numbers A_n, say A_m, is nonzero, then*

$$\left(\sum_{n=1}^{N} A_n B_n \right)^2 \leq \sum_{n=1}^{N} A_n^2 \sum_{n=1}^{N} B_n^2. \tag{3.1}$$

Proof *For nonzero real parameter λ we have*

$$0 \leq \sum_{n=1}^{N} (A_n \lambda + B_n)^2 = \lambda^2 \sum_{n=1}^{N} A_n^2 + 2\lambda \sum_{n=1}^{N} A_n B_n + \sum_{n=1}^{N} B_n^2.$$

DOI: 10.1201/9781003652540-3

Since the above expression is a quadratic equation in λ and non-negative, its discriminant is non-positive. In other words,

$$\left(2\sum_{n=1}^{N} A_n B_n\right)^2 - 4\sum_{n=1}^{N} A_n^2 \sum_{n=1}^{N} B_n^2 \leq 0,$$

otherwise it will have two distinct roots and part of the parabola will be negative. The equality allows the quadratic equation to have a double root. Therefore,

$$\left(\sum_{n=1}^{N} A_n B_n\right)^2 \leq \sum_{n=1}^{N} A_n^2 \sum_{n=1}^{N} B_n^2,$$

which is clearly valid even if all the numbers A_n are zero. □

Expression (3.1) is known as the *Cauchy's inequality for finite sums.*

Next we address the concept of uniform convergence. The next definition is a restatement of the third part of Definition 2.2.

Definition 3.1 *(Uniform Convergence) A sequence of functions $f_n(x) : [a,b] \to \mathbb{R}$, $n = 1,2\ldots$ is said to converge uniformly to $f(x) : [a,b] \to \mathbb{R}$ if and only if, for every positive real number ε, there is a natural number N^* depending on ε such that*

$$|f_n(x) - f(x)| < \varepsilon$$

holds for every natural number $n > N^*$ and every $x \in [a,b]$.

Later in this chapter we will show that if $f(x)$ is continuous and its Fourier coefficients are absolutely summable, then the Fourier series converges uniformly.

Definition 3.2 *(Pointwise Convergence) A sequence of functions $f_n(x) : [a,b] \to \mathbb{R}$, $n = 1,2\ldots$ is said to converge* pointwise *to f if for every $x \in [a,b]$ one has*

$$\lim_{n\to\infty} f_n(x) = f(x).$$

For example, the sequence of functions $f_n(x) = \frac{x^n}{n}$ converges pointwise to zero on the interval $[-1,1]$, because for each $x \in [-1,1]$ one has $|x^n/n| \leq 1/n$, and thus

$$\lim_{n\to\infty} \frac{x^n}{n} = 0.$$

In Example 2.2 we presented a scenario of pointwise convergence.

Uniform convergence implies pointwise convergence, but the converse is not true. For instance, the sequence $f_n(x) = x^n$ converges pointwise on the interval $[0,1]$, but it does not converge uniformly. We demonstrate this by assuming uniform convergence and arriving at a contradiction.

If $f_n(x)$ converges uniformly, then the limit function $f(x)$ is

$$f(x) = \begin{cases} 0 & \text{if } x \in [0,1) \\ 1 & \text{if } x = 1 \end{cases}$$

Uniform convergence means that for any $\varepsilon > 0$, there exists an $N^*(\varepsilon) \in \mathbb{N}$ such that $|f_n(x) - f(x)| < \varepsilon$ for all $n \geq N^*(\varepsilon)$ and for all $x \in [0,1]$. Now, consider $\varepsilon = \frac{1}{2}$. By uniform convergence, there must exist an $N^* \in \mathbb{N}$ such that for all $n \geq N^*$,

$$|f_n(x) - f(x)| < \frac{1}{2} \quad \text{for all} \quad x \in [0,1].$$

Choose $n = N^*$ and $x = \left(\frac{3}{4}\right)^{1/N^*}$. Since $x \in [0,1)$, the limit function satisfies $f(x) = 0$ and then

$$|f_N^*(x) - f(x)| = x^{N^*} - 0 = \frac{3}{4}.$$

However, $\frac{3}{4} > \frac{1}{2}$, which contradicts the assumption of uniform convergence. Thus, the sequence $f_n(x) = x^n$ does not converge uniformly on $[0,1]$.

Remark 6
- *The derivatives of a pointwise convergent sequence of functions do not have to converge.*
- *The integrals of a pointwise convergent sequence of functions do not have to converge.*

Example 3.1 Consider $f_n(x) : \mathbb{R} \to \mathbb{R}$ where $f_n(x) = \frac{1}{n}\sin(n^2 x)$. Then

$$\lim_{n\to\infty} f_n(x) = 0,$$

so the pointwise limit function is $f(x) = 0$. That is the sequence of functions converges to 0. On the other hand,

$$f_n'(x) = n\cos(n^2 x),$$

and for most $x \in \mathbb{R}$, the sequence $n\cos(n^2 x)$ is unbounded. Therefore, the sequence of derivatives $f_n'(x)$ does not converge pointwise.

□

Recall that if f_n is a sequence of functions defined on some set E, then we can consider the partial sums

$$S_N(x) = f_1(x) + \cdots + f_N(x) = \sum_{k=1}^{N} f_k(x).$$

If these converge as $N \to \infty$, and if this happens for every $x \in E$, then we say that the series **converges pointwise**. The sum of the series is

$$S(x) = \sum_{k=1}^{\infty} f_k(x) \stackrel{\text{def}}{=} \lim_{N\to\infty} \sum_{k=1}^{N} f_k(x) = \lim_{N\to\infty} S_N(x).$$

The sum $\sum_{k=1}^{\infty} f_k(x)$ is defined for each $x \in E$ and so it is a function on E.

The next criteria, known as Weierstrass M-test plays an important role in showing uniform convergence of Fourier series.

Lemma 7 *(Weierstrass M-test) Let $\{f_n\}$ be a sequence of functions defined on a set E. Suppose that for all n, $n = 1, \dots,$ there is a constant M_n such that $|f_n(t)| \leq M_n$, for all $t \in E$. If*

$$\sum_{n=1}^{\infty} M_n < \infty,$$

then

$$\sum_{n=1}^{\infty} f_n(t) \text{ converges absolutely and uniformly on the } E.$$

We remark that the Weierstrass M-test can be naturally generalized when the domain of the sequence of functions is a subset of Banach space equipped with an appropriate norm. For more on Banach spaces we refer the reader to [117].

Example 3.2 For $n = 1, 2, \dots$ define the sequence of functions $\{f_n\}$ on $\mathbb{R}$ by $f_n(t) = \dfrac{1}{t^2+n^2}$. Then $|f_n(t)| = \left|\dfrac{1}{t^2+n^2}\right| \leq \dfrac{1}{n^2} := M_n$, for all $t \in \mathbb{R}$, and $n \geq 1$. Since the series $\sum_{n=1}^{\infty} \dfrac{1}{n^2}$ converges, then by the Weierstrass M-test the series $\sum_{n=1}^{\infty} \dfrac{1}{t^2+n^2}$ converges uniformly on $\mathbb{R}$. Moreover we can say more. As each term of the series is continuous and the convergence is uniform the sum function is also continuous. (As the uniform limit of continuous functions is continuous.) □

Here is another example with a simple twist to it.

Example 3.3 We prove the following series

$$\sum_{n=1}^{\infty} \frac{n^2+x^4}{n^4+x^2}$$

converges to a continuous function $f : \mathbb{R} \to \mathbb{R}$.
Let c be a positive constant. Then for all $x \in [-c, c]$ we have that

$$\left|\frac{n^2+x^4}{n^4+x^2}\right| \leq \frac{n^2+x^4}{n^4} \leq \frac{1}{n^2} + \frac{c^4}{n^4} := M_n.$$

Additionally, the series

$$\sum_{n=1}^{\infty} M_n = \sum_{n=1}^{\infty} \frac{1}{n^2} + c^4 \sum_{n=1}^{\infty} \frac{1}{n^4}$$

converges, so Weierstrass M-test implies the series converges uniformly to a function f on the bounded interval $[-c, c]$. Each term in the series is continuous and since the uniform limit of continuous functions is continuous, we have the limit function f is continuous on $[-c, c]$ for every $c > 0$. Now since every $x \in \mathbb{R}$ lies in such an interval for sufficiently large c, it follows that f is continuous on $\mathbb{R}$. Note that the series does not converge uniformly on $\mathbb{R}$, so we cannot use the argument that the sum is continuous on $\mathbb{R}$ because the series converges uniformly on $\mathbb{R}$. □

3.1.1 Exercises

Exercise 3.1 *Show that* $f_n(x) = \frac{x^n}{n}$ *converges uniformly to* $f(x) = 0$ *on the interval* $(-1,1)$.

Exercise 3.2 *Give an example of a sequence of functions that converge pointwise but their derivative does not converge.*

Exercise 3.3 *Consider* $f_n(x) : [0,1] \to \mathbb{R}$, *where*

$$f_n(x) = \frac{2n^2 x}{(1+n^2x^2)^2}.$$

(a) Show that

$$\lim_{n\to\infty} f_n(x) = 0 \quad \textit{for all} \quad x \in [0,1].$$

(b) Show that

$$\lim_{n\to\infty} \int_0^1 f_n(x)\,dx = 1.$$

(c) What do you conclude?

Exercise 3.4 *Consider the following power series* $L(x)$, *which is also known as Euler's dilogarithm function*

$$L(x) = \sum_{n=1}^{\infty} \frac{x^n}{n^2}.$$

(a) Compute the domain of convergence for $L(x)$. *Be sure to give a full analysis of the endpoints.*
(b) Show that $L(x)$ *uniformly converges on its entire domain of convergence. (Use Weierstrass M-test)*
(c) Explain why the $L(x)$ *is continuous on its domain of convergence.*

Exercise 3.5 *Consider the sequence* $\{f_n\}$ *of functions* $f_n : \mathbb{R} \to \mathbb{R}$ *defined by*

$$f_n(x) = \frac{nx}{\sqrt{1+n^2x^2}}.$$

Find the pointwise limit of this sequence an $n \to \infty$. *Does the sequence converge uniformly on* $\mathbb{R}$?

3.2 Uniform and Absolute Convergence of Fourier Series

We have seen that the Fourier series of $f(x)$ could not converge to the exact function $f(x)$ on the whole interval of interest, as demonstrated by the Fourier pointwise convergence theorem. However, they remain identical across the interval of interest, with the exception of the few locations where the periodic extension of $f(x)$ exhibits a jump discontinuity. The same method is used to evaluate sine (cosine) series, except that the odd (even) periodic extension needs to be taken into account. The different extensions of $f(x)$ may induce a jump discontinuity in addition to the locations of jump discontinuity of $f(x)$ itself. We can see from previous examples that there are occasionally no jump discontinuities in the resulting series. In these situations, $f(x)$ will really equal $f(x)$ in the Fourier series. The next theorem is concerned with the uniform and absolute convergence of Fourier series. We intentionally proved the theorem for functions that are $2L$- periodic. In plain words, the theorem will say that for piecewise smooth $f(x)$, the Fourier series of $f(x)$ is continuous for $-L \leq x \leq L$, if and only if $f(x)$ is continuous and $f(-L) = f(L)$.

Theorem 3.1 (Uniform and Absolute Convergence on Arbitrary Intervals) *Suppose that $f(x) \in C_p'(-L,L)$. That is, $f(x)$ is piecewise smooth on the interval $-L < x < L$. If $f(x)$ is continuous on the entire interval $-L \leq x \leq L$, with $f(-L) = f(L)$, then the Fourier series of $f(x)$,*

$$\frac{a_0}{2} + \sum_{n=1}^{\infty} \left(a_n \cos \frac{n\pi x}{L} + b_n \sin \frac{n\pi x}{L} \right),$$

where a_0, a_n, and b_n are given by equations (1.14), (1.15), *and* (1.16), *respectively, converges uniformly and absolutely to $f(x)$ on $-L \leq x \leq L$.*

Proof Since $f(x)$ is piecewise smooth on the interval $-L \leq x \leq L$, we may define

$$A_n = \frac{1}{L} \int_{-L}^{L} f'(x) \cos \frac{n\pi x}{L} dx, \quad B_n = \frac{1}{L} \int_{-L}^{L} f'(x) \sin \frac{n\pi x}{L} dx.$$

Then

$$A_0 = \frac{1}{L} \int_{-L}^{L} f'(x)\, dx = \frac{1}{L}[f(L) - f(-L)] = 0.$$

Next we perform integration by parts. Let $u = \cos \dfrac{n\pi x}{L}$, $dv = f'(x)dx$. Then

$$\begin{aligned} A_n = \frac{1}{L} \int_{-\pi}^{\pi} \cos \frac{n\pi x}{L} f(x)\, dx &= \frac{1}{L} \left[\cos \frac{n\pi x}{L} f(x) \Big|_{-L}^{L} + \frac{n\pi}{L} \int_{-L}^{L} f(x) \sin \frac{n\pi x}{L} dx \right] \\ &= \frac{1}{L} \cos n\pi \Big(f(L) - f(-L) \Big) + \frac{n\pi}{L^2} \int_{-L}^{L} f(x) \sin \frac{n\pi x}{L} dx \\ &= \frac{n\pi}{L^2} \int_{-L}^{L} f(x) \sin \frac{n\pi x}{L} dx = \frac{n\pi}{L} b_n, \quad n = 1,2\ldots \end{aligned}$$

Thus,

$$b_n = \frac{L}{n\pi} A_n, \quad n = 1,2\ldots.$$

Along the same lines, for $u = \sin \dfrac{n\pi x}{L}$, $dv = f'(x)dx$ we arrive at

$$a_n = \frac{L}{n\pi} B_n, \quad n = 1,2\ldots$$

Let

$$S_N = \sum_{n=1}^{N} \sqrt{a_n^2 + b_n^2}$$

be the partial sum of the infinite series $\sum_{n=1}^{\infty} \sqrt{a_n^2 + b_n^2}$. Then

$$S_N = \sum_{n=1}^{N} \sqrt{a_n^2 + b_n^2} = \frac{L}{\pi} \sum_{n=1}^{N} \frac{1}{n} \sqrt{A_n^2 + B_n^2}.$$

Squaring both sides and using Cauchy's inequality given by (3.1) we arrive at

$$S_N^2 \leq \frac{L^2}{\pi^2} \sum_{n=1}^{N} \frac{1}{n^2} \sum_{n=1}^{N} \left(A_n^2 + B_n^2\right).$$

Now $\sum_{n=1}^{N} \frac{1}{n^2}$ is finite and hence it is bounded. Moreover, by (2.4) we have

$$S_N^2 \leq \frac{L^2}{\pi^2} \sum_{n=1}^{N} \frac{1}{n^2} \sum_{n=1}^{N} \left(A_n^2 + B_n^2\right) \leq D \frac{1}{L} \int_{-L}^{L} [f'(x)]^2 \, dx,$$

which is bounded since $f'(x)$ is square integrable on $-L \leq x \leq L$, where $\frac{L^2}{\pi^2} \sum_{n=1}^{\infty} \frac{1}{n^2} \leq D$, for a positive constant D. Thus, S_N^2 (or S_N) is bounded for all N. Since S_N is monotonically increasing and it is bounded, it must have a limit. That is

$$S = \lim_{N \to \infty} S_N(x) = \sum_{n=1}^{\infty} \sqrt{a_n^2 + b_n^2}.$$

This shows

$$\sum_{n=1}^{\infty} \sqrt{a_n^2 + b_n^2},$$

converges. Since

$$|a_n| = \sqrt{a_n^2} \quad \text{and} \quad |b_n| = \sqrt{b_n^2},$$

we have

$$|a_n| \leq \sqrt{a_n^2 + b_n^2} \quad \text{and} \quad |b_n| \leq \sqrt{a_n^2 + b_n^2}.$$

Moreover, using the comparison test we see that

$$\begin{aligned}
\left| \sum_{n=1}^{\infty} \left(a_n \cos \frac{n\pi x}{L} + b_n \sin \frac{n\pi x}{L} \right) \right| &\leq \sum_{n=1}^{\infty} \left(\left| a_n \cos \frac{n\pi x}{L} \right| + \left| b_n \sin \frac{n\pi x}{L} \right| \right) \\
&\leq \sum_{n=1}^{\infty} \left(|a_n| + |b_n| \right) \leq 2 \sum_{n=1}^{\infty} \sqrt{a_n^2 + b_n^2} < \infty.
\end{aligned}$$

Applying Weierstrass M-test we obtain that the convergence of the Fourier series is absolute and uniform on the interval $-L \leq x \leq L$. The conditions on f ensure the continuity of the periodic extension of f for all x. Hence it follows that the Fourier series converges absolutely and uniformly to $f(x)$ everywhere in the interval $-L \leq x \leq L$. □

We shall see in Chapter 4, that in most applications, we either encounter Fourier sine series or cosine Fourier series. Therefore, in the next two corollaries we customize Theorem 3.1 to sine and cosine.

Corollary 2 *Suppose that $f(x) \in C'_p(0,L)$. If $f(x)$ is continuous on the entire interval $0 \leq x \leq L$, with $f(0) = f(L) = 0$, then the Fourier sine series of $f(x)$,*

$$\sum_{n=1}^{\infty} b_n \sin \frac{n\pi x}{L},$$

where

$$b_n = \frac{2}{L} \int_0^L f(x) \sin \frac{n\pi x}{L} dx, \quad n = 1, 2, \ldots$$

converges uniformly and absolutely to $f(x)$ on $0 \leq x \leq L$.

Corollary 3 *Suppose that $f(x) \in C'_p(0,L)$. If $f(x)$ is continuous on the entire interval $0 \leq x \leq L$, then the Fourier cosine series of $f(x)$,*

$$\frac{a_0}{2} + \sum_{n=1}^{\infty} a_n \cos \frac{n\pi x}{L},$$

where

$$a_n = \frac{2}{L} \int_0^L f(x) \cos \frac{n\pi x}{L} dx, \quad n = 0, 1, \ldots$$

converges uniformly and absolutely to $f(x)$ on $0 \leq x \leq L$.

We lumped a_0 with a_n by assuming $n = 0$ is valid. Otherwise, compute a_0 separately, as we have been doing.

In Figs. 1.3 and 1.4, we displayed the periodic even extension and odd extension of

$$f(x) = \frac{x}{L} + 1, \quad 0 < x < L,$$

respectively. In Fig. 1.4, we have a discontinuous periodic odd extension due to the fact that we did not require $f(0) = f(L) = 0$. On the other hand, in Fig. 1.5, once we required $f(0) = f(L) = 0$ we ended up with continuous periodic odd extension. Observe that the periodic even extension does not introduce new jumps irrespective of the values of $f(0)$ or $f(L)$.

Theorem 3.1 and Corollaries 2 and 3 can be adjusted to intervals $-\pi \leq x \leq \pi$, or $0 \leq x \leq \pi$, by simply replacing L with π.

Example 3.4 In Example 1.5 we found that the Fourier series of

$$f(x) = |x|, \quad -\pi < x < \pi,$$

is

$$f(x) = \frac{\pi}{2} - \frac{4}{\pi}\sum_{n=1}^{\infty}\frac{\cos(2n-1)x}{(2n-1)^2}, \quad -\pi \leq x \leq \pi.$$

Clearly, $f(x) \in C_p'(-\pi,\pi)$ and $f(x)$ is continuous on $-\pi \leq x \leq \pi$. Moreover,

$$f(-\pi) = f(\pi) = \pi.$$

Hence, by Theorem 3.1

$$\frac{\pi}{2} - \frac{4}{\pi}\sum_{n=1}^{\infty}\frac{\cos(2n-1)x}{(2n-1)^2}$$

converges uniformly and absolutely to $f(x)$. This is evident from Figs. 2.2–2.4. Applying Weierstrass M-test, we see that

$$\left|\frac{\pi}{2} - \frac{4}{\pi}\sum_{n=1}^{\infty}\frac{\cos(2n-1)x}{(2n-1)^2}\right| \leq \frac{\pi}{2} + \frac{4}{\pi}\sum_{n=1}^{\infty}\frac{1}{(2n-1)^2} < \infty.$$

We conclude that

$$\frac{\pi}{2} - \frac{4}{\pi}\sum_{n=1}^{\infty}\frac{\cos(2n-1)x}{(2n-1)^2}$$

converges uniformly and absolutely. □

3.3 More on the Dirichlet Kernel

In this section we take a closer look at the Dirichlet kernel that is given by (2.10)

$$K_N(u) = \frac{\sin\left[\left(N+\frac{1}{2}\right)u\right]}{2\sin\left(\frac{u}{2}\right)} \quad \text{for} \quad u \neq 0, \pm 2\pi \ldots$$

The Dirichlet kernel arises in the computation of Fourier series. In the proof of Theorem 2.1, it is used in the partial sum of the Fourier series

$$S_N(x) = \frac{1}{\pi}\int_{-\pi}^{\pi} f(x-u)K_N(u)\,du. \tag{3.2}$$

For pointwise or uniform convergence of the Fourier series, we would ideally want the Dirichlet kernel to have the following property

$$\int_{-\pi}^{\pi} |K_N(x)|\,dx < \infty.$$

If $K_N(x)$ were absolutely integrable, the convolution $f * K_N$ defined by (3.2) would smooth the function $f(x)$, ensuring good convergence properties. However, the Dirichlet kernel is not absolutely integrable since

$$\int_{-\pi}^{\pi} |K_N(x)|\, dx \to \infty \quad \text{as} \quad N \to \infty.$$

The slow decay of $K_N(x)$'s oscillations (which lead to high peaks near $x = 0$) means that the Fourier coefficients might not decay fast enough for certain functions, which result in Gibb's phenomenon.

Lemma 8 *Let* $K_N(u) = \dfrac{\sin\left[\left(N+\frac{1}{2}\right)u\right]}{2\sin\left(\frac{u}{2}\right)}$ *be the Dirichlet kernel. Then*

$$\int_{-\pi}^{\pi} |K_N(u)|\, du \to \infty \quad as \quad N \to \infty.$$

Proof *Set*

$$L_N = \int_{-\pi}^{\pi} |K_N(u)|\, du = \int_{-\pi}^{\pi} \left| \frac{\sin\left[\left(N+\frac{1}{2}\right)u\right]}{2\sin\left(\frac{u}{2}\right)} \right| du = \int_{0}^{\pi} \left| \frac{\sin\left[\left(N+\frac{1}{2}\right)u\right]}{\sin\left(\frac{u}{2}\right)} \right| du,$$

due to symmetry. By letting $t = \frac{u}{2}$ *and utilizing*

$$\frac{1}{\sin t} > \frac{1}{t} \text{ for } t \in [-\frac{\pi}{2}, \frac{\pi}{2}],$$

we obtain, by setting $u = 2t$ *that*

$$L_N > \int_{0}^{\frac{\pi}{2}} \left| \frac{\sin\left[\left(N+\frac{1}{2}\right)2t\right]}{t} \right| dt.$$

Changing variables by letting $\theta = (N+\frac{1}{2})2t$, *we have* $d\theta = (N+\frac{1}{2})2dt$ *and the integral becomes*

$$L_N > \int_{0}^{(N+\frac{1}{2})\pi} \frac{|\sin(\theta)|}{\theta}\, d\theta.$$

Splitting the integral into intervals $[k\pi, (k+1)\pi]$ *for* $k \in \mathbb{Z}^+$ *gives*

$$L_N = \int_{0}^{(N+\frac{1}{2})\pi} \frac{|\sin\theta|}{\theta}\, d\theta > \sum_{k=0}^{N-1} \int_{k\pi}^{(k+1)\pi} \frac{|\sin\theta|}{\theta}\, d\theta. \tag{3.3}$$

Setting $u = \theta - k\pi$, *we have*

$$\sum_{k=0}^{N-1} \int_{k\pi}^{(k+1)\pi} \frac{|\sin\theta|}{\theta}\, d\theta = \sum_{k=0}^{N-1} \int_{0}^{\pi} \frac{|\sin(u+k\pi)|}{u+k\pi}\, du$$

$$= \sum_{k=0}^{N-1} \int_0^{\pi} \frac{\left|(-1)^k \sin(u)\right|}{u+k\pi} d\theta$$

$$= \sum_{k=0}^{N-1} \int_0^{\pi} \frac{\sin(u)}{u+k\pi} du.$$

By switching back from u to θ and substituting into (3.3)*, we obtain*

$$\begin{aligned} L_N &> \sum_{k=0}^{N-1} \int_0^{\pi} \frac{\sin(\theta)}{\theta+k\pi} d\theta > \sum_{k=0}^{N-1} \frac{1}{(k+1)\pi} \int_0^{\pi} \sin\theta d\theta \\ &= \frac{2}{\pi^2} \sum_{k=0}^{N-1} \frac{1}{k+1} > \frac{2}{\pi^2} \log N + \frac{2}{\pi^2} \gamma, \end{aligned}$$

where γ is Euler's constant. Thus

$$\int_{-\pi}^{\pi} |K_N(u)|\, du \sim \frac{2}{\pi^2} \log(N).$$

This shows

$$\lim_{N\to\infty} L_N = \lim_{N\to\infty} \int_{-\pi}^{\pi} \left|K_N(u)\right| du = \infty.$$

This complete the proof. □

The absolute integrability of the Dirichlet kernel would ensure smoother and more controlled convergence of the Fourier series. Its non-integrability, however, explains the oscillatory nature of the partial sums and why Fourier series may fail to converge uniformly or pointwise in general without imposing additional conditions on $f(x)$. To address the lack of absolute integrability of the Dirichlet kernel, mathematicians often use the Fejér kernel, (that we define below) which is a modification of the Dirichlet kernel. The Fejér kernel is absolutely integrable, and using it leads to better convergence properties, such as uniform convergence for continuous functions. It is derived from the Dirichlet kernel, but it has better properties for ensuring the convergence of Fourier series, particularly in the sense of Cesàro summation. For more reading on Fejér kernel, we refer to [141].

Let $K_k(x)$ be the Dirichlet kernel given by (2.10). Then the Fejér kernel $F_N(x)$ is defined as

$$F_N(x) = \frac{1}{N+1} \sum_{k=0}^{N} K_k(x). \tag{3.4}$$

where $K_k(x)$ is the Dirichlet kernel.

3.3.1 Exercises

Exercise 3.6 *Verify* $a_n = \dfrac{L}{n\pi} B_n$ *of Theorem 3.1.*

Exercise 3.7 *(a) Without finding the Fourier sine series, justify why the Fourier sine series of*

$$f(x) = \begin{cases} x, & 0 \le x \le \frac{L}{2} \\ -x+L, & \frac{L}{2} \le x \le L \end{cases}$$

converges uniformly and absolutely to $f(x)$ *and draw the function* f *over* $[-2L, 2L]$.
(b) Find the Fourier sine series and apply the Weierstrass M-test to it.

Exercise 3.8 *Assume the hypothesis of Theorem 3.1 holds. Let* $M = \max\{f'(x) : -L \le x \le L\}$. *Show that the Fourier coefficients* a_n *and* b_n *satisfy*

$$|a_n| \le \frac{2ML}{n\pi} \quad \textit{and} \quad |b_n| \le \frac{2ML}{n\pi}, \quad n = 1, 2, \ldots$$

Exercise 3.9 *Let* $M = \max\{f^{(k)}(x) : -L \le x \le L\}$. *Generalize Exercise 3.8 to functions* f *such that the derivatives* $f^{(k-1)}$ *are continuous as periodic functions and* $f^{(k)}$ *is piecewise continuous.*

Exercise 3.10 *Let* $F_N(x)$, $N = 0, 1 \ldots$ *be given by* (3.4). *Show that*
(a)

$$F_N(x) \ge 0 \quad \textit{for all} \quad x,$$

(b)

$$F_N(x) = \frac{1}{N+1} \frac{\sin^2\left(\frac{(N+1)x}{2}\right)}{2\sin^2\left(\frac{x}{2}\right)}, \quad x \ne 0,$$

and at $x = 0$, $F_N(x) = N+1$,
(c)

$$\int_{-\pi}^{\pi} F_N(x)\,dx = \frac{1}{2},$$

(d)

$$\int_{-\pi}^{\pi} |F_N(x)|\,dx < \infty.$$

3.4 Integrations and Differentiations of Fourier Series

Integrations and differentiations in Fourier series are powerful tools that allow the analysis and manipulation of functions in both time and frequency domains. Differentiating a Fourier series term by term enables the representation of the derivatives of a function in terms of its frequency components, providing insights into how the function's behavior changes at various frequencies. Similarly, integrating a Fourier series term by term helps in reconstructing antiderivatives or smoothing functions by attenuating high-frequency components. These techniques are particularly useful in solving differential equations, signal processing, and understanding the spectral properties of systems. However, differentiation is more delicate and requires careful handling. While integration tends to stabilize a function by reducing the impact of high-frequency terms, differentiation amplifies these components, which can lead to divergence or instability if the function is not sufficiently smooth. To avoid such issues, conditions like uniform convergence or additional smoothness of the original function must be ensured when differentiating a Fourier series. In contrast, integration is generally less sensitive, as it naturally averages out oscillations and avoids amplifying noise or discontinuities.

Integrations of Fourier Series

We begin by considering integrations of Fourier series. The integrations of a Fourier series may not be necessarily a Fourier series in the sense of periodicity. If $f(x)$ has a Fourier series on $[-L, L]$, then an integration of

$$f(x) \sim \frac{a_0}{2} + \sum_{n=1}^{\infty} \left(a_n \cos \frac{n\pi x}{L} + b_n \sin \frac{n\pi x}{L} \right),$$

may not lead to a periodic new Fourier series especially when $a_0 \neq 0$. We begin with the following proposition.

Proposition 1 *Suppose that the function $f(x)$ is piecewise continuous and is periodic with period $2L$. Then the function F given by*

$$F(x) = \int_{-L}^{x} f(t)\, dt, \tag{3.5}$$

is continuous and $2L$-periodic only if

$$F(-L) = F(L).$$

Proof *Using* (3.5) *we have*

$$F(x+2L) = \int_{-L}^{x+2L} f(t)\, dt = \int_{-L}^{L} f(t)\, dt + \int_{L}^{x+2L} f(t)\, dt.$$

By making the substitution $u = t - 2L$ in the second integral and utilizing the periodicity property $f(x+2L) = f(x)$, we arrive at

$$F(x+2L) = \int_{-L}^{L} f(t)\, dt + \int_{-L}^{x} f(t+2L)\, dt = \int_{-L}^{L} f(t)\, dt + \int_{-L}^{x} f(t)\, dt.$$

Consequently,

$$F(x+2L) = \int_{-L}^{L} f(t)\, dt + F(x).$$

But, $\int_{-L}^{L} f(t)\,dt = F(L) = F(-L) = 0$, *which implies that* $F(x+2L) = F(x)$.
□

Note that since $F(-L) = 0$, then the condition $F(-L) = F(L)$ implies that

$$0 = \int_{-L}^{-L} f(t)\,dt = \int_{-L}^{L} f(t)\,dt = La_0. \tag{3.6}$$

Notice that (3.6) implies that

$$a_0 = 0.$$

For the next theorem, a_0 is assumed to be the Fourier coefficient of a function $f(x)$ defined on the interval $[-L, L]$.

Theorem 3.2 *Suppose that the function f is $2L$-periodic and piecewise continuous on $[-L, L]$. Then the integral given by (3.5) determine a $2L$-periodic function if and only if (3.6) holds.*

Proof *Since $F(x) \in C_p(-L, L)$, it admits a convergent Fourier series expansion of the form*

$$F(x) = \frac{A_0}{2} + \sum_{n=1}^{\infty} \left(A_n \cos\frac{n\pi x}{L} + B_n \sin\frac{n\pi x}{L} \right).$$

Next we use integrations by parts to compute the coefficients, A_0, A_n, and B_n for $n = 1, 2, \ldots$ Letting $u = F(x)$ and $dv = dx$, we get

$$\begin{aligned} A_0 = \frac{1}{L}\int_{-L}^{L} F(x)\,dx &= \frac{1}{L}\left[xF(x)\Big|_{-L}^{L} - \int_{-L}^{L} xF'(x)\,dx \right] \\ &= F(L) + F(-L) - \frac{1}{L}\int_{-L}^{L} xF'(x)\,dx. \end{aligned}$$

Using (3.6) and the fact that $F'(x) = f(x)$ (from (3.5)), we see that

$$A_0 = -\frac{1}{L}\int_{-L}^{L} xf(x)\,dx. \tag{3.7}$$

For $n = 1, 2, \ldots$ we have for $dv = \cos\frac{n\pi x}{L}\,dx$, $u = F(x)$ that

$$\begin{aligned} A_n &= \frac{1}{L}\int_{-\pi}^{\pi} \cos\frac{n\pi x}{L} F(x)\,dx = \frac{1}{L}\left[\frac{L}{n\pi} \sin\frac{n\pi x}{L} F(x) \right]_{-L}^{L} - \frac{L}{n\pi}\int_{-L}^{L} F'(x)\sin\frac{n\pi x}{L}\,dx \\ &= -\frac{L}{n\pi}\left(\frac{1}{L}\int_{-L}^{L} f(x)\sin\frac{n\pi x}{L}\,dx \right) = -\frac{L}{n\pi} b_n, \quad n = 1, 2 \ldots \end{aligned}$$

Thus,

$$A_n = -\frac{L}{n\pi} b_n, \quad n = 1, 2, \ldots$$

where b_n is the Fourier sine coefficient of the function $f(x)$. Along the same lines we can find that

$$B_n = \frac{L}{n\pi}a_n, \quad n = 1,2\dots$$

where a_n is the Fourier cosine coefficient of the function $f(x)$. □

We have obtained the following theorem.

Theorem 3.3 *Suppose that the function f is 2L-periodic and piecewise continuous on $[-L,L]$ with $a_0 = 0$ so that*

$$f(x) \sim \sum_{n=1}^{\infty}\left(a_n \cos\frac{n\pi x}{L} + b_n \sin\frac{n\pi x}{L}\right). \tag{3.8}$$

Then, integrating both sides of (3.8) term by term yields

$$\int_{-L}^{x} f(s)\,ds \sim \frac{A_0}{2} + \sum_{n=1}^{\infty}\left(\frac{L}{n\pi}a_n \sin\frac{n\pi x}{L} - \frac{L}{n\pi}b_n \cos\frac{n\pi x}{L}\right), \tag{3.9}$$

where A_0 is given by (3.7). Moreover, the Fourier series in (3.9) converges for every $x \in \mathbb{R}$.

Since $F(-L) = F(L)$, this demonstrates that representation (3.9) is also valid at the endpoints of the open interval $-L < x < L$ and, therefore, at every point of the closed interval $-L \le x \le L$.

Example 3.5 In Exercise 1.5 we found the Fourier series of

$$f(x) = \begin{cases} 1, & 0 < x \le \pi \\ -1, & -\pi < x \le 0 \end{cases}, \quad f(x+2\pi) = f(x)$$

to be

$$f(x) \sim \frac{4}{\pi}\sum_{n=1}^{\infty}\frac{\sin(2n-1)x}{2n-1}.$$

Clearly, $a_0 = 0$, and hence Theorem 3.3 is applicable. For x in the interval $-\pi < x < \pi$,

$$A_0 = -\frac{1}{\pi}\int_{-\pi}^{\pi} x f(x)\,dx = -\frac{1}{\pi}\left[\int_{-\pi}^{0} x(-1)\,dx + \int_{0}^{\pi} x\,dx\right] = -\pi.$$

Next we evaluate the left-hand side of (3.9).
Case 1: $-\pi < x < 0$

$$\int_{-\pi}^{x} f(s)\,ds = \int_{-\pi}^{x}(-1)\,ds = -x - \pi.$$

Case 2: $0 < x < \pi$

$$\int_{-\pi}^{x} f(s)\,ds = \int_{-\pi}^{0}(-1)\,ds + \int_{0}^{x} ds = x - \pi.$$

Thus, the value of the integral is

$$\int_{-1}^{x} f(s)\,ds = \begin{cases} -x-\pi, & \text{if } -\pi < x < 0 \\ x-\pi, & \text{if } 0 < x < \pi \end{cases} = |x| - \pi.$$

Since $a_n = \frac{1}{2n-1}$ and $b_n = 0$, expression (3.9) becomes

$$|x| - \pi \sim -\frac{\pi}{2} - \frac{4}{\pi}\sum_{n=1}^{\infty} \frac{1}{(2n-1)^2} \cos(2n-1)x.$$

Solving for $|x|$ yields

$$|x| \sim \frac{\pi}{2} - \frac{4}{\pi}\sum_{n=1}^{\infty} \frac{1}{(2n-1)^2} \cos(2n-1)x, \quad -\pi < x < \pi, \tag{3.10}$$

which is the Fourier series of

$$f(x) = \begin{cases} -x, & \text{if } -\pi < x < 0 \\ x, & \text{if } 0 < x < \pi \end{cases} = |x|$$

or the Fourier cosine of $f(x) = x,\ 0 < x < \pi$, and hence its Fourier cosine converges to $f(x)$ uniformly and absolutely. □

As previously noted, the Fourier series of a function $f(x)$ may not converge uniformly to $f(x)$. However, when the series is integrated term by term, the resulting Fourier series always converges uniformly to the integral of $f(x)$ over the specified interval. For instance, in Example 3.5, the function $f(x)$ exhibits jump discontinuities at $0, \pm\pi, \ldots$ Despite these discontinuities, the Fourier series obtained through term-by-term integration converges uniformly to the function $|x|$, which corresponds to the integral of $f(x)$.

The above discussion focused on the case of a $2L$-periodic integrated Fourier series where the mean value is $\frac{a_0}{2} = 0$. But what happens if $\frac{a_0}{2} \neq 0$? Suppose that the function $f(x)$ is $2L$-periodic and has $\frac{a_0}{2} \neq 0$. To address this, we define a new function $h(x)$ as

$$h(x) = f(x) - \frac{a_0}{2}.$$

By this definition, $h(x)$ has a mean value of zero. Consequently, we can write

$$f(x) = h(x) + \frac{a_0}{2}.$$

Since $h(x)$ has zero mean, Theorem 3.3 applies to $h(x)$.

For any two points q, r in $[-L, L]$ we have

$$\int_q^r f(x)\,dx = \int_{-L}^{r} f(x)\,dx - \int_{-L}^{q} f(x)\,dx = \int_{-L}^{r} h(x)\,dx - \int_{-L}^{q} h(x)\,dx + \int_q^r \frac{a_0}{2}\,dx$$

Consequently,

$$\int_q^r f(x)\,dx = \int_q^r \frac{a_0}{2}\,dx + \sum_{n=1}^{\infty} \int_q^r \left[a_n \cos\frac{n\pi x}{L} + b_n \sin\frac{n\pi x}{L}\right] dx. \tag{3.11}$$

Hence the following result holds.

Theorem 3.4 *Suppose that the function f is $2L$-periodic and piecewise continuous on $[-L,L]$. Then the expression* (3.11) *is valid for any real numbers q,r in $[-L,L]$.*

As an application, we consider the following example.

Example 3.6 Integrate the Fourier series (3.10) of Example 3.5 from 0 to $\frac{\pi}{2}$. Then

$$\int_0^{\frac{\pi}{2}} |x|\,dx = \int_0^{\frac{\pi}{2}} \frac{\pi}{2}\,dx - \frac{4}{\pi}\sum_{n=1}^{\infty}\int_0^{\frac{\pi}{2}} \frac{1}{(2n-1)^2}\cos(2n-1)x\,dx.$$

After integrating, we obtain

$$\frac{\pi^2}{8} = \frac{\pi^2}{4} - \frac{4}{\pi}\sum_{n=1}^{\infty}\frac{(-1)^{n+1}}{(2n-1)^3}.$$

This relation produces

$$\frac{\pi^3}{32} = \sum_{n=1}^{\infty}\frac{(-1)^{n+1}}{(2n-1)^3}.$$

Differentiations of Fourier Series

In a Fourier series, the coefficients a_n and b_n typically decrease as n increases. When differentiating the series, the terms are multiplied by n (or higher powers of n for higher-order derivatives). This amplification can cause the differentiated series to grow rapidly, especially for functions with discontinuities or sharp transitions, leading to divergence or slower convergence. Thus, for a Fourier series to be differentiable term by term, the original function must be sufficiently smooth. Specifically, the function must be piecewise smooth, and its derivative must exist and be piecewise continuous. If the function lacks these properties, the differentiated series may fail to represent the derivative correctly or may not converge at all.

If $f(x)$ is continuous on $-L \le x \le L$ and $f'(x)$ is piecewise smooth, then the following statements are equivalent:

1. $f(L) = f(-L)$.
2. Fourier series of $f(x)$ is continuous.
3. $f(x) =$ Fourier series of $f(x)$ for $-L \le x \le L$.
4. Fourier series of $f(x)$ maybe differentiated term by term.

Theorem 3.5 *Let $f(x)$ be a 2L-periodic function that is continuous for all x, such that $f(-L) = f(L)$, and assume that its derivative $f'(x)$ is piecewise smooth for all x. If the Fourier series of $f(x)$ is given by*

$$f(x) \sim \frac{a_0}{2} + \sum_{n=1}^{\infty} \left(a_n \cos \frac{n\pi x}{L} + b_n \sin \frac{n\pi x}{L} \right),$$

then the Fourier series of $f'(x)$ is

$$f'(x) \sim \sum_{n=1}^{\infty} \left(-\frac{n\pi}{L} a_n \sin \frac{n\pi x}{L} + \frac{n\pi}{L} b_n \cos \frac{n\pi x}{L} \right), \tag{3.12}$$

which is obtained by term-by-term differentiation of the original series. Furthermore, the differentiated series converges to $f'(x)$ at every point where $f'(x)$ exists.

Proof *Since $f'(x)$ is assumed to be 2L-periodic and piecewise smooth, Theorem 2.1 ensures that the Fourier series of $f'(x)$ converges to the average of the left-hand and right-hand limits of $f'(x)$ at each x. Then*

$$f'(x) = \frac{A_0}{2} + \sum_{n=1}^{\infty} \left(A_n \cos \frac{n\pi x}{L} + B_n \sin \frac{n\pi x}{L} \right), \tag{3.13}$$

where the coefficients A_n and B_n are determined by

$$A_n = \frac{1}{L} \int_{-L}^{\pi} f'(x) \cos \frac{n\pi x}{L} dx, \quad B_n = \frac{1}{L} \int_{-L}^{L} f'(x) \sin \frac{n\pi x}{L} dx.$$

To prove the theorem, it suffices to show that the series in the equations for $f'(x)$ and the differentiated series are equivalent. We start with the calculation

$$A_0 = \frac{1}{L} \int_{-L}^{L} f'(x)\, dx = \frac{1}{L} [f(x)]_{-L}^{L} = f(L) - f(-L) = 0,$$

since $f(x)$ is continuous and 2L-periodic.

For $n \geq 1$, the Euler formulas for both f and f', along with integration by parts, yield

$$\begin{aligned} A_n &= \frac{1}{L} \int_{-L}^{L} f'(x) \cos \frac{n\pi x}{L} dx = \frac{1}{L} \left[f(x) \cos \frac{n\pi x}{L} \right]_{-L}^{L} + \frac{n\pi}{L^2} \int_{-\pi}^{\pi} f(x) \sin \frac{n\pi x}{L} dx \\ &= \frac{1}{L} [f(L) - f(-L)] \cos n\pi + \frac{n\pi}{L} \left(\frac{1}{L} \int_{-\pi}^{\pi} f(x) \sin \frac{n\pi x}{L} dx \right) \\ &= \frac{n\pi}{L} b_n, \end{aligned}$$

where we used the fact that $f(-L) = f(L)$ due to 2L-periodicity. Similarly,

$$B_n = -\frac{n\pi}{L} a_n, \quad n = 1, 2, \ldots$$

Thus, the series in Equations (3.12) and (3.13) are identical, thereby completing the proof. □

Example 3.7 It was shown in Exercise 1.29 that the Fourier series of

$$f(x) = |x|, \quad -\sqrt{3} < x < \sqrt{3}, \quad f(x+2\sqrt{3}) = f(x)$$

is given by

$$f(x) = \frac{\sqrt{3}}{2} - \frac{4\sqrt{3}}{\pi^2} \sum_{n=1}^{\infty} \frac{\cos\left((2n-1)\frac{\pi x}{\sqrt{3}}\right)}{(2n-1)^2}.$$

Note that

$$f(\sqrt{3}) = f(-\sqrt{3}) = \sqrt{3}.$$

However, the derivative of $f'(x)$ is the square wave and given by

$$f(x) = \begin{cases} 1, & \text{if } 0 < x < \sqrt{3} \\ -1, & \text{if } -\sqrt{3} < x < 0 \end{cases}$$

It is clear that $f'(x)$ is piecewise smooth for all x. Thus, by Theorem 3.5 the series of $f(x)$ can be differentiated term by term to obtain the Fourier series of $f'(x)$. That is

$$f'(x) = \frac{4}{\pi} \sum_{n=1}^{\infty} \frac{\sin\left((2n-1)\frac{\pi x}{\sqrt{3}}\right)}{(2n-1)}, \quad -\sqrt{3} < x < \sqrt{3}.$$

Note that $f'(x)$ has jump discontinuities at $x = j\sqrt{3},\ j = 0, \pm 1, \pm 2$, which is not the case for $f(x)$. □

3.4.1 Exercises

Exercise 3.11 *In Theorem 3.2 verify that the coefficient* $B_n = \frac{L}{n\pi} a_n,\ n = 1, 2, \ldots$

Exercise 3.12 *We already know that the Fourier series of* $f(x) = x$ *for* $-\pi < x < \pi$, *where* $f(x+2\pi) = f(x)$, *is*

$$f(x) = 2\sum_{n=1}^{\infty} \frac{(-1)^{n+1}}{n} \sin(nx).$$

Compute the Fourier series of each of the following 2π*-periodic functions.*

(a) $f_1(x) = x^2, \quad -\pi < x < \pi.$

(b) $f_2(x) = x^3, \quad -\pi < x < \pi.$

(c) $f_3(x) = x^4, \quad -\pi < x < \pi.$

Exercise 3.13 *Make use of Example 1.5 to compute the Fourier series for the function*

$$f(x) = x^2 \operatorname{sgn}(x) = \begin{cases} x^2, & \text{if } 0 < x < \pi \\ -x^2, & \text{if } -\pi < x < 0 \end{cases}$$

Exercise 3.14 Compute the Fourier series of $f(x) = 2x - |x|, \quad -\pi < x < \pi$, where $f(x+2\pi) = f(x)$.

Exercise 3.15 *Let $f(x)$ be given by (see Fig. 1.3).*

$$f(x) = \frac{x}{L} + 1, \quad 0 < x < L, \quad f(x+2L) = f(x).$$

(a) Find the Fourier cosine series of $f(x)$.
(b) Verify the hypotheses of Theorem 3.5 are satisfied and find the Fourier series of

$$f(x) = \begin{cases} \frac{1}{L}, & 0 < x < L \\ -\frac{1}{L}, & -L < x < 0 \end{cases}$$

(c) Do the hypotheses of Theorem 3.5 hold for $f''(x)$? Explain.

Exercise 3.16 *Repeat Exercise 3.15 for the function*

$$f(x) = \begin{cases} 0, & -\pi \le x < 0 \\ \sin x, & 0 \le x < \pi \end{cases}$$

$$f(x+2\pi) = f(x).$$

Exercise 3.17 *Given that the Fourier series of $f(x) = (x^2 - \pi^2)^2$ on $-\pi \le x \le \pi$ and $f(x+2\pi) = f(x)$ is*

$$f(x) = \frac{48\pi^4}{90} + 48 \sum_{n=1}^{\infty} \frac{(-1)^{n+1}}{n^4} \cos nx.$$

Show the following.
(a)

$$x(x^2 - \pi^2) = 12 \sum_{n=1}^{\infty} \frac{(-1)^n}{n^3} \sin nx, \quad -\pi \le x \le \pi.$$

(b)

$$x^2 = \frac{\pi^2}{3} + 4 \sum_{n=1}^{\infty} \frac{(-1)^n}{n^2} \cos nx, \quad -\pi \le x \le \pi.$$

(c)

$$x = 2 \sum_{n=1}^{\infty} \frac{(-1)^{n+1}}{n} \sin nx, \quad -\pi \le x \le \pi.$$

Exercise 3.18 *We already know that*

$$x = 2 \sum_{n=1}^{\infty} \frac{(-1)^{n+1}}{n} \sin nx, \quad -\pi < x < \pi.$$

(a) Integrate term by term three times in succession to obtain the series

$$\frac{x^4}{24} = \frac{2x^2}{12} - 2 \sum_{n=1}^{\infty} \frac{(-1)^n}{n^4} \cos nx + 2 \sum_{n=1}^{\infty} \frac{(-1)^n}{n^4}.$$

(b) Substitute $x = \frac{\pi}{2}$ and $x = \pi$ in the series to obtain

$$\sum_{n=1}^{\infty} \frac{1}{n^4} = \frac{\pi^4}{90}, \quad \sum_{n=1}^{\infty} \frac{(-1)^{n+1}}{n^4} = \frac{7\pi^4}{720}.$$

Exercise 3.19 *Use appropriate Fourier series to solve the following differential equations.*

(a) $y''(x) + 2y(x) = 1, \quad y(0) = y(\pi) = 0.$

(b) $y''(x) - 4y(x) = 1, \quad y(0) = y(\pi) = 0.$

(c) $y''(x) + y(x) = x, \quad y(0) = y(1) = 0.$

(d) $y''(x) + 2y(x) = x, \quad y'(0) = y'(\pi) = 0.$

3.5 Generalized Fourier Series and Orthonormal Functions

A generalized Fourier series extends the concept of the classical Fourier series to represent functions in terms of orthogonal basis functions beyond sines and cosines. While traditional Fourier series are ideal for periodic functions, many real-world problems involve functions defined on non-standard intervals or those requiring different basis functions (e.g., Legendre polynomials, Bessel functions). Generalized Fourier series provide a flexible framework to handle such cases, enabling solutions to a broader range of problems in fields like quantum mechanics, signal processing, and differential equations. This adaptability makes them essential for modeling complex systems and solving equations with varying boundary and initial conditions. We start with some preliminary work.

Definition 3.3 *Let the functions f and g be in $C_p(a,b)$. Then the inner product of f and g is defined as*

$$(f,g) = \int_a^b f(x)g(x)dx. \tag{3.14}$$

Thus, for any functions f, g, and h in $C_p(a,b)$, the following hold for a nonzero constant c.

(1) $(f,g) = (g,f)$,

(2) $(f, g+h = (f,g) + (f,h)$,

(3) $(cf,g) = c(f,g)$,

(4) $(f,f) \geq 0, \quad \text{and} \quad (f,f) = 0 \quad \text{if and only if} \quad f = 0.$

Next, we define the notion of a norm.

Definition 3.4 (Norm) *A* ***norm*** $\|\cdot\|$ *on a vector space* V *over a field* $\mathbb{F}$ *(typically* $\mathbb{R}$ *or* $\mathbb{C}$*) is a function* $\|\cdot\| : V \to \mathbb{R}$ *that satisfies the following conditions for all* $x, y \in V$ *and scalar* $\alpha \in \mathbb{F}$

(1) ***Non-negativity***

$$\|x\| \geq 0, \quad \text{and} \quad \|x\| = 0 \quad \text{if and only if} \quad x = 0.$$

(2) ***Absolute Homogeneity (or Positive Homogeneity)***

$$\|\alpha x\| = |\alpha|\|x\|, \quad \alpha \in \mathbb{R}.$$

(3) ***Triangle Inequality (Subadditivity)***

$$\|x+y\| \leq \|x\| + \|y\|.$$

If a function satisfies these three conditions, it is called a ***norm****, and the pair* $(V, \|\cdot\|)$ *is a* ***normed vector space****.*

Definition 3.5 *The* L^2 *space is defined as*

$$L^2[a,b] = \left\{ f : [a,b] \to \mathbb{R} \text{ such that } \int_a^b |f(x)|^2\, dx < \infty \right\}$$

or equivalently, the set of functions f *on* $[a,b]$ *such that the integral of its square is finite.*

Let

$$(f,f) = ||f||^2 = \int_a^b f^2(x)\, dx. \tag{3.15}$$

Then the L^2-norm of f, which represents the total energy of the function f over its domain is defined as

$$||f|| = \left(f,f\right)^{\frac{1}{2}}.$$

Definition 3.6 *The* L^2*-norm is induced by the* L^2*-inner product. That is*

$$(f,g) = \int_a^b f(x)\overline{g(x)}\, dx,$$

where $\overline{g(x)}$ *is the complex conjugate of* $g(x)$*. The norm is then*

$$\|f\|_{L^2} = \sqrt{(f,f)}.$$

Example 3.8

$$\|f\|_{L^2} = \left(\int_{-\infty}^{\infty} e^{-2x^2}\, dx\right)^{\frac{1}{2}} = \left(\frac{\pi}{2}\right)^{\frac{1}{2}}.$$

See Lemma 23 on how to evaluate the integral. □

The next definition is important in its own self.

Definition 3.7 *The* **span** *of the set of functions* $\{f_1, f_2, \ldots, f_n\}$ *consists of all functions that can be written as*

$$g(x) = c_1 f_1(x) + c_2 f_2(x) + \cdots + c_n f_n(x)$$

where $c_1, c_2, \ldots, c_n$ *are scalars (usually from* $\mathbb{R}$ *or* $\mathbb{C}$*). We may write*

$$g \in span(f_1, \ldots, f_n).$$

- *The span forms a* **vector space** *if the set of functions belongs to a vector space (e.g., the space of continuous functions or differentiable functions).*
- *If a set of functions' span covers the entire space, the functions are said to be* ***spanning*** *the space.*
- *If no function in the set can be written as a linear combination of the others, the functions are* ***linearly independent****.*

Example 3.9 Consider the functions

$$f_1(x) = 1, \quad f_2(x) = x, \quad \text{and} \quad f_3(x) = x^2.$$

Their span consists of all functions of the form

$$g(x) = c_1 + c_2 x + c_3 x^2,$$

which is the space of all polynomials of degree at most 2.

Definition 3.8 *A set of functions* $\varphi_1(x), \varphi_2(x), \ldots, \varphi_n(x)$ *is orthogonal on* (a,b) *if*

(1)

$$(\varphi_n, \varphi_m) = \begin{cases} 0, & n \neq m \\ \neq 0, & n = m \end{cases}$$

(2) orthonormal on (a,b) *if*

$$(\varphi_n, \varphi_m) = \delta_{nm} = \begin{cases} 0, & n \neq m \\ 1, & n = m \end{cases}$$

It is then clear that an orthogonal set of functions can be made into an orthonormal set by dividing each function in the set by its norm. That is,

$$\{\psi_n(x) = \frac{\varphi_n(x)}{||\varphi_n||}, \; n = 1, 2, \ldots\}$$

defines an orthonormal set of functions on (a,b). To see this, one can show that $||\psi_n||^2 = 1$ for all values of n by writing

$$(\psi_m, \psi_n) = \left(\frac{\varphi_m}{||\varphi_m||}, \frac{\varphi_n}{||\varphi_n||} \right) = \frac{(\varphi_m, \varphi_n)}{||\varphi_m|| \, ||\varphi_n||}$$

from which we arrive at

$$\|\psi_n\|^2 = (\psi_n, \psi_n) = \frac{(\varphi_n, \varphi_n)}{\|\varphi_n\|\|\varphi_n\|} = \frac{\|\varphi_n\|^2}{\|\varphi_n\|^2} = 1.$$

Example 3.10 The set of functions

$$\varphi_n(x) = \sin nx \quad (n = 1, 2, \ldots)$$

is orthogonal on the interval $0 < x < \pi$, since

$$(\varphi_n, \varphi_m) = \int_0^\pi \sin mx \sin nx\, dx = \begin{cases} 0, & \text{when } m \neq n \\ \frac{\pi}{2}, & \text{when } m = n \end{cases}$$

The norm $\|\varphi_n\|$ of each of these functions is

$$\|\varphi_n\| = \left(\int_0^\pi \sin^2 nx\, dx\right)^{1/2} = \left(\frac{\pi}{2}\right)^{1/2}.$$

Hence,

$$\psi_n(x) = \sqrt{\frac{2}{\pi}} \sin nx, \quad n = 1, 2, \ldots$$

form an orthonormal set on $(0, \pi)$. □

Now we explore *generalized Fourier series,* in which we represent a function using a set of orthogonal basis functions tailored to the problem's geometry and properties. Suppose that we have a set of orthogonal functions $\{\varphi_n(x) : n = 1, 2 \ldots\}$ on the interval (a, b). We wish to represent a function $f(x)$ that is piecewise continuous on (a, b) in the form

$$f(x) = \sum_{n=1}^{\infty} c_n \varphi_n(x), \tag{3.16}$$

where $\{\phi_n(x)\}$ are orthogonal basis functions on the interval (a, b). We are done once we find the coefficients c_n, $n = 1, 2, \ldots$ We assume the function f is well behaved so that

$$\int_a^b \left(\sum_{n=1}^{\infty} c_n \varphi_n(x)\right) dx = \sum_{n=1}^{\infty} \left(c_n \int_a^b \varphi_n(x)\, dx\right).$$

With this in mind, we multiply both sides of (3.16) with $\varphi_m(x)$ and integrate the resulting equation from $x = a$ to $x = b$. This leads us to

$$\int_a^b f(x)\varphi_m(x)\, dx = \sum_{n=1}^{\infty} c_n \int_a^b \varphi_n(x)\varphi_m(x)\, dx,$$

which is equivalent to

$$(f, \phi_n) = \sum_{m=1}^{\infty} c_m(\phi_m, \phi_n).$$

But $(\phi_m, \phi_n) = 0$ for all values of m except when $m = n$, and hence

$$(f, \varphi_n) = c_n(\phi_n, \phi_n) = c_n \int_a^b \varphi_n^2(x)\,dx.$$

Solving for c_n, we obtain that

$$c_n = \frac{(f, \varphi_n)}{(\phi_n, \phi_n)} = \frac{\int_a^b f(x)\phi_n(x)\,dx}{\int_a^b \varphi_n^2(x)\,dx}, \quad n = 1, 2, \ldots \tag{3.17}$$

The expression defined by (3.16) is known as the **Generalized Fourier series** and the corresponding Fourier coefficient c_n, $n = 1, 2, \ldots$ is given by expression (3.17). Note that if the basis function are orthonormal on the interval (a, b), then (3.17) reduces to

$$c_n = (f, \varphi_n) = \int_a^b f(x)\phi_n(x)\,dx, \quad n = 1, 2 \ldots \tag{3.18}$$

since

$$(\phi_n, \phi_n) = \int_a^b \phi_n^2(x)\,dx = 1.$$

Here is an example.

Example 3.11 (Legendre polynomials) Please note that, Legendre polynomials will be discussed in details in Chapter 7. Let $f(x) = x^2$ be defined on the interval $[-1, 1]$. The first three terms of Legendre polynomials $P_n(x)$ are

$$P_0(x) = 1, \quad P_1(x) = x, \quad P_2(x) = \frac{1}{2}(3x^2 - 1).$$

We are interested in writing $f(x)$ in terms of $P_0(x), P_1(x),$ and $P_2(x)$. That is

$$f(x) = \sum_{n=0}^{2} c_n P_n(x),$$

where the coefficients c_n are given by (3.17), for $n = 0, 1, 2$. First we compute

$$(P_n(x), P_n(x)) = \int_{-1}^{1} P_n^2(x)\,dx.$$

An easy calculations lead to

$$(P_0, P_0) = 2, \quad (P_1, P_1) = \frac{2}{3}, \quad (P_2, P_2) = \frac{2}{5}.$$

Thus,

$$c_0 = \frac{1}{2}\int_{-1}^{1} x^2 \cdot P_0(x)\,dx = \frac{1}{2}\int_{-1}^{1} x^2\,dx = \frac{1}{2} \cdot \frac{2}{3} = \frac{1}{3}.$$

$$c_1 = \frac{3}{2}\int_{-1}^{1} x^2 \cdot x\,dx = \frac{3}{2}\int_{-1}^{1} x^3\,dx = 0,$$

and

$$c_2 = \frac{5}{2}\int_{-1}^{1} x^2 \cdot \frac{1}{2}(3x^2 - 1)\,dx = \frac{5}{2}\int_{-1}^{1}\left(\frac{3}{2}x^4 - \frac{1}{2}x^2\right)dx = \frac{2}{3}.$$

The generalized Fourier series becomes

$$f(x) = \frac{1}{3}P_0(x) + 0 \cdot P_1(x) + \frac{2}{3}P_2(x) + \cdots,$$

or explicitly,

$$f(x) = \frac{1}{3} + \frac{2}{3}\cdot\frac{1}{2}(3x^2 - 1) = x^2.$$

This illustrates how the generalized Fourier series with Legendre polynomials can represent a given function over $[-1, 1]$.

3.5.1 Exercises

Exercise 3.20 Prove the following Schwartz inequality for integrals. For functions f and g in $C_p(a,b)$,

$$||(f,g)|| \leq ||f||\;||g||.$$

Exercise 3.21 Show the set of functions

$$\{f_n(x)\}_{n=1}^{\infty} = \left\{\frac{\sin(n\pi\ln(x))}{\sqrt{x}}\right\}, \quad x \in [1, e]$$

is orthogonal.

Exercise 3.22 Verify that

$$||\psi_n||^2 = 1,$$

where

$$\psi_n(x) = \sqrt{\frac{2}{\pi}}\sin(nx) \quad n = 1, 2, \ldots$$

for $x \in (0, \pi)$.

Exercise 3.23 Let $\{\varphi_n(x)\}_{n=1}^{\infty}$ be a sequence of orthogonal functions on (a,b). Show that if

$$\varphi_1(x) = 1,$$

then

$$(\varphi_n(x), 1) = 0, \quad n = 2, 3, \ldots$$

Exercise 3.24 In addition to the Legendre polynomials

$$P_0(x), \quad P_1(x), \quad P_2(x),$$

given in Example 3.11, consider

$$P_3(x) = \frac{x}{2}(5x^2 - 3) \quad \text{and} \quad P_4(x) = \frac{35x^4}{8} - \frac{15x^2}{4} + \frac{3}{8}.$$

(a) Show that $P_n(x)$ are orthogonal on $[-1,1]$, satisfying

$$\int_{-1}^{1} P_m(x)P_n(x)\,dx = \frac{2}{2n+1}\delta_{mn},$$

where δ_{mn} is the Kronecker delta and $n,m = 0,1,2,3,4$.

(b) Determine the generalized Fourier series of $f(x) = x(x-1)$, on $[-1,1]$.

(c) Determine the generalized Fourier series of $f(x) = x^{\frac{1}{3}}$, on $[-1,1]$.

Exercise 3.25 *Let* $\{\varphi_n(x)\}_{n=1}^{\infty}$ *be a sequence of functions that are orthogonal on* $[a,b]$. *Show that*

$$\left(\varphi_n(x)+\varphi_m(x), \varphi_n(x)+\varphi_m(x)\right) = \left(\varphi_n(x), \varphi_n(x)\right) + \left(\varphi_m(x), \varphi_m(x)\right).$$

In other words,

$$\int_a^b \left[\varphi_n(x)+\varphi_m(x)\right]^2 dx = \int_a^b \varphi_n^2(x)\,dx + \int_a^b \varphi_m^2(x)\,dx.$$

Exercise 3.26 *(Rademacher system) Assume* $f(x)$ *is an odd function of period* 2 *such that*

$$f_0(x) = 1, \quad \text{for } 0 < x < 1,$$

and that no integer belongs to the domain of $f_0(x)$. *Define*

$$f_n(x) = f_0(2^n x), \quad n = 1,2,\ldots$$

(a) Draw the graphs of $f_0(x), f_1(x), f_2(x)$, *and* $f_3(x)$ *on the interval* $(0,1)$.

(b) Show the set $\{f_n(x)\}$ *for* $n = 1,2,\ldots$ *is orthonormal on the interval* $(0,1)$.

3.6 Best Approximation in the Mean

We make use of the Generalized Fourier series that we discussed in Section 3.5 to obtain the best approximation in the mean of a piecewise function that is periodic. Given a function $f(x)$ and a set of orthogonal functions $\{\phi_n(x)\}$ with respect to a weight function $\sigma(x)$ over an interval $[a,b]$, where

$$f(x) = \sum_{n=1}^{\infty} \alpha_n \phi_n(x),$$

we show that the best approximation in the mean is the linear combination

$$f_M(x) = \sum_{n=1}^{M} \alpha_n \phi_n(x),$$

where the coefficients α_n are chosen to minimize the mean-square error

$$E = \int_a^b [f(x) - f_M(x)]^2 \sigma(x)\, dx.$$

The optimal coefficients α_n are the generalized Fourier coefficients

$$\alpha_n = \frac{\int_a^b f(x)\phi_n(x)\sigma(x)\, dx}{\int_a^b \phi_n^2(x)\sigma(x)\, dx}.$$

The concept of best approximation refers to the approximation that minimizes the error. However, the error can be defined in various ways. The difference between $f(x)$ and its approximation, $\sum_{n=1}^{M} \alpha_n \phi_n(x)$, is dependent on x. In some regions, the error $f(x) - \sum_{n=1}^{M} \alpha_n \phi_n(x)$ may be positive, while in others, it could be negative. One potential measure of the error is the maximum deviation over the entire interval

$$\max \left| f(x) - \sum_{n=1}^{M} \alpha_n \phi_n(x) \right|.$$

This definition of the error is reasonable, but it is rarely used because selecting the α_n values to minimize the maximum deviation is quite challenging. Instead, we typically define the error in terms of the **mean-square deviation**, given by

$$E \equiv \int_a^b \left[f(x) - \sum_{n=1}^{M} \alpha_n \phi_n(x) \right]^2 \sigma(x)\, dx. \tag{3.19}$$

In this formulation, large deviations even over small intervals incur significant penalties. We introduce a weight factor $\sigma(x)$ in the error definition to simplify minimization, as this specific weighting aligns with the orthogonality structure of the eigenfunctions $\phi_n(x)$. Crucially, $\sigma(x)$ matches the weight function used in the governing differential equation for $\phi_n(x)$, ensuring consistency between the error minimization framework and the eigenfunction orthogonality.

Theorem 3.6 *Let α_n ($n = 1, 2, \ldots$) denote the generalized Fourier coefficients of a function $f \in C_p(a,b)$ with respect to an orthogonal set $\{\phi_n(x)\}$ ($n = 1, 2, \ldots$) in this space. Then the specific combination*

$$\sum_{n=1}^{N} \alpha_n \phi_n(x)$$

provides the best approximation in the mean to $f(x)$ on the interval $a < x < b$. The corresponding mean square error E is given by equation (3.19).

Proof *To prove that the error E is actually minimized, we expand the square deviation in* (3.19) *and obtain*

$$E = \int_a^b \left(f^2 - 2\sum_{n=1}^{M} \alpha_n f \phi_n + \sum_{n=1}^{M} \sum_{l=1}^{M} \alpha_n \alpha_l \phi_n \phi_l \right) \sigma\, dx.$$

Due to the orthogonality of the eigenfunctions, the last expression derived form (3.19) *is simplified to*

$$\begin{aligned} E &= \int_a^b \left(f^2 - 2\sum_{n=1}^{M} \alpha_n f \phi_n + \sum_{n=1}^{M} \alpha_n^2 \phi_n^2 \right) \sigma\, dx \\ &= \sum_{n=1}^{M} \left[\alpha_n^2 \int_a^b \phi_n^2 \sigma\, dx - 2\alpha_n \int_a^b f \phi_n \sigma\, dx \right] + \int_a^b f^2 \sigma\, dx. \end{aligned} \tag{3.20}$$

Next, we attempt to complete the square to gain insight into selecting the coefficients α_n *in a way that minimizes the error. This involves rewriting the given expression in a squared form, allowing us to systematically determine the optimal values of* α_n *by minimizing the residual term. Thus,*

$$\alpha_n^2 \int_a^b \phi_n^2 \sigma\, dx - 2\alpha_n \int_a^b f\phi_n \sigma\, dx = \int_a^b \phi_n^2 \sigma\, dx \left[\alpha_n^2 - 2\alpha_n \frac{\int_a^b f\phi_n \sigma\, dx}{\int_a^b \phi_n^2 \sigma\, dx} \right]$$

$$= \int_a^b \phi_n^2 \sigma\, dx \left[\left(\alpha_n - \frac{\int_a^b f\phi_n \sigma\, dx}{\int_a^b \phi_n^2 \sigma\, dx} \right)^2 - \left(\frac{\int_a^b f\phi_n \sigma\, dx}{\int_a^b \phi_n^2 \sigma\, dx} \right)^2 \right]. \tag{3.21}$$

Substituting (3.21) *into* (3.20), *we obtain*

$$E = \sum_{n=1}^{M} \left\{ \int_a^b \phi_n^2 \sigma\, dx \left[\left(\alpha_n - \frac{\int_a^b f\phi_n \sigma\, dx}{\int_a^b \phi_n^2 \sigma\, dx} \right)^2 - \left(\frac{\int_a^b f\phi_n \sigma\, dx}{\int_a^b \phi_n^2 \sigma\, dx} \right)^2 \right] \right\} + \int_a^b f^2 \sigma\, dx. \tag{3.22}$$

The only term that depends on the unknowns α_n *appears in a nonnegative way. The minimum occurs only if that first term vanishes, determining the best coefficients*

$$\alpha_n = \frac{\int_a^b f\phi_n \sigma\, dx}{\int_a^b \phi_n^2 \sigma\, dx}, \quad n = 1, 2 \ldots \tag{3.23}$$

Plugging (3.23) *into* (3.22) *shows that the minimal error is*

$$\begin{aligned} E &= \sum_{n=1}^{M} \int_a^b \phi_n^2 \sigma\, dx \left[- \left(\frac{\int_a^b f\phi_n \sigma\, dx}{\int_a^b \phi_n^2 \sigma\, dx} \right)^2 \right] + \int_a^b f^2 \sigma\, dx \\ &= \int_a^b f^2 \sigma\, dx - \sum_{n=1}^{M} \alpha_n^2 \int_a^b \phi_n^2 \sigma\, dx. \end{aligned} \tag{3.24}$$

Equation (3.24) *shows that as M increases, the error decreases. Thus, we can think of a generalized Fourier series as an approximation scheme. The more terms in the truncated series that are used, the better the approximation.* □

Remark 7 *The best approximation (in the mean-square sense using the first M eigenfunctions) has two main advantages.*
(1) The coefficients α_n *are easy to determine, and*
(2) the coefficients are independent of M.

Example 3.12 For a Fourier sine series, where $\sigma(x) = 1$, $\phi_n(x) = \sin\frac{n\pi x}{L}$ and $\int_0^L \sin^2\frac{n\pi x}{L}\,dx = \frac{L}{2}$, it follows that

$$E = \int_0^L f^2(x)\,dx - \frac{L}{2}\sum_{n=1}^{M}\alpha_n^2,$$

where α_n are given by (3.23). Now set $L = \pi$ and $f(x) = x(\pi - x)$. Then, the function space V_N consists of the span of the first N eigenfunctions is

$$V_N = \text{span}\{\sin x, \sin 2x, \ldots, \sin Nx\}.$$

Using (3.23), the Fourier coefficients are given by (see Example 1.9)

$$\alpha_n = \frac{2}{\pi}\int_0^\pi x(\pi - x)\sin nx\,dx = \frac{8}{\pi(2n-1)^3}, \qquad n = 1, 2, \ldots.$$

Thus, the best approximation in V_N is

$$f_N(x) = \sum_{n=1}^{N}\frac{8}{\pi(2n-1)^3}\sin nx.$$

We are interested in bounding the error in this Best approximation. Since $f = \sum_{n=1}^{\infty}\alpha_n\phi_n$, we have that $f_N = \sum_{n=1}^{N}\alpha_n\phi_n$. The error in approximation is

$$f - f_N = \sum_{n=N+1}^{\infty}\alpha_n\phi_n.$$

The squared norm of the error is

$$\|f - f_N\|^2 = \left\|\sum_{n=N+1}^{\infty}\alpha_n\phi_n\right\|^2 = \sum_{n=N+1}^{\infty}\alpha_n^2\|\phi_n\|^2.$$

Since

$$\|\phi_n\|^2 = \frac{\pi}{2},$$

we have

$$\|f - f_N\|^2 = \sum_{n=N+1}^{\infty}\left(\frac{8}{\pi(2n-1)^3}\right)^2\frac{\pi}{2}.$$

Approximating the tail sum using integral bounds, we get

$$\|f - f_N\|^2 \leq \frac{32}{\pi}\sum_{n=N+1}^{\infty}\frac{1}{(2n-1)^6}.$$

Since the sum can be approximated by an integral,

$$\sum_{n=N+1}^{\infty}\frac{1}{(2n-1)^6} \approx \int_N^\infty \frac{dx}{(2x-1)^6} = \frac{1}{10(2N-1)^5},$$

we obtain

$$\|f - f_N\|^2 \le \frac{32}{\pi}\left(\frac{1}{10(2N-1)^5}\right) = \frac{16}{5\pi(2N-1)^5}.$$

□

To go deeper into the theory of Fourier series, one would need to take a significant detour into a more advanced mathematical framework known as Hilbert space, named after the German mathematician David Hilbert. This space provides the proper setting for a rigorous treatment of Fourier series and plays a fundamental role in modern analysis and physics. However, since a full exploration of Hilbert spaces is beyond the scope of this text, we will stop here and turn to the application of Fourier series in partial differential equations, particularly boundary value problems, using separation of variables and other accessible methods. For more on Hilbert spaces, we refer to [34].

3.6.1 Exercises

Exercise 3.27 (Gram-Schmidt process) *Suppose that we have a basis $\{f_j\}$ of functions and wish to convert it into an orthogonal basis $\{\phi_j\}$. The Gram-Schmidt process achieves this, ensuring that*

$$\phi_j \in span(f_0, \ldots, f_j).$$

The process is easy to apply. Start by taking f_j as the starting function, then subtract off the components of f_j in the direction of the previous ϕ's, so that the result is orthogonal to them. That is, we compute the sequence

$$\phi_0 = f_0$$

$$\phi_1 = f_1 - \frac{\langle f_1, \phi_0\rangle}{\langle \phi_0, \phi_0\rangle}\phi_0$$

$$\vdots$$

$$\phi_j = f_j - \sum_{k=0}^{j-1}\frac{\langle f_j, \phi_k\rangle}{\langle \phi_k, \phi_k\rangle}\phi_k.$$

It is easy to verify that this procedure generates the desired orthogonal basis.

Consider the interval $[-1,1]$ with the weight function $\sigma(x) = 1$ and $f_n(x) = x^n$, $n = 0, 1, \ldots$ Define

$$\langle h, g\rangle = \int_{-1}^{1} h(x)g(x)\sigma(x)dx.$$

Use the Gram-Schmidt process to construct the functions $\phi_i(x)$, $i = 0, 1, 2, 3, 4,$ starting with $\phi_0(x) = f_0(x) = 1$. These obtained functions are Legendre polynomials. These polynomials form an orthogonal basis, as we shall explore in detail in Chapter 7.

Exercise 3.28 *Find the best quadratic approximation in* $L_2([-1,1])$ *to* $f(x) = e^x$ *using the first three Legendre polynomials,*

$$P_0(x) = 1, \quad P_1(x) = x, \quad P_2(x) = \frac{1}{2}(3x^2 - 1).$$

Compute the coefficients and the resulting L_2*-error.*

Exercise 3.29 *Repeat Example 3.12 for*

$$f(x) = x(x^2 - \pi^2), \quad -\pi \le x \le \pi, \quad \sigma(x) = 1, \quad \phi_n(x) = \sin nx, \quad n = 1, 2\ldots$$

4

Heat Equation and Boundary Value Problems

The chapter on the heat equation and boundary value problems begins with an introduction to partial differential equations, providing a foundational understanding of their role in modeling various physical phenomena. The chapter delves into the derivation of the heat equation, highlighting its significance in describing the distribution of temperature over time in a given medium. A thorough analysis of boundary value problems follows, where the chapter examines the impact of different boundary conditions, such as Dirichlet, Neumann, and mixed types, on the solution of the heat equation. The methods for solving these problems, including separation of variables where Fourier series expansions are used.

4.1 Introduction to PDEs

A *partial differential equation*, short PDE, is an equation that contains the unknown function u and its partial derivatives. Recall that given a function of two variables, $f(x,y)$, the *partial derivative* of f with respect to x is the rate of change of f as x varies, keeping y constant and it is given by

$$\frac{\partial f}{\partial x} = \lim_{h \to 0} \frac{f(x+h,y) - f(x,y)}{h}.$$

Similarly, the *partial derivative* of f with respect to y is the rate of change of f as y varies, keeping x constant and it is given by

$$\frac{\partial f}{\partial y} = \lim_{h \to 0} \frac{f(x,y+h) - f(x,y)}{h}.$$

More often we write f_x, f_y to denote $\frac{\partial f}{\partial x}$ and $\frac{\partial f}{\partial y}$, respectively. Similar notations will be used to denote higher partial derivatives and mixed partial derivatives. Let D be a subset of $\mathbb{R}^2$ and $u = u(x,y)$ such that $u : D \to \mathbb{R}$. Then we may denote the general first order PDE in $u(x,y)$ by

$$F(x,\ y,\ u(x,y),\ u_x(x,y),\ u_y(x,y)) = 0$$

DOI: 10.1201/9781003652540-4

or

$$F(x,\ y,\ u,\ u_x,\ u_y) = 0, \tag{4.1}$$

for some function F. Below we list some important PDEs.

$$u_t + cu_x = 0, \quad c \in \mathbb{R} \quad \text{(Transport equation)}.$$

$$u_{xx} + u_{yy} = 0 \quad \text{(Laplace's equation)}.$$

$$u_t + uu_x = 0 \quad \text{(Burger's equation)}.$$

$$u_{tt} - c^2 u_{xx} = 0 \quad \text{(Wave equation)}.$$

$$u_t - ku_{xx} = 0 \quad \text{(Heat equation)}.$$

Loosely speaking, by a solution of (4.1), we mean a function $u(x,y) = \varphi(x,y)$ such that

$$F\big(x,\ y,\ \varphi,\ \varphi_x,\ \varphi_y\big) = 0$$

for all $x,\ y \in D \subseteq \mathbb{R}^2$. The solution of a first-order partial differential equation is generally a surface in three-dimensional space (or a hypersurface in higher dimensions). However, it can also be described in terms of a family of characteristic curves, which we will discuss in Chapter 8.

Example 4.1 For an arbitrary function f, $\varphi(x,t) = f(x-ct)$ is a general solution of the transport equation $u_t + cu_x = 0$ since

$$\varphi_t + c\varphi_x = -c\varphi'(x-ct) + c\varphi'(x-ct) = 0.$$

□

Order and linearity are two of the main properties of PDEs.

Definition 4.1 *The order of a partial differential equation is the highest order derivative in the given PDE.*

Definition 4.2 *A given partial differential equation is said to be linear if the unknown function and all of its derivatives enter linearly.*

For example, all the equations listed above are linear except Burger's equation. Now to better understand linearity we utilize the operator concept $\mathscr{L}$ on an appropriate space, where $\mathscr{L}$ is a *differential operator*. An operator is really just a function that takes a function as an argument instead of numbers as we are used to dealing with in functions. For example, $\mathscr{L}u$ assigns u a new function $\mathscr{L}u$. That is if we set

$$\mathscr{L} = \frac{\partial^2}{\partial t^2} - \frac{\partial^2}{\partial x^2},$$

then

$$\mathscr{L}u = u_{tt} - u_{xx}.$$

The next definition gives a precise and convenient way to test for linearity.

Definition 4.3 *An operator $\mathscr{L}$ is said to be linear if is satisfies*
(a)

$$\mathscr{L}(u_1 + u_2) = \mathscr{L}u_1 + \mathscr{L}u_2,$$

(b)

$$\mathscr{L}(cu_1) = c\mathscr{L}u_1,$$

for any functions u_1, u_2 and constant c. Moreover, the equation $\mathscr{L}u = 0$ is said to be linear if the operator $\mathscr{L}$ is linear.

Example 4.2 Consider $u_{tt} - u_{xx} = 0$. To show it is linear we let

$$\mathscr{L}u = u_{tt} - u_{xx}.$$

Then for any functions u, v, in the appropriate space of functions, we have

$$\mathscr{L}(u+v) = (u+v)_{tt} - (u+v)_{xx} = [u_{tt} - u_{xx}] + [v_{tt} - v_{xx}] = \mathscr{L}u + \mathscr{L}v,$$
$$\mathscr{L}(cu) = (cu)_{tt} - (cu)_{xx} = c(u_{tt}) - c(u_{xx}) = c[u_{tt} - u_{xx}] = c\mathscr{L}u.$$

This shows the PDE in question is linear. □

Example 4.3 The following PDE $uu_t - u_x = 0$ is not a linear. To see this, we let

$$\mathscr{L}u = uu_t - u_x.$$

Then for any functions u, v in the appropriate space of functions and a constant c, we have

$$\mathscr{L}(u+v) = (u+v)(u+v)_t - (u+v)_x = [uu_t - u_x] + [vv_t - v_x] + uv_t + vu_t \neq \mathscr{L}u + \mathscr{L}v.$$

This shows the PDE in question is not linear. □

Let $\mathscr{L}u = 0$ and assume $\mathscr{L}$ is a linear operator. If u is a solution, then so is cu, for some constant c since $\mathscr{L}(cu) = c\mathscr{L}u = 0$. Similarly, if v is another solution then so is $c_1u + c_2v$, since for any constants c_1 and c_2 we have $\mathscr{L}(c_1u + c_2v) = c_1\mathscr{L}u + c_2\mathscr{L}v$. Thereupon, we have the following theorem.

Theorem 4.1 *[Principle of Superposition] Assume $\mathscr{L}$ is a linear operator. If $u_1(x,y)$, $u_2(x,y)$, $\ldots$, $u_n(x,y)$ are solutions of $\mathscr{L}u = 0$, then so is*

$$c_1u_1(x,y) + c_2u_2(x,y) + \ldots + c_nu_n(x,y) = \sum_{i=1}^{n} c_iu_i(x,y).$$

Example 4.4 In this example we show that each of

$$u_n(x,t) = e^{-n^2\pi^2 t} \sin n\pi x, \quad n = 1, 2, \ldots, N$$

is a solution of the heat equation $u_{xx} - u_t = 0, \ t > 0$, for the temperature $u = u(x,t)$ in a rod, considered as a function of the distance x measured along the rod and of the time t. In addition, we show for constants c_i, $i = 1, 2, \ldots, N$ that

$$u(x,t) = \sum_{n=1}^{N} c_nu_n(x,t)$$

is also a solution.

To do so, we let $\mathscr{L}$ be the differential operator given by $\mathscr{L} = \frac{\partial^2}{\partial x^2} - \frac{\partial}{\partial t}$. It is clear that $\mathscr{L}$ is linear. Now for any fixed n, $n = 1,2,\ldots,N$ we have

$$\begin{aligned}
\mathscr{L}u_n &= \frac{\partial^2}{\partial x^2}\left(e^{-n^2\pi^2 t}\sin n\pi x\right) - \frac{\partial}{\partial t}\left(e^{-n^2\pi^2 t}\sin n\pi x\right)\\
&= \frac{\partial}{\partial x}\left(n\pi e^{-n^2\pi^2 t}\cos n\pi x\right) + n^2\pi^2 e^{-n^2\pi^2 t}\sin n\pi x\\
&= -n^2\pi^2 e^{-n^2\pi^2 t}\sin n\pi x + n^2\pi^2 e^{-n^2\pi^2 t}\sin n\pi x\\
&= 0.
\end{aligned}$$

This shows $\mathscr{L}u_n(x,t) = 0$, $n = 1,2,\ldots,N$. As for the rest of the work, we apply $\mathscr{L}$ to u.

$$\begin{aligned}
\mathscr{L}u &= \mathscr{L}\Big(\sum_{n=1}^{N} c_n u_n(x,t)\Big)\\
&= \mathscr{L}(c_1 u_1(x,t)) + \mathscr{L}(c_2 u_2(x,t)) + \mathscr{L}(c_N u_N(x,t))\\
&= c_1\mathscr{L}u_1(x,t) + c_2\mathscr{L}u_2(x,t) + c_N\mathscr{L}u_N(x,t)\\
&= 0.
\end{aligned}$$

□

Let $\mathscr{L}$ be a differential operator. Then the partial differential equation

$$\mathscr{L}u = f \tag{4.2}$$

for continuous function f in x and y is called *non-homogeneous partial differential equation*. Its corresponding *homogeneous partial differential equation* is

$$\mathscr{L}u = 0. \tag{4.3}$$

If we assume $\mathscr{L}$ is linear, then the construction of the general solution of the non-homogeneous PDE given by (4.2) is similar to its counterpart in ordinary differential equations. Let u_p be a *particular solution* of (4.2) and u_h be the *homogeneous solution* of (4.3). Then, due to the linearity of the differential operator $\mathscr{L}$ we see that

$$\mathscr{L}(u_h + u_p) = \mathscr{L}u_h + \mathscr{L}u_p = 0 + f = f.$$

Thus, it suffices to find u_p of (4.2) and add it to the homogeneous solution u_h of (4.3) to get the general solution

$$u = u_h + u_p.$$

When the PDE is linear and involves only simple derivatives of only one variable, it is more likely that it can be solved along the lines of an ordinary differential equation, as the next example shows.

Example 4.5 In this example we display various forms of the solution $u = u(x,y)$ for the following PDEs

(a) $u_y = 0$, b) $u_{yy} + u = 0$, (c) $u_{yyx} = 0$.
We begin with (a). The given PDE, $u_y = 0$ has no partial derivatives with respect to x, which indicates that the solution u is a function of x only. Thus, the solution is $u(x,y) = g(x)$, for some function g. Suppose that we impose the initial condition $u(x,a) = e^{2x}$, then

$$e^{2x} = u(x,a) = g(x),$$

which uniquely determines the function g.
(b) Now the PDE, $u_{yy} + u = 0$ can be thought of as the second-order ODE $z'' + z = 0$, which has the solution $z = c_1 \cos(y) + c_2 \sin(y)$. Or, $u(x,y) = g(x)\cos(y) + h(x)\sin(y)$, since the constants c_1 and c_2 may depend on the other variable x, where the functions g and h are differentiable.
(c) Let f be a twice differentiable function in y. Then, we integrate with respect to x and get

$$u_{yy} = f_{yy}(y),$$

and from which we have $u_y = f_y(y) + g(x)$. Integrating one more time we arrive at the general solution

$$u(x,y) = f(y) + yg(x) + h(x).$$

□

Example 4.6 Consider the PDE $u_{yy} + u = 3 + y$.
Since it is non-homogeneous we need to find u_h and u_p. From Example 4.5 we have

$$u_h(x,y) = g(x)\cos(y) + h(x)\sin(y).$$

Since $f = 3 + y$, we look for a particular solution in the form $u_p = A + By$, where A and B are to be determined by substituting, $(u_p)_{yy}$, and u_p into the original PDE. This implies that $A + By = 3 + y$, and therefore, $A = 3$, and $B = 1$. As a result, we obtain the general solution

$$u(x,y) = g(x)\cos(y) + h(x)\sin(y) + 3 + y.$$

□

4.2 Derivation of the Heat Equation

We consider the heat conduction problem of a thin rod and look at the solution. Part of our presentation is inspired and influenced by Strauss [136]. Let $u(x,t)$ represent the temperature at position x in a thin insulated rod at time t. The vertical axis will measure the temperature, and the rod will be positioned along the x-axis in the xu-coordinate system. As in the wave equation, we consider a small portion of the rod over an interval $[a,b]$. The heat or thermal energy of the rod situated at the interval $[a,b]$ is given by

$$D(x,t) = \int_a^b c\rho u dx,$$

where c denotes the specific heat capacity of the material of the rod and ρ is the mass density of the rod. The instantaneous change of the heat with respect to time will be the time derivative of the above equation

$$\frac{d}{dt}D(x,t) = \frac{d}{dt}\int_a^b c\rho u dx = \int_a^b c\rho u_t dx. \tag{4.4}$$

The above expression is true since a and b are constants and u is continuous. Recall that the thin rod is insulated. The change in heat must be balanced by the heat flux across the cross-section of the cylindrical piece around the interval $[a,b]$, as the heat cannot be gained or lost in the absence of an external heat source. Fourier's law states that the heat flux across the boundary will be inversely proportional to the temperature derivative in the direction of the boundary's outward normal, in this instance the x-derivative. The second way to compute the time rate of change of D is to notice that, in the the absence of heat sources within the rod, the quantity of heat in u can change only through the flow of heat across the boundaries of u at $x = a$ and $x = b$. The heat flux through a section of the rod is called the heat flux through the section.

Let κ denote the *thermal conductivity* of the rod. Recall that the thermal conductivity of a material is a measure of its ability to conduct heat. Then the heat flux into u at $x = a$ and $x = b$ is

$$-\kappa u_x(a,t) \quad \text{and} \quad \kappa u_x(b,t),$$

respectively. Thus, the total time rate of change of D is the sum of the rates at the two ends. Using the fundamental theorem of calculus, we may write

$$\frac{d}{dt}D(x,t) = \kappa\Big[u_x(b,t) - u_x(a,t)\Big] = \kappa\int_a^b u_{xx}dx. \tag{4.5}$$

A quick comparison of equations (4.4) and (4.5) yields

$$\int_a^b c\rho u_t dx = \kappa\int_a^b u_{xx}dx,$$

or

$$\int_a^b \Big(c\rho u_t - \kappa u_{xx}\Big)dx = 0.$$

Since the above integral must hold for all $x \in [a,b]$ with $a < b$ we must have

$$c\rho u_t - \kappa u_{xx} = 0,$$

throughout the material. Dividing by $c\rho$ and setting $k = \frac{\kappa}{c\rho}$, we arrive at the one-dimensional *heat equation*

$$u_t - ku_{xx} = 0, \tag{4.6}$$

where k is called the *thermal diffusivity* of the material.

If we consider the homogeneous heat equation in (4.6) on an interval $I \in \mathbb{R}$, then we have the heat problem, with initial and boundary conditions

$$\begin{cases} u_t = ku_{xx}, & x \in I,\ t > 0 \\ u(x,0) = f(x), & x \in I \\ u \text{ satisfies certain BCs} \end{cases} \tag{4.7}$$

Here, the initial condition $u(x,0) = f(x)$, means the lateral surface of the rod is insulated and parallel to the x-axis, and its initial temperatures are $f(x)$ for $x \in I$. In practice, the most common boundary conditions are the following:

(1) $u(0,t) = 0 = u(l,t) : \big(I = (0,l),\ \text{Dirichlet}\big)$. It is the case when both faces of the rod are kept at temperature zero.

(2) $u_x(0,t) = 0 = u_x(l,t) : \big(I = (0,l),\ \text{Neumann}\big)$. It is the case when both faces of the rod are insulated.

(3) $u_x(0,t) - a_0 u(0,t) = 0$ and $u_x(l,t) + a_l u(l,t) = 0 : \big(I = (0,l),\ \text{Robin}\big)$.

(4) $u(-l,t) = u(l,t) = 0$ and $u_x(-l,t) = u_x(l,t) = 0 : \big(I = (-l,l),\ \text{Periodic}\big)$.

The existence and uniqueness of the solution of the non-homogeneous heat equation will be proved in Chapter 7.

4.2.1 Exercises

Exercise 4.1 *Determine the order and use the operator $\mathscr{L}$ to decide linearity, non-linearity of the given equations.*
(a) $u_{xx} + 3u_y + 4u = 0$, *(b)* $u_{xy} + 3u_x = x$.
(c) $u_x + 3u_{yyyy} + u^{1/2} = 0$, *(d)* $u_x = 1$.

Exercise 4.2 *Show that $u(x,t) = e^{t-x}$ is a solution of $u_{tt} - u_{xx} = 0$.*

Exercise 4.3 *Show that $u(x,y) = f(bx - ay)$ for a differentiable function f is a solution of $au_x + bu_y = 0$.*

Exercise 4.4 *Show that $u(x,t) = \frac{1}{6}t^3 + (x - t^2/2)t + (x - t^2/2)^2$ solves $u_t + tu_x = x$, $-\infty < x < \infty$, $t > 0$, and satisfies the initial data $u(x,0) = x^2$.*

Exercise 4.5 *Solve $2u_x - 3u_y = 0$, subject to the condition $u(0,y) = e^y$.*

Exercise 4.6 *Find the general solution of each of the PDEs.*
(a) $u_{xx} + 4u = 0$, *(b)* $u_{xy} = 0$, *(c)* $u_{xy} + u_y = 0$.
(d) $u_{xx} + 3u_x + 2u = e^{5x}$, *(e)* $(x-y)u_{xy} - u_x + u_y = 0$. *(Hint: let $v = (x-y)u_x + u$.)*
(f) $u_{xxy} = 0$, *(g)* $u_{xy} + u_y = e^{2x+3y}$.

4.3 Separable Equations

The method of separating variables is a fundamental technique for solving partial differential equations, particularly those with boundary and initial value problems. It is important because it simplifies a complex PDE into a set of ordinary differential equations, which are often easier to solve. By assuming that the solution can be expressed as a product of functions, each depending only on a single variable, this method reduces the problem's dimensionality. This approach is particularly powerful for linear PDEs with separable boundary conditions, such as the heat equation, wave equation, and Laplace's equation. Additionally, the method often provides insight into the underlying physical processes by isolating independent contributions from each variable, leading to solutions that can be represented in terms of eigenfunctions and Fourier series expansions. Consider the general second-order partial differential equation given by

$$a(x,y)u_{xx} + c(x,y)u_{yy} + d(x,y)u_x + e(x,y)u_y + f(x,y)u = 0, \tag{4.8}$$

where the coefficients are assumed to be continuous on a subset of $\mathbb{R}^2$. We assume a nonzero separable solution of (4.8) in the form

$$u(x,y) = X(x)Y(y) \tag{4.9}$$

where $X \neq 0$ and $Y \neq 0$ are, respectively, functions of x and y alone, and are twice continuously differentiable. From now on, we suppress the independent variables x and y from all arguments. After performing partial differentiations and substituting equation (4.9) into equation (4.8), we obtain

$$aXY'' + cX''Y + dX'Y + eXY' + fXY = 0, \tag{4.10}$$

where the primes indicate differentiation with respect to their respective variables. Suppose that there is a function $\rho(x,y)$, such that dividing (4.10) by $\rho(x,y)$, we get

$$a_1(x)X''Y + b_1(y)X''Y + a_2(x)X'Y + b_2(y)XY' + [a_3(x) + b_3(y)]XY = 0. \tag{4.11}$$

Dividing both sides of (4.11) with XY, yields

$$\frac{a_1X''}{X} + \frac{a_2X'}{X} + a_3 = -\left(\frac{b_1Y''}{Y} + \frac{b_2Y'}{Y} + b_3\right). \tag{4.12}$$

The left side of equation (4.12) depends only on x, while the right side depends only on y. Differentiating equation (4.12) with respect to x, we have

$$\frac{d}{dx}\left(\frac{a_1X''}{X} + \frac{a_2X'}{X} + a_3\right) = 0. \tag{4.13}$$

Integration of (4.13) yields

$$\frac{a_1X''}{X}+\frac{a_2X'}{X}+a_3=-\lambda, \tag{4.14}$$

where λ is a separation constant. From equations (4.13) and (4.14), we also have

$$\frac{b_1Y''}{Y}+\frac{b_2Y'}{Y}+b_3=\lambda. \tag{4.15}$$

Rewriting equations (4.14) and (4.15), we obtain

$$a_1X''+a_2X'+(a_3+\lambda)X=0, \tag{4.16}$$

$$b_1Y''+b_2Y'+(b_3-\lambda)Y=0. \tag{4.17}$$

Thus, $u(x,y)$ is the solution of equation (4.8) if $X(x)$ and $Y(y)$ satisfy the ordinary differential equations (4.16) and (4.17), respectively.

4.3.1 Exercises

Exercise 4.7 *Determine whether each equation is separable, and justify your answer.*
(a) $c^2u_{xx}=u_{tt}+2\gamma(x)u_t$, *(b)* $u_{xx}+2u_{xy}+u_{yy}=0$.

4.4 BVP-Heat Equation

We examine the one-dimensional heat equation with both insulated (Neumann boundary conditions) and non-insulated (Dirichlet boundary conditions). In our analysis, we will utilize the concept of *separation of variables* after we recast the PDE into an ODE. Before we consider our heat flow problem(s), we make the following definition.

Definition 4.4 *Consider the Sturm–Liouville problem*

$$\Big(r(x)X'(x)'\Big)'+\big[q(x)+\lambda p(x)\big]X(x)=0, \tag{4.18}$$

where p,q,r and r' are continuous for all $x\in[a,b]$, and $p(x)>0$, and $r(x)>0$ for $x\in[a,b]$. For constants α_1, α_2, β_1, β_2, we impose the boundary conditions

$$\alpha_1y(a)+\beta_1y'(a)=0,\quad \alpha_2y(b)+\beta_2y'(b)=0, \tag{4.19}$$

with

$$\alpha_1^2+\beta_1^2\neq 0,\quad \alpha_2^2+\beta_2^2\neq 0.$$

A value of λ for which problem (4.18) *and* (4.19) *has* **nontrivial solution** *is called an* **eigenvalue***; and the* **nontrivial solution** *is called an* **eigenfunction**.

Note that the Sturm–Liouville problem will be discussed in detail in Chapter 7, along with other closely related concepts.

4.5 Insulated Homogeneous Heat Problem with Homogeneous Boundaries

Consider the homogeneous insulated heat problem on a finite slab

$$u_t - ku_{xx} = 0, \quad 0 < x < c, \quad t > 0, \tag{4.20}$$

subject to the initial condition

$$u(x,0) = f(x), \quad 0 < x < c, \tag{4.21}$$

and to the boundary conditions

$$u_x(0,t) = 0, \;\; u_x(c,t) = 0, \quad 0 < x < c, \tag{4.22}$$

that models the diffusion of heat in a one-dimensional rod of length c with insulated boundaries. The function $u(x,t)$ represents the temperature at position x and time t, while k is the thermal diffusivity of the material that is always positive. The boundary conditions $u_x(0,t) = 0$ and $u_x(c,t) = 0$ signify that there is no heat flux through the ends of the rod, meaning the boundaries are perfectly insulated. Physically, this implies that the heat energy remains confined within the rod, and no energy is lost or gained at its boundaries. The initial condition $u(x,0) = f(x)$ specifies the initial temperature distribution along the rod. The problem seeks to determine how the temperature evolves over time while satisfying the given constraints. We refer to this problem as homogeneous because the heat flow described by (4.20) is homogeneous, and its boundary conditions, given by (4.22), are also homogeneous. We seek a nontrivial solution $u(x,t)$ of (4.20) that satisfies the initial condition (4.21) along with the boundary conditions given by (4.22).

• Method of Solution

We seek separated solutions of the form

$$u(x,t) = X(x)T(t), \tag{4.23}$$

where X is a function of x alone, and T is a function of t alone. Note that both X and T must be nontrivial to ensure a nontrivial solution, i.e., $X \neq 0$ and $T \neq 0$. By differentiating (4.23) with respect to t and x and substituting into (4.20), we obtain the relation

$$X(x)T'(t) = kX''(x)T(t).$$

Since $X(x) \neq 0$ and $T(t) \neq 0$, we may divide by the term $X(x)T(t)$ to separate the variables. Thus,

$$\frac{X''(x)}{X(x)} = \frac{T'(t)}{kT(t)}.$$

Since the left-hand side is a function of x alone and the right-hand side is a function of t alone, the two sides must have a common constant value $-\lambda$ (we choose $-\lambda$ for convenience). Therefore,

$$\frac{X''(x)}{X(x)} = \frac{T'(t)}{kT(t)} = -\lambda.$$

The Neumann boundary conditions given by (4.22) yield

$$0 = u_x(0,t) = X'(0)T(t),$$

which must hold for all $t > 0$. Thus, we have $X'(0) = 0$. Similarly

$$0 = u_x(c,t) = X'(c)T(t),$$

implies $X'(c) = 0$. Consequently, we arrive at the second-order differential equation

$$X''(x) + \lambda X(x) = 0, \quad X'(0) = 0, \quad X'(c) = 0, \tag{4.24}$$

which is a subset of the Sturm–Liouville problem (4.18). Additionally, we obtain the first-order ordinary differential equation

$$T'(t) + \lambda kT(t) = 0, \quad t > 0. \tag{4.25}$$

We begin by solving (4.24), as its solution is well-defined due to the presence of two boundary conditions.

- $\lambda = 0$. In this case, (4.24) reduces to $X''(x) = 0$, which has the solution

  $$X(x) = Ax + B,$$

  where A and B are constants. Applying $X'(0) = 0$ gives $A = 0$. Applying $X'(c) = 0$ imposes no restriction on B. Hence, B is arbitrary, and we set it equal to 1. Thus, for $\lambda = 0$, the eigenfunction is

  $$X_0(x) = 1.$$

- $\lambda < 0$. If $\lambda < 0$, we set $\lambda = -\alpha^2$ with $\alpha > 0$. Then, (4.24) becomes

  $$X''(x) - \alpha^2 X(x) = 0,$$

  whose general solution is

  $$X(x) = c_1 e^{\alpha x} + c_2 e^{-\alpha x}.$$

  Differentiating, we get

  $$X'(x) = \alpha c_1 e^{\alpha x} - \alpha c_2 e^{-\alpha x} = \alpha\left(c_1 e^{\alpha x} - c_2 e^{-\alpha x}\right).$$

  Applying $X'(0) = 0$ gives $c_1 = c_2$. Substituting this into $X(x)$, we get

  $$X(x) = 2c_1 \cosh(\alpha x).$$

  The condition $X'(c) = 0$ implies $c_1 \sinh(\alpha c) = 0$, which is not possible for $\alpha > 0$. Thus, we must have $c_1 = 0$, and consequently, there is no nontrivial solution in this case.

- $\lambda > 0$. If $\lambda > 0$, we set $\lambda = \alpha^2$ with $\alpha > 0$. Then, (4.24) becomes

$$X''(x) + \alpha^2 X(x) = 0,$$

whose general solution is

$$X(x) = A\cos\alpha x + B\sin\alpha x.$$

Differentiating, we get

$$X'(x) = -A\alpha\sin\alpha x + B\alpha\cos\alpha x.$$

Applying $X'(0) = 0$ gives $B = 0$. Substituting $B = 0$ into $X(x)$, we have

$$X(x) = A\cos\alpha x.$$

Applying $X'(c) = 0$ yields $A\sin\alpha c = 0$, which implies

$$\alpha c = n\pi, \quad n = 1, 2, \ldots,$$

and therefore

$$\alpha = \frac{n\pi}{c}, \quad n = 1, 2, \ldots$$

Note that, $A = 0$ would have led to the trivial solution. Thus, for $\lambda_n = \left(\frac{n\pi}{c}\right)^2$, the corresponding eigenfunctions are

$$X_n(x) = \cos\frac{n\pi x}{c}, \quad n = 1, 2, \ldots,$$

where we have set $A = 1$.

Turning to (4.25), we solve it for the already determined eigenvalues

$$\lambda_0 = 0 \quad \text{and} \quad \lambda_n = \left(\frac{n\pi}{c}\right)^2, \quad n = 1, 2, \ldots.$$

For $\lambda_0 = 0$, (4.25) has the solution

$$T_0(t) = 1.$$

For $\lambda_n = \left(\frac{n\pi}{c}\right)^2$, $n = 1, 2, \ldots$, the first-order differential equation

$$T'(t) + \left(\frac{n\pi}{c}\right)^2 kT(t) = 0, \quad t > 0,$$

has the solution

$$T_n(t) = e^{-\frac{n^2\pi^2 k}{c^2}t}, \quad n = 1, 2, \ldots.$$

Setting $u_n(x,t) = T_n(t)X_n(x)$, the function satisfies the heat equation and its boundary conditions. Therefore, by the superposition principle, for any constants a_n, the general solution can be written as

$$u(x,t) = \frac{a_0}{2} + \sum_{n=1}^{\infty} a_n e^{-\frac{n^2\pi^2 k}{c^2}t} \cos\frac{n\pi x}{c}. \tag{4.26}$$

Applying the initial condition $u(x,0) = f(x)$ to (4.26), we obtain the Fourier cosine series

$$f(x) = \frac{a_0}{2} + \sum_{n=1}^{\infty} a_n \cos \frac{n\pi x}{c},$$

where

$$a_0 = \frac{2}{c}\int_0^c f(x)\,dx \quad \text{and} \quad a_n = \frac{2}{c}\int_0^c f(x) \cos \frac{n\pi x}{c}\,dx, \quad n = 1,2,\ldots. \tag{4.27}$$

In conclusion, (4.26) and (4.27) provide the desired solution. Let's take a closer look at (4.26) and (4.27). Suppose that $f(x)$ is piecewise continuous on $0 < x < c$. Then there is a positive constant D such that

$$\left| \frac{1}{c}\int_0^c f(x) \cos \frac{n\pi x}{c}\,dx \right| \leq D,$$

and hence

$$|a_n| \leq 2D.$$

So, from (4.26), we obtain

$$|u(x,t)| \leq \left|\frac{a_0}{2}\right| + 2D \sum_{n=1}^{\infty} e^{-\frac{n^2\pi^2 k}{c^2}t}$$

Let $d_n = e^{-\frac{n^2\pi^2 k}{c^2}t} \leq e^{-\frac{n^2\pi^2 k}{c^2}t_{min}}$, where $t_{min} > 0$ is a lower bound for t. Set $M_n = e^{-\frac{n^2\pi^2 k}{c^2}}$. Then the sequence M_n forms a geometric series with terms that decay exponentially. Therefore,

$$\sum_{n=1}^{\infty} M_n = \sum_{n=1}^{\infty} e^{-\frac{n^2\pi^2 k}{c^2}} < \infty.$$

Because

$$|d_n(t)| \leq M_n = e^{-\frac{n^2\pi^2 k}{c^2}}$$

for all $t > 0$, and because

$$\sum_{n=1}^{\infty} M_n < \infty,$$

the Weierstrass M-test ensures that the series converges uniformly for $t > 0$. Next we show the obtained solution given by (4.26) satisfies the initial condition (4.21). We already showed that for any fixed x $(0 < x < c)$, the series in expression (4.26) converges uniformly with respect to t when $t \geq 0$. Consequently, it represents a function that is continuous in t for $t \geq 0$. This establishes that the solution $u(x,t)$ is continuous in t when $t \geq 0$, particularly at $t = 0$. Specifically, we have

$$\lim_{\substack{t\to 0\\ t>0}} u(x,t) = u(x,0),$$

or equivalently, $u(x,0^+) = u(x,0) = f(x)$, for each fixed $x \in (0,c)$.

Example 4.7 For $0 < x < \pi$, $t > 0$, consider the heat flow problem

$$u_t(x,t) - 4u_{xx}(x,t) = 0, \quad u(x,0) = (x^2 - \pi^2)^2, \quad u_x(0,t) = u_x(\pi,t) = 0.$$

Using (4.26), the solution takes the form

$$u(x,t) = \frac{a_0}{2} + \sum_{n=1}^{\infty} a_n e^{-4n^2 t} \cos nx.$$

Replacing x with zero, we obtain

$$f(x) = \frac{a_0}{2} + \sum_{n=1}^{\infty} a_n \cos nx,$$

which is the same cosine Fourier series of Exercise 3.17 on $0 < x < \pi$, where

$$\frac{a_0}{2} = \frac{48\pi^4}{90} \quad \text{and} \quad a_n = 48\frac{(-1)^{n+1}}{n^4}, \quad n = 1,2\ldots$$

As a result, the complete solution of the heat problem takes the form

$$u(x,t) = \frac{48\pi^4}{90} + 48\sum_{n=1}^{\infty} \frac{(-1)^{n+1}}{n^4} e^{-4n^2 t} \cos nx.$$

□

4.6 Insulated Homogeneous Heat Problem with Non-Homogeneous Boundaries

The insulated homogeneous heat problem with non-homogeneous boundaries involves the diffusion of heat in a homogeneous medium where the heat equation governs the temperature distribution, and the boundaries have specified, non-zero heat flux or temperature values. This setup models scenarios where the material properties are uniform, but external conditions, such as fixed temperatures or heat inputs, create non-uniform effects at the boundaries.

Consider the homogeneous insulated heat problem with non-homogeneous boundary conditions on a finite slab

$$u_t - ku_{xx} = 0, \quad 0 < x < c, \quad t > 0, \tag{4.28}$$

subject to the initial condition

$$u(x,0) = f(x), \quad 0 < x < c, \tag{4.29}$$

and to the boundary conditions

$$u_x(0,t) = T_1, \quad u_x(c,t) = T_2, \quad 0 < x < c. \tag{4.30}$$

• Method of Solution

Due to the presence of non-homogeneous boundary conditions, we must adapt our method for finding the solution. In this case the method of separation of variables does not work since the boundary conditions are non-homogeneous. We use to concept of *equilibrium temperature, or steady-state temperature.* In a one-dimensional heat flow problem, the **equilibrium temperature** is the steady-state solution to the heat equation, where the temperature no longer changes with time. That is

$$u_t(x,t) = 0 \quad \text{for all} \quad t > 0.$$

Thus, setting $u_t(x,t) = 0$ in (4.28) leads to the second-order differential equation

$$\frac{d^2u(x)}{dx^2} = 0, \tag{4.31}$$

subject to the given boundary conditions. Denoting the solution of (4.31) by $u^E(x)$, we have

$$u^E(x) = Ax + B.$$

Using the boundary conditions

$$\frac{du(x)}{dx}\Big|_{x=0} = T_1 \quad \text{and} \quad \frac{du(x)}{dx}\Big|_{x=c} = T_2,$$

we get

$$u_x^E(0) = A = T_1, \quad u_x^E(c) = A = T_2,$$

if and only if

$$T_1 = T_2 = T.$$

We see that $u^E(x) = Tx + B$, for arbitrary constant B. Next, we define a new function $v(x,t)$ that represents the deviation of the temperature $u(x,t)$ from the equilibrium temperature $u^E(x)$, given by

$$v(x,t) = u(x,t) - u^E(x).$$

Then,

$$v_t(x,t) = u_t(x,t), \quad v_x(x,t) = u_x(x,t) - T, \quad v_{xx}(x,t) = u_{xx}(x,t),$$

and consequently,

$$v_x(0,t) = u_x(0,t) - T = T - T = 0, \quad \text{and} \quad v_x(c,t) = u_x(c,t) - T = T - T = 0.$$

Moreover,

$$v(x,0) = u(x,0) - u^E(x) = f(x) - Tx - B := g(x).$$

By recasting our original problem in terms of $v(x,t)$ we arrive at

$$v_t - ku_{xx} = 0, \; 0 < x < c, \; t > 0, \; v(x,0) = g(x), \; v_x(0,t) = 0, \quad v_x(c,t) = 0,$$

that match with (4.20)– (4.22) whose solution is

$$v(x,t) = \frac{a_0}{2} + \sum_{n=1}^{\infty} a_n e^{-\frac{n^2\pi^2 k}{c^2}t} \cos\frac{n\pi x}{c},$$

where

$$a_0 = \frac{2}{c}\int_0^c g(x)\,dx \quad \text{and} \quad a_n = \frac{2}{c}\int_0^c g(x)\cos\frac{n\pi x}{c}\,dx, \quad n = 1,2,\ldots$$

Since, $v(x,t) = u(x,t) - u^E(x)$, the solution to the original problem takes the form

$$u(x,t) = u^E(x) + \frac{a_0}{2} + \sum_{n=1}^{\infty} a_n e^{-\frac{n^2\pi^2 k}{c^2}t} \cos\frac{n\pi x}{c}, \tag{4.32}$$

where

$$a_0 = \frac{2}{c}\int_0^c \left[f(x) - u^E(x)\right] dx \quad \text{and} \quad a_n = \frac{2}{c}\int_0^c \left[f(x) - u^E(x)\right]\cos\frac{n\pi x}{c}\,dx, \quad n = 1,2,\ldots$$

The heat equation with given initial and Neumann boundary conditions has a unique solution. Since we have transformed the problem into a homogeneous boundary value problem for $v(x,t)$, and the steady-state solution $u^E(x)$ is uniquely determined if $T_1 = T_2$, the overall solution is unique. If $T_1 \neq T_2$, a steady-state solution does not exist, indicating that the problem as posed is not well-posed. Thus, the solution is unique provided that $T_1 = T_2$.

Remark 8 *The arbitrary constant B in $u^E(x)$ does not affect the solution given by* (4.32)*, as it either cancels out or appears in integrals that evaluate to zero. To see this,*

$$\begin{aligned} u^E(x) + \frac{a_0}{2} &= Tx + B + \frac{1}{c}\int_0^c \left[f(x) - u^E(x)\right] dx \\ &= Tx + B + \frac{1}{c}\int_0^c [f(x) - Tx]\,dx - \frac{1}{c}\int_0^c B\,dx = Tx + \frac{1}{c}\int_0^c [f(x) - Tx]\,dx. \end{aligned}$$

Along the same lines,

$$\begin{aligned} a_n &= \frac{2}{c}\int_0^c \left[f(x) - Tx\right]\cos\frac{n\pi x}{c}\,dx + \frac{2}{c}\int_0^c B\cos\frac{n\pi x}{c}\,dx \\ &= \frac{2}{c}\int_0^c \left[f(x) - Tx\right]\cos\frac{n\pi x}{c}\,dx. \end{aligned}$$

4.6.1 Exercises

Exercise 4.8 *Find the eigenvalues and eigenfunctions for*

$$x^2X''(x) + xX'(x) + \lambda X(x) = 0, \quad X'(1) = 0, \quad X(b) = 0.$$

Exercise 4.9 *Let $u(x,t)$ be given by* (4.26)*. Show that* (*a*) $\lim_{t\to\infty} u(x,t) = \frac{a_0}{2} = \frac{1}{c}\int_0^c f(x)\,dx.$

(*b*) $u_x(0^+,t) = \lim_{\substack{x\to 0\\ x>0}} u_x(x,t) = 0, \quad u_x(c^-,t) = \lim_{\substack{x\to c\\ x<c}} u_x(x,t) = 0.$

Exercise 4.10 *Solve the heat flow problem.*

(*a*) $u_t(x,t) - u_{xx}(x,t) = 0, \quad u(x,0) = \sin x, \quad u_x(0,t) = u_x(\pi,t) = 0, \quad 0 < x < \pi.$

(*b*) $u_t(x,t) - u_{xx}(x,t) = 0, \quad u(x,0) = \pi - x, \quad u_x(0,t) = u_x(\pi,t) = 0, \quad 0 < x < \pi$

(*c*) $u_t(x,t) - 4u_{xx}(x,t) = 3u, \quad u(x,0) = x, \quad u_x(0,t) = u_x(1,t) = 0, \quad 0 < x < 1.$

Exercise 4.11 *For a constant h, use the substitution* $u(x,t) = e^{-ht}v(x,t)$ *to solve*

$$u_t = ku_{xx} - hu, \quad -\pi < x < \pi, \quad t > 0,$$

$$u(x,0) = f(x), \quad -\pi \le x \le \pi,$$

$$u_x(-\pi,t) = u_x(\pi,t), \quad u(-\pi,t) = u(\pi,t).$$

Exercise 4.12 *(See Example 1.10). Solve the heat flow problem.*

$$u_t(x,t) - u_{xx}(x,t) = 0, \quad 0 < x < 8,$$

$$u(x,0) = f(x), \quad u_x(0,t) = u_x(8,t) = 0,$$

where

$$f(x) = \begin{cases} 2 - x, & 0 < x < 4 \\ x - 6, & 4 < x < 8. \end{cases}$$

Exercise 4.13 *From Example 4.7, set*

$$u_n(x,t) = \frac{48\pi^4}{90} + 48\sum_{n=1}^{\infty} \frac{(-1)^{n+1}}{n^4} e^{-4n^2t}\cos nx.$$

For a positive constant B, show that the error between $u(x,t)$ *and the first term approximation* $u_1(x,t)$ *satisfies*

$$\left|u_n(x,t) - u_1(x,t)\right| \le B\frac{e^{-4t}}{1 - e^{-4t}}, \quad t > 0.$$

Conclude that

$$\lim_{t\to\infty} u_n(x,t) = u_1(x,t).$$

Hint: Use the geometric series result $\sum_{n=1}^{\infty} r^n = \frac{r}{1-r}$, *for* $|r| < 1$.

Exercise 4.14 *Solve the heat flow problem.*

$$u_t(x,t) - u_{xx}(x,t) = 0, \quad 0 < x < c, \quad t > 0$$

$$u(x,0) = f(x), \quad u(0,t) = u_x(c,t) = 0, \quad 0 < x < c.$$

Exercise 4.15 *Solve the heat flow problem.*

$$u_t(x,t) - \alpha^2 u_{xx}(x,t) = 0, \quad 0 \le x \le \pi, \quad t > 0,$$
$$u(0,t) = 0, \quad u_x(\pi,t) = 0, \quad 0 < x < c,$$
$$u(x,0) = 3\sin\frac{5x}{2}.$$

Ans: $u(x,t) = 3\sin\frac{5x}{2}\exp\left(-\left(\frac{5\alpha}{2}\right)^2 t\right)$.

Exercise 4.16 *(Non-homogeneous heat flow) Let* $u = u(x,t)$ *and consider the non-homogeneous heat flow problem*

$$\begin{aligned} &u_t - ku_{xx} = pu, \quad 0 < x < c, \quad t > 0, \\ &u(x,0) = f(x), \\ &u_x(0,t) = T_1, \ u_x(c,t) = T_2. \end{aligned} \tag{4.33}$$

We studied the case $p = 0$. *If* $u^E(x)$ *is the steady-state solution for the equilibrium temperature, then it satisfies*

$$-k\frac{d^2u^E(x)}{dx^2} = pu^E(x), \quad \frac{du^E(x)}{dx}|_{x=0} = T_1 \quad \text{and} \quad \frac{du^E(x)}{dx}|_{x=c} = T_2.$$

Find $u^E(x)$ *for the cases*

(a) $p > 0$, *(b)* $p < 0$

(c) Make use of the steady-state solution $u^E(x)$ *combined with the method of separation of variables and solve (4.33) for the specific values* $k = p = T_1 = 1, T_2 = 0$ *and for initial function* $f(x)$.

4.7 Non-Insulated Homogeneous Heat Problem with Homogeneous Boundaries

Consider the non-insulated homogeneous heat flow with Homogeneous Dirichlet conditions

$$u_t - ku_{xx} = 0, \quad 0 < x < c, \quad t > 0, \tag{4.34}$$

subject to the initial condition

$$u(x,0) = f(x), \quad 0 < x < c, \tag{4.35}$$

and to the boundary conditions

$$u(0,t) = 0, \ u(c,t) = 0, \quad 0 < x < c. \tag{4.36}$$

The boundary conditions given by (4.36) indicate that the ends of the rod are maintained at a constant temperature of zero, representing perfectly homogeneous Dirichlet boundary conditions. This setup models the diffusion of heat in a rod with fixed endpoints, emphasizing the effects of conduction while neglecting external sources or sinks of heat. The problem seeks to determine how the temperature evolves over time while satisfying the given constraints. We refer to this problem as homogeneous because the heat flow described by (4.34) is homogeneous, and its boundary conditions, given by (4.36), are also homogeneous. We seek a nontrivial solution $u(x,t)$ of (4.34) that satisfies the initial condition (4.35) along with the boundary conditions given by (4.36).

• Method of Solution

We seek separated solutions of the form

$$u(x,t) = X(x)T(t), \tag{4.37}$$

where X is a function of x alone, and T is a function of t alone. Note that both X and T must be nontrivial to ensure a nontrivial solution, i.e., $X \neq 0$ and $T \neq 0$. Therefore,

$$\frac{X''(x)}{X(x)} = \frac{T'(t)}{kT(t)} = -\lambda.$$

The Dirichlet boundary conditions given by (4.36) yield

$$0 = u(0,t) = X(0)T(t),$$

which must hold for all $t > 0$. Thus, we have $X(0) = 0$. Similarly,

$$0 = u(c,t) = X(c)T(t),$$

implies $X(c) = 0$. Consequently, we arrive at the second-order differential equation

$$X''(x) + \lambda X(x) = 0, \quad X(0) = 0, \quad X(c) = 0, \tag{4.38}$$

which is a subset of the Sturm–Liouville problem (4.18). Additionally, we obtain the first-order ordinary differential equation

$$T'(t) + \lambda k T(t) = 0, \quad t > 0. \tag{4.39}$$

We begin by solving (4.38), as its solution is well-defined due to the presence of two boundary conditions.

- $\lambda = 0$. In this case, (4.38) reduces to $X''(x) = 0$, which has the solution

  $$X(x) = Ax + B,$$

  where A and B are constants. Applying $X(0) = 0$ gives $B = 0$. Applying $X(c) = 0$ gives $A = 0$. We conclude that $\lambda = 0$ leads to a trivial solution.

- **$\lambda < 0$.** If $\lambda < 0$, we set $\lambda = -\alpha^2$ with $\alpha > 0$. Then, (4.38) becomes

$$X''(x) - \alpha^2 X(x) = 0,$$

whose general solution is

$$X(x) = c_1 e^{\alpha x} + c_2 e^{-\alpha x}.$$

Applying $X(0) = 0$ gives $c_1 = -c_2$. Substituting this into $X(x)$, we get

$$X(x) = 2c_1 \sinh(\alpha x).$$

The condition $X(c) = 0$ implies $c_1 \sinh(\alpha c) = 0$, which is not possible for $\alpha > 0$. Hence, we must have $c_1 = 0$, which leads to the trivial (zero) solution.

- **$\lambda > 0$.** If $\lambda > 0$, we set $\lambda = \alpha^2$ with $\alpha > 0$. Then, equation (4.38) becomes

$$X''(x) + \alpha^2 X(x) = 0,$$

whose general solution is given by

$$X(x) = A\cos(\alpha x) + B\sin(\alpha x).$$

Applying the boundary condition $X(0) = 0$ yields $A = 0$. Substituting $A = 0$ into $X(x)$, we obtain

$$X(x) = B\sin(\alpha x).$$

Applying the second boundary condition $X(c) = 0$ gives

$$B\sin(\alpha c) = 0.$$

Here, we have two possibilities; either $B = 0$, which leads to the trivial solution, or

$$\sin(\alpha c) = 0.$$

To avoid the trivial solution, we require

$$\alpha c = n\pi, \quad n = 1, 2, \ldots$$

which implies

$$\alpha = \frac{n\pi}{c}.$$

Thus, for $\lambda_n = \left(\frac{n\pi}{c}\right)^2$, the corresponding eigenfunctions are

$$X_n(x) = \sin\frac{n\pi x}{c}, \quad n = 1, 2, \ldots,$$

where we have set $B = 1$.

Equation (4.39), has the solution

$$T_n(t) = e^{-\frac{n^2\pi^2 k}{c^2}t}, \quad n = 1, 2, \ldots.$$

Setting $u_n(x,t) = T_n(t)X_n(x)$, the function satisfies the heat equation and its boundary conditions. Therefore, by the superposition principle, for any constants b_n, the general solution can be written as

$$u(x,t) = \sum_{n=1}^{\infty} b_n e^{-\frac{n^2\pi^2 k}{c^2}t} \sin\frac{n\pi x}{c}. \tag{4.40}$$

Applying the initial condition $u(x,0) = f(x)$ to (4.40), we obtain the Fourier sine series

$$f(x) = \sum_{n=1}^{\infty} b_n \sin\frac{n\pi x}{c},$$

where

$$b_n = \frac{2}{c}\int_0^c f(x)\sin\frac{n\pi x}{c}\,dx, \quad n = 1, 2, \ldots. \tag{4.41}$$

In conclusion, (4.40) and (4.41) provide the desired solution.

Example 4.8 For $0 < x < \pi$, $t > 0$, consider the heat flow problem

$$u_t(x,t) - 4u_{xx}(x,t) = 0, \quad u(x,0) = \frac{\pi}{2}, \quad u(0,t) = u(\pi,t) = 0.$$

Then the solution takes the form

$$u(x,t) = \sum_{n=1}^{\infty} b_n e^{-4n^2 t} \sin nx.$$

Replacing x with zero, we obtain

$$f(x) = \sum_{n=1}^{\infty} b_n \sin nx.$$

This is the Fourier sine series of $f(x)$ and hence

$$b_n = \frac{2}{\pi}\int_0^{\pi} \frac{\pi}{2} \sin nx\,dx = -\frac{1}{n}[(-1)^n - 1].$$

Consequently, $b_n = \frac{2}{n}$ for n being odd, and $b_n = 0$ for n being even. Thus the solution is

$$u(x,t) = \sum_{n=1}^{\infty} \frac{2}{(2n-1)} e^{-4(2n-1)^2 t} \sin(2n-1)x.$$

□

4.8 Non-Insulated Homogeneous Heat Problem with Non-Homogeneous Boundaries

Consider the homogeneous heat problem with non-homogeneous boundary conditions on a finite slab

$$u_t - ku_{xx} = 0, \quad 0 < x < c, \quad t > 0, \tag{4.42}$$

subject to the initial condition

$$u(x,0) = f(x), \quad 0 < x < c, \tag{4.43}$$

and to the boundary conditions

$$u(0,t) = T_1, \quad u(c,t) = T_2, \quad 0 < x < c. \tag{4.44}$$

The above homogeneous heat problem on a finite slab, describes the diffusion of heat in a medium without internal heat sources. The temperature distribution evolves over time according to the heat equation, where the thermal diffusivity determines the rate of heat transfer. The initial condition specifies the temperature profile across the slab at the starting time. The boundary conditions are Dirichlet conditions that specify fixed temperatures at both ends of the domain. This setup is typical in heat conduction problems where the ends of a rod are maintained at constant, possibly different, temperatures T_1 and T_2. Again, due to the presence on the non-homogeneous boundary conditions, we need to improvise our method of finding the solution. The method of separation of variables does not work since the boundary conditions are non-homogeneous. As before, we use to concept of *equilibrium temperature.* Thus, setting $u_t(x,t) = 0$ in (4.42) leads to the second-order differential equation

$$-k\frac{d^2u(x)}{dx^2} = 0, \tag{4.45}$$

subject to the given boundary conditions

$$u(0) = T_1 \quad \text{and} \quad u(c) = T_2.$$

We denote the solution of (4.45) by $u^E(x)$. Then

$$u^E(x) = Ax + B.$$

Applying both boundary conditions gives,

$$A = \frac{T_2 - T_1}{c} \quad \text{and} \quad B = T_1.$$

Then

$$u^E(x) = \frac{T_2 - T_1}{c}x + T_1.$$

Next we introduce a new function $v(x,t)$ that measures the displacement of the temperature $u(x,t)$ from the equilibrium temperature $u^E(x)$

$$v(x,t) = u(x,t) - u^E(x).$$

Then,

$$v_t(x,t) = u_t(x,t) \quad \text{and} \quad v_{xx}(x,t) = u_{xx}(x,t),$$

and consequently,

$$v(0,t) = u(0,t) - T_1 = 0 \quad \text{and} \quad v(c,t) = u(c,t) - T_2 = 0.$$

Moreover,

$$v(x,0) = u(x,0) - u^E(x) = f(x) - u^E(x) := g(x).$$

By recasting our original problem in terms of $v(x,t)$ we arrive at

$$\begin{aligned}
&v_t - ku_{xx} = 0, \quad 0 < x < c, \quad t > 0,\\
&v(x,0) = g(x),\\
&v(0,t) = 0, \quad v(c,t) = 0,
\end{aligned}$$

that match with (4.34)– (4.36) whose solution is

$$v(x,t) = \sum_{n=1}^{\infty} b_n e^{-\frac{n^2\pi^2 k}{c^2}t} \sin\frac{n\pi x}{c},$$

where

$$b_n = \frac{2}{c}\int_0^c g(x)\sin\frac{n\pi x}{c}\,dx, \quad n = 1,2,\ldots$$

Since, $v(x,t) = u(x,t) - u^E(x)$, the solution to the original problem takes the form

$$\begin{aligned}
u(x,t) &= u^E(x) + \sum_{n=1}^{\infty} b_n e^{-\frac{n^2\pi^2 k}{c^2}t} \sin\frac{n\pi x}{c}\\
&= \frac{T_2 - T_1}{c}x + T_1 + \sum_{n=1}^{\infty} b_n e^{-\frac{n^2\pi^2 k}{c^2}t} \sin\frac{n\pi x}{c},
\end{aligned}$$

where

$$b_n = \frac{2}{c}\int_0^c \left[f(x) - u^E(x)\right]\sin\frac{n\pi x}{c}\,dx, \quad n = 1,2,\ldots$$

Now we consider the non-homogeneous heat flow with homogeneous boundary conditions

$$\begin{aligned}
&u_t - ku_{xx} = pu, \quad 0 < x < c, \quad t > 0,\\
&u(x,0) = f(x),\\
&u(0,t) = u(c,t) = 0.
\end{aligned} \tag{4.46}$$

Using the method of separation of variables of the form $u(x,t) = X(x)T(t)$, we arrive at the equations along with the corresponding boundary conditions for $X(x)$,

$$X''(x) + \frac{p}{k}X(x) = -\lambda X(x), \quad X(0) = X(c) = 0 \quad \text{and} \quad T'(t) + k\lambda T(t) = 0.$$

Now we may proceed as before to find the solution. In the next example we give a sketch on how to use the steady-state solution, equilibrium solution to recast the problem so that the method of separation of variables can be applied.

Example 4.9 *(Non-homogeneous)* Consider the non-homogeneous heat flow with non-homogeneous boundary conditions

$$\begin{aligned} &u_t - ku_{xx} = pu, \quad 0 < x < c, \quad t > 0, \\ &u(x,0) = f(x), \\ &u(0,t) = T_1, \quad u(c,t) = T_2. \end{aligned} \tag{4.47}$$

This problem describes the flow of heat in a material rod where heat conduction, represented by u_{xx}, is influenced by both internal heat production or absorption and an exponential growth or decay factor pu. The term u_t represents the rate of change of temperature $u(x,t)$ over time, while the diffusion term ku_{xx} models the spread of heat along the rod. The factor pu accounts for temperature changes caused by sources or sinks that are proportional to the current temperature. The Neumann boundary conditions $u_x(0,t) = T_1$ and $u_x(c,t) = T_2$ describe the heat flux (rate of heat transfer) at the ends of the rod, while the initial condition $u(x,0) = f(x)$ specifies the initial temperature distribution. This setup reflects how heat evolves in a rod subjected to combined effects of conduction, proportional heat sources or sinks, and specified flux at the boundaries.

First find the steady-state solution or the equilibrium temperature, $u^E(x)$ for

$$-k\frac{d^2u^E(x)}{dx^2} = pu^E(x), \quad u^E(0) = T_1 \quad \text{and} \quad u^E(c) = T_2.$$

Next we let the solution $u(x,t)$ be written as the sum of a steady-state component $u^E(x)$ and a transient component $w(x,t)$ such that

$$u(x,t) = u^E(x) + w(x,t).$$

Then,

$$u_t(x,t) = w_t(x,t) \quad \text{and} \quad u_{xx}(x,t) = \left(u^E(x)\right)_{xx} + w_{xx}(x,t).$$

Therefore, $u_t - ku_{xx} = pu$, becomes

$$w_t(x,t) - k\left[\left(u^E(x)\right)_{xx} + w_{xx}(x,t)\right] = p\left[u^E(x) + w(x,t\right].$$

Using the fact that $u^E(x)$ satisfies the steady-state equation, or $-k\left(u^E(x)\right)_{xx} = pu^E(x)$ the transient equation becomes

$$w_t - kw_{xx} = pw.$$

As for the boundary conditions, we have

$$w(0,t) = u(0,t) - u^E(0) = T_1 - T_1 = 0, \quad \text{and} \quad w(c,t) = u(c,t) - u^E(c) = T_2 - T_2 = 0.$$

Hence (4.33) can be recasted as

$$\begin{aligned} &w_t - kw_{xx} = pw, \quad 0 < x < c, \quad t > 0, \\ &w(x,0) = f(x) - u^E(x), \\ &w(0,t) = w(c,t) = 0. \end{aligned}$$

Now, the method of separation of variables can be applied, as outlined above, to solve the reformulated problem with homogeneous boundary conditions (see Exercise 4.19). □

4.8.1 Exercises

Exercise 4.17 *Find the eigenvalues and eigenfunctions for*

$$X''(x) + (\lambda + 1)X(x) = 0, \quad X(0) = 0, \quad X(1) = 0.$$

Exercise 4.18 *Find the eigenvalues and eigenfunctions for*

$$(1+x)^2 X''(x) + 2(1+x)X(x) + 3\lambda X(x) = 0, \quad X(0) = 0, \quad X(1) = 0.$$

Exercise 4.19 *Derive the solution of* (4.46).

Exercise 4.20 *Find the steady-state solution of the problem*

$$\frac{\partial}{\partial x}\left(\kappa \frac{\partial u}{\partial x}\right) = c\rho \frac{\partial u}{\partial t}, \quad 0 < x < a, \quad t > 0,$$

$$u(0,t) = T_0, \quad u(a,t) = T_1, \quad t > 0,$$

if the conductivity varies in a linear fashion with x

$$\kappa(x) = \kappa_0 + \beta x,$$

where κ_0 and β are constants.

Exercise 4.21 *Solve the heat flow problem.*

(a) $u_t(x,t) - u_{xx}(x,t) = 0, \quad u(x,0) = 0, \quad u(0,t) = 1, \quad u(3,t) = 3, \quad 0 < x < 3.$

(b) $u_t(x,t) - 5u_{xx}(x,t) = 0, \quad u(x,0) = 2x, \quad u(0,t) = 0, \quad u(\pi,t) = 1, \quad 0 < x < \pi.$

(c) $u_t(x,t) - ku_{xx}(x,t) = 0, \quad u(x,0) = x(\ell - x), \quad u(0,t) = 0, \quad u(\ell,t) = 0, \quad 0 < x < \ell$. *(See Example 1.9)*

Exercise 4.22 *Suppose that f in (4.41) is piecewise continuous and let $u(x,t)$ be given by (4.40). For a positive constant B show that for all $t > 0$ and $0 < x < c$,*

$$u(x,t) \leq B \frac{e^{-\frac{\pi^2 k}{c^2} t}}{1 - e^{-\frac{\pi^2 k}{c^2} t}}.$$

Then conclude that

$$\lim_{t \to \infty} u(x,t) = 0.$$

Exercise 4.23 *Solve*

$$u_t(x,t) - k u_{xx}(x,t) = 0, \quad u(x,0) = f(x), \quad u(0,t) = 0, \quad u(a,t) = 0, \quad 0 < x < a,$$

where

$$f(x) = \begin{cases} \frac{2T_0 x}{a}, & 0 < x < \frac{a}{2} \\ \frac{2T_0(a-x)}{a}, & \frac{a}{2} < x < a \end{cases}$$

Exercise 4.24 *Solve using separation of variables.*

$$u_t(x,t) - u_{xx}(x,t) = u, \quad 0 < x < \pi,$$

$$u(x,0) = 3, \quad u(0,t) = u(\pi,t) = 0.$$

Exercise 4.25 (A.N. Virkar, T.B. Jackson, and R.A. Cutler) *[138] Let C be the oxygen concentration, D the diffusion constant, a the thickness of a grain, and C_0 and C_1 known concentrations. Then, the kinetics of oxygen removal from a grain of aluminum nitride by diffusion is governed by the following boundary value problem*

$$\frac{\partial C}{\partial t} = D \frac{\partial^2 C}{\partial x^2}, \quad 0 < x < a, \quad 0 < t,$$

$$C(0,t) = C_1, \quad C(a,t) = C_1, \quad 0 < t,$$

$$C(x,0) = C_0, \quad 0 < x < a.$$

(a) Find the steady-state solution, $C^E(x)$.

(b) State the problem (partial differential equation, boundary conditions, and initial condition) for the transient $w(x,t) = C(x,t) - C^E(x)$.

(c) Solve the problem for $w(x,t)$, and write out the complete solution $C(x,t)$.

(d) The concentration in the center of the grain, $C(a/2,t)$, varies from C_0 at time $t = 0$ toward C_1 as t increases. Suppose that we want to find out how long it takes for this concentration to complete 90% of the change it will make from C_0 to C_1; that is, we want to solve this equation for t

$$C\left(\frac{a}{2},t\right) - C_0 = 0.9(C_1 - C_0).$$

Show that this equation is equivalent to the equation

$$w\left(\frac{a}{2},t\right) = -0.1(C_1 - C_0).$$

Find an approximate formula for the solution by using just the first term of the series for $w(x,t)$.

Exercise 4.26 *A fluid flows between two parallel plates held at temperature* 0. *At the inlet, the fluid temperature is* T_0 *and initially the fluid is at temperature* T_1. *If* V *is the speed of the fluid in the x-direction, a problem describing the temperature* $u(x,y,t)$ *is*

$$k\frac{\partial^2 u}{\partial y^2} = \frac{\partial u}{\partial t} + V\frac{\partial u}{\partial x}, \quad 0 < y < b, \quad x > 0, \quad t > 0,$$

$$u(x,0,t) = 0, \quad u(x,b,t) = 0, \quad x > 0, \quad t > 0,$$

$$u(0,y,t) = T_0, \quad 0 < y < b, \quad t > 0,$$

$$u(x,y,0) = T_1, \quad x > 0, \quad 0 < y < b.$$

Use the separation of variables

$$u(x,y,t) = \psi(x+Vt)\phi(x-Vt)Y(y)$$

and solve the boundary value problem in Y. *Obtain the solution*

$$u_n(x,y,t) = \varphi_n(x-Vt)\exp\left(-\frac{\lambda_n^2 k(x+Vt)}{2V}\right)\sin(\lambda_n y)$$

and show that is satisfies the partial differential equation and boundary conditions at $y = 0$ *and* $y = b$, *without restriction of* φ_n *(except differentiability).*

Exercise 4.27 *Consider* $u_t = u_{xx} + u$ *on* $0 < x < \pi$, $t > 0$, *with* $u(0,t) = u(\pi,t) = 0$ *and* $u(x,0) = \sin x$. *Show that* $u(x,t) \le e^t \sin x$.

4.9 Heat Source

Consider the simplest form of the heat flow problem with heat source $q = q(x)$ (independent of t) in a finite rod. The heat problem is modeled by

$$u_t = ku_{xx} + q(x), \quad 0 < x < c, \tag{4.48}$$

$$u(0,t) = b_1, \quad u(c,t) = b_2, \quad t > 0, \tag{4.49}$$

$$u(x,0) = f(x), \quad 0 < x < c. \tag{4.50}$$

This describes a heat conduction problem in a rod with an additional constant source term, $q(x)$, representing heat generation or absorption. The equation (4.48) models how the temperature $u(x,t)$ evolves over time. The term u_{xx} accounts for heat diffusion, while $q(x)$ adds or removes heat at each point x. The boundary conditions given by (4.49) specify fixed temperatures at the two ends of the rod, and the initial condition (4.50) sets the temperature distribution at the start.

If $q(x) > 0$, heat is being generated inside the rod at the point x, increasing the temperature. Conversely, if $q(x) < 0$, heat is being absorbed at x, lowering the temperature.

• Method of Solution

Following the same approach as in Example 4.9, determine the equilibrium solution $u^E(x)$ that satisfies the given PDE and boundary conditions,

$$0 = k\frac{du^E}{dx^2} + q(x), \quad 0 < x < c,$$

$$u^E(0) = b_1, \quad u^E(c) = b_2.$$

Then let

$$w(x,t) = u(x,t) - u^E(x).$$

Substituting $u(x,t) = w(x,t) + u^E(x)$ into the original PDE gives a problem for $w(x,t)$ of the form

$$w_t = kw_{xx}, \quad 0 < x < c,$$

$$w(0,t) = 0, \quad w(c,t) = 0, \quad t > 0,$$

$$w(x,0) = f(x) - u^E(x), \quad 0 < x < c.$$

Now $w(x,t)$ satisfies the homogeneous heat equation PDE and homogeneous boundary conditions, i.e., the basic heat problem. Solve the heat problem for $w(x,t)$, and then obtain

$$u(x,t) = w(x,t) + u^E(x).$$

Note that things become more complicated for non-constant sources $q(x)$, but the principle remains the same.

We provide a working example.

Example 4.10 *(Heat source with non-homogeneous BCs)* Consider the heat flow with a source and non-homogeneous boundary conditions

$$u_t - u_{xx} = \sin x, \quad 0 < x < \pi,$$

$$u(0,t) = 1, \quad u(\pi,t) = 3, \quad t > 0,$$

$$u(x,0) = \sin x, \quad 0 < x < \pi.$$

Let $u^E(x)$ be steady-state solution of the equilibrium temperature. Then $u^E(x)$ must solve

$$0 = \frac{du^E}{dx^2} + \sin x, \quad 0 < x < \pi$$

along with the boundary conditions

$$u^E(0) = 1, \quad u^E(\pi) = 3.$$

Integrating twice and then applying the boundary conditions yield

$$u^E(x) = \sin x + \frac{2}{\pi}x + 1.$$

Set

$$u(x,t) = u^E(x) + w(x,t).$$

Then,

$$u_t(x,t) = w_t(x,t) \text{ and } u_{xx}(x,t) = \left(u^E(x)\right)_{xx} + w_{xx}(x,t) = -\sin x + w_{xx}(x,t).$$

Therefore, $u_t - u_{xx} = \sin x$ becomes

$$w_t(x,t) - \left[-\sin x + w_{xx}(x,t)\right] = \sin x.$$

Thus, the transient equation becomes

$$w_t - w_{xx} = 0$$

As for the boundary conditions, we have

$$w(0,t) = u(0,t) - u^E(0) = 1 - 1 = 0 \quad \text{and} \quad w(\pi,t) = u(\pi,t) - u^E(\pi) = 3 - 3 = 0.$$

Hence the original problem can be recasted as

$$\begin{aligned} &w_t - w_{xx} = 0, \quad 0 < x < \pi, \quad t > 0, \\ &w(x,0) = \sin x - u^E(x) = -\frac{2}{\pi}x - 1, \\ &w(0,t) = w(\pi,t) = 0, \end{aligned}$$

that align with (4.34)– (4.36) with $c = \pi$. Thus, we have the corresponding solution

$$w(x,t) = \sum_{n=1}^{\infty} b_n e^{-n^2kt} \sin nx,$$

where

$$b_n = -\frac{2}{\pi}\int_0^{\pi} \left[\frac{2}{\pi}x + 1\right] \sin nx dx, \quad n = 1,2,\ldots$$

For the sake of completeness let's compute the sine coefficient b_n. By splitting the integral, we get

$$b_n = -\frac{2}{\pi}\left(\int_0^{\pi} \frac{2}{\pi}x \sin nx dx + \int_0^{\pi} \sin nx dx\right).$$

Let $u = x$ and $dv = \sin nx\,dx$. An integration by parts yields,

$$\int_0^{\pi} x \sin nx\,dx = \left[-\frac{x\cos nx}{n}\right]_0^{\pi} + \frac{1}{n}\int_0^{\pi} \cos nx\,dx = -\frac{\pi(-1)^n}{n}.$$

Hence,

$$-\frac{2}{\pi}\cdot\frac{2}{\pi}\int_0^{\pi} x\sin nx\,dx = \frac{4(-1)^n}{n\pi}.$$

Additionally,

$$-\frac{2}{\pi}\int_0^{\pi} \sin nx\,dx = \left[-\frac{\cos nx}{n}\right]_0^{\pi} = \frac{2}{n\pi}\left((-1)^n - 1\right).$$

Combining the terms leads to

$$b_n = \frac{4(-1)^n}{n\pi} + \frac{2}{n\pi}\left((-1)^n - 1\right) = \frac{6(-1)^n - 2}{n\pi}, \quad n = 1, 2, \ldots$$

Since, $w(x,t) = u(x,t) - u^E(x)$, the solution to the original problem takes the form

$$u(x,t) = u^E(x) + \sum_{n=1}^{\infty} b_n e^{-n^2kt} \sin nx,$$

or

$$u(x,t) = \sin x + \frac{2}{\pi}x + \frac{2}{\pi}\sum_{n=1}^{\infty} \frac{3(-1)^n - 1}{n} e^{-n^2kt} \sin nx.$$

Let

$$M_n = e^{-n^2kt_0}, \quad t \geq t_0 \geq 0.$$

Then

$$\begin{aligned}\frac{M_{n+1}}{M_n} &= \frac{e^{-(n+1)^2kt_0}}{e^{-n^2kt_0}} = e^{-\left((n+1)^2-n^2\right)kt_0} \\ &= e^{-2nkt_0-kt_0} \leq e^{-kt_0} < 1, \quad n = 1, 2, 3, \ldots.\end{aligned}$$

Thus by the ratio test the sum

$$\sum_{n=1}^{\infty} M_n,$$

converges absolutely. Therefore,

$$\begin{aligned}\left|\frac{2}{\pi}\sum_{n=1}^{\infty} \frac{3(-1)^n - 1}{n\pi} e^{-n^2kt} \sin nx\right| &\leq \frac{8}{\pi}\sum_{n=1}^{\infty} e^{-n^2kt} \leq \frac{8}{\pi}\sum_{n=1}^{\infty} e^{-nkt} \\ &\leq \frac{8}{\pi}\sum_{n=1}^{\infty}\left(e^{-kt}\right)^n = \frac{8}{\pi}\frac{e^{-kt}}{1-e^{-kt}}, \quad t > 0.\end{aligned}$$

where we have used a brand of the geometric series

$$\sum_{n=1}^{\infty} r^n = \frac{r}{1-r}, \quad \text{for} \quad |r| < 1.$$

Letting $t \to \infty$ in the solution, we see that

$$\lim_{t\to\infty}\left(u(x,t) - u^E(x)\right) \le \lim_{t\to\infty} \frac{8}{\pi}\frac{e^{-kt}}{1-e^{-kt}} = 0.$$

Hence

$$\lim_{t\to\infty} u(x,t) = u^E(x) = \sin x + \frac{2}{\pi}x + 1,$$

as expected. That is, as t grows larger, the solution approaches the steady-state solution of the equilibrium temperature. □

4.10 Heat Flow with Variable Sources and Time-Dependent Boundaries

Now, we analyze the problem of a heat source with both spatial and temporal dependence, considering non-homogeneous boundary conditions that vary over time. In particular we consider the following non-homogeneous problem

$$u_t = ku_{xx} + Q(x,t), \quad 0 < x < L, \quad t > 0, \tag{4.51}$$

$$u(0,t) = A(t), \quad u(L,t) = B(t), \tag{4.52}$$

$$u(x,0) = f(x). \tag{4.53}$$

This heat flow problem models the temperature distribution, $u(x,t)$, in a one-dimensional medium over time, accounting for both internal heat generation and non-homogeneous boundary conditions. The heat equation describes the evolution of temperature, where k is the thermal diffusivity and $Q(x,t)$ represents heat generation, or thermal energy that varies with position and time. The Dirichlet boundary conditions specify that the ends of the medium are maintained at time-dependent temperatures, $A(t)$ and $B(t)$, rather than fixed values. Additionally, the initial condition provides the temperature distribution at $t = 0$. Together, these conditions lead to a non-homogeneous problem that may require analytical or numerical methods to solve, depending on the complexity of $Q(x,t)$, $A(t)$, $B(t)$, and $f(x)$.

- **Method of Solution**

We begin with finding the simplest function, say, $p(x,t)$ that satisfies $p(0,t) = A(t)$ and $p(L,t) = B(t)$. The simplest form of $p(x,t)$ is a linear path in x between $(0,A(t))$ and $(L,B(t))$. Consequently,

$$p(x,t) = A(t) + \frac{x}{L}[B(t) - A(t)]. \tag{4.54}$$

We assume the functions $A(t), B(t)$ are continuous and uniformly bounded for all $t \ge 0$. Next, we seek to convert the non-homogeneous boundary conditions into homogeneous ones. To accomplish this, we introduce

$$v(x,t) = u(x,t) - p(x,t).$$

Then

$$u_t = v_t + p_t, \quad u_{xx} = v_{xx} + p_{xx} = v_{xx}.$$

Substituting into (4.51) we arrive at

$$v_t + p_t = kv_{xx} + Q(x,t).$$

Letting

$$\overline{Q}(x,t) = Q(x,t) - p_t$$

we arrive at

$$v_t = kv_{xx} + \overline{Q}(x,t).$$

As for the boundary conditions, one obtains

$$v(0,t) = u(0,t) - p(0,t) = A(t) - A(t) = 0,$$

$$v(L,t) = u(L,t) - p(L,t) = B(t) - B(t) = 0.$$

Finally, we transfer our initial condition. That is

$$v(0,t) = u(0,t) - p(0,t) = f(x) - p(x,0) := g(x). \tag{4.55}$$

Thus, our original problem has been reformulated in terms of v as a non-homogeneous heat flow problem with homogeneous boundary conditions. That is,

$$v_t = kv_{xx} + \overline{Q}(x,t), \quad 0 < x < L, \quad t > 0, \tag{4.56}$$

$$v(0,t) = 0, \quad v(L,t) = 0, \tag{4.57}$$

$$v(x,0) = g(x). \tag{4.58}$$

Let λ_n and φ_n be the eigenvalues and corresponding eigenfunctions of the homogeneous problem

$$v_t = kv_{xx}, \quad v(0,t) = 0, \quad v(L,t) = 0.$$

We express the solution $v(x,t)$ of (4.56)–(4.58) as a generalized Fourier series of the eigenfunctions φ_n with time dependent coefficients $a_n(t)$ and obtain,

$$v(x,t) = \sum_{n=1}^{\infty} a_n(t)\varphi_n(x). \tag{4.59}$$

We now represent $Q(x,t)$ as a generalized Fourier series, where the eigenfunctions form the spatial basis and the coefficients depend on time. The expansion is expressed as

$$\overline{Q}(x,t) = \sum_{n=1}^{\infty} \overline{q}_n(t)\varphi_n(x),$$

where the time-dependent Fourier coefficients $\overline{q}_n(t)$ are determined by the following integral (see Section 3.5)

$$\overline{q}_n(t) = \frac{\int_0^L \overline{Q}(x,t)\varphi_n(x)\,dx}{\int_0^L \varphi_n^2(x)\,dx}. \tag{4.60}$$

With this in mind, expression (4.56) takes the form

$$v_t = kv_{xx} + \sum_{n=1}^{\infty} \bar{q}_n(t)\varphi_n(x), \quad 0 < x < L, \quad t > 0. \tag{4.61}$$

Let $'$ denote differentiation with respect to the argument of the corresponding function. Assuming the generalized Fourier series in (4.59) can be differentiated term by term, substituting (4.59) into (4.61) leads to

$$\sum_{n=1}^{\infty} a_n'(t)\varphi_n(x) = k\sum_{n=1}^{\infty} a_n(t)\varphi_n''(x) + \sum_{n=1}^{\infty} \bar{q}_n(t)\varphi_n(x).$$

Since

$$\varphi_n''(x) = -\lambda_n \varphi_n(x),$$

the above expression reduces to

$$\sum_{n=1}^{\infty} \Big[a_n'(t) + k\lambda_n a_n(t) - \bar{q}_n(t)\Big]\varphi_n(x) = 0.$$

However, the above expression must hold for all $0 < x < L$, and so we must have

$$a_n'(t) + \lambda_n k a_n(t) - \bar{q}_n(t) = 0, \quad t > 0. \tag{4.62}$$

Making use of the variation of parameters formula, the solution $a_n(t)$ of (4.62) is found to be

$$a_n(t) = a_n(0)e^{-\int_0^t \lambda_n k\,du} + \int_0^t \bar{q}_n(s)e^{-\int_s^t \lambda_n k\,du}ds, \quad t \geq 0,$$

which reduces to

$$a_n(t) = a_n(0)e^{-\lambda_n kt} + \int_0^t \bar{q}_n(s)e^{-\lambda_n k(t-s)}ds, \quad t \geq 0. \tag{4.63}$$

We are done once we determine the initial value $a_n(0)$. Setting $x = 0$ in (4.59) and making use of (4.58) we arrive at

$$g(x) = \sum_{n=1}^{\infty} a_n(0)\varphi_n(x), \quad 0 < x < L.$$

Then by Section 3.5 we have

$$a_n(0) = \frac{\int_0^L g(x)\varphi_n(x)\,dx}{\int_0^L \varphi_n^2(x)\,dx}. \tag{4.64}$$

Putting everything together, we arrive at using (4.59) that

$$v(x,t) = \sum_{n=1}^{\infty} \Big[\frac{\int_0^L g(x)\varphi_n(x)\,dx}{\int_0^L \varphi_n^2(x)\,dx}e^{-\lambda_n kt} + \int_0^t \bar{q}_n(s)e^{-\lambda_n k(t-s)}ds\Big]\varphi_n(x), \tag{4.65}$$

where $\overline{q}_n(t)$ is given by (4.60). Finally, the solution to our original problem (4.51)–(4.53) is

$$u(x,t) = v(x,t) + p(x,t), \tag{4.66}$$

where $p(x,t)$ is given by (4.54).

Remark 9 *If the boundary conditions specified in* (4.52) *are homogeneous, i.e.,* $A(t) = B(t) = 0$, *then* $p(x,t) = 0$. *Consequently, the solution of* (4.51)–(4.53) *simplifies to*

$$u(x,t) = v(x,t),$$

as given in (4.66) *and* $v(x,t)$ *is given by* (4.65).

□

Example 4.11 Consider the following non-homogeneous problem with a source

$$\begin{aligned} &u_t = ku_{xx} + q(t), \quad 0 < x < L, \quad t > 0, \\ &u(0,t) = 0, \quad u(L,t) = 0, \quad u(x,0) = f(x), \end{aligned}$$

where the heat source depends only on time. In this case, the function $p(x,t) = 0$. Hence, there is no need to reformulate the problem, and we can proceed directly with the original formulation. Let λ_n and φ_n be the eigenvalues and corresponding eigenfunctions of the homogeneous problem

$$u_t = ku_{xx}, \quad u(0,t) = 0, \quad u(L,t) = 0.$$

Since this is the same problem that was studied in Section 4.7, the eigenvalues and eigenfunctions are given by

$$\lambda_n = \left(\frac{n\pi}{L}\right)^2, \quad \varphi_n(x) = \sin\frac{n\pi x}{L}, \quad n = 1, 2, \ldots.$$

We compute some of the terms

$$\int_0^L \varphi_n^2(x)\,dx = \int_0^L \sin^2\frac{n\pi x}{L}\,dx = \frac{1}{2}\int_0^L \Big(1 - \cos\frac{2n\pi x}{L}\Big)\,dx = \frac{L}{2}.$$

Using this result, (4.64) becomes

$$a_n(0) = \frac{2}{L}\int_0^L f(x)\sin\frac{n\pi x}{L}\,dx.$$

Finally, applying (4.65), we obtain the solution

$$u(x,t) = \sum_{n=1}^{\infty}\left[\frac{2}{L}\int_0^L f(x)\sin\frac{n\pi x}{L}\,dx\, e^{-\lambda_n kt} + \int_0^t \overline{q}_n(s)e^{-\lambda_n k(t-s)}\,ds\right]\varphi_n(x), \tag{4.67}$$

where for our problem $\overline{Q}(x,t) = q(t)$. Adjusting (4.60) to this specific case, we find

$$\overline{q}_n(t) = \frac{\int_0^L q(t)\varphi_n(x)\,dx}{\int_0^L \varphi_n^2(x)\,dx} = \frac{2q(t)}{L}\int_0^L \sin\frac{n\pi x}{L}\,dx = \frac{2q(t)}{L}\cdot\frac{L}{n\pi}\left[1-(-1)^n\right].$$

Thus

$$\overline{q}_n(t) = \frac{4q(t)}{\pi(2n-1)}, \quad n = 1, 2, \ldots$$

Using (4.67), the solution takes the form

$$u(x,t) = \frac{2}{L}\sum_{n=1}^{\infty}\left[\sin\frac{n\pi x}{L}\int_0^L f(x)\sin\frac{n\pi x}{L}\,dx\right]e^{-\left(\frac{n\pi}{L}\right)^2 kt}$$
$$+\sum_{n=1}^{\infty}\left[\sin\frac{(2n-1)\pi x}{L}\int_0^t \frac{4q(s)}{\pi(2n-1)}e^{-\left(\frac{(2n-1)\pi}{L}\right)^2 k(t-s)}\,ds\right]. \quad (4.68)$$

4.11 Method of Eigenfunction Expansion Using Green's Formula

We begin this section by stating and proving *Green's formula* that we utilize for eigenfunctions expansion.

Lemma 9 *(Green's Formula) Let the functions u and v be twice continuously differentiable. Then*

$$\int_a^b \left(u\frac{d^2v}{dx^2} - v\frac{d^2u}{dx^2}\right)dx = \left(u\frac{dv}{dx} - v\frac{du}{dx}\right)\Big|_a^b. \quad (4.69)$$

Proof *To simplify notations, we use the notion of operator. Define the operator L by*

$$L = \frac{d}{dx}[p(x)\frac{d}{dx}] + q(x),$$

where $p(x)$ and $q(x)$ are continuous functions. Then L acting on u and v separately gives,

$$L(u) = \frac{d}{dx}\left(p\frac{du}{dx}\right) + qu \quad \text{and} \quad L(v) = \frac{d}{dx}\left(p\frac{dv}{dx}\right) + qv.$$

Thus,

$$\begin{aligned} uL(v) - vL(u) &= u\frac{d}{dx}\left(p\frac{dv}{dx}\right) + uqv - v\frac{d}{dx}\left(p\frac{du}{dx}\right) - vqu \\ &= u\frac{d}{dx}\left(p\frac{dv}{dx}\right) - v\frac{d}{dx}\left(p\frac{du}{dx}\right). \end{aligned}$$

Simple manipulations leads to

$$u\frac{d}{dx}\left(p\frac{dv}{dx}\right) = \frac{d}{dx}\left(up\frac{dv}{dx}\right) - p\frac{dv}{dx}\frac{du}{dx},$$

and

$$v\frac{d}{dx}\left(p\frac{du}{dx}\right) = \frac{d}{dx}\left(vp\frac{du}{dx}\right) - p\frac{du}{dx}\frac{dv}{dx}.$$

Consequently,

$$uL(v) - vL(u) = \frac{d}{dx}\left[p\left(u\frac{dv}{dx} - v\frac{du}{dx}\right)\right], \tag{4.70}$$

known as differential form of Lagrange's identity. The right-hand side of (4.70) *shows* $uL(v) - vL(u)$ *is an* **exact differential**. *The* **integral form of Lagrange's identity** *is also known as* **Green's formula** *is obtained by integrating* (4.70)

$$\int_a^b [uL(v) - vL(u)]\,dx = p\left(u\frac{dv}{dx} - v\frac{du}{dx}\right)\Bigg|_a^b \tag{4.71}$$

for any functions u and v. This is a very useful formula. To prove (4.69) *we set* $p = 1$ *and* $q = 0$*. Then* $L = \frac{d^2}{dx^2}$ *and the integrand in* (4.71) *becomes*

$$uL(v) - vL(u) = u\frac{d^2v}{dx^2} - v\frac{d^2u}{dx^2}.$$

Hence both sides of (4.71) *simplify to*

$$\int_a^b \left(u\frac{d^2v}{dx^2} - v\frac{d^2u}{dx^2}\right)dx = \left(u\frac{dv}{dx} - v\frac{du}{dx}\right)\Bigg|_a^b.$$

This completes the proof. □

We consider the heat equation with a source and non-homogeneous time-dependent boundary conditions

$$\frac{\partial u}{\partial t} = k\frac{\partial^2 u}{\partial x^2} + Q(x,t), \quad 0 < x < L, \quad t > 0, \tag{4.72}$$

$$u(0,t) = A(t), \quad t > 0, \tag{4.73}$$

$$u(L,t) = B(t), \quad t > 0, \tag{4.74}$$

$$u(x,0) = f(x), \quad 0 \le x \le L. \tag{4.75}$$

In the previous section, we addressed the same problem by introducing a transition function $p(x,t)$ to transform the boundary conditions into homogeneous ones. This allowed us to reformulate the heat flow problem in terms of a new variable $v(x,t)$, defined as $v(x,t) = u(x,t) - p(x,t)$, which satisfied homogeneous boundary conditions. In the final step, the solution for $u(x,t)$ was obtained by reconstructing it as $u(x,t) = v(x,t) + p(x,t)$. Next, we demonstrate how the method of eigenfunction expansion can be applied directly to solve the problem (4.72)–(4.75). It is crucial to note that term-by-term differentiation of the Fourier series is not generally valid when dealing with non-homogeneous boundary conditions, and additional care must be taken to ensure the solution's consistency and validity. Consider the homogeneous problem with boundary conditions

$$\frac{d^2\Phi}{dx^2} + \lambda_n\Phi = 0, \quad 0 < x < L, \quad \Phi(0) = 0, \quad \Phi(L) = 0, \tag{4.76}$$

whose corresponding eigenvalues and eigenfunctions are known to be

$$\lambda_n = \left(\frac{n\pi}{L}\right)^2, \quad \Phi_n(x) = \sin\frac{n\pi x}{L}, \quad n = 1,2,\ldots$$

As any piecewise smooth function can be represented as a series of these eigenfunctions, we set

$$u(x,t) = \sum_{n=1}^{\infty} b_n(t)\Phi_n(x), \tag{4.77}$$

and its generalized Fourier coefficients $b_n(t)$ are given by

$$b_n(t) = \frac{\int_0^L u(x,t)\Phi_n(x)\,dx}{\int_0^L \Phi_n^2(x)\,dx}. \tag{4.78}$$

Since $u(x,t)$ is not continuous at $x = 0$ and $x = L$, term-by-term differentiation in (4.77) with respect to x is not valid. However, term-by-term differentiation with respect to t is feasible and can be applied. Consequently,

$$\frac{\partial u}{\partial t} = \sum_{n=1}^{\infty} \frac{db_n}{dt}\Phi_n(x). \tag{4.79}$$

Substituting (4.79) into (4.72) results in

$$\sum_{n=1}^{\infty} \frac{db_n}{dt}\Phi_n(x) = k\frac{\partial^2 u}{\partial x^2} + Q(x,t).$$

Due to the orthogonality of the eigenfunctions, it follows that

$$\frac{db_n}{dt} = \frac{\int_0^L \left[k\frac{\partial^2 u}{\partial x^2} + Q(x,t)\right]\Phi_n(x)\,dx}{\int_0^L \Phi_n^2(x)\,dx}. \tag{4.80}$$

As done previously, we represent the source term $Q(x,t)$ in terms of its generalized Fourier series

$$Q(x,t) = \sum_{n=1}^{\infty} q_n(t)\Phi_n(x).$$

Again, due to the orthogonality of $\Phi_n(x)$, we obtain

$$q_n(t) = \frac{\int_0^L Q(x,t)\Phi_n(x)\,dx}{\int_0^L \Phi_n^2(x)\,dx}. \tag{4.81}$$

Substituting (4.81) into (4.80), yields

$$\frac{db_n}{dt} = q_n(t) + k\frac{\int_0^L \frac{\partial^2 u}{\partial x^2}\Phi_n(x)\,dx}{\int_0^L \Phi_n^2(x)\,dx}. \tag{4.82}$$

To express the last term on the right-hand side of (4.82) in terms of $b_n(t)$, we use the result of Lemma 9,

$$\int_0^L \left(u\frac{\partial^2 v}{\partial x^2} - v\frac{\partial^2 u}{\partial x^2}\right) dx = \left[u\frac{\partial v}{\partial x} - v\frac{\partial u}{\partial x}\right]_0^L. \tag{4.83}$$

Let $v = \phi_n(x)$. Since $\dfrac{d^2\Phi_n}{dx^2} = -\lambda_n\Phi$, we have

$$\int_0^L u\frac{\partial^2 v}{\partial x^2} = \int_0^L u\frac{d^2\Phi_n}{dx^2}\,dx = -\lambda_n\int_0^L u\Phi_n\,dx. \tag{4.84}$$

Now we work with the rightside of (4.83) and obtain

$$\begin{aligned} u\frac{d\Phi_n}{dx}\bigg|_0^L &= u\frac{n\pi}{L}\cos\frac{n\pi}{L}\bigg|_0^L = u(L,t)n\pi\cos n\pi - u(0,t)n\pi \\ &= \frac{n\pi}{L}\left[B(t)(-1)^n - A(t)\right] \end{aligned} \tag{4.85}$$

and

$$\Phi_n\frac{\partial u}{\partial x}\bigg|_0^L = \Phi_n(L)\frac{\partial u}{\partial x}(L,t) - \Phi_n(0)\frac{\partial u}{\partial x}(0,t) = 0.$$

Then, from (4.83) we have

$$\begin{aligned} \int_0^L \Phi_n(x)\frac{\partial^2 u}{\partial x^2}\,dx &= \int_0^L u\frac{d^2\Phi_n}{dx^2}\,dx - \left[u\frac{d\Phi_n}{dx} - \Phi_n\frac{\partial u}{\partial x}\right]_0^L \\ &= -\lambda_n\int_0^L u\Phi_n\,dx - u\frac{d\Phi_n}{dx}\bigg|_0^L \\ &= -\lambda_n\int_0^L u\Phi_n\,dx - \frac{n\pi}{L}\left[B(t)(-1)^n - A(t)\right]. \end{aligned} \tag{4.86}$$

After substituting (4.86) and (4.78) into (4.82), we obtain a non-homogeneous first-order differential equation for the generalized Fourier coefficients b_n. Specifically, we arrive at

$$\frac{db_n}{dt} + k\lambda_n b_n = q_n(t) + k\frac{n\pi}{L}\frac{A(t) - (-1)^n B(t)}{\int_0^L \Phi_n^2(x)\,dx}. \tag{4.87}$$

To explicitly determine the solution, we require $b_n(0)$. By applying the initial condition (4.75) together with (4.77), we obtain the following relation

$$f(x) = \sum_{n=1}^{\infty} b_n(0)\Phi_n(x),$$

from which we conclude that

$$b_n(0) = \frac{\int_0^L f(x)\Phi_n(x)\,dx}{\int_0^L \Phi_n^2(x)\,dx}. \tag{4.88}$$

Thus, the solution of (4.72)–(4.75) is given by (4.76), where the coefficients b_n are determined by solving (4.87) with the initial condition (4.88). □

4.11.1 Exercises

Exercise 4.28 *Find the steady-state solution, or equilibrium solution to solve the non-homogeneous heat flow with non-homogeneous boundary conditions*

$$u_t - u_{xx} = x, \quad 0 < x < \pi, \quad t > 0,$$
$$u(x,0) = f(x),$$
$$u(0,t) = A, \quad u(\pi,t) = B.$$

Exercise 4.29 *Solve*

$$u_t - u_{xx} = x, \quad 0 < x < \pi, \quad t > 0,$$
$$u(x,0) = e^x,$$
$$u(0,t) = 1, \quad u(\pi,t) = 5,$$

and then show

$$\lim_{t\to\infty} u(x,t) = u^E(x).$$

Exercise 4.30 *For positive constants a and u_0 solve the non-homogeneous problem*

$$u_t - u_{xx} = e^{-at}\sin x, \quad 0 < x < \pi, \quad t > 0,$$
$$u(x,0) = u_0 \quad 0 \le x \le \pi,$$
$$u(0,t) = 0, \quad u(\pi,t) = 1, \quad t > 0,$$

and then show

$$\lim_{t\to\infty} u(x,t) = u^E(x).$$

Exercise 4.31 *For positive constant q_0 solve the non-homogeneous problem*

$$u_t = ku_{xx} + q_0, \quad 0 < x < \pi, \quad t > 0,$$
$$u(0,t) = 0, \quad u(\pi,t) = 0, \quad u(x,0) = 0.$$

Ans:

$$u(x,t) = \frac{4q_0}{\pi k}\sum_{n=1}^{\infty}\frac{1-\exp\left[-(2n-1)^2kt\right]}{(2n-1)^3}\sin\left[(2n-1)x\right].$$

Exercise 4.32 *For positive constant q_0 and negatives constants a and b, solve the non-homogeneous problem*

$$u_t = ku_{xx} + q_0, \quad 0 < x < \pi, \quad t > 0,$$
$$u(0,t) = e^{at}, \quad u(\pi,t) = e^{bt}, \quad u(x,0) = 1.$$

Exercise 4.33 *Solve*

$$u_t = 2tu_{xx} + e^{-t^2}\sin x, \quad 0 < x < \pi, \quad t > 0,$$
$$u(0,t) = u(\pi,t) = 0, \quad t > 0,$$
$$u(x,0) = 3\sin 2x.$$

Ans: $u(x,t) = te^{-t^2}\sin x + 3e^{-t^2}\sin 2x.$

Exercise 4.34 *Solve the inhomogeneous problem*

$$u_t = u_{xx} + 1, \quad 0 < x < 1, \quad t > 0,$$
$$u(0,t) = 0, \quad u(1,t) = 0, \quad t > 0,$$
$$u(x,0) = f(x).$$

Ans: $u(x,t) = \frac{1}{2}x(1-x) + \sum_{n=1}^{\infty} b_n e^{-n^2\pi^2 t} \sin n\pi x, \quad b_n = 2\int_0^1 \left[f(x) - \frac{1}{2}x(1-x)\right] \sin n\pi x \, dx.$

Exercise 4.35 *(Neumann type boundary condition) Develop a theory for*

$$u_t = ku_{xx} + Q(x,t), \quad 0 < x < L, \quad t > 0,$$
$$u_x(0,t) = A(t), \quad u(L,t) = B(t),$$
$$u(x,0) = f(x).$$

Hint: Let $p(x,t) = c_1x^2 + c_2x$. *Use the appropriate boundary conditions that* $p(x,t)$ *must satisfy to determine* c_1 *and* c_2.

Exercise 4.36 *With the aid of Exercise 4.35 solve*

$$u_t = ku_{xx} + ax^2, \quad 0 < x < c, \quad t > 0,$$
$$u_x(0,t) = 0, \quad u(c,t) = 0,$$
$$u(x,0) = 0.$$

Ans: $u(x,t) = \frac{ac^2 t}{3} + \frac{4ac^4}{\pi^4 k} \sum_{n=1}^{\infty} \frac{(-1)^n}{n^4} \left[1 - e^{-\frac{n^2\pi^2 k}{c^2}t}\right] \cos \frac{n\pi x}{c}$

Exercise 4.37 *Use the method of eigenfunction expansion using Green's formula to solve*

$$u_t = ku_{xx} + x\cos t, \quad 0 < x < \pi, \quad t > 0,$$
$$u(0,t) = t^2, \quad u(\pi,t) = 2t,$$
$$u(x,0) = \sin x.$$

Exercise 4.38 *Let* $L = \frac{d^4}{dx^4}$ *be a fourth-order differential operator.*

(a) Show that $uL(v) - vL(u)$ *is an exact differential.*

(b) $\int_0^1 [uL(v) - vL(u)]\,dx = 0$, *if* u *and* v *are any two functions satisfying the boundary conditions*

$$u(0) = u(1) = \frac{du}{dx}(0) = \frac{d^2u}{dx^2}(1) = 0,$$
$$v(0) = v(1) = \frac{dv}{dx}(0) = \frac{d^2v}{dx^2}(1) = 0.$$

4.12 Laplace Equation on Rectangles and Unbounded Domains

In this section, we are interested in the Laplacian equation in 2-D on the rectangle

$$\mathscr{R} = \{(x,y) : 0 \leq x \leq b,\ 0 \leq y \leq b\}.$$

Thus, we consider (see Fig. 4.1)

$$u_{xx}(x,y) + u_{yy}(x,y) = 0, \quad 0 < x < a, \quad 0 < y < b, \tag{4.89}$$
$$u(0,y) = 0, \quad u(a,y) = 0, \tag{4.90}$$
$$u(x,0) = f_1(x), \quad u(x,b) = f_2(x). \tag{4.91}$$

Equation (4.89) describes a situation where the temperature $u(x,y)$ in the rectangle does not change with time (steady-state). The heat flows within the rectangle are governed by the principle that there is no net accumulation of heat at any point inside (as indicated by the Laplace equation). The temperature inside the rectangle is entirely determined by the boundary values $f_1(x), f_2(x)$ and the fixed temperatures on the vertical edges. In electrostatics, $u(x,y)$ can represent the electric potential in a region where there are no free charges ($\nabla^2 u = 0$). The boundary conditions specify the potential on the edges of the rectangle. In fluid dynamics, $u(x,y)$ can represent the stream function in a two-dimensional incompressible flow. The Laplacian equation ensures that the flow is steady and irrotational. In each context, the solution $u(x,y)$ represents the equilibrium distribution of the physical quantity (temperature, potential, or stream function) determined entirely by the boundary values. The boundary conditions given by (4.90) are called Dirichlet boundary conditions.

• Method of Solution

We aim to find a nontrivial solution by assuming that the function $u(x,y)$ can be expressed as the product of two separate functions, one depending on x and the other on y, such that

$$u(x,y) = X(x)Y(y).$$

Substituting into the PDE given by (4.89), we obtain that

$$X''(x)Y(y) + X(x)Y''(y) = 0.$$

Since $X(x) \neq 0$ and $Y(y) \neq 0$, we may divide by the term $X(x)Y(y)$ to separate the variables. That is,

$$\frac{X''(x)}{X(x)} = -\frac{Y''(y)}{Y(y)}.$$

Since the left-hand side is a function of x alone, it does not vary with y. However, it is equal to a function of y alone, and so it can not vary with x. Hence the two sides must have some constant value $-\lambda$ in common. That is,

$$\frac{X''(x)}{X(x)} = -\frac{Y''(y)}{Y(y)} = -\lambda.$$

This gives the Sturm–Liouville problem in X

$$X''(x) + \lambda X(x) = 0, \quad X(0) = 0, \quad X(a) = 0, \tag{4.92}$$

and

$$Y''(y) - \lambda Y(y) = 0. \tag{4.93}$$

Using arguments similar to those used earlier, we conclude that for values of $\lambda \leq 0$, the only solution to (4.93) is the **trivial solution**. For $\lambda > 0$, we set $\lambda = \alpha^2, \alpha > 0$. Then the corresponding eigenvalues and eigenfunctions of (4.92) are

$$\lambda_n = \left(\frac{n\pi}{a}\right)^2, \quad X_n(x) = \sin\frac{n\pi}{a}x, \quad n = 1, 2\ldots$$

Next we solve (4.92) for $\lambda = \alpha^2$ whose solution is

$$Y_n(y) = C_1 e^{\alpha y} + C_2 e^{-\alpha y}.$$

We aim to express this solution in terms of hyperbolic sine and cosine functions. Thus, for nonzero constants a_1 and b_1 we may set

$$C_1 = \frac{a_1 + b_1}{2} \text{ and } C_2 = \frac{a_1 - b_1}{2}.$$

Then

$$\begin{aligned} Y(y) &= \frac{a_1 + b_1}{2} e^{\alpha y} + \frac{a_1 - b_1}{2} e^{-\alpha y} = a_1 \frac{e^{\alpha y} + e^{-\alpha y}}{2} + b_1 \frac{e^{\alpha y} - e^{-\alpha y}}{2} \\ &= a_1 \cosh(\alpha y) + b_1 \sinh(\alpha y). \end{aligned}$$

Setting $\lambda_n = \left(\frac{n\pi}{a}\right)^2 = \alpha^2$, into the above equation, we obtain

$$Y_n(y) = a_n \cosh\left(\frac{n\pi y}{a}\right) + b_n \sinh\left(\frac{n\pi y}{a}\right), \quad n = 1, 2, \ldots$$

where a_n and b_n are to be determined using (4.91). Thus, the general solution is

$$u(x,y) = \sum_{n=1}^{\infty} \left[a_n \cosh\left(\frac{n\pi y}{a}\right) + b_n \sinh\left(\frac{n\pi y}{a}\right)\right] \sin\left(\frac{n\pi x}{a}\right). \tag{4.94}$$

Using the boundary condition $u(x,0) = f_1(x)$ in (4.94) gives

$$f_1(x) = \sum_{n=1}^{\infty} a_n \sin\frac{n\pi x}{a}.$$

This implies that the coefficients a_n are the Fourier sine series coefficients of $f_1(x)$, and therefore

$$a_n = \frac{2}{a}\int_0^a f_1(x)\sin\frac{n\pi x}{a}\,dx, \quad n = 1,2,\ldots. \tag{4.95}$$

The second boundary condition $u(x,b) = f_2(x)$ leads to

$$f_2(x) = \sum_{n=1}^{\infty}\left[a_n\cosh\left(\frac{n\pi b}{a}\right) + b_n\sinh\left(\frac{n\pi b}{a}\right)\right]\sin\frac{n\pi x}{a}.$$

This implies that

$$B_n = a_n\cosh\left(\frac{n\pi b}{a}\right) + b_n\sinh\left(\frac{n\pi b}{a}\right),$$

where B_n must be the Fourier sine coefficients of $f_2(x)$. Thus,

$$B_n = \frac{2}{a}\int_0^a f_2(x)\sin\left(\frac{n\pi x}{a}\right)dx, \quad n = 1,2,\ldots. \tag{4.96}$$

Solving for b_n in (4.96) yields

$$b_n = \frac{B_n - a_n\cosh\left(\frac{n\pi b}{a}\right)}{\sinh\left(\frac{n\pi b}{a}\right)}, \quad n = 1,2,\ldots. \tag{4.97}$$

Finally, the general solution of (4.89)–(4.91) is obtained by substituting a_n and b_n into (4.94), where a_n, B_n, and b_n are given by the equations (4.95), (4.96), and (4.97), respectively. □

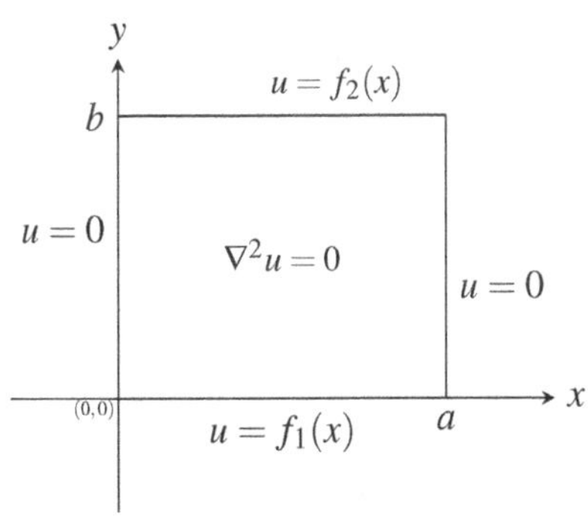

FIGURE 4.1: Dirichlet problem in rectangular coordinates.

• Half-unbounded Rectangle

We consider

$$u_{xx}(x,y) + u_{yy}(x,y) = 0, \quad 0 < x < a, \quad 0 < y < \infty, \tag{4.98}$$
$$u(0,y) = 0, \quad u(a,y) = 0, \tag{4.99}$$

$$u(x,0) = f(x), \tag{4.100}$$

$$u(x,y) \quad \text{remains bounded as} \quad y \to \infty. \tag{4.101}$$

This problem describes the steady-state heat distribution $u(x,y)$ in a semi-infinite strip $0 < x < a$, $0 < y < \infty$. Equation (4.98) is the Laplace equation, ensuring the absence of any internal heat sources or sinks. The boundary conditions (4.99) impose that the temperature is zero at the vertical edges $x = 0$ and $x = a$. The condition (4.100) specifies the temperature distribution $f(x)$ along the bottom edge $y = 0$. Finally, (4.101) ensures that the solution remains bounded as $y \to \infty$, modeling physical behavior where the temperature stabilizes far from the boundary.

- **Method of Solution**

Assuming a nonzero solution of the form $u(x,y) = X(x)Y(y)$, and performing separation of variables as before, we arrive at the following equations

$$X''(x) + \lambda X(x) = 0, \quad X(0) = 0, \quad X(a) = 0, \tag{4.102}$$

and

$$Y''(y) - \lambda Y(y) = 0. \tag{4.103}$$

Using arguments similar to those used earlier, we conclude that for values of $\lambda \leq 0$, the only solution to (4.102) is the **trivial solution**. For $\lambda > 0$, we set $\lambda = \alpha^2, \alpha > 0$. The corresponding eigenvalues and eigenfunctions of (4.102) are

$$\lambda_n = \left(\frac{n\pi}{a}\right)^2, \quad X_n(x) = \sin\frac{n\pi x}{a}, \quad n = 1, 2 \ldots$$

Next, we solve (4.103) for $\lambda = \alpha^2$, whose general solution is given by

$$Y(y) = C_1 e^{\alpha y} + C_2 e^{-\alpha y}.$$

Since the independent variable y is unbounded, we must require $C_1 = 0$, otherwise the term

$$|C_1| e^{\alpha y} \to \infty, \quad \text{as } y \to \infty.$$

Thus, we are left with $Y(y) = e^{-\alpha y}$, where we set $C_2 = 1$. Setting $\alpha = \frac{n\pi}{a}$, we obtain the solution

$$Y_n(y) = e^{\frac{n\pi y}{a}}.$$

Finally,

$$u(x,y) = \sum_{n=1}^{\infty} b_n e^{\frac{n\pi y}{a}} \sin\frac{n\pi x}{a}. \tag{4.104}$$

Applying the initial condition (4.100), we obtain the relation

$$f(x) = \sum_{n=1}^{\infty} b_n \sin\frac{n\pi x}{a},$$

where b_n is the Fourier sine series of $f(x)$ given by

$$b_n = \frac{2}{a}\int_0^a f(x)\sin\frac{n\pi x}{a}\,dx, \quad n = 1, 2, \ldots$$

4.12.1 Exercises

Exercise 4.39 *Solve*

$$\begin{aligned} &u_{xx}(x,y)+u_{yy}(x,y)=0, \quad 0<x<a, \quad 0<y<b,\\ &u(0,y)=0, \quad u(a,y)=0,\\ &u(x,0)=0, \quad u(x,b)=f(x). \end{aligned}$$

Ans: $u(x,y)=\sum_{n=1}^{\infty} b_n \sinh(\frac{n\pi y}{a}) \sin\frac{n\pi x}{a}, \ b_n=\frac{2}{a\sinh(\frac{n\pi b}{a})}\int_0^a f(x)\sin\frac{n\pi x}{a}dx, \ n=1,2,\ldots.$

Exercise 4.40 *Solve*

$$\begin{aligned} &u_{xx}(x,y)+u_{yy}(x,y)=0, \quad 0<x<a, \quad 0<y<b,\\ &u(0,y)=0, \quad u(a,y)=0,\\ &u(x,0)=f(x), \quad u(x,b)=0. \end{aligned}$$

Ans: $u(x,y)=\sum_{n=1}^{\infty} b_n \sinh(\frac{n\pi y}{a}) \sin\frac{n\pi x}{a}, \ b_n=\frac{2}{a\sinh(\frac{n\pi b}{a})}\int_0^a f(x)\sin\frac{n\pi x}{a}dx, \ n=1,2,\ldots.$

Exercise 4.41 *Solve*

$$\begin{aligned} &u_{xx}(x,y)+u_{yy}(x,y)=0, \quad 0<x<1, \quad 0<y<\pi,\\ &u(0,y)=0, \quad u(1,y)=0,\\ &u(x,0)=0, \quad u(x,\pi)=1. \end{aligned}$$

Ans: $u(x,y)=\sum_{n=1}^{\infty}\frac{2}{n\pi}[1+(-1)^{n+1}]\frac{\sinh n\pi y}{\sinh n\pi^2}\sin n\pi x.$

Exercise 4.42 *Solve*

$$\begin{aligned} &u_{xx}(x,y)+u_{yy}(x,y)=0, \quad 0<x<1, \quad 0<y<\pi,\\ &u(x,0)=x, \quad u(x,\pi)=0,\\ &u_x(0,y)=0, \quad u_x(\pi,y)=0. \end{aligned}$$

Exercise 4.43 *Solve*

$$\begin{aligned} &u_{xx}(x,y)+u_{yy}(x,y)=0, \quad 0<x<a, \quad 0<y<b,\\ &u(x,0)=0, \quad u(x,b)=0,\\ &u(0,y)=f_1(y), \quad u(a,y)=f_2(y). \end{aligned}$$

Exercise 4.44 *We consider the Poisson problem (heat source)*

$$u_{xx}(x,y)+u_{yy}(x,y)+q(x)=0, \quad 0<x<\pi, \quad 0<y<\infty,$$

$$u(0,y) = 0, \quad u(\pi,y) = 0,$$
$$u(x,0) = 0,$$
$$u(x,y) \text{ remains bounded as } y \to \infty.$$

(a) Assume a bounded solution of the form

$$u(x,y) = \sum_{n=1}^{\infty} B_n(y) \sin nx,$$

show that

$$B_n(y) = \frac{q_n}{n^2}(1 - e^{-ny}), \quad n = 1,2,\ldots$$

where the q_n *are the Fourier sine series for* $q(x)$, $0 < x < \pi$.

Hint: Write $q(x) = \sum_{n=1}^{\infty} q_n \sin nx$.

(b) Show that when $q(x) = q_0$, *where* q_0 *is constant, then the solution from part (a) becomes*

$$u(x,t) = \frac{4q_0}{\pi} \sum_{n=1}^{\infty} \frac{1 - e^{-(2n-1)y}}{(2n-1)^3} \sin(2n-1)x.$$

Exercise 4.45 *We consider the Laplace equation with two sides are insulated and one side is maintained at temperature zero. Solve*

$$u_{xx}(x,y) + u_{yy}(x,y) = 0, \quad 0 < x < a, \quad 0 < y < b,$$
$$u_x(0,y) = 0, \quad u_x(a,y) = 0,$$
$$u(x,b) = 0, \quad u(x,0) = f(x)$$

Ans: $u(x,y) = \frac{1}{2}(\pi - y) + \sum_{n=1}^{\infty} \frac{2}{\pi n^2}[(-1)^n - 1] \frac{\sinh n(\pi - y)}{\sinh n\pi} \cos nx.$

Exercise 4.46 *Consider the Laplacian problem described by* (4.89)–(4.91) *and assume the functions* f_1 *and* f_2 *are bounded. Show that*

$$u(x,y), \quad u_{xx}(x,y), \quad u_{yy}(x,y)$$

all converge uniformly on the rectangle.

Exercise 4.47 *Obtain the solution of the following Dirichlet problem on a rectangle.*

The general Dirichlet problem with boundary conditions:

$$u_{xx}(x,y) + u_{yy}(x,y) = 0, \quad 0 < x < a, \quad 0 < y < b,$$
$$u(x,0) = f_1(x), \quad u(x,a) = f_2(x), \quad 0 \le x \le a,$$
$$u(0,y) = f_3(y), \quad u(b,y) = f_4(y), \quad 0 \le y \le b,$$

can be solved by decomposing it into four separate problems, each with one non-homogeneous boundary condition while the others are set to zero. By solving each subproblem individually, as done in the preceding example, and then summing the four solutions, the final solution to the Dirichlet problem for a rectangle is obtained.

Exercise 4.48 *Let $u(x,y)$ be a function that is harmonic everywhere on the entire plane, meaning that it satisfies the Laplace equation*

$$\Delta u = u_{xx} + u_{yy} = 0 \quad \textit{for all } (x,y) \in \mathbb{R}^2.$$

Show that if $u(x,y)$ is bounded either above or below, then $u(x,y)$ must be a constant function.

Exercise 4.49 *Let*

$$D = \{x \in \mathbb{R}^2 : |x| < 1\}.$$

Suppose u is harmonic in D, continuous on $\overline{D}$, and satisfies

$$u(x) = \sin^2\theta \quad \textit{on } |x| = 1,$$

where θ is the polar angle. Show that

$$0 < u(x) < 1 \quad \textit{for all } x \in D.$$

4.13 Laplacian in Polar, Cylindrical and Spherical Coordinates

In this section we will discuss the Laplacian $\nabla^2 u$, in *polar, cylindrical and spherical coordinates.* When solving boundary value problems in more than one dimension, it is often necessary to use other coordinate systems than the cartesian. It is then important to be able to express the Laplacian operator in these coordinate systems.

In two dimension the Laplacian can be written

$$\nabla^2 u = \frac{\partial^2 u(x,y)}{\partial x^2} + \frac{\partial^2 u(x,y)}{\partial y^2},$$

while in three dimensions we write

$$\nabla^2 u = \frac{\partial^2 u(x,y,z)}{\partial x^2} + \frac{\partial^2 u(x,y,z)}{\partial y^2} + \frac{\partial^2 u(x,y,z)}{\partial z^2}.$$

Either equation may be written

$$\nabla^2 u = 0$$

and we will have to learn how to express the **Laplacian**

$$\nabla^2 = \frac{\partial^2}{\partial x^2} + \frac{\partial^2}{\partial y^2} + \frac{\partial^2}{\partial z^2},$$

in different ways such as in polar, cylindrical or spherical coordinates.
We begin by considering the Laplacian in polar coordinates. Thus, we make use of the transformation

$$x = r\cos\theta, \quad y = r\sin\theta.$$

Laplace's equation in this coordinate system can be shown to be,

$$\nabla^2 u(r,\theta) = \frac{\partial^2 u(r,\theta)}{\partial r^2} + \frac{1}{r}\frac{\partial u(r,\theta)}{\partial r} + \frac{1}{r^2}\frac{\partial^2 u(r,\theta)}{\partial \theta^2}. \tag{4.105}$$

In *cylindrical coordinates,* we set

$$x = \rho\cos\phi, \quad y = \rho\sin\phi, \quad z = z,$$

and it can also be shown that Laplace's equation in cylindrical coordinates takes the form

$$\nabla^2 u(\rho,\phi,z) = \frac{\partial^2 u(\rho,\phi,z)}{\partial \rho^2} + \frac{1}{\rho}\frac{\partial u(\rho,\phi,z)}{\partial \rho} + \frac{1}{\rho^2}\frac{\partial^2 u(\rho,\phi,z)}{\partial \phi^2} + \frac{\partial^2 u(\rho,\phi,z)}{\partial z^2}. \tag{4.106}$$

Note that expression (4.105) is a special case of (4.106) by simply holding z constant.

In **spherical coordinates**

$$x = \rho\sin\theta\cos\phi, \quad y = \rho\sin\theta\sin\phi, \quad z = \rho\cos\theta$$

we have

$$\begin{aligned}\nabla^2 u(\rho,\theta,z) &= \frac{1}{\rho^2}\left[\frac{\partial}{\partial\rho}(\rho^2\frac{\partial u(\rho,\theta,\phi)}{\partial\rho})\right.\\ &+ \left.\frac{1}{\sin\theta}\frac{\partial}{\partial\theta}(\sin\theta\frac{\partial u(\rho,\theta,\phi)}{\partial\theta}) + \frac{1}{\sin^2\theta}\frac{\partial^2 u(\rho,\theta,\phi)}{\partial\phi^2}\right]\end{aligned} \tag{4.107}$$

• **Derivation of** (4.106) $\left(\nabla^2 u(\rho,\phi,z)\right.$ **in Cylindrical Coordinates**$\left.\right)$

Next, we give a brief derivation of (4.106). We already know that

$$\rho = \sqrt{x^2+y^2} \quad \text{and} \quad \phi = \arctan\left(\frac{y}{x}\right).$$

Using the chain rule we have

$$\frac{\partial u}{\partial x} = \frac{\partial u}{\partial \rho}\frac{\partial \rho}{\partial x} + \frac{\partial u}{\partial \phi}\frac{\partial \phi}{\partial x} + \frac{\partial u}{\partial z}\frac{\partial z}{\partial x}.$$

However,

$$\frac{\partial \rho}{\partial x} = \frac{x}{\sqrt{x^2+y^2}} = \frac{x}{\rho} = \frac{\rho\cos\phi}{\rho} = \cos\phi.$$

In similar fashions, we arrive at

$$\frac{\partial \rho}{\partial y} = \sin\phi, \quad \frac{\partial \phi}{\partial x} = -\frac{\sin\phi}{\rho}, \quad \text{and} \quad \frac{\partial \phi}{\partial y} = \frac{\cos\phi}{\rho}. \tag{4.108}$$

Substituting into $\frac{\partial u}{\partial x}$ we have

$$\frac{\partial u}{\partial x} = \cos\phi \frac{\partial u}{\partial \rho} - \frac{\sin\phi}{\rho}\frac{\partial u}{\partial \phi}, \tag{4.109}$$

since $\frac{\partial z}{\partial x} = 0$. To obtain $\frac{\partial^2 u}{\partial x^2}$, we replace the function u in (4.109) by $\frac{\partial u}{\partial x}$. That is,

$$\begin{aligned}
\frac{\partial^2 u}{\partial x^2} &= \cos\phi \frac{\partial}{\partial \rho}\left(\frac{\partial u}{\partial x}\right) - \frac{\sin\phi}{\rho}\frac{\partial}{\partial \phi}\left(\frac{\partial u}{\partial x}\right) \\
&= \cos\phi \frac{\partial}{\partial \rho}\left(\cos\phi \frac{\partial u}{\partial \rho} - \frac{\sin\phi}{\rho}\frac{\partial u}{\partial \phi}\right) - \frac{\sin\phi}{\rho}\frac{\partial}{\partial \phi}\left(\cos\phi \frac{\partial u}{\partial \rho} - \frac{\sin\phi}{\rho}\frac{\partial u}{\partial \phi}\right) \\
&= \cos\phi\left(\cos\phi \frac{\partial^2 u}{\partial \rho^2} + \frac{\sin\phi}{\rho^2}\frac{\partial u}{\partial \phi} - \frac{\sin\phi}{\rho}\frac{\partial^2 u}{\partial \rho \partial \phi}\right) \\
&- \frac{\sin\phi}{\rho}\left(-\sin\phi \frac{\partial u}{\partial \rho} + \cos\phi \frac{\partial^2 u}{\partial \phi \partial \rho} - \frac{\cos\phi}{\rho}\frac{\partial u}{\partial \phi} - \frac{\sin\phi}{\rho}\frac{\partial^2 u}{\partial \phi^2}\right).
\end{aligned}$$

Using the fact that

$$\frac{\partial^2 u}{\partial \rho \partial \phi} = \frac{\partial^2 u}{\partial \phi \partial \rho},$$

the above expression simplifies to

$$\begin{aligned}
\frac{\partial^2 u}{\partial x^2} &= \cos^2\phi \frac{\partial^2 u}{\partial \rho^2} - \frac{2\sin\phi\cos\phi}{\rho}\frac{\partial^2 u}{\partial \phi \partial \rho} + \frac{\sin^2\phi}{\rho^2}\frac{\partial^2 u}{\partial \phi^2} \\
&+ \frac{\sin^2\phi}{\rho}\frac{\partial u}{\partial \rho} + \frac{2\sin\phi\cos\phi}{\rho^2}\frac{\partial u}{\partial \phi}.
\end{aligned} \tag{4.110}$$

By similar steps, we see that

$$\begin{aligned}
\frac{\partial u}{\partial y} &= \frac{\partial u}{\partial \rho}\frac{\partial \rho}{\partial y} + \frac{\partial u}{\partial \phi}\frac{\partial \phi}{\partial y} + \frac{\partial u}{\partial z}\frac{\partial z}{\partial x} \\
&= \sin\phi \frac{\partial u}{\partial \rho} + \frac{\cos\phi}{\rho}\frac{\partial u}{\partial \phi}.
\end{aligned}$$

Moreover,

$$\frac{\partial^2 u}{\partial y^2} = \sin^2\phi \frac{\partial^2 u}{\partial \rho^2} + \frac{2\sin\phi\cos\phi}{\rho}\frac{\partial^2 u}{\partial \phi \partial \rho} + \frac{\cos^2\phi}{\rho^2}\frac{\partial^2 u}{\partial \phi^2}$$

$$+\quad \frac{\cos^2\phi}{\rho}\frac{\partial u}{\partial \rho} - 2\frac{2\sin\phi\cos\phi}{\rho^2}\frac{\partial u}{\partial \phi}. \tag{4.111}$$

A substitution of (4.110) and (4.111) into

$$\nabla^2 u = \frac{\partial^2 u(x,y,z)}{\partial x^2} + \frac{\partial^2 u(x,y,z)}{\partial y^2} + \frac{\partial^2 u(x,y,z)}{\partial z^2},$$

leads to (4.106).

4.13.1 Exercises

Exercise 4.50 *Verify all the terms in expression* (4.108).

Exercise 4.51 *Provide all the detail in obtaining expression* (4.111).

Exercise 4.52 *Show that* (4.106) *can be manipulated to obtain,*

(a)

$$\nabla^2 u = \frac{1}{\rho}(\rho u_\rho)_\rho + \frac{1}{\rho^2}u_{\phi\phi} + u_{zz}.$$

(b) When u is independent of z, so that $u = u(\rho,\phi)$, the expression reduces to the two-dimensional Laplacian in polar coordinates

$$\nabla^2 u = u_{\rho\rho} + \frac{1}{\rho}u_\rho + \frac{1}{\rho^2}u_{\phi\phi}.$$

(c) Laplace's equation $\nabla^2 u = 0$ in polar coordinates can, therefore, be written as

$$\rho^2 u_{\rho\rho} + \rho u_\rho + u_{\phi\phi} = 0.$$

4.14 The Laplacian on an Annulus and a Disk: Analysis and Applications

This section is devoted to steady temperatures, or Laplacian on an annulus and on disc.

• Laplacian on an Annulus

For an application, we consider the Laplacian given by equation (4.105) on an annulus D defined by

$$D = \{(x,y) : 1 \le x^2 + y^2 \le 2\}.$$

We aim to determine the potential in the annular region D. By holding the coordinate z constant in (4.105), the equation simplifies, leading to the following notation

$$u_{\rho\rho} + \frac{1}{\rho}u_\rho + \frac{1}{\rho^2}u_{\phi\phi} = 0, \quad 1 < \rho < 2, \quad 0 < \phi < \pi, \tag{4.112}$$

along with boundary conditions

$$u(\rho,0) = 0, \quad u(\rho,\pi) = u_0, \tag{4.113}$$

$$u(1,\phi) = 0, \quad u(2,\phi) = 0, \tag{4.114}$$

as depicted in Fig. 4.2. The problem involves solving the Laplacian equation in polar coordinates to determine the potential $u(\rho,\phi)$ within the annular region D. This geometry, known as an annulus, arises in various physical applications such as heat conduction, fluid flow, and electrostatics, where symmetry about a central axis simplifies the problem to two dimensions in polar coordinates. By assuming the z-coordinate is constant, the Laplacian in cylindrical coordinates reduces to the simplified form in equation (4.112). The region of interest lies between two concentric circles with radii $\rho = 1$ and $\rho = 2$, and the potential satisfies boundary conditions on both the radial edges ((4.114)) and the angular edges ((4.113)). The task is to find $u(\rho,\phi)$ subject to these constraints, which describe a steady-state solution where the potential remains bounded and respects the imposed physical conditions at the boundaries. We seek a nontrivial solution of the form

$$u(\rho,\phi) = R(\rho)\Theta(\phi).$$

First, differentiate and substitute the result into (4.112). Then, divide the resulting equation by $R(\rho)\Theta(\phi)$. Finally, by separating variables and equating both sides to $-\lambda$, we obtain two second-order differential equations.

$$\Theta'' - \lambda\Theta = 0, \quad \Theta(0) = 0, \tag{4.115}$$

and

$$\rho^2 R''(\rho) + \rho R'(\rho) + \lambda R(\rho) = 0, \quad \rho > 0, \tag{4.116}$$
$$R(1) = R(2) = 0.$$

Equation (4.116) is of the form of second-order Cauchy-Euler equation and it can be reduced in the following manner. Let

$$\rho = e^t, \text{ or equivalently } \quad t = \ln(\rho).$$

Then,

$$\frac{dR}{d\rho} = \frac{dR}{dt}\frac{dt}{d\rho} = \frac{dR}{dt}\frac{1}{\rho} = e^{-t}\frac{dR}{dt}.$$

Moreover,

$$R''(\rho) = \frac{d^2R}{d\rho^2} = \frac{d}{d\rho}\left(\frac{dR}{d\rho}\right) = e^{-t}\frac{d}{dt}\left(e^{-t}\frac{dR}{dt}\right)$$

$$= e^{-t}\left[-e^{-t}\frac{dR}{dt}+e^{-t}\frac{d^2R}{dt^2}\right]$$
$$= -e^{-2t}\frac{dR}{dt}+e^{-2t}\frac{d^2R}{dt^2}=\frac{1}{\rho^2}\left[\frac{d^2R}{dt^2}-\frac{dR}{dt}\right].$$

A substitution of $\frac{dR}{d\rho}$ and $\frac{d^2R}{d\rho^2}$ into (4.116) gives

$$\rho^2\left[\frac{1}{\rho^2}\left(\frac{d^2R}{dt^2}-\frac{dR}{dt}\right)\right]+\rho\left(\frac{1}{\rho}\frac{dR}{dt}\right)+\lambda\rho = \frac{d^2R}{dt^2}-\frac{dR}{dt}+\frac{dR}{dt}+\lambda\rho$$
$$= R''(t)+\lambda R(t).$$

Thus, we consider

$$R''(t)+\lambda R(t)=0 \quad \text{where} \quad \rho=e^t.$$

It is easy to verify that $R(t)=0$ for $\lambda \le 0$. For $\lambda > 0$, we assume $\lambda=\alpha^2$, $\alpha>0$. Then

$$R''(t)+\alpha^2 R(t)=0,$$

has the general solution

$$R(t)=A\cos(\alpha t)+B\sin(\alpha t).$$

In term of ρ, the solution takes the form

$$R(\rho)=A\cos(\alpha\ln(\rho))+B\sin(\alpha\ln(\rho)).$$

Applying $0=R(1)$, we obtain $A=0$. Applying the second boundary condition and keeping in mind that $A=0$, we arrive at $B\sin(\alpha\ln(2))=0$, from which we obtain $\alpha\ln(2)=n\pi$, $n=1,2,\ldots$. This yields

$$\alpha_n=\frac{n\pi}{\ln(2)}, \quad n=1,2,\ldots.$$

Consequently, the corresponding eigenfunctions are given by

$$R_n(\rho)=\sin(\alpha_n\ln(\rho)), \quad n=1,2,\ldots.$$

Next we normalize the eigenfunctions with respect the weight function $p=\frac{1}{\rho}$, by setting

$$\zeta_n=\frac{R_n(\rho)}{||R_n||}, \quad n=1,2,\ldots, \quad \text{where}$$

$$||R_n||^2=\int_1^2\frac{\sin^2(\alpha\ln(\rho))}{\rho}d\rho.$$

For learning purposes, we show the necessary steps for evaluating the integral.

$$\int_1^2\frac{\sin^2(n\pi\ln(\rho))}{\rho}d\rho = \int_1^2\frac{1-\cos(2\alpha\ln(\rho))}{2\rho}d\rho=\frac{1}{2}\ln(\rho)\Big|_1^2-\int_1^2\frac{\cos(2\alpha\ln(\rho))}{2\rho}d\rho$$

$$= \frac{\ln(2)}{2} - \int_1^2 \frac{\cos(2\alpha\ln(\rho))}{2\rho} d\rho.$$

To evaluate the integral we make the substitution $u = 2\alpha\ln(\rho)$. Then, $\frac{du}{2\alpha} = \frac{d\rho}{\rho}$ and

$$\int_1^2 \frac{\cos(2\alpha\ln(\rho))}{2\rho} d\rho = \int_0^{2\alpha\ln(2)} \frac{\cos(u)}{4\alpha} du = \frac{\sin(2\alpha\ln(2))}{4\alpha}$$
$$= \frac{\sin(2\frac{n\pi}{\ln(2)}\ln(2))}{4\alpha} = 0.$$

Thus,

$$||R_n|| = \sqrt{\frac{\ln(2)}{2}}.$$

Hence, the normalized eigenfunctions are given by

$$\zeta_n(\rho) = \sqrt{\frac{2}{\ln(2)}} \sin(\alpha_n \ln(\rho)), \quad n = 1, 2, \ldots.$$

Similarly, for the same eigenvalues $\lambda_n = \alpha_n^2$, the system

$$\Theta'' - \lambda\Theta = 0, \quad \Theta(0) = 0,$$

has the general solution

$$\Theta_n(\phi) = c_1 e^{\alpha_n \phi} + c_2 e^{-\alpha_n \phi}.$$

An application of the boundary condition, leads to $c_2 = -c_1$. By setting $c_1 = \frac{1}{2}$, we then have $c_2 = -\frac{1}{2}$, and the solution takes the form

$$\Theta_n(\phi) = \sinh(\alpha_n \phi), \quad n = 1, 2, \ldots.$$

So, the solution of (4.112) can be written as

$$u(\rho, \phi) = \sum_{n=1}^{\infty} d_n \sinh(\alpha_n \phi) \zeta_n(\rho).$$

Using $u(\rho, \pi) = u_0$, we get

$$u_0 = \sum_{n=1}^{\infty} d_n \sinh(\alpha_n \pi) \zeta_n(\rho). \tag{4.117}$$

Recall that the set of functions given by ζ_n are normalized with respect to the weight function $p = \frac{1}{\rho}$ and therefore,

$$\int_1^2 \zeta_n(\rho)\zeta_m(\rho) d\rho = \begin{cases} 1, & m = n \\ 0, & m \neq n \end{cases}$$

Hence, by multiplying both sided of (4.117) with $\zeta_m(\rho)$ and then integrate with respect to ρ from 1 to 2 we arrive at

$$u_0 \int_1^2 \zeta_n(\rho)d\rho = d_n \sinh(\alpha_n \pi),$$

or,

$$u_0 \int_1^2 \frac{1}{\rho}\sqrt{\frac{2}{\ln(2)}} \sin(\alpha_n \ln(\rho))d\rho = d_n \sinh(\alpha_n \pi).$$

From which we obtain after integrating,

$$d_n = \frac{1}{\sinh(\alpha_n \pi)}\Big(\frac{u_0\sqrt{2\ln(2)}}{\pi}\frac{1-(-1)^n}{n}\Big).$$

Accordingly, the solution takes the form

$$u(\rho,\phi) = \frac{2u_0}{\pi}\sum_{n=1}^{\infty}\frac{1-(-1)^n}{n}\frac{\sinh(\alpha_n\phi)}{\sinh(\alpha_n\pi)}\sin(\alpha_n\ln(\rho)),$$

where

$$\alpha_n = \frac{n\pi}{\ln(2)}, \quad n = 1,2,\ldots.$$

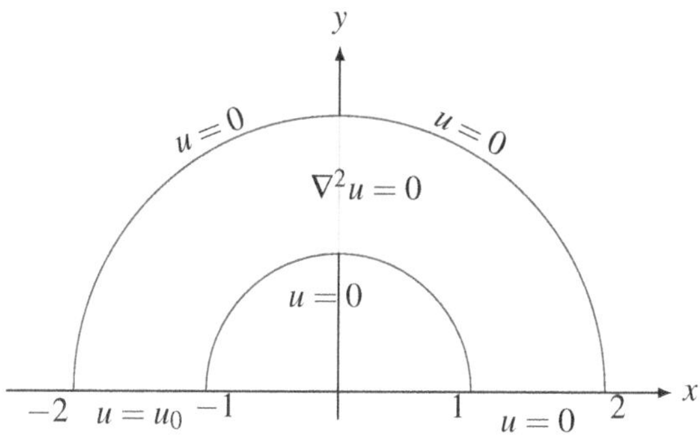

FIGURE 4.2: Laplacian on an annulus.

• Laplacian on a Disk

Now we turn our attention to steady temperature, or Laplacian on a disk D of the form

$$D = \{(x,y) : 0 < x^2 + y^2 \le r^2\}.$$

As before, the Laplacian takes the form

$$u_{\rho\rho} + \frac{1}{\rho}u_\rho + \frac{1}{\rho^2}u_{\phi\phi} = 0, \quad 0 < \rho < r, \quad 0 < \phi < 2\pi, \tag{4.118}$$

$$u(r,\phi) = f(\phi), \quad 0 \le \phi < 2\pi,$$
$$u(\rho,\phi+2\pi) = u(\rho,\phi), \tag{4.119}$$
$$u(\rho,\phi) \quad \text{is bounded as} \quad \rho \to 0^+.$$

as depicted in Fig. 4.3. We remark that the periodicity condition on the solution given by (4.119) is satisfied when

$$\Theta(\pi) = \Theta(-\pi) \quad \text{and} \quad \Theta'(\pi) = \Theta'(-\pi).$$

We seek a nontrivial solution of the form

$$u(\rho,\phi) = R(\rho)\Theta(\phi)$$

and obtain, as we did before

$$\Theta'' - \lambda\Theta = 0, \quad \Theta(\pi) = \Theta(-\pi) \quad \text{and} \quad \Theta'(\pi) = \Theta'(-\pi), \tag{4.120}$$

and

$$\rho^2 R''(\rho) + \rho R'(\rho) + \lambda R(\rho) = 0, \quad R(\rho) \quad \text{is bounded as} \quad \rho \to 0^+. \tag{4.121}$$

If $\lambda < 0$, we set $\lambda = -\alpha^2, \alpha > 0$. Then from (4.120) we have

$$\Theta'' - \alpha^2\Theta = 0,$$

whose solution is

$$\Theta(\phi) = c_1 e^{-\alpha\phi} + c_2 e^{\alpha\phi}.$$

Applying the periodicity conditions, one obtains

$$c_1 e^{-\alpha\pi} + c_2 e^{\alpha\pi} = c_1 e^{\alpha\pi} + c_2 e^{-\alpha\pi},$$

and

$$-\alpha c_1 e^{-\alpha\pi} + \alpha c_2 e^{\alpha\pi} = -\alpha c_1 e^{\alpha\pi} + \alpha c_2 e^{-\alpha\pi}.$$

Dividing both sides of the second equation by α, and then adding the resulting equation to the first equation, we obtain

$$4c_2 \sinh(\alpha\pi) = 0.$$

Since $\alpha > 0$, it follows that $\sinh(\alpha\pi) \neq 0$. Therefore, the above equation holds if and only if $c_2 = 0$. Thus, we conclude that there is only the trivial solution for $\lambda < 0$. For $\lambda = 0$, the solution is $\Theta(\phi) = c_1\phi + c_2$. Applying the periodicity condition, we obtain $c_1 = 0$, and c_2 is an arbitrary, and hence $\Theta(\phi) = c$, where c is a nonzero constant. If $\lambda > 0$, we set $\lambda = \alpha^2$, $\alpha > 0$. Then (4.120) has the general solution

$$\Theta(\phi) = c_1 \cos \alpha\phi + c_2 \sin \alpha\phi.$$

Applying the periodicity conditions, we arrive at

$$c_1 \cos \alpha\pi + c_2 \sin \alpha\pi = c_1 \cos \alpha\pi - c_2 \sin \alpha\pi,$$

and

$$-\alpha c_1 \sin \alpha\pi + \alpha c_2 \cos \alpha\phi = \alpha c_1 \sin \alpha\pi + \alpha c_2 \cos \alpha\phi.$$

Either equation gives

$$\sin \alpha\pi = 0,$$

which is satisfied for $\alpha_n = n$, $n = 1, 2, \ldots$, or equivalently, $\lambda_n = n^2$. Therefore, the corresponding eigenfunctions are

$$\Theta_n(\phi) = a_n \cos n\phi + b_n \sin n\phi, \quad n = 1, 2 \ldots$$

Let $\rho = e^t$, as we did before. Then, equation (4.121) takes the form

$$R''(t) + \lambda R(t) = 0.$$

We are only interested in its solution for $\lambda \geq 0$.

For $\lambda = 0$, the solution is

$$R(t) = c_1 t + c_2.$$

In terms of ρ, we have

$$R(\rho) = c_1 \ln(\rho) + c_2.$$

Since

$$\lim_{\rho \to 0^+} R(\rho) = -\infty,$$

the solution becomes unbounded. Thus, we must take $c_1 = 0$. Consequently, for any nonzero constant d, we have

$$R(\rho) = d.$$

For $\lambda_n = n^2$, the solution of the equation

$$R''(t) + \lambda R(t) = 0$$

takes the form

$$R(t) = c_1 e^{nt} + c_2 e^{-nt}.$$

In terms of ρ, the solution becomes

$$R(\rho) = c_1 \rho^n + c_2 \rho^{-n} = c_1 \rho^n + \frac{c_2}{\rho^n}.$$

Hence,

$$\lim_{\rho \to 0^+} R(\rho) = \lim_{\rho \to 0^+} \left(c_1 \rho^n + \frac{c_2}{\rho^n} \right).$$

Since

$$\lim_{\rho \to 0^+} \frac{c_2}{\rho^n} = \infty,$$

we must set $c_2 = 0$. As a result, the solution is

$$R(\rho) = \rho^n, \quad n = 1, 2, \ldots,$$

where we set $c_1 = 1$. The general solution of (4.118) along with its initial and boundary conditions is

$$u(\rho,\phi) = \frac{a_0}{2} + \sum_{n=1}^{\infty} \rho^n \Big(a_n \cos n\phi + b_n \sin n\phi \Big), \quad 0 < \phi < 2\pi. \tag{4.122}$$

Applying the initial condition to the solution gives

$$f(\phi) = \frac{a_0}{2} + \sum_{n=1}^{\infty} r^n \Big(a_n \cos n\phi + b_n \sin n\phi \Big),$$

where $a_0, r^n a_n$ and $r^n b_n$ are the Fourier coefficients of f. Thus,

$$a_0 = \frac{1}{\pi}\int_0^{2\pi} f(\phi)\,d\phi,\ r^n a_n = \frac{1}{\pi}\int_0^{2\pi} f(\phi)\cos n\phi\,d\phi,\ r^n b_n = \frac{1}{\pi}\int_0^{2\pi} f(\phi)\sin n\phi\,d\phi.$$

Next, we put the solution given by (4.122) in the form of **Poisson integral.** We have the following corollary.

Corollary 4 *The solution of* (4.118) *can be put in the form of Poisson integral given by*

$$u(\rho,\phi) = \frac{1-(\frac{\rho}{r})^2}{2\pi}\int_0^{2\pi} \frac{f(t)}{1-2\frac{\rho}{r}\cos(\phi-t)+(\frac{\rho}{r})^2}\,dt. \tag{4.123}$$

Proof *In the coefficients a_0, a_n and b_n change the dummy variable ϕ to t and then substitute the coefficients into the solution* (4.118), *given by* (4.122). *In doing so, we obtain*

$$\begin{aligned}
u(\rho,\phi) &= \frac{1}{2\pi}\int_0^{2\pi} f(t)\,dt + \frac{1}{\pi}\sum_{n=1}^{\infty}\Big(\frac{\rho}{r}\Big)^n \Big[\int_0^{2\pi} f(t)\cos(nt)\,dt\cos(n\phi) \\
&\quad + \int_0^{2\pi} f(t)\sin(nt)\,dt\sin(n\phi)\Big] \\
&= \frac{1}{2\pi}\int_0^{2\pi} f(t)\,dt + \frac{1}{\pi}\sum_{n=1}^{\infty}\Big(\frac{\rho}{r}\Big)^n \Big[\int_0^{2\pi} f(t)\Big(\cos(nt)\cos(n\phi)+\sin(nt)\sin(n\phi)\Big)dt\Big] \\
&= \frac{1}{2\pi}\int_0^{2\pi} f(t)\Big[1+2\sum_{n=1}^{\infty}\Big(\frac{\rho}{r}\Big)^n \cos n(\phi-t)\Big]dt.
\end{aligned} \tag{4.124}$$

Notice that

$$\begin{aligned}
1+2\sum_{n=1}^{\infty}\Big(\frac{\rho}{r}\Big)^n \cos n(\phi-t) &= 1+2\Re\sum_{n=1}^{\infty}\Big(\frac{\rho}{r}\Big)^n e^{in(\phi-t)} \\
&= 1+2\Re\frac{\frac{\rho}{r}e^{i(\phi-t)}}{1-\frac{\rho}{r}e^{i(\phi-t)}} \\
&= 1+\frac{\frac{\rho}{r}e^{i(\phi-t)}}{1-\frac{\rho}{r}e^{i(\phi-t)}}+\frac{\frac{\rho}{r}e^{-i(\phi-t)}}{1-\frac{\rho}{r}e^{-i(\phi-t)}} \\
&= \frac{1-(\frac{\rho}{r})^2}{1-2\frac{\rho}{r}\cos(\phi-t)+(\frac{\rho}{r})^2}.
\end{aligned} \tag{4.125}$$

Substituting (4.125) *into* (4.124) *gives* (4.123). *This completes the proof.* □

Next, we analyze the solution (4.124) in terms of its convergence and its adherence to the given boundary conditions. Let $R = \dfrac{\rho}{r}$. Then $0 < R < 1$, and hence, by (4.124) we have that

$$\begin{aligned}
\left|u(\rho,\phi)\right| &\leq \frac{1}{2\pi}\int_0^{2\pi}\left|f(t)\right|dt + \frac{1}{\pi}\int_0^{\pi}\sum_{n=1}^{\infty}\left|R^n\cos n(\phi-t)\right|dt \\
&\leq M_1 + \sum_{n=1}^{\infty} M_2 R^n < \infty,
\end{aligned}$$

for positive constants M_1 and M_2. Taking partial derivatives in (4.124) with respect to ρ we may easily conclude that for a positive constant M_3

$$\begin{aligned}
\left|\frac{\partial^2 u(\rho,\phi)}{\partial\rho^2}\right| &\leq \int \frac{1}{\pi}\int_0^{\pi}\sum_{n=1}^{\infty} n(n-1)\left(\frac{\rho}{r}\right)^{n-2}\left|\cos n(\phi-t)\right|\Big] dt \\
&\leq \sum_{n=1}^{\infty} n(n-1)\left(\frac{\rho}{r}\right)^{n-2} < \infty.
\end{aligned}$$

Along the same lines, it can be shown that

$$\left|\frac{\partial^2 u(\rho,\phi)}{\partial\phi^2}\right| < \infty.$$

Moreover,

$$\begin{aligned}
\nabla^2 u &= u_{\rho\rho} + \frac{1}{\rho}u_\rho + \frac{1}{\rho^2}u_{\phi\phi} \\
&= \frac{1}{\pi}\int_0^{2\pi}\sum_{n=1}^{\infty} r^{-2}R^{n-2}\left(n(n-1)+n-n^2\right)\cos n(\phi-t)\,dt = 0, \quad 0 \leq R \leq R_0 < 1.
\end{aligned}$$

As each term in the series represents a harmonic function and the series converges uniformly, it follows that $u(\rho,\phi)$ remains harmonic at any interior point within the region $0 \leq R < 1$. The final step is to verify that u meets the prescribed boundary condition $f(\phi)$. Set $f(\phi) = 1$. Then from (4.124) we have that

$$\begin{aligned}
u(\rho,\phi) &= \frac{1}{2\pi}\int_0^{2\pi}\left[1 + 2\sum_{n=1}^{\infty}\left(\frac{\rho}{r}\right)^n \cos n(\phi-t)\right]dt \\
&= 1 + \frac{1}{\pi}\sum_{n=1}^{\infty}\left(\frac{\rho}{r}\right)^n \int_0^{2\pi}\cos n(\phi-t)dt = 1,
\end{aligned}$$

since

$$\int_0^{2\pi}\cos n(\phi-t)dt = 0, \quad n = 1,2,\ldots$$

Thus, $u(\rho,\phi) = 1$, or

$$1 = \frac{1}{2\pi}\int_0^{2\pi}\frac{1-R^2}{1-2R\cos(\phi-t)+R^2}dt.$$

Therefore,

$$f(\phi) = \frac{1}{2\pi}\int_0^{2\pi} \frac{1-R^2}{1-2R\cos(\phi - t) + R^2} f(\phi)\,dt, \quad 0 \le R < 1.$$

Consequently, we have the important relation

$$u(R,\phi) - f(\phi) = \frac{1}{2\pi}\int_0^{2\pi} \frac{(1-R^2)\,[f(t)-f(\phi)]}{1-2R\cos(\phi - t) + R^2}\,dt. \tag{4.126}$$

Since $f(\phi)$ is uniformly continuous on $[0,2\pi]$, for any given $\varepsilon > 0$, there exists a positive number $\delta(\varepsilon)$ such that whenever $|\phi - t| < \delta$, it follows that $|f(\phi) - f(t)| < \varepsilon$. On the other hand, if $|\phi - t| \ge \delta$ and $\phi - t \ne 2n\pi$ for $n = 0,1,2,\ldots$, then

$$\lim_{R\to 1^-} \frac{1-R^2}{1-2R\cos(\phi - t)+R^2} = 0.$$

In other words, there exists ρ_0 such that if $|\phi - t| \ge \delta$, then

$$\frac{1-R^2}{1-2R\cos(\phi - t)+R^2} < \varepsilon,$$

for $0 \le R \le R_0 < 1$. Hence, equation (4.126) yields

$$\begin{aligned} |u(\rho,\phi) - f(\phi)| &\le \frac{1}{2\pi}\int\limits_{|\phi - t|\ge\delta}^{2\pi} \frac{(1-R^2)|f(t)-f(\phi)|}{1-2R\cos(\phi - t)+R^2}dt \\ &+ \frac{1}{2\pi}\int\limits_{|\phi - t|<\delta}^{2\pi} \frac{(1-R^2)|f(\phi)-f(t)|}{1-2R\cos(\phi - t)+R^2}dt \\ &\le \frac{1}{2\pi}(2\pi\varepsilon)\big[2\max_{0\le\phi\le 2\pi}|f(\phi)|\big] + \frac{\varepsilon}{2\pi}(2\pi) \\ &= \varepsilon\Big[\Big(1 + 2\max_{0\le\phi\le 2\pi}|f(\phi)|\Big)\Big]. \end{aligned}$$

Consequently, we have

$$u(r,\phi) = \lim_{R\to 1^-} u(\rho,\phi) = f(\phi).$$

Thus, we have shown that the solution of (4.118) exists and is unique, and it is either given by (4.122) or (4.123).

Notice that, for $\rho = 0$, relation (4.123) implies that

$$u(0,\phi) = u(0) = \frac{1}{2\pi}\int_0^{2\pi} f(t)\,dt,$$

and this gives the formal result.

Theorem 4.2 (Mean Value Theorem) *If u is harmonic in a circle, then the value of u at the center is equal to the mean value of u on the boundary of the circle.*

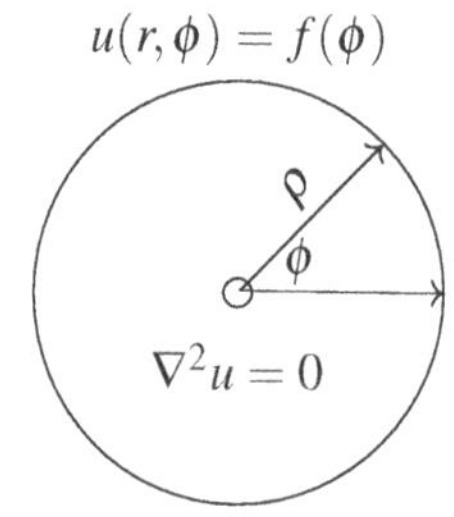

FIGURE 4.3: Laplacian on a disk.

4.14.1 Exercises

Exercise 4.53 *Redo* (4.112)- (4.114) *over the annulus*

$$D = \{(x,y) : b^2 \leq x^2 + y^2 \leq d^2\},$$

where $d > b$.

Exercise 4.54 *For constant* u_0, *solve*

$$\begin{aligned}
&u_{\rho\rho} + \frac{1}{\rho}u_\rho + \frac{1}{\rho^2}u_{\phi\phi} = 0, \quad 0 < \rho < 2, \quad 0 < \phi < 2\pi, \\
&u(2,\phi) = u_0, \quad 0 \leq \phi < 2\pi, \\
&u(\rho,\phi + 2\pi) = u(\rho,\phi), \quad u(\rho,\phi) \quad \textit{is bounded as} \quad \rho \to 0^+.
\end{aligned} \tag{4.127}$$

Exercise 4.55 *Repeat Exercise 4.54 by changing* (4.127) *to* $u(2,\phi) = \phi$.
Ans: $u(\rho,\phi) = \pi - 2\sum_{n=1}^{\infty} \left(\frac{r}{2}\right)^n \frac{\sin n\phi}{n}$.

Exercise 4.56 *For constant* u_0, *solve*

$$\begin{aligned}
&u_{\rho\rho} + \frac{1}{\rho}u_\rho + \frac{1}{\rho^2}u_{\phi\phi} = 0, \quad 1 < \rho < b, \quad 0 < \phi < \pi, \\
&u(\rho,0) = 0, \quad u(\rho,\pi) = 0, \quad 1 < \rho < b, \\
&u(1,\phi) = 0, \quad u(b,\phi) = u_0.
\end{aligned}$$

Ans: $u(\rho,\phi) = \frac{4u_0}{\pi}\sum_{n=1}^{\infty} \frac{\rho^{2n-1} - \rho^{-(2n-1)}}{b^{2n-1} - b^{-(2n-1)}} \cdot \frac{\sin((2n-1)\phi)}{2n-1}$.

Exercise 4.57 *(Partially Insulated) Solve*

$$\begin{aligned} &u_{\rho\rho} + \frac{1}{\rho}u_\rho + \frac{1}{\rho^2}u_{\phi\phi} = 0, \quad 0 < \rho < a, \quad 0 < \phi < \pi, \\ &u_\phi(\rho, 0) = 0, \quad u_\phi(\rho, \pi) = 0, \quad 1 < \rho < b, \\ &u(a, \phi) = f(\phi). \end{aligned}$$

Ans: $u(\rho,\phi) = \frac{1}{\pi}\int_0^\pi f(s)\,ds + \frac{2}{\pi}\sum_{n=1}^{\infty}\left(\frac{\rho}{a}\right)^n \cos(n\phi)\int_0^\pi f(s)\cos(ns)\,ds.$

Exercise 4.58 *(Heat flow in a sphere) Consider the heat flow equation* $u_t = \nabla^2 u$ *where* $\nabla^2 u$ *is given by (4.107). Assume* u *is independent of* θ *and* ϕ*. Assume the sphere is initially at temperature* $f(\rho)$ *and its surface at* $\rho = 1$ *is kept at temperature zero. Thus, we must solve the heat flow problem*

$$u_t(\rho,t) = \nabla^2 u(\rho,t), \quad 0 < \rho < 1, \quad t > 0,$$

$$u(1,t) = 0, \quad u(\rho,0) = f(\rho).$$

Hint: Make use of the substitution $v(\rho,t) = \rho u(\rho,t)$*, and then make use of known result.*

Ans: $u(\rho,t) = \frac{1}{\rho}\sum_{n=1}^{\infty} b_n \exp\left(-n^2\pi^2 kt\right)\sin(n\pi r)$*, where* $b_n = 2\int_0^1 r f(r)\sin(n\pi r)\,dr.$

Exercise 4.59 *(Heat flow in a sphere) Suppose that in Exercise 4.58 we take*

$$u(1,t) = g(t) \quad \text{and} \quad u(\rho,0) = 0.$$

Show that the solution is given by

$$u(\rho,t) = \frac{2\pi k}{\rho}\sum_{n=1}^{\infty}(-1)^{n+1} n\sin(n\pi\rho)\int_0^t e^{-n^2\pi^2 k(t-s)}g(s)\,ds.$$

Exercise 4.60 *(Heat flow in a sphere) Let* $g(t) = t$*, in Exercise 4.59. Demonstrate that the solution is then given by*

$$u(\rho,t) = t - \frac{2}{\pi^3 k\rho}\sum_{n=1}^{\infty}\frac{(-1)^{n+1}}{n^3}\left(1 - e^{-n^2\pi^2 kt}\right)\sin n\pi\rho.$$

Exercise 4.61 *Solve*

$$\begin{aligned} &u_{\rho\rho} + \frac{1}{\rho}u_\rho + \frac{1}{\rho^2}u_{\phi\phi} = 0, \quad a < \rho < b, \quad 0 < \phi < \alpha, \\ &u(\rho, \alpha) = 0, \quad u(\rho, 0) = f(\rho), \quad a \le \rho \le b, \\ &u(a, \phi) = f(\phi), \quad u(b, \phi) = 0, \quad 0 \le \phi \le \alpha. \end{aligned}$$

Exercise 4.62 (Dirichlet problem for a circular annulus) *Solve*

$$u_{\rho\rho} + \frac{1}{\rho}u_\rho + \frac{1}{\rho^2}u_{\phi\phi} = 0, \quad a < \rho < b, \quad 0 < \phi < 2\pi,$$
$$u(a,\phi) = f(\phi), \quad u(b,\phi) = g(\phi), \quad 0 \le \phi \le 2\pi.$$

In addition, $u(\rho,\phi)$ must satisfy the periodicity condition. Accordingly, $f(\phi)$ and $g(\phi)$ must also be periodic with period 2π.

4.15 Poisson's Equation with a Source

In this section we look into the two-dimensional Poisson equation with a source $Q(x,y)$ that is independent of time t given by

$$\nabla^2 u = \frac{\partial^2 u}{\partial x^2} + \frac{\partial^2 u}{\partial y^2} = Q(x,y), \quad 0 < x < L, \quad 0 < y < H. \tag{4.128}$$

The equation describes the distribution of u in a two-dimensional domain in response to the source $Q(x,y)$. Boundary conditions must be specified to solve this equation, such as Dirichlet conditions (u specified on the boundary) or Neumann conditions (the normal derivative of u specified on the boundary). We assume the boundary conditions on the rectangle

$$u(0,y) = f_1(y), \quad u(L,y) = f_2(y), \tag{4.129}$$

$$u(x,0) = 0, \quad u(x,H) = 0. \tag{4.130}$$

Recall that the solution oscillates in one direction. Since we have two homogeneous boundary conditions at $y = 0$ and $y = L$, the method of eigenfunction expansion consists in expanding $u(x,y)$ in a series of these eigenfunctions. In other words, the eigenfunctions satisfy

$$Y''(y) - \lambda Y(y) = 0, \quad Y(0) = Y(H) = 0.$$

Hence the eigenvalues and corresponding eigenfunctions are

$$\lambda_n = \frac{n^2\pi^2}{H^2}, \quad \varphi_n(y) = \sin\frac{n\pi y}{H}, \quad n = 1, 2, \ldots$$

As any piecewise smooth function can be represented as a series of these eigenfunctions, we set

$$u(x,y) = \sum_{n=1}^{\infty} b_n(x)\varphi_n(y). \tag{4.131}$$

Similarly, by letting

$$Q(x,y) = \sum_{n=1}^{\infty} q_n(x)\varphi_n(y)$$

we obtain

$$q_n(x) = \frac{\int_0^H Q(x,y)\varphi_n(y)\,dy}{\int_0^H \varphi_n^2(y)\,dy}. \tag{4.132}$$

Taking partial derivatives in (4.131) with respect to x and y and then substituting into (4.128) yields

$$\sum_{n=1}^{\infty} b_n''(x)\varphi_n(y) + \sum_{n=1}^{\infty} b_n(x)\varphi_n''(y) = \sum_{n=1}^{\infty} q_n(x)\varphi_n(y).$$

Since $\varphi_n''(y) = -\lambda_n\varphi_n$, the above expression provides us with

$$\sum_{n=1}^{\infty} \left[b_n''(x) - \lambda_n b_n(x) - q_n(x)\right]\varphi_n(y) = 0$$

for all x and y in the designated rectangle. We obtain the non-homogeneous second-order differential equation in $b_n(x)$ that we must solve based on two initial conditions,

$$b_n''(x) - \lambda_n b_n(x) = q_n(x), \tag{4.133}$$

where $q_n(x)$ is given by (4.132). Making use of (4.129), we obtain

$$f_1(y) = \sum_{n=1}^{\infty} b_n(0)\varphi_n(y),$$

from which we arrive at

$$b_n(0) = \frac{2}{H}\int_0^H f_1(y)\varphi_n(y)\,dy. \tag{4.134}$$

Along the same lines we have

$$b_n(H) = \frac{2}{H}\int_0^H f_2(y)\varphi_n(y)\,dy. \tag{4.135}$$

Now that we have two initial conditions for $b_n(x)$, we can solve (4.133) to obtain its unique solution, $b_n(x)$. Finally, the solution to (4.128)–(4.130) is given by (4.131), along with (4.132)–(4.135) and the unique solution $b_n(x)$ of (4.133).

4.15.1 Exercises

Exercise 4.63 *Find the general solution of*

$$\nabla^2 u = \frac{\partial^2 u}{\partial x^2} + \frac{\partial^2 u}{\partial y^2} = Q(x,y), \quad 0 < x < L, \quad 0 < y < H,$$

$$u(0,y) = 0, \quad u(L,y) = 0,$$

$$u(x,0) = f_1(x), \quad u(x,H) = f_2(x).$$

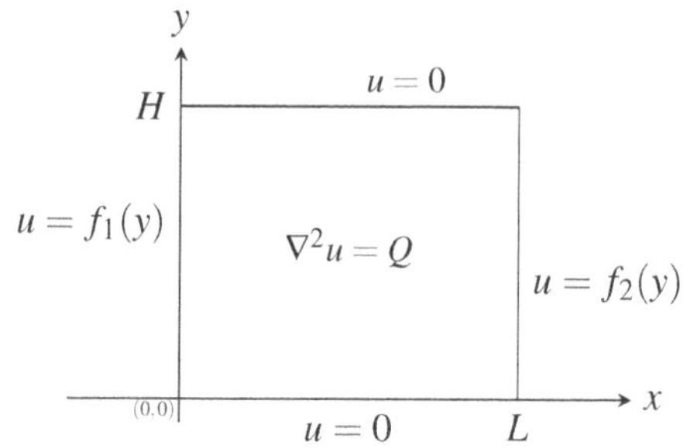

FIGURE 4.4: Poisson's problem with a source.

Exercise 4.64 *Find the general solution of*

$$\nabla^2 u = \frac{\partial^2 u}{\partial x^2} + \frac{\partial^2 u}{\partial y^2} + Q(y), \quad 0 < x < L, \quad 0 < y < H,$$

$$u(0,y) = 0 = u(L,y),$$
$$u(x,0) = u(x,H) = 0.$$

Use a Fourier sine series in x.

Exercise 4.65 *Solve Equation* (4.133), *given that*

$$f_1(y) = 1 \quad and \quad f_2(y) = \sin y.$$

Exercise 4.66 *Solve*

$$\nabla^2 u = \frac{\partial^2 u}{\partial x^2} + \frac{\partial^2 u}{\partial y^2} = e^{2y} \sin x, \quad 0 < x < \pi, \quad 0 < y < L,$$

$$u(0,y) = u(\pi,y) = 0,$$
$$u(x,0) = 0, \quad u(x,L) = f(x).$$

Exercise 4.67 *Solve the Poisson equation with a source term on the rectangular domain* $\Omega = (0,1) \times (0,1)$.

$$\frac{\partial^2 u}{\partial x^2} + \frac{\partial^2 u}{\partial y^2} = -\sin \pi x \sin \pi y,$$

subject to homogeneous Dirichlet boundary conditions

$$u(0,y) = u(1,y) = u(x,0) = u(x,1) = 0.$$

Hint: Assume a solution of the form

$$u(x,y) = A \sin \pi x \sin \pi y.$$

Substitute into the equation to solve for A.

Exercise 4.68 *Consider the time-dependent heat equation with a source term*

$$\frac{\partial u}{\partial t} = \nabla^2 u - \sin \pi x \sin \pi y$$

on the domain $\Omega = (0,1) \times (0,1)$, *with initial condition and boundary condition*

$$u(x,y,0) = 0, \quad u = 0 \quad \text{on} \quad \partial\Omega,$$

(a) Show that as $t \to \infty$, *the solution* $u(x,y,t)$ *approaches the steady-state solution from Exercise 4.67.*

(b) Solve the time-dependent problem using separation of variables and verify the steady-state limit.

5

The Wave Equation and Boundary Value Problems

This chapter begins with the derivation of the wave equation, a fundamental partial differential equation that describes the propagation of waves in various physical systems, such as vibrating strings, sound waves, and electromagnetic waves. Starting from basic physical principles like Newton's laws of motion, Hooke's law, and the continuity of motion, the wave equation is formulated to describe small transverse displacements of a stretched string. Once the wave equation is derived, attention turns to solving it for specific boundary value problems that arise in the study of vibrating strings. Using the method of separation of variables, the wave equation is decomposed into simpler ordinary differential equations, one governing spatial behavior and the other governing time evolution.

The solutions are expressed in terms of sine and cosine functions to satisfy boundary conditions, leading to discrete eigenvalues corresponding to the natural frequencies of vibration. The general solution is then constructed as a superposition of these harmonic modes, revealing the physical behavior of the string in response to initial conditions.

Some of the applications include determining the normal modes of vibration of a fixed string and analyzing the effect of initial displacement or velocity profiles on the motion of the string. We extend the method to solve more complex boundary conditions, such as strings with one free end or with damping effects and transverse vibration of a beam.

5.1 Derivation of the Wave Equation

In this section we will discuss the derivation of the wave equation. Let $l > 0$ be the length of a thin string that is stretched between two points on the x-axis. A string vibrates only if it is tightly stretched. Assume the string undergoes relatively small transverse vibrations (think of the string of a musical instrument, say a violin string). Let ρ be the linear density of the string, measured in units of mass per unit of length. We will assume that the string is made of homogeneous material and its density is

DOI: 10.1201/9781003652540-5

constant along the entire length of the string. The displacement of the string from its equilibrium state at time t and position x will be denoted by $u(x,t)$. We assume the string is positioned in such a way that its left endpoint coincides with the origin of the xu coordinate system. Consider the motion of a small portion of the string sitting atop the interval $[a,b]$. Then the corresponding mass is $\rho(b-a)$, and acceleration u_{tt}. Using Newton's second law of motion we have

$$\rho(b-a) = \text{Total force}. \tag{5.1}$$

Since the mass of the string is negligible, we may discard the effect of gravity on the string. In addition, we may as well ignore air resistance, and other external forces. Thus, the only force that is acting on the string is the tension force $\mathbb{T}(x,t)$. Assuming that the string is perfectly flexible, the tension force will have the direction of the tangent vector along the string. At a fixed time t the position of the string is given by the parametric equations, $x = x$, $u = u(x,t)$, where x is a parameter. Then, the tangent vector is $\langle 1, u_x \rangle$, with corresponding unit vector $\langle \frac{1}{\sqrt{1+u_x^2}}, \frac{u_x}{\sqrt{1+u_x^2}} \rangle$. Under this set up the tension force takes the form

$$\mathbb{T}(x,t) = \langle \frac{T(x,t)}{\sqrt{1+u_x^2}}, \frac{T(x,t)u_x}{\sqrt{1+u_x^2}} \rangle \tag{5.2}$$

where $T(x,t)$ is the magnitude of the tension force. Due to the assumption of a small vibration, it is safe to assume that u_x is small, and thus, via Taylor's expansion we have

$$\sqrt{1+u_x^2} = 1 + \frac{1}{2}u_x^2 + o(u_x^4) \approx 1.$$

Substituting this approximation into (5.2), we arrive at an equivalent form of the tension force

$$\mathbb{T}(x,t) = \langle T(x,t), T(x,t)u_x \rangle.$$

Since there is no longitudinal displacement, we arrive at the following identities for the balances of forces (5.1) in the x, respectively u directions

$$0 = T(b,t) - T(a,t)$$

$$\rho(b-a)u_{tt} = T(b,t)u_x(b,t) - T(a,t)u_x(a,t).$$

Simply stated, the first equation shows that the tensions from the two edges of the little portion of the string balance each other out in the x direction (no longitudinal motion). From this, we can also infer that the position of the string has no impact on the tension force. Hence, the second equation might be rewritten as

$$\rho u_{tt} = T\frac{u_x(b,t) - u_x(a,t)}{b-a}.$$

Taking the limit in the above equation we arrive at the wave equation

$$\rho u_{tt} = \lim_{b \to a} T\frac{u_x(b,t) - u_x(a,t)}{b-a} = Tu_{xx},$$

or

$$u_{tt} - c^2 u_{xx} = 0,$$

with $c^2 = \frac{T}{\rho}$.

The above derivation can be generalized to incorporate effects of other forces, as displayed below. The wave equation

$$u_{tt} - c^2 u_{xx} + r u_t = 0, \quad r > 0,$$

reflects air resistance as a force proportional to the speed u_t. On the other hand, the wave equation

$$u_{tt} - c^2 u_{xx} + ku = 0, \quad k > 0,$$

incorporates transverse elastic force that is proportional to the displacement u. Finally, the wave equation

$$u_{tt} - c^2 u_{xx} = f(x,t), \quad r > 0,$$

incorporates externally applied forces. Such equation is referred to as inhomogeneous wave equation, which we will discuss in Section 5.3 and Chapter 10.

5.2 Vibrating String with Fixed Ends

Consider the simplest form of a vibrating string with fixed ends. The vibration is modeled by

$$u_{tt} - c^2 u_{xx} = 0, \quad 0 < x < l, \quad t > 0, \tag{5.3}$$

$$u(x,0) = f(x), \quad 0 \leq x \leq l, \tag{5.4}$$

$$u_t(x,0) = g(x), \quad 0 \leq x \leq l, \tag{5.5}$$

$$u(0,t) = u(l,t) = 0, \quad t \geq 0. \tag{5.6}$$

The physical properties of a vibrating string with fixed ends are governed by the wave equation (5.3) where $u(x,t)$ represents the transverse displacement of the string at position x and time t, and c is the wave speed determined by the tension and mass density of the string. The fixed boundary conditions given by (5.6) , reflect the fact that the string is anchored at its endpoints $x = 0$ and $x = l$. The initial conditions defined by (5.4) and (5.5) specify the string's initial displacement and velocity profile, respectively.

Together, these conditions model the propagation of waves along the string while ensuring that the vibration is constrained to the fixed ends. The wave equation describes how energy is distributed along the string, resulting in wave-like motion that can exhibit standing waves or harmonic oscillations depending on the initial and boundary conditions.

• Method of Solution

We are seeking a non-trivial solution, and hence, if we assume the solution $u(x,t)$ is the product of two functions, one in x and the other in t, such that

$$u(x,t) = X(x)T(t).$$

Substituting into (5.3), we obtain

$$X''(x)T(t) - X(x)T''(t) = 0.$$

Since $X(x) \neq 0$ and $T(t) \neq 0$, we may divide by the term $X(x)T(t)$ to separate the variables. That is,

$$\frac{X''(x)}{T(t)} = \frac{T''(t))}{T(t)}.$$

Since the left-hand side is a function of x alone, it does not vary with t. However, it is equal to a function of t alone, and so it can not vary with x. Hence the two sides must have some constant value $-\lambda$ in common. That is,

$$\frac{X''(x)}{X(x)} = \frac{T''(t))}{T(t)} = -\lambda.$$

This gives the Sturm–Liouville problem

$$X''(x) + \lambda X(x) = 0, \quad X(0) = 0, \quad X(l) = 0, \tag{5.7}$$

and

$$T''(t) + \lambda c^2 T(t) = 0. \tag{5.8}$$

Using arguments similar to those used earlier, we conclude that for values of $\lambda \leq 0$, the only solution to (5.7) is the **trivial solution**. If $\lambda > 0$, we set $\lambda = \alpha^2$ with $\alpha > 0$. Then (5.7) becomes

$$X''(x) + \alpha^2 X(x) = 0,$$

whose general solution is

$$X(x) = A\cos\alpha x + B\sin\alpha x.$$

Applying $X(0) = 0$ gives $A = 0$. Substituting $A = 0$ into $X(x)$, we have

$$X(x) = B\sin\alpha x.$$

Applying $X(l) = 0$ yields $\sin\alpha l = 0$, which implies

$$\alpha l = n\pi, \quad n = 1, 2, \ldots$$

and therefore

$$\alpha = \frac{n\pi}{l}, \quad n = 1, 2, \ldots$$

Thus, for $\lambda_n = \left(\frac{n\pi}{l}\right)^2$, the corresponding eigenfunctions are

$$X_n(x) = \sin\frac{n\pi x}{l}, \quad n = 1, 2, \ldots$$

On the other hand, given $\lambda_n = \left(\frac{n\pi}{l}\right)^2$, the solution of (5.8) is

$$T_n(t) = a_n \cos\frac{n\pi ct}{l} + b_n \sin\frac{n\pi ct}{l}, \quad n = 1, 2, \ldots$$

for arbitrary constants a_n and b_n. Therefore, the solution takes the general form

$$u(x,t) = \sum_{n=1}^{\infty}\left[a_n \cos\frac{n\pi ct}{l} + b_n \sin\frac{n\pi ct}{l}\right]\sin\frac{n\pi x}{l}. \tag{5.9}$$

Differentiating the series (5.9) with respect to t gives

$$u_t = \sum_{n=1}^{\infty}\frac{n\pi c}{l}\left[-a_n \sin\frac{n\pi ct}{l} + b_n \cos\frac{n\pi ct}{l}\right]\sin\frac{n\pi x}{l}.$$

Next, by applying the initial conditions (5.4) and (5.5), we obtain

$$u(x,0) = f(x) = \sum_{n=1}^{\infty} a_n \sin\frac{n\pi x}{l},$$

$$u_t(x,0) = g(x) = \sum_{n=1}^{\infty} b_n \frac{n\pi c}{l}\sin\frac{n\pi x}{l}.$$

These equations are satisfied if $f(x)$ and $g(x)$ can be represented by Fourier sine series. The coefficients are given by

$$a_n = \frac{2}{l}\int_0^l f(x)\sin\frac{n\pi x}{l}\,dx, \quad b_n = \frac{2}{n\pi c}\int_0^l g(x)\sin\frac{n\pi x}{l}\,dx.$$

Now we try to write the solution in (5.9) in compact form when the initial velocity is zero, or $g(x) = 0$. Suppose that f is continuous on $0 \le x \le l$ such that $f(0) = f(l) = 0$. Then the solution

$$u(x,t) = \sum_{n=1}^{\infty} B_n \sin\frac{n\pi x}{l}\cos\frac{n\pi ct}{l}, \tag{5.10}$$

where

$$B_n = \frac{2}{l}\int_0^l f(x)\sin\frac{n\pi x}{l}dx, \quad n = 1, 2, \ldots$$

Solution (5.10) is easily written in a closed form that does not involve infinite series. We can do this with the aid of the trigonometric identity:

$$2\sin A\cos B = \sin(A+B) + \sin(A-B),$$

by first noting that

$$\sin\frac{n\pi x}{l}\cos\frac{n\pi ct}{l} = \frac{1}{2}\left[\sin\frac{n\pi(x+ct)}{l} + \sin\frac{n\pi(x-ct)}{l}\right],$$

series (5.9) then becomes

$$u(x,t) = \frac{1}{2}\sum_{n=1}^{\infty} B_n \sin\frac{n\pi(x+ct)}{l} + \frac{1}{2}\sum_{n=1}^{\infty} B_n \sin\frac{n\pi(x-ct)}{l}. \tag{5.11}$$

The series

$$\sum_{n=1}^{\infty} a_n \sin\frac{n\pi x}{l},$$

represents the odd periodic extension of $f(x)$. By calling this extension $F(x)$ then (5.11) may take the compact form

$$u(x,t) = \frac{1}{2}\left[F(x-ct) + F(x+ct)\right],$$

which we will develop in Chapter 10 and is referred to as d'Alembert's solution.

Example 5.1 *(Plucked string)* Consider the spring problem given by (5.3)–(5.6) with the assumptions that $g(x) = 0$ and

$$u(x,0) = f(x) = \begin{cases} \frac{hx}{a}, & 0 \le x \le a \\ \frac{h(L-x)}{L-a}, & a \le x \le L \end{cases}$$

A plucked string refers to a string that is set into motion by being pulled away from its resting position, here by an amount h and then released (see Fig. 5.1). This is a common way to generate vibrations and produce sound in stringed musical instruments, such as guitars, harps, or sitars.
The coefficients b_n are identically zero, since $g(x) = 0$. On the other hand,

$$a_n = \frac{2}{L}\int_0^L f(x)\sin\frac{n\pi x}{L}\,dx = \frac{2}{L}\int_0^a \frac{hx}{a}\sin\frac{n\pi x}{L}\,dx + \frac{2}{L}\int_a^L \frac{h(L-x)}{L-a}\sin\frac{n\pi x}{L}\,dx.$$

An integration by parts yields after simplification to

$$a_n = \frac{2hl^2}{\pi^2 a(L-a)}\frac{1}{n^2}\sin\frac{n\pi a}{L}. \tag{5.12}$$

Consequently, the displacement of the string is

$$u(x,t) = \frac{2hL^2}{\pi^2 a(L-a)}\sum_{n=1}^{\infty}\frac{1}{n^2}\sin\frac{n\pi a}{L}\sin\frac{n\pi x}{L}\cos\frac{n\pi ct}{L}.$$

Let us examine the physical meaning relative to music where instruments with vibrating strings are used. The obtained solution is a sum of harmonics (vibrational modes), each corresponding to a frequency. The term

$$\cos\left(\frac{n\pi ct}{L}\right)$$

represents oscillation in time with frequency

$$f_n = \frac{nc}{2L},$$

The first frequency, or the fundamental frequency (first harmonic)

$$f_1 = \frac{c}{2L} = \frac{1}{2L}\sqrt{\frac{T}{\rho}}$$

where $c = \sqrt{\frac{T}{\rho}}$ is the wave speed, T is the tension in the string, ρ is the linear density (mass per unit length), and L is the length of the string determines the musical pitch of the note you hear when the string is plucked. For example, the note A_4 on a guitar string vibrates primarily at 440 Hz–this is its fundamental. The other terms in the series (with $n > 1$) are overtones or higher harmonics. Their frequencies are integer multiples of the fundamental. These shape the *timbre* or tone quality, but the first frequency is the base.

□

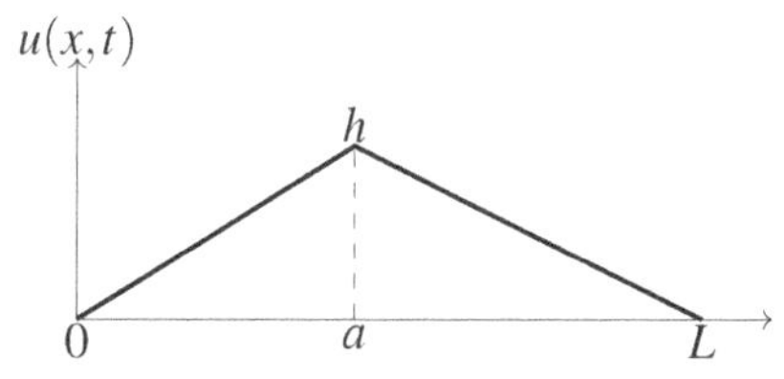

FIGURE 5.1: The string is raised to a height of h at $x = a$ and then released with zero initial velocity.

Example 5.2 *(Struck string)* Consider the spring problem given by (5.3)–(5.6) with the assumptions that the initial displacement $f(x) = 0$ and initial velocity

$$u_t(x,0) = g(x) = \begin{cases} \frac{v_0 x}{a}, & 0 \le x \le a \\ \frac{v_0(L-x)}{L-a}, & a \le x \le L \end{cases}$$

A struck string refers to a string that is set into motion by being hit or struck at a specific point, rather than being plucked or displaced. This action typically occurs in musical instruments such as pianos, harpsichords, or certain percussion instruments. The coefficients a_n are identically zero, since $f(x) = 0$. On the other hand,

$$b_n = \frac{2}{n\pi c}\int_0^a v_0 \frac{x}{a}\sin\frac{n\pi x}{L}\,dx + \frac{2}{n\pi c}\int_a^L v_0 \frac{(L-x)}{(L-a)}\sin\frac{n\pi x}{L}\,dx = \frac{2v_0 L^3}{\pi^3 ca(L-a)}\frac{1}{n^3}\sin\frac{n\pi a}{L}, \tag{5.13}$$

and consequently, the solution is given by

$$u(x,t) = \frac{2v_0 l^3}{\pi^3 ac(L-a)}\sum_{n=1}^{\infty}\frac{1}{n^3}\sin\frac{n\pi a}{L}\sin\frac{n\pi x}{L}\cos\frac{n\pi ct}{L}.$$

□

In the Exercises 5.6 and 5.7 we look at the **damped wave equation**

$$u_{tt} + bu_t = c^2 u_{xx},$$

due to air resistance that is proportional to the spring's velocity, where $b > 0$ is the damping constant.

5.2.1 Air Pressure in Thin Pipe

Let $p(x,t)$ be the air pressure perturbation as a function of position x and time t. Also, suppose that $u(x,t)$ is particle velocity along the axis of the pipe. Let ρ_0 = equilibrium density of the air, and c = speed of sound in the air. Then the equations that govern the system are the "Continuity Equations" given by the mass conservation equation

$$\frac{\partial p}{\partial x} = -\rho_0 \frac{\partial u}{\partial t},$$

and the momentum equation, or Newton's second law

$$\frac{\partial u}{\partial x} = -\frac{1}{\rho_0}\frac{\partial p}{\partial t}.$$

Combining the above equations to eliminate $u(x,t)$, we obtain the wave equation for air pressure $p(x,t)$

$$\frac{\partial^2 p}{\partial t^2} = c^2 \frac{\partial^2 p}{\partial x^2},$$

where

$$c = \sqrt{\frac{\gamma p_0}{\rho_0}}$$

is the speed of sound in the pipe, with γ being the adiabatic index, p_0 the equilibrium pressure, and ρ_0 the equilibrium density. Since the pipe is sealed at both ends, there is no air motion at the boundaries. This implies that the velocity of air particles at the ends must be zero. Using the acoustic relationship between pressure and particle velocity, this means

$$\frac{\partial p}{\partial x} = 0 \quad \text{at} \quad x = 0 \quad \text{and} \quad x = L,$$

where L is the length of the pipe. These are Neumann boundary conditions, indicating that the spatial derivative of pressure is zero at the pipe ends.

5.2.2 Exercises

Exercise 5.1 *Work out the details for* (5.12) *and* (5.13).

In Exercises 5.2–5.5, solve for the solution $u(x,t)$.

Exercise 5.2 $u_{tt} - 4u_{xx} = 0, \quad u(x,0) = x, \quad u_t(x,0) = 0, \quad u(0,t) = u(\pi,t) = 0.$

Exercise 5.3 $u_{tt} - 4u_{xx} = 0, \quad u(x,0) = x(x-\pi), \quad u_t(x,0) = 0, \quad u(0,t) = u(\pi,t) = 0.$

Exercise 5.4 $u_{tt} - c^2 u_{xx} = 0, \quad u(x,0) = 3\sin x, \quad u_t(x,0) = 0, \quad u(0,t) = u(\pi,t) = 0.$

Exercise 5.5 $u_{tt} - c^2 u_{xx} = 0, \quad u(x,0) = 0, \quad u_t(x,0) = x\sin x, \quad u(0,t) = u(1,t) = 0.$

In Exercises 5.6–5.7 use separation of variables to solve.

Exercise 5.6 $u_{tt} + 2u_t = u_{xx}, \quad u(x,0) = x, \quad u_t(x,0) = 0, \quad u(0,t) = u(\pi,t) = 0.$

Exercise 5.7 $u_{tt} + 2u_t = 4u_{xx}, \quad u(x,0) = 0, \quad u_t(x,0) = g(x), \quad u(0,t) = u(\pi,t) = 0,$

$$g(x) = \begin{cases} 0, & 0 < x < 1 \\ 1, & 1 < x < \pi \end{cases}$$

Exercise 5.8 *Find the general solution of*

$$u_{tt} - c^2 u_{xx} = 0, \quad 0 < x < \pi, \quad t > 0,$$

$$u(x,0) = f(x), \quad 0 \le x \le \pi,$$

$$u_t(x,0) = g(x), \quad 0 \le x \le \pi,$$

$$u_x(0,t) = u_x(\pi,t) = 0, \quad t \ge 0.$$

Exercise 5.9 *Solve* $u_{tt} - 4u_{xx} = 0, \ u_t(x,0) = 0, \quad u(0,t) = u(4,t) = 0.$

$$u(x,0) = f(x) = \begin{cases} hx, & 0 \le x \le 2 \\ -h(x-4), & 2 < x \le 4 \end{cases}$$

Exercise 5.10 *(Timoshenko, S., and J.N. Goodier. Theory of Elasticity)[138] In a musical instrument, a string is generally less flexible and for such a string, the displacement u may satisfy the following partial differential equation*

$$\frac{\partial^2 u}{\partial x^2} - \eta \frac{\partial^4 u}{\partial x^4} = \frac{1}{c^2}\frac{\partial^2 u}{\partial t^2}, \quad 0 < x < a, \quad t > 0,$$

where $c^2 = \frac{T}{\rho}$, *and* $\eta = \frac{EI}{T}$, *with E as Young's modulus for the material, and I as the second moment of area. Assuming that* $u(x,t) = \phi(x)T(t)$, *carry out a separation of variables and find the eigenvalue problem for* ϕ. *Assume boundary conditions of the form*

$$u(0,t) = 0, \quad u(a,t) = 0,$$

$$\frac{\partial^2 u}{\partial x^2}(0,t) = 0, \quad \frac{\partial^2 u}{\partial x^2}(a,t) = 0, \quad t > 0.$$

Show that $\phi(x) = \sin(\mu x)$ *satisfies the eigenvalue problem found provided that* $\mu = \frac{n\pi}{a}$ *and*

$$\lambda^2 = \mu^2 + \eta\mu^4, \quad \text{where} \quad -\lambda^2 = \frac{T''(t)}{c^2 T(t)}.$$

Exercise 5.11 *(Love, A.E.H)[98] The longitudinal vibration of a thin rod has been described by the equations*

$$c^2 \frac{\partial^2 u}{\partial x^2} = \frac{\partial^2 u}{\partial t^2} - \eta \frac{\partial^4 u}{\partial x^2 \partial t^2}, \quad 0 < x < a, \quad t > 0.$$

Here, $u(x,t)$ *is the displacement of the points at position x when there is no motion;* $\eta = \nu R^2$, *where* ν *is Poisson's ratio and R is the radius of gyration of a cross section of the rod. Assume boundary conditions of the form*

$$u(0,t) = 0, \quad u(a,t) = 0,$$

and use separation of variables $u(x,t) = \phi(x)T(t)$. *The calculations are simplified if you let* $\frac{\phi''}{\phi} = -\lambda^2$.

Exercise 5.12 *Rework Exercises 5.6–5.7 using the transformation*

$$u(x,t) = e^{-bt} v(x,t),$$

where $b > 0$ *is the damping constant.*

5.3 Forced Vibrating String at Resonance

Now we analyze a vibrating string subject to outside forces. In particular, we consider the forced vibrating string with non-homogeneous conditions of the form

$$u_{tt} - c^2 u_{xx} = Q(x,t), \quad 0 < x < L, \quad t > 0, \tag{5.14}$$

$$u(0,t) = A, \ u(L,t) = B, \tag{5.15}$$

$$u(x,0) = f(x), \tag{5.16}$$

$$u_t(x,0) = g(x), \quad 0 < x < L. \tag{5.17}$$

This describes a vibrating string influenced by external forces, modeled by the non-homogeneous wave equation with boundary and initial conditions. Equation (5.14) represents the wave propagation, where $Q(x,t)$ accounts for external forces acting on the string. The boundary conditions (5.15)ensure the string's displacement is fixed at $x = 0$ and $x = L$. The initial conditions given by (5.16) and (5.17) specify the string's initial shape and velocity. Together, these conditions describe the evolution of the string's motion under the influence of external forces.

• Method of Solution

We begin with finding the simplest function, say, $p(x)$ that satisfies $p(0) = A$ and $p(L) = B$. The simplest form of $p(x)$ is a straight path between $(0,A)$ and (L,B). Hence,

$$p(x) = A + \frac{x}{L}[B - A]. \tag{5.18}$$

Next, we aim to transform the non-homogeneous boundary conditions into homogeneous ones. To achieve this, we define

$$v(x,t) = u(x,t) - p(x).$$

Then

$$u_{tt} = v_{tt}, \quad u_{xx} = v_{xx} - p_{xx} = v_{xx}.$$

Substituting into (5.14) we arrive at

$$v_{tt} = c^2 v_{xx} + Q(x,t).$$

As for the boundary conditions, one obtains

$$v(0,t) = u(0,t) - p(0) = A - A = 0,$$

$$v(L,t) = u(L,t) - p(L) = B - B = 0.$$

Finally, we transfer our initial condition. That is

$$v(x,0) = u(x,0) - p(0) = f(x) - A. \tag{5.19}$$

Similarly,

$$v_t(x,0) = u_t(x,0) = g(x). \tag{5.20}$$

Thus, our original problem has been reformulated in terms of v as a non-homogeneous vibrating string problem with homogeneous boundary conditions. Consequently,

$$v_{tt} = c^2 v_{xx} + Q(x,t), \quad 0 < x < L, \quad t > 0, \tag{5.21}$$
$$v(0,t) = 0, \quad v(L,t) = 0, \tag{5.22}$$
$$v(x,0) = f(x) - A, \tag{5.23}$$
$$v_t(x,0) = g(x). \tag{5.24}$$

Let λ_n and φ_n be the eigenvalues and corresponding eigenfunctions of the homogeneous problem

$$v_{tt} = c^2 v_{xx}, \quad v(0,t) = 0, \quad v(L,t) = 0.$$

We express the solution $v(x,t)$ of (5.21)–(5.23) as a generalized Fourier series of the eigenfunctions φ_n with time dependent coefficients $a_n(t)$ and obtain,

$$v(x,t) = \sum_{n=1}^{\infty} a_n(t)\varphi_n(x). \tag{5.25}$$

We now represent $Q(x,t)$ as a generalized Fourier series, where the eigenfunctions form the spatial basis and the coefficients depend on time. The expansion is expressed as

$$Q(x,t) = \sum_{n=1}^{\infty} q_n(t)\varphi_n(x),$$

where the time-dependent Fourier coefficients $q_n(t)$ are determined by the following integral (see Section 3.5)

$$q_n(t) = \frac{\int_0^L Q(x,t)\varphi_n(x)\,dx}{\int_0^L \varphi_n^2(x)\,dx}. \tag{5.26}$$

With this in mind, expression (5.21) takes the form

$$v_{tt} = c^2 v_{xx} + \sum_{n=1}^{\infty} q_n(t)\varphi_n(x), \quad 0 < x < L, \quad t > 0. \tag{5.27}$$

Let $'$ denote differentiation with respect to the argument of the corresponding function. Assuming the generalized Fourier series in (5.25) can be differentiated term by term, substituting (5.25) into (5.27)) leads to

$$\sum_{n=1}^{\infty} a_n''(t)\varphi_n(x) = c^2 \sum_{n=1}^{\infty} a_n(t)\varphi_n''(x) + \sum_{n=1}^{\infty} q_n(t)\varphi_n(x).$$

Since

$$\varphi_n''(x) = -\lambda_n \varphi_n(x),$$

the above expression reduces to

$$\sum_{n=1}^{\infty} \left[a_n''(t) + c^2\lambda_n a_n(t) - q_n(t)\right]\varphi_n(x) = 0.$$

However, the above expression must hold for all $0 < x < L$, and so we must have

$$a_n''(t) + \lambda_n c^2 a_n(t) - q_n(t) = 0, \quad t > 0. \tag{5.28}$$

Equation (5.31) is of second order and non-homogeneous. In most situations, it is considered that the vibrations of a string is subjected to a periodic forcing

$$Q(x,t) = q(x)\cos(\omega t).$$

In this special case, (5.26) implies that

$$q_n(t) = \frac{\int_0^L q(x)\varphi_n(x)\,dx}{\int_0^L \varphi_n^2(x)\,dx}\cos\omega t = q_n\cos\omega t, \tag{5.29}$$

where

$$q_n = \frac{\int_0^L q(x)\varphi_n(x)\,dx}{\int_0^L \varphi_n^2(x)\,dx}. \tag{5.30}$$

Then the general second-order differential equation in t reduces to

$$a_n''(t) + \lambda_n c^2 a_n(t) - q_n \cos \omega t = 0, \quad t > 0. \tag{5.31}$$

The homogeneous part is

$$a_n''(t) + \lambda_n c_n^2 a(t) = 0,$$

and it has the solution

$$a_n^h(t) = C_1 \cos(\sqrt{\lambda_n} ct) + C_2 \sin(\sqrt{\lambda_n} ct),$$

where C_1 and C_2 are constants determined by initial conditions. Based on the term $q_n \cos \omega t$, we assume a particular solution of the form

$$a_n^p(t) = A \cos \omega t + B \cos \omega t.$$

Differentiating $a_n^p(t)$ twice and substituting into (5.31), we obtain after some simplifications that,

$$A(\lambda_n c^2 - \omega^2) \cos \omega t + B(\lambda_n c^2 - \omega^2) \sin \omega t = q_n \cos \omega t.$$

Equating both sides we arrive at

$$A = \frac{q_n}{\lambda_n c^2 - \omega^2}, \qquad B = 0.$$

Since the general solution is $a_n(t) = a_n^h(t) + a_n^p(t)$, we obtain

$$a_n(t) = C_1 \cos(\sqrt{\lambda_n} ct) + C_2 \sin(\sqrt{\lambda_n} ct) + \frac{q_n}{\lambda_n c^2 - \omega^2} \cos \omega t. \tag{5.32}$$

Making use of (5.23) and (5.24) into (5.25) leads to

$$f(x) - A = \sum_{n=1}^{\infty} a_n(0) \varphi_n(x),$$

and

$$g(x) = \sum_{n=1}^{\infty} a_n'(0) \varphi_n(x).$$

Since φ_n are orthogonal, we arrive at using Section 3.5 that

$$a_n(0) = \frac{\int_0^L \left[f(x) - A\right] \varphi_n(x)\, dx}{\int_0^L \varphi_n^2(x)\, dx}, \quad a_n'(0) = \frac{\int_0^L g(x) \varphi_n(x)\, dx}{\int_0^L \varphi_n^2(x)\, dx}. \tag{5.33}$$

Using the two relations in (5.33), we can determine the constants C_1 and C_2 in (5.32), thereby fully solving for $a_n(t)$. Finally, substituting $a_n(t)$ into (5.25), we obtain the complete solution. In order for (5.32) to be valid, we must assume that

$$c^2 \lambda_n - \omega^2 \neq 0.$$

In other words, *the forcing frequency ω is different from the natural frequency $c\sqrt{\lambda_n}$.* If the forcing frequency ω is the same as one of the natural frequencies $c\sqrt{\lambda_n}$, **resonance occurs.** Technically speaking, if the forcing frequency aligns with one of the eigenfrequencies of the string, energy is efficiently transferred into the string, resulting in amplified oscillations. This phenomenon leads to standing wave patterns with well-defined nodes and antinodes. □

Example 5.3 Consider the following non-homogeneous problem

$$\begin{aligned} &u_{tt} = u_{xx} + x\cos\,\omega t \quad 0 < x < \pi, \quad t > 0, \\ &u(0,t) = 4, \quad u(\pi,t) = 5, \quad u(x,0) = x, \quad u_t(x,0) = 2. \end{aligned}$$

Then

$$p(x) = 4 + \frac{x}{\pi}$$

and

$$v(x,t) = u(x,t) - p(x) = u(x,t) - 4 - \frac{x}{\pi}.$$

Consequently, we have the new vibration problem

$$\begin{aligned} &v_{tt} = v_{xx} + x\cos\,\omega t, \quad 0 < x < \pi, \quad t > 0, \\ &v(0,t) = 0, \quad v(\pi,t) = 0, \quad v(x,0) = x - 4, \quad v_t(x,0) = 2. \end{aligned}$$

Let λ_n and φ_n be the eigenvalues and corresponding eigenfunctions of the homogeneous problem

$$v_{tt} = v_{xx}, \quad v(0,t) = 0, \quad v(\pi,t) = 0.$$

That is they are solutions pairs to

$$X''(x) + \lambda X(x) = 0, \quad X(0) = X(\pi) = 0.$$

The eigenvalues and corresponding eigenfunctions can be easily determined and are given by

$$\lambda_n = n^2, \quad \varphi_n(x) = \sin\,nx, \quad n = 1,2,\ldots.$$

We compute some of the terms.

$$\int_0^\pi \varphi_n^2(x)\,dx = \int_0^\pi \sin^2 nx\,dx = \frac{1}{2}\int_0^\pi (1 - \cos\,2nx)\,dx = \frac{\pi}{2}.$$

Using this result, (5.33) becomes

$$a_n(0) = \frac{2}{\pi}\int_0^\pi [x-4]\sin\,nx\,dx = \frac{2(4-\pi)}{\pi n}\big[1 + (-1)^{n+1}\big]. \tag{5.34}$$

Along the same lines,

$$a_n'(0) = \frac{2}{\pi}\int_0^\pi 2\sin(nx)\,dx = \begin{cases} \frac{8}{\pi n}, & \text{if } n \text{ is odd} \\ 0, & \text{if } n \text{ is even} \end{cases} \tag{5.35}$$

Using (5.32) we have

$$a_n(t) = C_1 \cos nt + C_2 \sin nt + \frac{q_n}{n^2 - \omega^2} \cos \omega t. \tag{5.36}$$

Moreover, from (5.30) we have

$$q_n = \frac{2}{\pi} \int_0^{\pi} x \sin nx\, dx = \frac{2(-1)^{n+1}}{n}, \quad n = 1, 2, \ldots.$$

Henceforth, $a_n(0) = C_1 + \frac{q_n}{n^2 - \omega^2}$, from which we have

$$C_1 = a_n(0) - \frac{q_n}{n^2 - \omega^2}. \tag{5.37}$$

From (5.36) we have

$$a_n'(t) = -nC_1 \sin nt + nC_2 \cos nt - \omega \frac{q_n}{n^2 - \omega^2} \sin \omega t.$$

Substituting $t = 0$ into both sides gives $a_n'(0) = nC_2$, which implies

$$C_2 = \frac{a_n'(0)}{n} = \begin{cases} \frac{8}{\pi n^2}, & \text{if } n \text{ is odd} \\ 0, & \text{if } n \text{ is even} \end{cases} \tag{5.38}$$

Let $a_n(0), a_n'(0)$ be given by (5.34) and (5.35), respectively. The solution of $v(x,t)$ is given by (5.25) and hence using (5.36)–(5.38) we arrive at

$$\begin{aligned} v(x,t) &= \sum_{n=1}^{\infty} \left[\left(a_n(0) - \frac{2(-1)^{n+1}}{n(n^2 - \omega^2)} \right) \cos nt + \frac{2(-1)^{n+1}}{n(n^2 - \omega^2)} \cos \omega t \right] \sin nx \\ &+ \sum_{n=1}^{\infty} \frac{8}{\pi (2n-1)^2} \sin (2n-1)x, \quad n \neq \omega. \end{aligned}$$

Thus, the general solution to the original problem is

$$u(x,t) = v(x,t) + p(x),$$

which leads to the conclusion that

$$\begin{aligned} u(x,t) &= 4 + \frac{x}{\pi} + \sum_{n=1}^{\infty} \left[\left(a_n(0) - \frac{2(-1)^{n+1}}{n(n^2 - \omega^2)} \right) \cos nt + \frac{2(-1)^{n+1}}{n(n^2 - \omega^2)} \cos \omega t \right] \sin nx \\ &+ \sum_{n=1}^{\infty} \frac{8}{\pi (2n-1)^2} \sin (2n-1)x, \quad n \neq \omega, \end{aligned}$$

where $a_n(0)$ is given by (5.34). □

5.3.1 Exercises

Exercise 5.13 *Show that if the forcing frequency ω is the same as one of the natural frequencies $c\sqrt{\lambda_n}$, then the general solution to (5.32) becomes*

$$a_n(t) = c_1 \cos(c\sqrt{\lambda_n} t) + c_2 \sin(c\sqrt{\lambda_n} t) + \frac{q_n}{2\omega} t \sin \omega t.$$

Exercise 5.14 *Solve*

$$\begin{aligned} &u_{tt} = u_{xx} + \sin x \sin t, \quad 0 < x < \pi, \quad t > 0, \\ &u(0,t) = 4, \quad u(\pi,t) = 5, \quad u(x,0) = x, \quad u_t(x,0) = 2. \end{aligned}$$

Exercise 5.15 *Solve*

$$\begin{aligned} &u_{tt} = u_{xx} + \sin x \cos 2t, \quad 0 < x < \pi, \quad t > 0, \\ &u(0,t) = 4, \quad u(\pi,t) = 5, \quad u(x,0) = x, \quad u_t(x,0) = 2. \end{aligned}$$

Exercise 5.16 *Develop similar theory for*

$$\begin{aligned} &u_{tt} - c^2 u_{xx} = Q(x,t), \quad 0 < x < L, \quad t > 0, \\ &u_x(0,t) = 0, \quad u_x(L,t) = 0, \\ &u(x,0) = f(x), \quad u_t(x,0) = g(x), \quad 0 < x < L. \end{aligned}$$

Exercise 5.17 *Redo Example 5.3 for*

$$Q(x,t) = \sin \omega t.$$

Exercise 5.18 *For a positive constant h solve*

$$\begin{aligned} &u_{tt} = u_{xx} + h, \quad 0 < x < L, \quad t > 0, \\ &u(0,t) = 0, \quad u(L,t) = 0, \quad u(x,0) = 0, \quad u_t(x,0) = 0. \end{aligned}$$

Ans: $u(x,t) = -\sum_{n=1}^{\infty} \frac{4l^2 h}{c^2\pi^3} \frac{\cos\left((2n-1)\frac{\pi c t}{l}\right)}{(2n-1)^3} \sin\left((2n-1)\frac{\pi x}{l}\right).$

Exercise 5.19 *For a positive constant h solve*

$$\begin{aligned} &u_{tt} = u_{xx} + h, \quad 0 < x < 1, \quad t > 0, \\ &u(0,t) = 0, \quad u(1,t) = 0, \\ &u_t(x,0) = 0, \quad u(x,0) = x(1-x), \quad 0 \le x \le 1. \end{aligned}$$

Ans: $u(x,t) = \sum_{n=1}^{\infty} \left[\frac{4}{n^3\pi^3}[1-(-1)^n] \cos n\pi t + \frac{2h}{n^3\pi^3}[1-(-1)^n](1-\cos n\pi t) \right] \sin n\pi x.$

Exercise 5.20 *Consider the non-homogeneous string problem with time-dependent non-homogeneous boundary conditions*

$$\begin{aligned} &u_{tt} - c^2 u_{xx} = Q(x,t), \quad 0 < x < L, \quad t > 0, \\ &u(0,t) = A(t), \quad u(L,t) = B(t), \\ &u(x,0) = f(x), \quad u_t(x,0) = g(x), \quad 0 < x < L. \end{aligned}$$

(a) Find its general solution. Hint: see Section 4.10.

(b) For a positive constant h use part (a) to solve

$$\begin{aligned} &u_{tt} = u_{xx} + h, \quad 0 < x < 1, \quad t > 0, \\ &u(0,t) = t, \quad u(1,t) = \sin t, \\ &u(x,0) = x(1-x), \quad u_t(x,0) = 0, \quad 0 \le x \le 1. \end{aligned}$$

Ans: $u(x,t) = -\frac{2}{h}x^2 + (\sin t - t + \frac{2}{h})x + t + \sum_{n=1}^{\infty} C_n \cos n\pi t \sin n\pi x$

$C_n = 2\int_0^1 \left(x(1-x) + \frac{2}{h}x^2 - \frac{2}{h}x \right) \sin n\pi x\, dx.$

Exercise 5.21 *Consider a string that is subjected to air resistance proportional to its velocity. Then we are to solve*

$$\begin{aligned} &u_{tt} = a^2 u_{xx} - cu_t + F(x)\sin \omega t, \quad 0 < x < L, \quad t > 0, \\ &u(0,t) = 0, \quad u(L,t) = 0, \quad u(x,0) = 0, \quad u_t(x,0) = 0, \quad 0 \le x \le L. \end{aligned}$$

Show that the solution is given by

$$u(x,t) = \sum_{n=1}^{\infty} \beta_n F_n \sin\left(\frac{n\pi x}{L}\right) \sin(\omega t + \alpha_n),$$

where

$$\beta_n = \left[(\omega_n^2 - \omega^2)^2 + \omega^2 c^2\right]^{-1/2},$$

$$\alpha_n = \tan^{-1}\left(\frac{\omega c}{\omega_n^2 - \omega^2}\right),$$

and F_n are the Fourier sine coefficients of $F(x)$.

5.4 Vibrating String and Double Fourier Series

Double Fourier series provide a powerful tool for solving problems in two dimensions, particularly in mathematical physics and engineering. The choice of sine, cosine, or mixed bases depends on the boundary conditions of the domain and the nature of the problem.

A double Fourier series is an extension of the Fourier series to functions of two variables, typically x and y. It represents a function $f(x,y)$ defined on a rectangular domain $[0,L_x] \times [0,L_y]$ as an infinite sum of sines and cosines in both variables. Double Fourier series are used in solving partial differential equations, especially in problems involving heat, wave, or Laplace equations in two dimensions.

Let $f(x,y)$ be a continuously differentiable periodic function with a period of $2L$ in both variables. That is

$$f(x+2L,y) = f(x,y+2L) = f(x,y).$$

For each value of y, we can expand $f(x,y)$ in a uniformly convergent Fourier series of the form

$$f(x,y) = \frac{1}{2}a_0(y) + \sum_{n=1}^{\infty}\left[a_n(y)\cos\frac{n\pi x}{L} + b_n(y)\sin\frac{n\pi x}{L}\right].$$

The coefficients are given by

$$a_n(y) = \frac{1}{L}\int_{-L}^{L} f(x,y)\cos\frac{n\pi x}{L}\,dx,$$

$$b_n(y) = \frac{1}{L}\int_{-L}^{L} f(x,y)\sin\frac{n\pi x}{L}\,dx,$$

and these coefficients are continuously differentiable in y. The coefficients can be expanded in uniformly convergent Fourier series of the form

$$a_n(y) = \frac{1}{2}a_{n0} + \sum_{m=1}^{\infty}\left(a_{nm}\cos\frac{m\pi y}{L} + b_{nm}\sin\frac{m\pi y}{L}\right),$$

$$b_n(y) = \frac{1}{2}c_{n0} + \sum_{m=1}^{\infty}\left(c_{nm}\cos\frac{n\pi y}{L} + d_{nm}\sin\frac{n\pi y}{L}\right),$$

where

$$a_{nm} = \frac{1}{L^2}\int_{-L}^{L}\int_{-L}^{L} f(x,y)\cos\frac{n\pi x}{L}\cos\frac{m\pi y}{L}\,dx\,dy,$$

$$b_{nm} = \frac{1}{L^2}\int_{-L}^{L}\int_{-L}^{L} f(x,y)\cos\frac{n\pi x}{L}\sin\frac{m\pi y}{L}\,dx\,dy,$$

$$c_{nm} = \frac{1}{L^2}\int_{-L}^{L}\int_{-L}^{L} f(x,y)\sin\frac{n\pi x}{L}\cos\frac{m\pi y}{L}\,dx\,dy,$$

$$d_{nm} = \frac{1}{L^2}\int_{-L}^{L}\int_{-L}^{L} f(x,y)\sin\frac{n\pi x}{L}\sin\frac{m\pi y}{L}\,dx\,dy.$$

Putting the series for the coefficients into the series for $f(x,y)$, we have

$$\begin{aligned} f(x,y) \quad &\sim \quad \frac{1}{4}a_{00} + \frac{1}{2}\sum_{m=1}^{\infty}\left[a_{0m}\cos\frac{m\pi y}{L} + b_{0m}\sin\frac{m\pi y}{L}\right] \\ &+ \quad \frac{1}{2}\sum_{n=1}^{\infty}\left[a_{n0}\cos\frac{n\pi x}{L} + c_{n0}\sin nx\right] \\ &+ \quad \sum_{n=1}^{\infty}\sum_{m=1}^{\infty}\left[a_{nm}\cos\frac{n\pi x}{L}\cos\frac{m\pi y}{L} + b_{nm}\cos\frac{n\pi x}{L}\sin\frac{m\pi y}{L}\right.\end{aligned}$$

$$+ \quad c_{nm} \sin\frac{n\pi x}{L}\cos\frac{m\pi y}{L} + d_{nm}\sin\frac{n\pi x}{L}\sin\frac{m\pi y}{L}\Big] \tag{5.39}$$

and hence the name of *double Fourier series.* Expression (5.39) can be reduced depending on the type of functions we have. For example, if

$$f(-x,y) = -f(x,y) \text{ and } f(x,-y) = -f(x,y),$$

then

$$f(x,y) \sim \sum_{n=1}^{\infty}\sum_{m=1}^{\infty} d_{nm}\sin\frac{n\pi x}{L}\sin\frac{m\pi y}{L}, \tag{5.40}$$

where

$$d_{nm} = \frac{4}{L^2}\int_0^L\int_0^L f(x,y)\sin\frac{n\pi x}{L}\sin\frac{m\pi y}{L}\,dx\,dy.$$

Example 5.4 Consider the function

$$f(x,y) = xy, \quad -1 < x < 1, \quad -1 < y < 1.$$

Then it is obvious that

$$f(-x,y) = -xy = -f(x,y) \quad \text{and} \quad f(x,-y) = -xy = -f(x,y).$$

Then

$$\begin{aligned} d_{nm} &= 4\int_0^1\int_0^1 f(x,y)\sin n\pi x\sin m\pi y\,dx\,dy = 4\int_0^1\int_0^1 xy\sin n\pi x\sin m\pi y\,dx\,dy \\ &= 4\left(\int_0^1 x\sin n\pi x\,dx\right)\left(\int_0^1 y\sin m\pi y\,dy\right) = 4\left[\frac{-\cos n\pi}{n\pi}\right]\left[\frac{-\cos m\pi}{m\pi}\right] \\ &= \frac{4(-1)^{n+m}}{nm\pi^2}. \end{aligned}$$

Therefore,

$$xy \sim 4\sum_{n=1}^{\infty}\sum_{m=1}^{\infty}\frac{(-1)^{n+m}}{nm\pi^2}\sin n\pi x\sin m\pi y.$$

Remark 10 *Above, we discussed the double Fourier series of functions defined on a square. However, this concept can be extended to rectangles as well. For example, if we assume the periodicity conditions $f(x+2L,y) = f(x,y)$ and $f(x,y+2H) = f(x,y)$, then the discussion can be generalized to a function $f(x,y)$ defined on the rectangle $-L < x < L$ and $-H < y < H$. In this case, the arguments of the trigonometric terms would be $\frac{n\pi x}{L}$ and $\frac{m\pi y}{H}$ for x and y, respectively.*

□

Example 5.5 *(The Vibrating Rectangular Membrane)* The problem is given by the two-dimensional wave equation in Cartesian coordinates

$$u_{tt} = c^2\left(u_{xx} + u_{yy}\right), \quad t > 0, \quad 0 < x < L, \quad 0 < y < H, \tag{5.41}$$

subject to the following set of boundary conditions

$$u(0,y,t) = 0, \quad u(L,y,t) = 0, \quad t > 0, \quad 0 < y < H, \tag{5.42}$$

$$u(x,0,t) = 0, \quad u(x,H,t) = 0, \quad t > 0, \quad 0 < x < L, \tag{5.43}$$

and a pair of initial conditions

$$u(x,y,0) = f(x,y), \quad u_t(x,y,0) = g(x,y). \tag{5.44}$$

We assume that the functions f and g are piecewise continuous on the rectangle of their domain and possess continuous partial derivatives of the first order. The two-dimensional wave equation describes the vibration of a rectangular membrane under tension. The term $u(x,y,t)$ represents the vertical displacement of the membrane at position (x,y) and time t. The equation (5.41) expresses the balance between the inertial forces of the membrane (represented by u_{tt}) and the restoring forces due to tension (represented by $u_{xx} + u_{yy}$). The speed of wave propagation, c, is determined by the physical properties of the membrane, such as its tension and mass per unit area. The boundary conditions ensure that the edges of the membrane remain fixed, as indicated by $u = 0$ at $x = 0, x = L$ and $y = 0, y = H$. The initial conditions given by (5.44), specify the initial shape and velocity distribution of the membrane, which determine how it begins to vibrate. Together, the wave equation, boundary conditions, and initial conditions model the physical behavior of the membrane's oscillations over time.

We use the method of separation of variables by assuming a nonzero solution of the form

$$u(x,y,t) = X(x)Y(y)T(t). \tag{5.45}$$

Hence (5.45) satisfies (5.41) if

$$\frac{T'(t)}{c^2T(t)} = \frac{X''(x)}{X(x)} + \frac{Y''(y)}{Y(y)} = -\lambda.$$

Additionally, $\frac{X''(x)}{X(x)} + \frac{Y''(y)}{Y(y)} = -\lambda$, implies that for another separation constant μ,

$$\frac{Y''(y)}{Y(y)} = -\lambda - \frac{X''(x)}{X(x)} = -\mu.$$

So we are led to two Sturm–Liouville problems

$$X''(x) + (\lambda - \mu)X(x) = 0, \quad X(0) = 0, \quad X(L) = 0$$

and

$$Y''(y) + \mu Y(y) = 0, \quad Y(0) = 0, \quad Y(H) = 0,$$

and to the second-order differential equation in $T(t)$,

$$T''(t) + \lambda c^2 T(t) = 0.$$

We begin by solving the Sturm–Liouville problem for $Y(y)$, as it involves only one separation constant. From previous work we already know that this problem has eigenvalues $\mu = \frac{m^2\pi^2}{H^2}$, where $m = 1,2,\ldots$, and the corresponding eigenfunctions are

$$Y_m(y) = \sin\frac{m\pi y}{H}, \quad m = 1,2,\ldots.$$

Examining the equations in X, we observe that, due to its boundary conditions, it has the eigenvalues

$$\lambda - \mu = \frac{n^2\pi^2}{L^2}, \quad n = 1,2,\ldots$$

with the corresponding eigenfunctions

$$X_n(x) = \sin\frac{n\pi x}{L}, \quad n = 1,2,\ldots.$$

By substituting

$$\lambda = \mu + \frac{n^2\pi^2}{L^2} = \frac{m^2\pi^2}{H^2} + \frac{n^2\pi^2}{L^2},$$

into the equation in T, we obtain

$$T''(t) + (\frac{m^2\pi^2}{H^2} + \frac{n^2\pi^2}{L^2})c^2T(t) = 0,$$

whose general solution is

$$T_{nm}(t) = \cos(\omega_{nm}t) + \sin(\omega_{nm}t),$$

where

$$\omega_{nm} = c\sqrt{\frac{m^2\pi^2}{H^2} + \frac{n^2\pi^2}{L^2}},$$

is the natural frequency of the system. Consequently, the solution to our boundary value problem takes the form

$$u(x,y,t) = \sum_{n=1}^{\infty}\sum_{m=1}^{\infty}(d_{nm}\cos\omega_{nm}t + b_{nm}\sin\omega_{nm}t)\sin\frac{n\pi x}{L}\sin\frac{m\pi y}{H}, \tag{5.46}$$

where ω_{nm} is the natural frequency of the system. Applying the first initial condition of (5.44) yields the relation

$$f(x,y) = \sum_{n=1}^{\infty}\sum_{m=1}^{\infty} d_{nm}\sin\left(\frac{n\pi x}{L}\right)\sin\left(\frac{m\pi y}{H}\right),$$

which is (5.40). Therefore,

$$d_{nm} = \frac{4}{LH}\int_0^H\int_0^L f(x,y)\sin\frac{n\pi x}{L}\sin\frac{m\pi y}{L}\,dxdy.$$

Differentiating (5.46) with respect to t and then applying the second initial condition of (5.44) gives

$$g(x,y) = \sum_{n=1}^{\infty}\sum_{m=1}^{\infty} b_{nm}\omega_{nm} \sin\frac{n\pi x}{L} \sin\frac{m\pi y}{H}.$$

Again, we have a double Fourier sine series, where its coefficient is provided by

$$b_{nm} = \frac{4}{\omega_{nm}LH}\int_0^H \int_0^L g(x,y)\sin\frac{n\pi x}{L}\sin\frac{m\pi y}{H}dxdy.$$

Substituting the coefficients d_{nm} and b_{nm} into (5.46) gives the full solution to the vibrating rectangular membrane. □

5.4.1 Exercises

Exercise 5.22 *Provide a special form of the solution* (5.39) *in the following cases when,*

(a) $f(-x,y) = f(x,y)$ *and* $f(x,-y) = f(x,y)$,

(b) $f(-x,y) = f(x,y)$ *and* $f(x,-y) = -f(x,y)$,

(b) $f(-x,y) = -f(x,y)$ *and* $f(x,-y) = f(x,y)$.

Exercise 5.23 *Determine the double Fourier series of the following functions.*

$$\begin{array}{lll}
(a)\quad f(x,y) = 1, & -L < x < L, & -H < y < H.\\
(b)\quad f(x,y) = x^2y, & 0 < x < \pi, & 0 < y < \pi.\\
(c)\quad f(x,y) = x^2y^2, & 0 < x < \pi, & 0 < y < \pi.\\
(d)\quad f(x,y) = x+y, & -\pi < x < \pi, & -\pi < y < \pi.\\
(e)\quad f(x,y) = x\sin y, & -\pi < x < \pi, & -\pi < y < \pi.\\
(f)\quad f(x,y) = xy, & 0 < x < 1, & 0 < y < 2.
\end{array}$$

Exercise 5.24 *Redo Example 5.5 for* $L = H = \pi$ *and* $g(x) = 0$.

Exercise 5.25 *Consider the two-dimensional heat equation*

$$u_t = c^2\nabla^2 u = c^2(u_{xx} + u_{yy}), \quad 0 < x < a, \quad 0 < y < b.$$

At the edges of the plate, we impose homogeneous Dirichlet boundary conditions

$$u(0,y,t) = u(a,y,t) = 0, \quad 0 \le y \le b,\ t \ge 0,$$
$$u(x,0,t) = u(x,b,t) = 0, \quad 0 \le x \le a,\ t \ge 0.$$

The initial condition is given by

$$u(x,y,0) = f(x,y), \quad (x,y) \in R,$$

where $R = [0,a] \times [0,b]$.

(a) Find the general solution.

(b) Find the solution when a 2×2 square plate with thermal diffusivity $c = \frac{1}{3}$ is heated in such a way that the temperature in the lower half is 50, while the temperature in the upper half is 0. After that, it is insulated laterally, and the temperature at its edges is held at 0. Find an expression that gives the temperature $u(x,y,t)$ in the plate for $t > 0$.

Hint: the initial condition $f(x,y)$ is given by

$$f(x,y) = \begin{cases} 50, & \text{if } 0 \le y \le 1 \\ 0, & \text{if } 1 < y < 2 \end{cases}$$

Ans (b): $u(x,y,t) = \dfrac{200}{\pi^2} \sum_{m=1}^{\infty} \sum_{n=1}^{\infty} \dfrac{[1+(-1)^{m+1}][1-\cos\frac{n\pi}{2}]}{mn} \sin\dfrac{m\pi x}{2} \sin\dfrac{n\pi y}{2} e^{-\frac{\pi^2(m^2+n^2)t}{36}}$.

Exercise 5.26 *Solve*

$$\nabla^2 u = \frac{\partial^2 u}{\partial x^2} + \frac{\partial^2 u}{\partial y^2} + \frac{\partial^2 u}{\partial z^2} = 0, \quad 0 < x < \pi, \quad 0 < y < \pi, \quad 0 < z < \pi$$

$$u(0,y,z) = u(\pi,y,t) = 0, \quad u(x,0,z) = u(x,\pi,z) = 0,$$
$$u(x,y,\pi) = 0, \quad u(x,y,0) = g(x,y).$$

This problem arises in electrostatics when u is the potential whose value g is given on the face $z = 0$, while the other faces are perfect conductors kept at zero potential. u can also be interpreted as an equilibrium temperature distribution when the faces are maintained at temperatures 0 and g, respectively.

Ans: $u(x,y,z) = \sum_{n=1}^{\infty} \sum_{m=1}^{\infty} d_{nm} \dfrac{\sinh\left(\sqrt{n^2+m^2}\,\pi\right)}{\sinh\left(\sqrt{n^2+m^2}\,(\pi - z)\right)} \sin nx \sin my.$

5.5 Vibration of a Beam

In this section, we examine the transverse vibration of a beam, which involves oscillatory motion perpendicular to its longitudinal axis. This phenomenon occurs when external forces or disturbances, such as impacts or dynamic loads, induce deflection and oscillation in the beam. First we prove the following lemma.

Lemma 10 *Let $X(x)$ be a nonzero solution to the fourth-order boundary value problem,*

$$X^{(4)}(x) - \mu X(x) = 0, \quad X(0) = X''(0) = X(b) = X''(b) = 0,$$

then

$$\mu \ge 0.$$

Proof *Consider the integral*

$$\int_0^b X(x)X^{(4)}(x)\,dx.$$

A first integration by parts, with $u = X(x)$ and $dv = X^{(4)}(x)\,dx$, yields

$$\int_0^b X(x)X^{(4)}(x)\,dx = X(x)X'''(x)\big|_0^b - \int_0^b X'(x)X'''(x)\,dx = -\int_0^b X'(x)X'''(x)\,dx.$$

A second integration by parts, with $u = X'(x)$ and $dv = X'''(x)\,dx$, gives

$$-\int_0^b X'(x)X'''(x)\,dx = -X'(x)X''(x)\big|_0^b + \int_0^b X''(x)X''(x)\,dx = \int_0^b \left(X''(x)\right)^2 dx.$$

Thus, we have shown that

$$\int_0^b X(x)X^{(4)}(x)\,dx = \int_0^b \left(X''(x)\right)^2 dx. \tag{5.47}$$

Now, substituting $X^{(4)}(x) = \mu X(x)$ into the left-hand side of (5.47), *we obtain*

$$\int_0^b \mu\left(X(x)\right)^2 dx = \int_0^b \left(X''(x)\right)^2 dx \geq 0,$$

or equivalently

$$\mu \int_0^b \left(X(x)\right)^2 dx \geq 0,$$

from which we conclude that $\mu \geq 0$. This conclude the proof.

□

Transverse Vibration of a Beam

Consider

$$u_{tt} + c^2 u_{xxxx} = 0, \quad 0 < x < l, \quad t > 0, \tag{5.48}$$

subject to the following set of boundary conditions

$$u(0,t) = u(l,t) = 0, \tag{5.49}$$

$$u_{xx}(0,t) = u_{xx}(l,t) = 0, \tag{5.50}$$

and a pair of initial conditions

$$u(x,0) = f(x), \quad u_t(x,0) = g(x), \quad 0 < x < l. \tag{5.51}$$

The above problem is known as the transverse vibration of a beam. The transverse vibration of a beam is governed by the wave equation given by (5.48) which describes the oscillatory motion of the beam's displacement $u(x,t)$ as a function of position x and time t. The parameter c represents the wave speed, which depends on the material properties and geometric characteristics of the beam. The equation captures the dynamic balance of forces acting on an infinitesimal segment of the beam, combining inertia and internal elastic restoring forces.

The beam is subject to boundary conditions that reflect its constraints. Specifically, the conditions (5.49) indicate that the beam is clamped or fixed at both ends, preventing displacement at $x=0$ and $x=L$. Additionally, the conditions given by (5.50) imply that the bending moment at the fixed ends is zero, a typical condition for a beam with free rotational ends or no applied moment. To complete the problem formulation, initial conditions are specified by (5.51). These conditions define the initial displacement $f(x)$ and initial velocity $g(x)$ of the beam at time $t=0$. Together, the wave equation, boundary conditions, and initial conditions establish a well-posed problem, enabling the determination of the beam's motion for all $t>0$. The interplay between the wave-like dynamics and the boundary conditions leads to characteristic vibrational modes and frequencies, central to understanding the beam's response to external excitations.

• Method of Solution

Again, we make use of separation of variables by assuming a nonzero solution $u(x,t)=X(x)T(t)$. A substitution into (5.48) leads to

$$\frac{T''(t)}{c^2T(t)} = -\frac{X^{(4)}(x)}{X(x)} = -\mu.$$

Thus, we are lead to the fourth-order boundary value problem

$$X^{(4)}(x)-\mu X(x)=0, \quad X(0)=0,\ X(l)=0, \quad X''(0)=X''(l)=0,$$

and to the second-order differential equation

$$T''(t)+\mu c^2 T(t)=0.$$

We already know from Lemma 10 that $\mu \geq 0$. To avoid the use of square root we let

$$\mu=\alpha^4,\ \alpha>0,$$

instead of $\mu=\alpha^2$. We assume a solution of $X^{(4)}(x)-\alpha^4X(x)=0$ of the form $X(x)=e^{rx}$. Then, we arrive at the characteristic equation

$$r^4-\alpha^4=(r^2-\alpha^2)(r^2+\alpha^2)=0.$$

Consequently, the general solution is

$$X(x)=c_1e^{\alpha x}+c_2e^{-\alpha x}+C_3\cos\alpha x+C_4\sin\alpha x.$$

We prefer to write the part of the solution

$$c_1e^{\alpha x}+c_2e^{-\alpha x}$$

in terms of cosine and sine hyperbolic. By letting

$$c_1 = \frac{C_1 + C_2}{2} \quad \text{and} \quad c_2 = \frac{C_1 - C_2}{2},$$

we have

$$c_1 e^{\alpha x} + c_2 e^{-\alpha x} = C_1 \cosh \alpha x + C_2 \sinh \alpha x.$$

Finally, the solution takes the form

$$X(x) = C_1 \cosh \alpha x + C_2 \sinh \alpha x + C_3 \cos \alpha x + C_4 \sin \alpha x.$$

Our next job is determining the constants. A differentiation of $X(x)$ twice with respect to x leads to

$$X''(x) = \alpha^2 C_1 \cosh \alpha x + \alpha^2 C_2 \sinh \alpha x - C_3 \alpha^2 \cos \alpha x - C_4 \alpha^2 \sin \alpha x.$$

Applying the conditions $X(0) = X''(0) = 0$, implies $C_1 + C_3 = 0$, and $\alpha^2(C_1 - C_3) = 0$. Since $\alpha \neq 0$, we obtain $C_1 = C_3 = 0$. Combine this with $C_1 + C_3 = 0$, yields $C_1 = 0$. On the other hand, the conditions $X(L) = X''(L) = 0$ give

$$C_2 \sinh(\alpha l) + C_4 \sin(\alpha l) = 0, \quad C_2 \sinh(\alpha l) - C_4 \sin(\alpha l) = 0.$$

By adding the two equations, we obtain

$$C_2 \sinh(\alpha l) = 0,$$

Subtracting one from the other we have

$$C_4 \sin(\alpha l) = 0.$$

Since $\sinh(\alpha l) \neq 0$, we must have $C_2 = 0$. On the other hand, for a nontrivial solution, we have the relation

$$C_4 \sin(\alpha l) = 0.$$

From this, we arrive at

$$\alpha = \frac{n\pi}{l}, \quad n = 1, 2, \ldots$$

Thus, the corresponding eigenfunctions are given by

$$X_n(x) = A_n \sin\left(\frac{n\pi x}{l}\right), \quad n = 1, 2, \ldots$$

It can be shown that for $\mu = 0$, $X(x) = 0$ is the only solution. Replacing $\mu = \alpha^4$, in the equation for $T(t)$, we obtain the general solution

$$T_n(t) = B_n \cos\left(c\left(\frac{n\pi}{l}\right)^2 t\right) + C_n \sin\left(c\left(\frac{n\pi}{l}\right)^2 t\right).$$

Thus, the general solution for the transverse vibrations of a beam is

$$u(x,t) = \sum_{n=1}^{\infty} \left[a_n \cos\left(c\left(\frac{n\pi}{l}\right)^2 t \right) + b_n \sin\left(c\left(\frac{n\pi}{l}\right)^2 t \right) \right] \sin\left(\frac{n\pi x}{l}\right). \tag{5.52}$$

To satisfy the initial condition $u(x,0) = f(x)$, we must have

$$u(x,0) = f(x) = \sum_{n=1}^{\infty} a_n \sin\frac{n\pi x}{l},$$

from which we find

$$a_n = \frac{2}{l}\int_0^l f(x)\sin\frac{n\pi x}{l}dx. \tag{5.53}$$

Now the application of the second initial condition gives

$$u_t(x,0) = g(x) = \sum_{n=1}^{\infty} b_n c\left(\frac{n\pi}{l}\right)^2 \sin\frac{n\pi x}{l},$$

and hence,

$$b_n = \frac{2}{cl\left(\frac{n\pi}{l}\right)^2}\int_0^l g(x)\sin\frac{n\pi x}{l}dx. \tag{5.54}$$

Thus, the solution to (5.48)–(5.51) is given by (5.52) where a_n and b_n are given by (5.53) and (5.54), respectively.

5.5.1 Exercises

Exercise 5.27 *Solve the fourth order non-homogeneous differential equation that govern an elastic beam*

$$\frac{d^4w(x)}{dx^4} + 4\beta^4 w(x) = q(x),$$

where $\beta^4 = \frac{k}{4EI}$ *is a constant, subject to the boundary conditions*

$$w(0) = 0, \quad w'(0) = 0, \quad w'(l) = 0, \quad \textit{and} \quad EIw'''(l) = -V.$$

Exercise 5.28 *Solve and provide physical explanation of*

$$u_{tt} + u_{xxxx} = 0, \quad 0 < x < 1, \quad t > 0,$$

subject to the following set of boundary conditions

$$u(0,t) = u_x(0,t) = 0,$$

$$u(1,t) = u_{xx}(1,t) = 0,$$

and a pair of initial conditions

$$u(x,0) = 0.1\sin\pi x, \quad u_t(x,0) = 0, \quad 0 < x < 1.$$

Exercise 5.29 *Suppose that we have a uniform beam with linear density ρ and Young's modulus E. Let I denote the moment of inertia of the cross-section of the beam. Then, the boundary value problem that governs the transverse vibrations of the uniform beam or an elastic bar of length l is given by the position function $u(x,t)$, which satisfies the following equations*

$$u_{tt} + a^4 u_{xxxx} = 0, \quad 0 < x < l, \quad t > 0,$$

where $a^4 = \frac{EI}{\rho}$ with initial conditions

$$u(x,0) = f(x), \quad u_t(x,0) = 0, \quad 0 < x < l.$$

Find the general solutions in each of the following cases.

(a) $u_{xx}(0,t) = u_{xxx}(0,t) = u_{xx}(l,t) = u_{xxx}(l,t) = 0$. (**Free at each ends.**)

(b) $u(0,t) = u_x(0,t) = u_{xx}(l,t) = u_{xxx}(l,t) = 0$. (**Fixed at** $x = 0$, **free at** $x = l$.)

6

Sturm–Liouville Systems and Applications

Studying the eigenvalues and eigenfunctions of Sturm–Liouville problems is essential in mathematical physics and engineering, as they arise naturally in the analysis of a wide range of physical phenomena, including heat conduction, vibrations, quantum mechanics, and wave propagation. These problems provide a framework for understanding how certain differential operators act on functions within specific boundary conditions, revealing key properties like stability and resonance. Bessel functions and Legendre polynomials are integral to this study because they are solutions to specific types of Sturm–Liouville problems that appear in cylindrical and spherical coordinate systems, respectively. For example, Bessel functions model wave behavior in circular or cylindrical domains, such as vibrations of a circular drumhead or electromagnetic waves in a cylindrical waveguide. Similarly, Legendre polynomials describe potentials and fields in spherical domains, such as gravitational or electric fields around planets. These functions not only provide analytical solutions to complex problems but also form orthogonal sets, enabling the expansion of arbitrary functions in terms of these eigenfunctions for practical computations.

6.1 Sturm–Liouville Problem

Consider the second-order differential equation with parameter $\lambda \in \mathbb{R}$,

$$y''(x) + P(x)y'(x) + Q(x)y(x) + \lambda R(x)y(x) = 0, \quad a \leq x \leq b,$$

where P, Q, R are continuous and R is positive on $[a,b]$. Multiply both sides of the above equation with

$$r(x) = e^{\int P(x)dx} > 0,$$

and then by observing that

$$\left(r(x)y'\right)' = r'(x)y' + r(x)y'',$$

DOI: 10.1201/9781003652540-6

the above equation may take the form

$$\left(r(x)y'\right)' + q(x)y + \lambda p(x)y = 0, \tag{6.1}$$

where $q(x) = r(x)Q(x)$ and $p(x) = r(x)R(x) > 0$ for all $x \in [a,b]$. For constants α_1, α_2, β_1, β_2, we impose the boundary conditions

$$\alpha_1 y(a) + \beta_1 y'(a) = 0, \quad \alpha_2 y(b) + \beta_2 y'(b) = 0, \tag{6.2}$$

with

$$\alpha_1^2 + \beta_1^2 \neq 0, \quad \alpha_2^2 + \beta_2^2 \neq 0.$$

Definition 6.1 *The Sturm–Liouville problem given by* (6.1) *and* (6.2) *is said to be* **regular** *when*
(i) $p(x), q(x), r(x)$, *and* r' *are continuous on the closed interval* $a \le x \le b$;
(ii) $p(x) > 0$ *and* $r(x) > 0$ *when* $a \le x \le b$.

A value of λ for which problem (6.1) and (6.2) has **nontrivial solution** is called an **eigenvalue** and the **nontrivial solution** is called an **eigenfunction**.

Throughout this book, we have been working with Sturm–Liouville problems. Now, we aim to formalize and refine the concept, enabling us to tackle more complex problems effectively.

Example 6.1 The eigenvalue problem

$$3y'' + 6xy' + x^2 y + \lambda y = 0, \quad 0 < x < 1,$$

can be put in the form of (6.1) by setting $r(x) = e^{\int 2x dx} = e^{x^2}$. Then we have

$$q(x) = \frac{x^2}{3}e^{x^2}, \quad p(x) = \frac{1}{3}e^{x^2},$$

and consequently, the problem takes the form

$$(e^{x^2}y')' + \left(\frac{x^2}{3}e^{x^2} + \lambda \frac{1}{3}e^{x^2}\right)y = 0.$$

□

The orthogonality of eigenfunctions in the Sturm–Liouville problem with respect to a weight function is crucial because it facilitates the expansion of arbitrary functions in terms of these eigenfunctions. This property ensures that the eigenfunctions form an orthogonal basis in the weighted inner product space. Consequently, it simplifies the computation of coefficients in series expansions (e.g., Fourier-like series) and aids in solving partial differential equations by reducing them to simpler, solvable forms.

Using a differential linear operator in Sturm–Liouville problem proofs is important because it provides a structured and general framework to analyze the problem. The operator formalism highlights key properties like self-adjointness, which guarantees real eigenvalues and orthogonal eigenfunctions with respect to the weight function. Additionally, it simplifies the manipulation of differential equations and boundary conditions, making proofs of fundamental results, such as orthogonality and completeness of eigenfunctions, more concise and systematic. We define the Sturm–Liouville differential operator by

$$\mathscr{L}z := \frac{d}{dx}\left(r(x)\frac{dz}{dx}\right) + q(x)z, \tag{6.3}$$

Then, equation (6.1) can be written as

$$\mathscr{L}z + \lambda p(x)z = 0. \tag{6.4}$$

Definition 6.2 *The differential operator* $\mathscr{L}$ *given by* (6.3) *is self-adjoint if there exists a continuously differentiable function g such that*

$$\left(w\mathscr{L}z - z\mathscr{L}w\right)dx = dg, \tag{6.5}$$

where z *and* w *satisfy* $\mathscr{L}z = \mathscr{L}w = 0$. *In other words,* $\left(w\mathscr{L}z - z\mathscr{L}w\right)dx$ *must be exact.*

We have the following lemma.

Lemma 11 *The differential operator* $\mathscr{L}$ *given by* (6.3) *is self-adjoint.*

Proof *Let* z *and* w *satisfy* $\mathscr{L}z = \mathscr{L}w = 0$. *Then*

$$\begin{aligned} w\mathscr{L}z - z\mathscr{L}w &= w\frac{d}{dx}\left(r(x)\frac{dz}{dx}\right) + wq(x)z \\ &- z\frac{d}{dx}\left(r(x)\frac{dw}{dx}\right) - zq(x)w \\ &= w\frac{d}{dx}\left(r(x)\frac{dz}{dx}\right) - z\frac{d}{dx}\left(r(x)\frac{dw}{dx}\right) \\ &= wrz'' + wr'z' - zrw'' - zr'w' \\ &= r[wz'' - zw''] + r'[wz' - zw']. \end{aligned} \tag{6.6}$$

On the other-hand, after simple calculation we find that

$$\frac{d}{dx}\{r(x)[wz' - zw']\} = r[wz'' - zw''] + r'[wz' - zw'],$$

which is the right-hand side of (6.6). *Hence,*

$$w\mathscr{L}z - z\mathscr{L}w = \frac{d}{dx}\{r(x)[wz' - zw']\}. \tag{6.7}$$

Or,

$$\left(w\mathscr{L}z - z\mathscr{L}w\right)dx = d\{r(x)[wz' - zw']\} := dg. \tag{6.8}$$

Note that an integration of (6.7) *gives Green's formula that we obtained in* (4.69). *This completes the proof.*

Lemma 11 is pivotal in establishing the orthogonality of the eigenfunctions. To maintain clarity and ensure a smooth progression of the argument, we will restate the following definition.

Definition 6.3 *If for any two functions f and g we have*

$$\int_a^l f(x)g(x)p(x)dx = 0, \tag{6.9}$$

then we say f and g are orthogonal on $[a,b]$ with respect to the weight function $p(x) > 0$.

Theorem 6.1 *For the Sturm–Liouville problem given by* (6.1) *and* (6.2) *the following statements hold.*

(i) The eigenvalues are real and to each eigenvalue there corresponds a single eigenfunction up to a constant multiple.

(ii) The eigenvalues form an infinite sequence $\lambda_1, \lambda_2, \ldots, \lambda_n, \ldots,$ and can be ordered in a manner that $0 < \lambda_1 < \lambda_2 < \lambda_3 < \ldots$ with

$$\lim_{n\to\infty} (\lambda_n) = \infty.$$

(iii) If λ_m and λ_n are two distinct eigenvalues, then the corresponding eigenfunctions $y_m(x)$ and $y_n(x)$ are orthogonal on the interval $[0,l]$ with respect to the weight function $p(x)$.

Proof *We will prove items (i) and (iii), while item (ii) lies beyond the scope of this course. We begin with the proof of (iii) by assuming two distinct eigenvalues, λ_n and λ_m, such that $\lambda_n \neq \lambda_m$ for $n \neq m$, and their corresponding eigenfunctions $\varphi_n(x)$ and $\varphi_m(x)$. Then*

$$\mathscr{L}\varphi_n + \lambda_n p(x)\varphi_n = 0, \quad \mathscr{L}\varphi_m + \lambda_m p(x)\varphi_m = 0.$$

Multiplying the first equation with $\varphi_m(x)$ and the second with $\varphi_n(x)$ and then subtracting the resulting equations gives

$$\varphi_m(x)\mathscr{L}\varphi_n - \varphi_n(x)\mathscr{L}\varphi_m = (\lambda_m - \lambda_n)p(x)\varphi_n(x)\varphi_m(x).$$

Making use of (6.7) *with $w = \varphi_m(x)$, $z = \varphi_n(x)$ in Lemma 11 we arrive at*

$$\begin{aligned}(\lambda_m - \lambda_n)p(x)\varphi_n(x)\varphi_m(x) &= \varphi_m(x)\mathscr{L}\varphi_n - \varphi_n(x)\mathscr{L}\varphi_m \\ &= \frac{d}{dx}\{r(x)[\varphi_m(x)\varphi_n'(x) - \varphi_n(x)\varphi_m'(x)]\}. \end{aligned} \tag{6.10}$$

Since the eigenfunctions satisfy the boundary conditions given by (6.2), *we have that*

$$\varphi_n'(a) = -\frac{\alpha_1 \varphi_n(a)}{\beta_1},\ \varphi_n'(b) = -\frac{\alpha_2 \varphi_n(b)}{\beta_2},\ \varphi_m'(a) = -\frac{\alpha_1 \varphi_m(a)}{\beta_1},\ \varphi_m'(b) = -\frac{\alpha_2 \varphi_m(b)}{\beta_2}. \tag{6.11}$$

Integrating both sides of (6.10) *from a to b and making use of the equations in* (6.11) *gives*

$$\begin{aligned}(\lambda_m - \lambda_n)\int_a^b p(x)\varphi_n(x)\varphi_m(x)\,dx &= \left[r(x)\Big(\varphi_m(x)\varphi_n'(x) - \varphi_n(x)\varphi_m'(x)\Big)\right]_a^b \\ &= r(b)\Big(\varphi_m(b)\varphi_n'(b) - \varphi_n(b)\varphi_m'(b)\Big) \\ &- r(a)\Big(\varphi_m(a)\varphi_n'(a) - \varphi_n(a)\varphi_m'(a)\Big) = 0.\end{aligned} \tag{6.12}$$

Since $\lambda_m - \lambda_n \neq 0$, *we obtain from expression* (6.12) *that*

$$\int_a^b p(x)\varphi_n(x)\varphi_m(x)\,dx = 0. \tag{6.13}$$

This complete the proof of item (iii). It is clear from (6.12) *that* (6.13) *holds,*

1. *if* $r(a) = 0$, *and the second boundary condition in* (6.2) *is not needed,*
2. *if* $r(b) = 0$, *and the first boundary condition in* (6.2) *is not needed.*

Now we prove (i). Suppose that λ *is a complex eigenvalue, and let* $\phi(x)$ *be the corresponding eigenfunction. Since the differential equation defining the eigenfunction may be complex,* $\phi(x)$ *is also allowed to be complex. Let* $\overline{z}$ *denote the complex conjugate of* z. *That is if* $z = x + iy$, *then* $\overline{z} = x - iy$. *Note that if* $z = 0$, *then* $\overline{z} = 0$. *Thus, the complex conjugate of* $\mathscr{L}\phi + \lambda p(x)\phi = 0$ *is also valid and*

$$\overline{\mathscr{L}\phi} + \overline{\lambda} p(x)\overline{\phi} = 0,$$

assuming that the coefficient $p(x)$ *is real and hence* $\overline{p(x)} = p(x)$. *The complex conjugate of* $\mathscr{L}\phi$ *is exactly* $\mathscr{L}$ *operating on the complex conjugate of* ϕ. *That is,* $\overline{\mathscr{L}\phi} = \mathscr{L}\overline{\phi}$. *Thus, we have shown that if* λ *is a complex eigenvalue with corresponding eigenfunction* ϕ, *then* $\overline{\lambda}$ *is also an eigenvalue with corresponding eigenfunction* $\overline{\phi}$. *Now, we show that* λ *cannot be complex. As we have shown, if* λ *is an eigenvalue, then so is* $\overline{\lambda}$. *According to Theorem 6.1, the corresponding eigenfunctions (*ϕ *and* $\overline{\phi}$*) must be orthogonal (with weight* $p(x)$*). Since* $\phi\overline{\phi} = |\phi|^2 \geq 0$ *and* $p(x) > 0$, *we have fom* (6.12) *that*

$$(\lambda - \overline{\lambda})\int_a^b \phi\overline{\phi}p(x)\,dx = (\lambda - \overline{\lambda})\int_a^b p(x)|\phi|^2\,dx = 0.$$

This can only hold when $\lambda - \overline{\lambda} = 0$, *and hence* λ *is real.* □

Example 6.2 Consider the Sturm–Liouville equation

$$\mathscr{L}y = (xy')' + 2xy = -\lambda x^{-1}y$$

subject to a set of homogeneous boundary conditions

$$y'(1) = 0, \quad y'(2) = 0.$$

Expanding the derivative, we have $xy'' + y' + 2xy = -\lambda x^{-1}y$. Multiply through by x to obtain

$$x^2y'' + xy' + (\lambda + 2)y = 0,$$

which is Cauchy-Euler equation. Letting $x = e^t$, we reduce the problem to

$$y''(t) + (\lambda + 2)y(t) = 0,$$

(see Section 4.14). If $\lambda = -2$, then the solution is $y(t) = c_1 + c_2t$, and in terms of x, $y(x) = c_1 + c_2 \ln x$. Either boundary condition, gives $c_2 = 0$, and we are left with a constant solution for $\lambda = -2$. If $\lambda + 2 > 0$, then the general solution is

$$y(x) = c_1 \cos\left(\sqrt{\lambda + 2}\ln|x|\right) + c_2 \sin\left(\sqrt{\lambda + 2}\ln|x|\right).$$

Next, we apply the boundary conditions. The condition $y'(1) = 0$ forces $c_2 = 0$. This leaves

$$y(x) = c_1 \cos\left(\sqrt{\lambda + 2}\ln x\right).$$

The second condition, $y'(2) = 0$, yields

$$\sin\left(\sqrt{\lambda + 2}\ln 2\right) = 0.$$

This will give nontrivial solutions when

$$\sqrt{\lambda + 2}\ln 2 = n\pi, \quad n = 1, 2, \ldots$$

In summary, the eigenfunctions for this eigenvalue problem are

$$y_n(x) = \cos\left(n\pi \frac{\ln x}{\ln 2}\right), \quad 1 \le x \le 2, \tag{6.14}$$

and the eigenvalues are given by

$$\lambda_n = \left(\frac{n\pi}{\ln 2}\right)^2 - 2, \quad n = 1, 2, \ldots$$

□

The following corollary provides a useful criterion for determining whether a Sturm–Liouville problem has negative eigenvalues, streamlining the process of analyzing the nature of the eigenvalues.

Corollary 5 *If λ is an eigenvalue of the regular Sturm–Liouville problem* (6.1) *and* (6.2)*, then $\lambda \geq 0$ provided that*

$$q(x) \leq 0 \quad (a \leq x \leq b) \quad \text{and} \quad \alpha_1\alpha_2 \leq 0, \quad \beta_1\beta_2 \geq 0.$$

We have the following definition concerning *singular Sturm–Liouville problem*

Definition 6.4 *The Sturm–Liouville problem given by* (6.1) *and* (6.2) *is said to be* **singular** *if*

(i) at least one of the regularity conditions mentioned in Definition 6.1 is failed to hold;

(ii) or the interval on which the problem is defined is unbounded.

Example 6.3 Here are two equations that are singular Sturm–Liouville problem,

(a) **Legendre's equation**

$$\frac{d}{dx}\left[(1-x^2)\frac{dy}{dx}\right] + \lambda y = 0, \quad -1 < x < 1,$$

with the conditions that y and $\frac{dy}{dx}$ are finite as $x \to \pm 1$. In this case, $r(x) = 1 - x^2$ and $p(x) = 1$, and $r(x)$ vanishes at $x = \pm 1$.

(b) The **Bessel equation for fixed** ν

$$\frac{d}{dx}\left(x\frac{dy}{dx}\right) + \left(\lambda x - \frac{\nu^2}{x}\right)y = 0, \quad 0 < x < a,$$

with the boundary conditions $y(a) = 0$ and y and $\frac{dy}{dx}$ are finite as $x \to 0^+$ is a singular Sturm–Liouville problem. Here

$$r(x) = x, \quad q(x) = -\frac{\nu^2}{x}, \quad p(x) = x.$$

We observe that

$$r(0) = 0, \quad q(x) \text{ becomes infinite as } x \to 0^+, \quad p(0) = 0$$

therefore, the system is singular.

□

6.2 Expansion in Series of Eigenfunctions

The eigenfunctions that arise from the Sturm–Liouville problem (6.1) and (6.2) are orthogonal with respect to a weight function $p(x)$. In particular, suppose that we have a set of orthogonal functions $\{\varphi_n(x) : n = 1, 2\ldots\}$ on the interval (a,b). We wish to represent a function $f(x)$ that is piecewise continuous on (a,b) in the form

$$f(x) = \sum_{n=1}^{\infty} c_n \varphi_n(x), \tag{6.15}$$

where $\{\phi_n(x)\}$ are orthogonal basis functions on the interval (a,b). We are done once we find the coefficients c_n, $n = 1, 2, \ldots$ With this in mind, we multiply both sides of (6.15) with $p(x)\varphi_m(x)$ and integrate the resulting equation from $x = a$ to $x = b$. This leads us to

$$\int_a^b p(x) f(x) \varphi_m(x)\, dx = \sum_{n=1}^{\infty} c_n \int_a^b p(x) \varphi_n(x) \varphi_m(x)\, dx,$$

which is equivalent to

$$(f, \phi_m) = \sum_{m=1}^{\infty} c_m (\phi_m, \phi_n).$$

But $(\phi_m, \phi_m) = 0$ for all values of m except when $m = n$, and hence

$$(f, \varphi_m) = c_m (\phi_m, \phi_m).$$

Replacing m with n, the above relation gives

$$c_n = \frac{(f, \varphi_n)}{(\phi_n, \phi_n)} = \frac{\int_a^b p(x) f(x) \phi_n(x)\, dx}{\int_a^b p(x) \phi_n^2(x)\, dx}, \; n = 1, 2, \ldots \tag{6.16}$$

The expression defined by (6.15) is known as the **Generalized Fourier series** and the corresponding Fourier coefficient c_n, $n = 1, 2, \ldots$ is given by expression (6.16). Note that if the basis function are orthonormal on the interval (a,b), then (6.16) reduces to

$$c_n = (f, \varphi_n) = \int_a^b p(x) f(x) \phi_n(x)\, dx, \; n = 1, 2\ldots \tag{6.17}$$

since

$$(\phi_n, \phi_n) = \int_a^b p(x) \phi_n^2(x)\, dx = 1.$$

We have the following theorem

Theorem 6.2 *Let $\varphi_1, \varphi_2, \ldots$ be eigenfunctions of a regular Sturm–Liouville problem satisfying equations* (6.1) *and* (6.2).*If $f(x)$ is piecewise smooth on the interval $a < x < b$, then the following expansion holds*

$$\sum_{n=1}^{\infty} c_n \varphi_n(x) = \frac{f(x^+) + f(x^-)}{2}, \quad a < x < b, \tag{6.18}$$

where

$$c_n = \frac{\int_a^b f(x) \varphi_n(x) p(x)\, dx}{\int_a^b \varphi_n^2(x) p(x)\, dx}.$$

Moreover, if the series

$$\sum_{n=1}^{\infty} |c_n| \left(\int_a^b \varphi_n^2(x) p(x)\, dx \right)^{1/2}$$

converges, then the series in equation (6.18) *converges uniformly for* $a \le x \le b$.

6.2.1 Exercises

Exercise 6.1 *What boundary conditions are needed if* $r(a) = r(b) = 0$, *for* (6.13) *to hold?*

Exercise 6.2 *Show the eigenfunctions of Example 6.2 given by* (6.14) *are orthogonal on the interval* $[1,2]$ *with respect to the weight function* $p(x) = \frac{1}{x}$.

Exercise 6.3 *Prove Corollary 5.*

Exercise 6.4 *Find the eigenvalues and corresponding eigenfunctions of the given Sturm–Liouville problems.*

(a) $y'' + \lambda y = 0, \quad y'(0) = 0, \quad y'(\pi) = 0.$

(b) $(xy')' + \dfrac{\lambda}{x} y = 0, \quad y(1) = 0, \quad y(e^2) = 0.$

Exercise 6.5 *Find the eigenvalues and corresponding eigenfunctions of the given Sturm–Liouville problems.*

(a) $x^2 y'' + 3xy' + \lambda y = 0, \quad 1 \le x \le e \quad y(1) = 0, \quad y(e) = 0.$

(b) $(1+x)^2 y'' + 2(1+x)y' + 3\lambda y = 0, \quad 0 \le x \le 1, \quad y(0) = 0, \quad y(1) = 0.$

(c) $\dfrac{d}{dx}\left(e^{2x}\dfrac{dy}{dx}\right) + e^{2x}(1+\lambda)y = 0, \quad y(0) = y(\pi) = 0.$

Exercise 6.6 *The eigenvalue problem*

$$x^2 y'' - \lambda x y' + \lambda y = 0, \quad y(1) = y(2) = 0,$$

is not a Sturm–Liouville eigenvalue problem. Show that none of the eigenvalues are real by solving the equation.

Exercise 6.7 *Let the operator* $\mathscr{L}$ *be defined by*

$$\mathscr{L} z = \frac{d^4}{dx^4}(z).$$

Consider the fourth order differential equation

$$\mathscr{L} z = \lambda z$$

with two distinct eigenvalues λ_n and λ_m, and corresponding eigenfunctions $\varphi_n(x)$ and $\varphi_m(x)$. Show that

$$\int_a^b \left[\varphi_m(x)\mathscr{L}\varphi_n \quad - \quad \varphi_n(x)\mathscr{L}\varphi_m\right]dx = (\lambda_n - \lambda_m)\int_a^b \varphi_n(x)\varphi_m(x)\,dx$$
$$= \left[\varphi_n(x)\varphi_m'''(x) - \varphi_m(x)\varphi_n'''(x) - \varphi_n'(x)\varphi_m''(x) + \varphi_m'(x)\varphi_n''(x)\right]_a^b.$$

Exercise 6.8 *Set $X(x) = \frac{Y(x)}{x}$, $0 < x < a$ to solve the eigenvalue problem*

$$\frac{1}{x^2}\frac{d}{dx}\left(x^2\frac{dX}{dx}\right) + \lambda X(x) = 0, \quad 0 < x < a,$$

$$X(0) \text{ (finite)} \quad \text{and} \quad X(a) = 0.$$

Is this a regular Sturm–Liouville problem? Are the eigenfunctions orthogonal?

Exercise 6.9 *Decide if whether or not $\lambda = 0$ is an eigenvalue of the problem*

$$\frac{d^2X(x)}{dx^2} + \lambda x X(x) = 0, \quad X(0) = 0, \quad X(a) = 0, \quad 0 < x < a.$$

Exercise 6.10 *Show that the eigenvalues and corresponding eigenfunctions of*

$$(e^x\phi')' + e^x\gamma\phi = 0, \quad 0 < x < a,$$

with boundary conditions

$$\phi(0) = 0, \quad \phi(a) = 0.$$

are given by

$$\gamma_n^2 = \frac{n^2\pi^2}{a^2} + \frac{1}{4}, \quad \phi_n(x) = e^{-x/2}\sin\frac{n\pi x}{a}.$$

Additionally, expand the function $f(x) = 1$ in terms of the eigenfunctions.

Exercise 6.11 *Show that the eigenvalues and corresponding eigenfunctions of*

$$(x\phi')' + \lambda\frac{1}{x}\phi = 0, \quad 1 < x < b,$$

with boundary conditions

$$\phi(1) = 0, \quad \phi(b) = 0$$

are given by

$$\lambda_n^2 = \left(\frac{n\pi}{\ln(b)}\right)^2, \quad \phi_n(x) = \sin(\lambda_n\ln(x)), \quad n = 1, 2, \ldots.$$

Additionally, expand the function $f(x) = x$ in terms of the eigenfunctions.

Exercise 6.12 *Recall that the adjoint, $\mathscr{L}^\dagger$, of operator $\mathscr{L}$ satisfies*

$$\langle u, \mathscr{L}v\rangle = \langle \mathscr{L}^\dagger u, v\rangle$$

for all v in the domain of $\mathscr{L}$ and u in the domain of $\mathscr{L}^\dagger$. Here, the domain of a differential operator $\mathscr{L}$ is the set of all $u \in L^2(a,b)$ satisfying a given set of homogeneous boundary conditions. Moreover,

$$\langle u, \mathscr{L}v\rangle = \int_a^b u\mathscr{L}v\,dx.$$

For $D = \frac{d}{dx}$, let

$$\mathscr{L} = a_2(x)D^2 + a_1(x)D + a_0(x).$$

(a) Find the adjoint $\mathscr{L}^\dagger$ of $\mathscr{L}$.

Ans: $\mathscr{L}^\dagger = \dfrac{d^2}{dx^2}a_2(x) - \dfrac{d}{dx}a_1(x) + a_0(x).$

(b) Determine $\mathscr{L}^\dagger$ and its domain for the operator $\mathscr{L}u = \frac{du}{dx}$, where u satisfies the boundary conditions $u(0) = 2u(1)$ on $[0,1]$.

Ans:

1. $\mathscr{L}^\dagger = -\frac{d}{dx}$

2. $uv\Big|_{x=1}^{0} = 0 \Rightarrow 0 = u(1)[v(1) - 2v(0)] \Rightarrow v(1) = 2v(0).$

6.3 Rayleigh Quotient

The Rayleigh quotient provides an approximation for the eigenvalues of a Sturm–Liouville problem. The Rayleigh quotient is particularly useful in obtaining estimates for the eigenvalues because it allows one to use approximate or trial functions, even if they do not exactly solve the differential equation. By minimizing the Rayleigh quotient with respect to the trial function, one can estimate the smallest eigenvalue, providing insights into the behavior of the system before solving the full eigenvalue problem. It serves as a tool for refining approximations in problems where the exact solutions are difficult to obtain.

We begin by considering the linear operator $\mathscr{L}$ defined in equation (6.3)

$$\mathscr{L}z = \frac{d}{dx}\left(r(x)\frac{dz}{dx}\right) + q(x)z.$$

In terms of $\mathscr{L}$, the second-order differential equation for the Sturm–Liouville problem is given by

$$\mathscr{L}z = -\lambda p(x)z.$$

Let λ_n be the eigenvalues of the regular Sturm–Liouville problem with corresponding eigenfunctions ϕ_n. Then we have

$$\mathscr{L}\phi_n = -\lambda_n p(x)\phi_n.$$

Multiply both sides of the equation by ϕ_n and integrate over the interval $[a,b]$. This leads to the equation

$$\int_a^b \phi_n \frac{d}{dx}\left(r\frac{d\phi_n}{dx}\right) dx + \int_a^b q\phi_n^2 dx = -\lambda_n \int_a^b p\phi_n^2 dx,$$

Solving for the eigenvalues λ_n, we get

$$\lambda_n = -\frac{\int_a^b \left[\phi_n \frac{d}{dx}\left(r\frac{d\phi_n}{dx}\right) + q\phi_n^2\right] dx}{\int_a^b p\phi_n^2 dx}.$$

This formulation provides a convenient way to calculate or approximate eigenvalues based on the corresponding eigenfunctions $\phi_n(x)$. We cannot directly compute the eigenvalue using this formula since the eigenfunctions, $\phi_n(x)$, remain unknown. Still, we can explore further insights based on this result. To refine the expression, consider rewriting the numerator by applying integration by parts to the first term. Specifically, let $u = \phi_n$ and $dv = \frac{d}{dx}\left(r\frac{d\phi_n}{dx}\right) dx$. Then an integration by parts yields

$$\int_a^b \phi_n \frac{d}{dx}\left(r\frac{d\phi_n}{dx}\right) dx = \left[r\phi_n \frac{d\phi_n}{dx}\right]_a^b - \int_a^b r\left(\frac{d\phi_n}{dx}\right)^2 dx.$$

Substituting this result into the eigenvalue formula leads to the Rayleigh Quotient

$$\lambda_n = \frac{-\left[r\phi_n \frac{d\phi_n}{dx}\right]_a^b + \int_a^b \left[r\left(\frac{d\phi_n}{dx}\right)^2 - q\phi_n^2\right] dx}{\int_a^b p\phi_n^2 dx}. \tag{6.19}$$

This representation is particularly useful in many physical problems, where the sign of the eigenvalue plays a critical role. For example, in solving the heat equation, we encounter the first-order differential equation

$$T'(t) + k\lambda T(t) = 0, \quad t > 0$$

whose solution is

$$T(t) = e^{-k\lambda t}.$$

We expect heat to dissipate, which is evident from the solution indicating that $\lambda > 0$. In fact, heat in this case dissipates exponentially for positive values of λ. Similarly, in the wave equation, vibrations occur only for positive eigenvalues. Therefore, ensuring nonnegative eigenvalues imposes the following conditions on the coefficients:

(a) $q(x) \le 0$, and

(b) $-\left[r\phi_n \frac{d\phi_n}{dx}\right]_a^b \ge 0.$

These conditions guarantee physically meaningful solutions for the underlying differential equations. As far as q is concerned, we consider the non-homogeneous heat flow equation

$$u_t = ku_{xx} + \alpha u, \quad 0 < x < L, \quad t > 0.$$

In the context of our problem, we identify $q = \alpha$. When $\alpha \le 0$, this corresponds to an energy-absorbing (endothermic) reaction. Therefore, the condition $q \le 0$ is physically meaningful in such scenarios.

6.4 Reformulation of the Zero Eigenvalue and Minimum Eigenvalue Concepts

If λ is a zero eigenvalue, it follows that $q(x) \equiv 0$, and the constants α_1 and α_2 must both be zero in the homogeneous boundary conditions. This conclusion arises by setting the numerator in the Rayleigh quotient to zero, which implies $q(x) = 0$ and $\phi_n'(x) = 0$. Substituting $\phi_n'(x) = 0$ into the boundary conditions enforces specific constraints on the allowable types of boundary conditions.

Recall item (ii) of Theorem 6.1 states that the eigenvalues of Sturm–Liouville problems with homogeneous boundary conditions can be ordered as $\lambda_1 < \lambda_2 < \cdots$ This implies the existence of a smallest eigenvalue. Thus, we state the following theorem.

Theorem 6.3 *The minimum value of the Rayleigh quotient for all continuous functions $y(x)$ satisfying the boundary conditions given by* (6.2) *is the lowest eigenvalue λ_1, where*

$$\lambda_1 = \frac{-\left[ry\frac{dy}{dx}\right]_a^b + \int_a^b \left(r\left(\frac{dy}{dx}\right)^2 - qy^2\right)dx}{\int_a^b py^2 dx}. \tag{6.20}$$

In other words, the minimum value of this expression is attained when $y(x) = \phi_1(x)$, the eigenfunction corresponding to the smallest eigenvalue.

This result is particularly useful for approximating the smallest eigenvalue by selecting trial functions $y(x)$ that are continuous and satisfy the boundary conditions, even if they do not solve the differential equation exactly. As mentioned earlier, the eigenfunctions are sometimes unknown, yet the expression in (6.20) can still be used to provide an upper bound on the smallest eigenvalue. This is achieved by selecting a trial function $y_T(x)$ that satisfies the specified boundary conditions and then computing

$$R(y_T) = \frac{-\left[r y_T \frac{dy_T}{dx} \right]_a^b + \int_a^b \left(r \left(\frac{dy_T}{dx} \right)^2 - q y_T^2 \right) dx}{\int_a^b p y_T^2 \, dx}, \tag{6.21}$$

which leads to the result

$$\lambda_1 \leq R(y_T),$$

where R_T serves as an upper bound for the first eigenvalue λ_1.

Example 6.4 The boundary value problem

$$\phi'' + \lambda \phi = 0, \quad \phi(0) = 0, \quad \phi(1) = 0,$$

has the eigenvalues and corresponding eigenfunctions

$$\lambda_n = n^2 \pi^2, \quad y_n(x) = \sin nx, \quad n = 1, 2 \ldots$$

Then the first or lowest eigenvalue is when $n = 1$, or $\lambda_1 = \pi^2$. For a trial function $y_T(x)$ we set $y_T(x) = x - x^2$, then $y_T(0) = y_T(1) = 0$, and hence it satisfies the given boundary conditions. Note that $y_T(x)$ may not necessarily satisfy the differential equation. Since $r = 1, p = 1, q = 0$ we have from (6.21) that

$$y_T(x) \frac{dy_T}{dx}\Big|_0^1 = y_T(1) \frac{dy_T}{dx}(1) - y_T(0) \frac{dy_T}{dx}(0) = 0,$$

$$\int_0^1 \left[r \left(\frac{dy_T}{dx} \right)^2 - q y_T^2 \right] dx = \int_0^1 (1 - 2x)^2 \, dx.$$

In a similar fashion we arrive at

$$\int_0^1 p y_T^2 \, dx = \int_0^1 (x - x^2)^2 \, dx.$$

Thus,

$$R(y_T) = \frac{\int_0^1 (1 - 2x)^2 \, dx}{\int_0^1 (x - x^2)^2 \, dx} = 10.$$

It is clear that the result is confirmed, as

$$\pi^2 = \lambda_1 < R(y_T) = 10.$$

□

6.4.1 Exercises

Exercise 6.13 *Consider the boundary value problem*

$$y'' + \lambda y = 0, \quad y(0) = 0, \quad y(1) = 0.$$

(a) Compute $R(y_1)$ and show it is equal to λ_1.

(b) Compute $R(y_T)$ for the trial function

$$y_T = \begin{cases} x, & 0 \le x \le \frac{1}{2} \\ 1-x, & \frac{1}{2} \le x \le 1 \end{cases}$$

(c) Redo part (b) for the trial function

$$y_T = x(1-x).$$

(d) Which function is a better estimate of λ_1?

(d) Use the trial function

$$y_T = x\left(x - \frac{1}{2}\right)(x-1)$$

to estimate λ_2. How close is the estimate of y_T to λ_2?

Exercise 6.14 *Consider the boundary value problem*

$$y'' + \lambda\left(1 + \frac{1}{10}x\right)y = 0, \quad y(0) = y(\pi) = 0.$$

Compute $R(y_T)$ for the following trial functions.

(a)

$$y_T = \begin{cases} x, & 0 \le x \le \frac{\pi}{2} \\ \pi - x, & \frac{\pi}{2} \le x \le \pi \end{cases}$$

(b)

$$y_T = x(\pi - x).$$

Exercise 6.15 *Find the lowest eigenvalue of the problem*

$$e^{\alpha x}\frac{d^2\varphi}{dx^2} + \lambda e^{\alpha x}\varphi = 0, \quad 0 < x < 1,$$

$$\varphi(0) = 0, \quad \varphi(1) = 0.$$

Compare your answer to the first eigenvalue by solving the boundary value problem.

Exercise 6.16 *Estimate the lowest eigenvalue of the problem*

$$\frac{d^2\varphi}{dx^2} - x\varphi + \lambda\varphi = 0, \quad 0 < x < 1,$$

$$\varphi(0) = 0, \quad \varphi(1) = 0.$$

6.5 Applications to Nonuniform String

The general form of the wave equation for transverse vibrations of a string is derived using Newton's second law

$$\text{Force} = \text{Mass} \times \text{Acceleration}.$$

For a string under constant tension T_0, the transverse displacement $u(x,t)$ satisfies

$$T_0 \frac{\partial^2 u}{\partial x^2} = \rho(x) \frac{\partial^2 u}{\partial t^2}.$$

Rearranging, we get the **nonuniform wave equation**

$$\frac{\partial^2 u}{\partial t^2} = \frac{T_0}{\rho(x)} \frac{\partial^2 u}{\partial x^2}. \tag{6.22}$$

Here, $\frac{T_0}{\rho(x)}$ is the position-dependent wave speed squared. The wave speed at position x is

$$v(x) = \sqrt{\frac{T_0}{\rho(x)}}.$$

Since $\rho(x)$ varies, the wave speed $v(x)$ also varies, which results in nonuniform wave propagation. This is different from a uniform string, where the wave speed is constant. The boundary conditions depend on the setup (e.g., fixed ends, free ends). For a string fixed at both ends $x = 0$ and $x = L$,

$$u(0,t) = 0 \quad \text{and} \quad u(L,t) = 0. \tag{6.23}$$

If we add initial conditions

$$u(x,0) = f(x), \quad u_t(x,0) = g(x), \tag{6.24}$$

the solution to the wave equation is determined by the specified initial displacement $f(x)$ and initial velocity $g(x)$. In conclusion, The spatial modes and frequencies depend on the specific variation of $\rho(x)$. Analytical solutions are generally more complex than for a uniform string, and numerical methods are often needed to study such systems in detail. In this study, we would like to use Rayleigh Quotient to have a lower and upper bounds on the first eigenvalue.

- **Analyzing the Problem**

We assume a nonzero solution of the form $u(x,t) = X(x)T(t)$. Substituting into (6.22) and separating the variables, we obtain

$$T''(t) + \lambda T(t) = 0. \tag{6.25}$$

On the other hand, the spatial part satisfies the boundary value problem

$$T_0X''(x) + \lambda\rho(x)X(x) = 0, \quad X(0) = X(L) = 0. \tag{6.26}$$

The boundary value problem given by system (6.26) cannot be solved in closed form. To gain insight into the eigenvalues λ_n, we will use the Rayleigh Quotient. Naturally, we expect oscillatory solutions to (6.26), which are only possible if $\lambda > 0$. Since finding the eigenvalues and their corresponding eigenfunctions explicitly is infeasible, we will rely on the Rayleigh Quotient to determine the sign of λ and to estimate its lower and upper bounds. Let $\phi_n(x)$ be the eigenfunctions of (6.26). Then the corresponding ($r = T_0, q = 0, p = \rho$) Rayleigh quotient to our problem is

$$\lambda = \frac{T_0\int_0^L \left(\frac{d\phi}{dx}\right)^2 dx}{\int_0^L \phi^2\rho(x)dx}. \tag{6.27}$$

Clearly, from (6.27) we conclude that $\lambda \geq 0$. However, if $\lambda = 0$, then (6.27) has only the trivial solution. Thus, we conclude that $\lambda > 0$. This is a big relief since now we know the solution of (6.25) is of linear combination of

$$\sin(\gamma t) \text{ and } \cos(\gamma t),$$

where we let $\lambda = \gamma^2$. According to the principle of superposition, the solution to (6.22) is

$$u(x,t) = \sum_{n=1}^{\infty} a_n \sin(\gamma_n t)\phi_n(x) + \sum_{n=1}^{\infty} b_n \cos(\gamma_n t)\phi_n(x).$$

Applying the initial conditions given by (6.24) we have

$$f(x) = \sum_{n=1}^{\infty} b_n\phi_n(x) \quad \text{and} \quad g(x) = \sum_{n=1}^{\infty} a_n\sqrt{\lambda_n}\phi_n(x),$$

from which we obtain the generalized Fourier coefficients a_n and b_n. Due to the orthogonality of the eigenfunctions with respect to the weight function $\rho(x)$, we obtain that

$$b_n = \frac{\int_a^b f(x)\phi_n(x)\rho(x)\,dx}{\int_a^b \phi_n^2(x)\rho(x)\,dx},$$

$$a_n = \frac{\int_a^b g(x)\phi_n(x)\rho(x)\,dx}{\gamma_n\int_a^b \phi_n^2(x)\rho(x)dx}.$$

We need to find upper and lower estimations of the first eigenvalue λ_1. Set

$$T_0 = \rho(x) = 1.$$

Then the boundary value problem

$$\phi''(x) + \overline{\lambda}\phi(x) = 0, \quad \phi(0) = \phi(L) = 0,$$

has eigenvalues and corresponding eigenfunctions

$$\overline{\lambda}_n = \frac{n^2\pi^2}{L^2}, \quad \phi_n(x) = \sin\frac{n\pi x}{L}, \quad n = 1, 2, \ldots$$

Hence the first eigenvalue is

$$\overline{\lambda}_1 = \frac{\pi^2}{L^2}.$$

The mass density function $\rho(x)$ is assumed to be continuous on the closed and bounded interval $[0, L]$. So by the Extreme Value Theorem, $\rho(x)$ achieves it absolute minimum and absolute maximum for some points $x \in [0, L]$. Let $\rho_{\min}$ and $\rho_{\max}$ denote the minimum value and maximum value of ρ on $[0, L]$, respectively. Then

$$\rho_{\min} \leq \rho(x) \leq \rho_{\max}.$$

We also know that

$$\overline{\lambda}_1 = \frac{\int_0^L \left(\frac{d\phi}{dx}\right)^2 dx}{\int_0^L \phi^2(x)dx} = \frac{\pi^2}{L^2}.$$

With the aid of (6.27) we know that

$$\lambda_1 = \min \frac{T_0 \int_0^L \left(\frac{d\phi}{dx}\right)^2 dx}{\int_0^L \phi^2 \rho(x)dx}.$$

Therefore,

$$\lambda_1 = \min \frac{T_0 \int_0^L \left(\frac{d\phi}{dx}\right)^2 dx}{\int_0^L \phi^2 \rho(x)dx} \leq \frac{T_0}{\rho_{\min}} \frac{\int_0^L \left(\frac{d\phi}{dx}\right)^2 dx}{\int_0^L \phi^2(x)dx} = \frac{T_0}{\rho_{\min}} \frac{\pi^2}{L^2}.$$

Along the same lines, we obtain

$$\lambda_1 = \min \frac{T_0 \int_0^L \left(\frac{d\phi}{dx}\right)^2 dx}{\int_0^L \phi^2 \rho(x)dx} \geq \frac{T_0}{\rho_{\max}} \frac{\int_0^L \left(\frac{d\phi}{dx}\right)^2 dx}{\int_0^L \phi^2(x)dx} = \frac{T_0}{\rho_{\max}} \frac{\pi^2}{L^2}.$$

The above two inequalities give the lower and upper bounds of the smallest eigenvalue. That is,

$$\frac{T_0}{\rho_{\max}} \frac{\pi^2}{L^2} \leq \lambda_1 \leq \frac{T_0}{\rho_{\min}} \frac{\pi^2}{L^2}.$$

Taking the square root on both sides of the above double inequality, we obtain

$$\sqrt{\frac{T_0}{\rho_{\max}}} \frac{\pi}{L} \leq \sqrt{\lambda_1} \leq \sqrt{\frac{T_0}{\rho_{\min}}} \frac{\pi}{L}. \tag{6.28}$$

The inequality (6.18) provides bounds for the square root of the smallest eigenvalue λ_1, which corresponds to the lowest frequency of oscillation of a variable-density string. This inequality implies that the lowest frequency of oscillation of the variable string lies between the lowest frequencies of vibration of two hypothetical strings with constant densities:

- One string has a constant density equal to the maximum density of the variable string, $\rho_{\max}$.
- The other string has a constant density equal to the minimum density of the variable string, $\rho_{\min}$.

Since the wave speed $v(x)$ depends on the local density $\rho(x)$ via $v(x) = \sqrt{\frac{T_0}{\rho(x)}}$, the oscillations of the variable string are influenced by the variation in density. The strings with constant densities $\rho_{\max}$ and $\rho_{\min}$ serve as bounds for this behavior:

- The string with $\rho_{\max}$ vibrates more slowly, leading to a lower bound on $\sqrt{\lambda_1}$.
- The string with $\rho_{\min}$ vibrates more quickly, leading to an upper bound on $\sqrt{\lambda_1}$.

Thus, the inequality highlights that the lowest frequency of the variable string is bounded by the frequencies of two uniform strings with the extreme densities of the variable string.

6.5.1 Exercises

Exercise 6.17 *Obtain an upper and lower bound for* $\rho(x)$, *when* $\rho(x) = 1 + xe^{-x}$, $0 \leq x \leq 2$.

Exercise 6.18 *Derive a similar inequality to* (6.18) *when the boundaries conditions given by* (6.23) *are changed to* $u(0,t) = 0$ *and* $u_x(L,t) = 0$.

Exercise 6.19 *Use separation of variables to solve the wave propagation in a string with variable mass density*

$$
\begin{aligned}
&u_{tt} = (1+x)^2 u_{xx}, && 0 < x < 1, \quad t > 0,\\
&u(0,t) = 0, \quad u(1,t) = 0, && t > 0,\\
&u(x,0) = f(x), && 0 < x < 1,\\
&u_t(x,0) = g(x), && 0 < x < 1.
\end{aligned}
$$

Here the mass density of the string at the point x is given by $\frac{1}{(1+x)^2}$.

Exercise 6.20 *Use separation of variables to solve the heat flow problem in a finite strip with a spatially varying diffusion coefficient* $(1+x)^2$ *variable conductivity*

$$
\begin{aligned}
&u_t = (1+x)^2 u_{xx}, && 0 < x < 1, \quad t > 0,\\
&u(0,t) = 0, \quad u(1,t) = 0, && t > 0,\\
&u(x,0) = x(1-x)\sqrt{1+x}, && 0 < x < 1.
\end{aligned}
$$

The initial temperature distribution is given by

$$x(1-x)\sqrt{1+x},$$

which is a function that is zero at $x = 0$ *and* $x = 1$ *and reaches a peak somewhere in the middle of the interval. This means that initially, the system is not in thermal* *equilibrium, and heat will flow over time. In conclusion, the problem describes heat conduction in a material where thermal diffusivity increases with x.*

6.6 Robin's Boundary Conditions in Heat Flow

Up to this point, we have considered physical problems, such as heat flow or vibrations of a string, where the boundary conditions were either Dirichlet or Neumann. Consequently, the associated mathematics was relatively straightforward. In this section, we turn our attention to problems involving Robin boundary conditions (see Section 4.2).

We begin by considering a problem that models the cooling of a slab over time, with heat transfer mechanisms. Specifically, we consider

$$u_t(x,t) = ku_{xx}(x,t), \quad 0 < x < b, \quad t > 0, \tag{6.29}$$

$$u_x(0,t) = 0, \quad u_x(b,t) = -hu(b,t), \tag{6.30}$$

$$u(x,0) = f(x), \tag{6.31}$$

where h is a positive constant. The PDE given by (6.29) represents the diffusion of heat within the slab. The term $u_t(x,t)$ denotes the rate of change of temperature with time, and $ku_{xx}(x,t)$ represents the spatial diffusion of heat, where k is the thermal diffusivity of the material. Heat flows from regions of higher temperature to regions of lower temperature within the slab. The boundary condition $u_x(0,t) = 0$, states that there is no heat flux at $x = 0$, implying that the face $x = 0$ is insulated. As a result, the temperature gradient u_x at $x = 0$ is zero, and no heat can flow through this boundary. On the other hand, the condition at $x = b$

$$u_x(b,t) = -hu(b,t)$$

incorporates Newton's law of cooling, which states that the heat flux at the boundary $x = b$ is proportional to the temperature difference between the surface and the surrounding medium. Since the medium is assumed to be at zero temperature, the heat flux is proportional to $u(b,t)$. The proportionality constant $h > 0$ is the heat transfer coefficient, which quantifies the rate of heat transfer between the slab's surface and the surrounding medium. Physically, this boundary condition models the cooling effect of the surrounding medium on the face $x = b$. Heat flows out of the slab into the medium, leading to a decrease in temperature at $x = b$ over time.

The initial condition given by (6.31) specifies the initial temperature distribution of the slab. The function $f(x)$ provides the temperature profile across the slab at time $t = 0$. The evolution of the temperature $u(x,t)$ for $t > 0$ is determined by the heat diffusion within the slab and the boundary conditions at $x = 0$ and $x = b$.

When addressing boundary value problems with Robin boundary conditions, it is often advantageous to normalize the eigenfunctions to streamline notation and calculations. We ask the reader to consult with Section 3.5.

• Method of Solution

We apply the method of separation of variables once more, assuming a nonzero solution of the form

$$u(x,t) = X(x)T(t).$$

Substituting this into equation (6.29), we obtain

$$\frac{X''(x)}{X(x)} = \frac{T'(t)}{kT(t)} = -\lambda.$$

The Robin boundary condition $u_x(b,t) = -hu(b,t)$ leads to

$$X'(b)T(t) + hX(b)T(t) = \left(X'(b) + hX(b)\right)T(t) = 0,$$

which must hold for all $t > 0$. Hence, we have the condition

$$X'(b) + hX(b) = 0.$$

The transformations for the other boundary conditions follow similarly. Consequently, we arrive at the second-order differential equation

$$X''(x) + \lambda X(x) = 0, \quad X'(0) = 0, \quad X'(b) + hX(b) = 0, \tag{6.32}$$

which is a specific form of the Sturm–Liouville problem (6.1). Additionally, we obtain the first-order ordinary differential equation

$$T'(t) + \lambda kT(t) = 0, \quad t > 0. \tag{6.33}$$

We begin by solving (6.32). Given that $q = 0$, $\alpha_1\alpha_2 = 0$, and $\beta_1\beta_2 > 0$, Corollary 5 implies that $\lambda \geq 0$.

- **$\lambda = 0$.** In this case, (6.32) simplifies to $X''(x) = 0$ whose solution is given by

  $$X(x) = Ax + B,$$

  where A and B are constants. Applying the condition $X'(0) = 0$ yields $A = 0$. Using the Robin boundary condition,

  $$X'(b) + hX(b) = hB = 0,$$

  we find that $B = 0$. Therefore, the only solution is the trivial solution.

- **$\lambda > 0$.** We set $\lambda = \alpha^2$ with $\alpha > 0$. Then, (6.32) becomes

  $$X''(x) + \alpha^2 X(x) = 0,$$

  whose general solution is

  $$X(x) = A\cos\alpha x + B\sin\alpha x.$$

Differentiating, we get

$$X'(x) = -A\alpha \sin \alpha x + B\alpha \cos \alpha x.$$

Applying $X'(0) = 0$ gives $B = 0$. Substituting $B = 0$ into $X(x)$, we obtain

$$X(x) = A\cos \alpha x.$$

Using the Robin boundary condition

$$X'(b) + hX(b) = -A\alpha \sin \alpha b + Ah\cos \alpha b = A\big(-\alpha \sin \alpha b + h\cos \alpha b\big) = 0.$$

Since $A \neq 0$, we must have

$$-\alpha \sin \alpha b + h\cos \alpha b = 0,$$

which implies

$$\tan \alpha b = \frac{h}{\alpha}. \tag{6.34}$$

The transcendental equation given by (6.34) cannot be solved algebraically because it involves both the transcendental function $\tan \alpha b$ and the rational term $\frac{h}{\alpha}$, making it impossible to isolate α explicitly. The equation (6.34) can be visualized as the intersection of the graph of the tangent function $\tan \alpha b$ with the hyperbolic curve $\frac{h}{\alpha}$. The solutions correspond to the α-values, denoted as α_n for $n = 1, 2, \ldots$, where these two curves intersect. The function $\tan \alpha b$ has vertical asymptotes at $\alpha b = \frac{\pi}{2}, \frac{3\pi}{2}, \ldots$ creating repeating branches, while the hyperbolic term $\frac{h}{\alpha}$ approaches infinity as $\alpha \to 0$ and decreases monotonically as α increases. This difference in behavior makes graphical methods particularly effective, as the intersections occur in specific regions based on the periodicity of $\tan \alpha b$. See Fig. 6.1 for all the relevant information. Based on this discussion, the eigenfunctions of (6.32) are $\lambda_n = \alpha_n^2$ with corresponding eigenfunctions

$$X_n(x) = \cos \alpha_n x, \quad n = 1, 2, \ldots$$

where the eigenvalues $\lambda_n = \alpha_n^2$, $n = 1, 2, \ldots$ satisfy the relation

$$\tan \alpha_n b = \frac{h}{\alpha_n}, \quad n = 1, 2, \ldots \tag{6.35}$$

For clarity, we normalize the eigenfunctions to simplify the process of calculating the Fourier coefficients.

$$\begin{aligned} ||X_n||^2 &= \int_0^b X_n^2(x)\,dx = \int_0^b \cos^2 \alpha_n x\,dx \\ &= \frac{1}{2}\int_0^b (1 + \cos 2\alpha_n x)\,dx = \frac{1}{2}\left(b + \frac{\sin 2\alpha_n b}{2\alpha_n}\right). \end{aligned}$$

Using the identity $\sin 2\alpha_n b = 2\sin \alpha_n b \cos \alpha_n b$, we arrive at the simplified form

$$||X_n||^2 = \frac{1}{2}\left(b + \frac{\sin \alpha_n b \cos \alpha_n b}{\alpha_n}\right).$$

On the other hand, from (6.35) we have

$$\frac{\tan \alpha_n b}{h} = \frac{1}{\alpha_n},$$

which when substituted in the right-hand side of $||X_n^2||^2$ yield

$$||X_n|| = \sqrt{\frac{hb + \sin^2 \alpha_n b}{2h}}.$$

Finally, the normalized eigenfunctions are

$$\psi_n(x) = \frac{X_n(x)}{||X_n||} = \sqrt{\frac{2h}{hb + \sin^2 \alpha_n b}} \cos \alpha_n x, \quad n = 1, 2 \ldots \tag{6.36}$$

Equation (6.33) has the solution

$$T_n(t) = e^{-\alpha_n^2 kt}, \quad n = 1, 2, \ldots$$

where α_n satisfy (6.35). Thus, the general solution is

$$u(x,t) = \sum_{n=1}^{\infty} a_n e^{-\alpha_n^2 kt} \psi_n(x), \tag{6.37}$$

where ψ_n are given by (6.36). Applying the initial condition $u(x,0) = f(x)$ to (6.37), we obtain the Fourier cosine series

$$f(x) = \sum_{n=1}^{\infty} a_n \psi_n(x),$$

from which we have

$$a_n = (\psi_n(x), f(x)) = \int_0^b \psi_n(x) f(x)\, dx = \sqrt{\frac{2h}{hb + \sin^2 \alpha_n b}} \int_0^b f(x) \cos \alpha_n x dx, \; n = 1, 2 \ldots \tag{6.38}$$

Finally, the general solution of (6.29)–(6.31) is given by (6.37), where $\psi_n(x)$ and a_n are given by (6.36) and (6.38), respectively. For the sake of completeness, we present the form of the solution, which is given by

$$\begin{aligned} u(x,t) &= \sum_{n=1}^{\infty} \sqrt{\frac{2h}{hb + \sin^2 \alpha_n b}} \int_0^b f(x) \cos \alpha_n x dx e^{-\alpha_n^2 kt} \sqrt{\frac{2h}{hb + \sin^2 \alpha_n b}} \cos \alpha_n x \\ &= \sum_{n=1}^{\infty} \frac{2h}{hb + \sin^2 \alpha_n b} \left(\int_0^b f(x) \cos \alpha_n x dx \right) e^{-\alpha_n^2 kt} \cos \alpha_n x. \end{aligned}$$

□

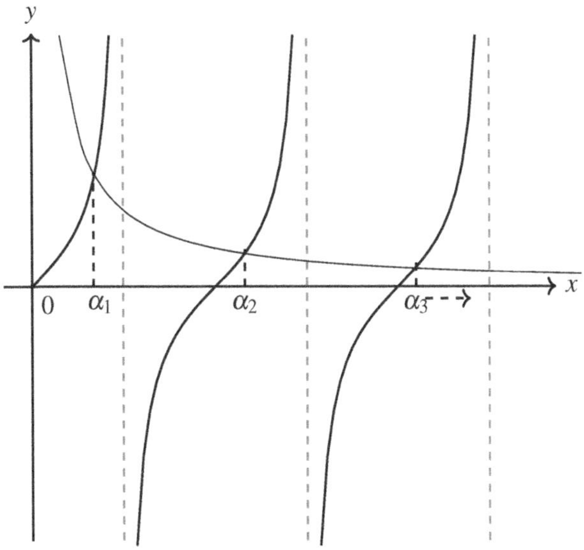

FIGURE 6.1: $y = \tan \alpha c$; $y = \frac{h}{\alpha}$.

Example 6.5 *We want to find the temperature $u(x,t)$ of the slab at position x at time t. The function $u(x,t)$ satisfies the boundary value problem*

$$u_t = k u_{xx}, \quad 0 < x < L, \quad t > 0,$$

$$u(0,t) = 0, \quad hu(L,t) + u_x(L,t) = 0, \quad u(x,0) = U_0.$$

Using separation of variables, we obtain

$$\frac{X''}{X} = \frac{T'}{kT} = -\lambda,$$

$$X'' + \lambda X = 0, \quad T' + \lambda kT = 0.$$

Transforming the boundary conditions, we get

$$X(0) = 0, \quad hX(L) + X'(L) = 0.$$

For $\lambda = 0$, we have the trivial solution. For $\lambda > 0$ with $\lambda = \beta^2$, $\beta > 0$, we have the eigenvalues and associated eigenfunctions

$$\lambda_n = \beta_n^2, \quad X_n(x) - \sin \beta_n x$$

for $n = 1, 2, 3, \ldots$, where β_n satisfies the relation

$$\tan \beta_n L = -\frac{\beta_n}{h}.$$

When we substitute $\lambda = \beta_n^2$ in the right-hand equation in $T(t)$, we get the first-order equation

$$T_n'(t) = -\beta_n^2 k T_n(t),$$

whose solution is

$$T_n(t) = \exp\left(-\beta_n^2 kt\right).$$

Consequently, the general solution is

$$u(x,t) = \sum_{n=1}^{\infty} c_n \exp\left(-\beta_n^2 kt\right) \sin \beta_n x.$$

Applying the non-homogeneous condition we arrive at

$$u(x,0) = \sum_{n=1}^{\infty} c_n \sin \beta_n x = U_0,$$

where

$$c_n = \frac{\int_0^L U_0 \sin \beta_n x \, dx}{\int_0^L \sin^2 \beta_n x \, dx}.$$

We have

$$\int_0^L U_0 \sin \beta_n x \, dx = U_0 \frac{(1-\cos \beta_n L)}{\beta_n},$$

and

$$\int_0^L \sin^2 \frac{\beta_n x}{L} \, dx = \frac{1}{2h}\left(hL + \cos^2 \beta_n L\right).$$

Hence

$$c_n = \frac{2U_0 h(1-\cos \beta_n L)}{\beta_n (hL + \cos^2 \beta_n L)}.$$

Finally, the solution is

$$u(x,t) = 2U_0 h \sum_{n=1}^{\infty} \frac{(1-\cos \beta_n L)}{\beta_n (hL + \cos^2 \beta_n L)} \exp\left(-\beta_n^2 kt\right) \sin \beta_n x.$$

□

6.6.1 Exercises

In Exercises 6.21–6.23 assume the constant $h > 0$ and find the eigenvalues and normalized eigenfunctions.

Exercise 6.21 $X''(x) + \lambda X(x) = 0, \quad X(1) = 0, \quad hX(0) - X'(0) = 0.$

Exercise 6.22 $X''(x) + \lambda X(x) = 0, \quad X(0) = 0, \quad hX(1) - X'(1) = 0.$

Exercise 6.23 $X''(x) + \lambda X(x) = 0, \quad X(0) = 0, \quad X(L) + hX'(L) = 0.$

Exercise 6.24 *Find the eigenvalues and corresponding normalized eigenfunctions for*

$$y''(x) + y'(x) + (1+\lambda)y(x) = 0, \quad y(0) = 0, \quad y(1) = 0.$$

Exercise 6.25 *Show that the eigenvalues and eigenfunctions of the Sturm–Liouville problem*

$$X''(x) + \lambda X(x) = 0, \quad X(0) + X'(0) = 0, \quad X(1) = 0,$$

are given by

$$\lambda_0 = 0, \quad X_0(x) = x - 1,$$

and

$$\lambda_n = \alpha_n^2, \quad X_n(x) = \alpha_n \cos(\alpha_n x) - \sin(\alpha_n x),$$

for $n = 1$, *where* $\{\alpha_n\}_{n=1}^{\infty}$ *are the positive roots of* $\tan x = x$.

Exercise 6.26 *Consider*

$$X''(x) + \lambda X(x) = 0, \quad hX(0) - X'(0) = 0 = hX(L) + X'(L) \quad (h > 0),$$

where

$$\lambda_n = \frac{\beta_n^2}{L^2}, \quad X_n(x) = \beta_n \cos\frac{\beta_n x}{L} + hL\sin\frac{\beta_n x}{L}, \quad (n \geq 1),$$

where β_n *is the nth positive root of the equation*

$$\tan x = \frac{2hLx}{x^2 - h^2L^2}.$$

To estimate β_n *for large n, sketch the graphs of*

$$y = 2hL\cot x, \quad y = \frac{x^2 - h^2L^2}{x}.$$

(When you let $\lambda = \alpha^2$, *then* $\beta = \alpha L$.*)*

Exercise 6.27 *Solve, make sure to normalize the eigenfunctions.*

(a) $u_t = u_{xx}, \quad u(0,t) = 0, \quad u_x(\pi,t) + u(\pi,t) = 0, \quad u(x,0) = 1 - \frac{x}{\pi}$.

(b) $u_t = 3u_{xx}, \quad u(0,t) = 0, \quad u_x(2,t) + u(2,t) = 0, \quad u(x,0) = x$.

Exercise 6.28 *Consider*

$$u_t = ku_{xx}, \quad u(0,t) = 0, \quad u_x(L,t) + hu(L,t) = 0, \quad u(x,0) = f(x).$$

(a) Solve if $hL > -1$.

(b) Solve if $hL = -1$.

Exercise 6.29 *For a positive constant h, solve the wave equation problem with Robin boundary conditions*

$$\begin{cases} u_{tt} - c^2 u_{xx} = 0, & 0 \leq x \leq l,\ t > 0 \\ u_x(0,t) - hu(0,t) = 0, \quad u_x(l,t) = 0 & \\ u(x,0) = f(x), & 0 \leq x \leq l \\ u_t(x,0) = g(x), \quad 0 \leq x \leq l & \end{cases}$$

for given and properly behaved functions f and g.

Exercise 6.30 *Solve the wave propagation problem with Robin boundary condition*

$$\begin{cases} u_{tt} - c^2 u_{xx} = 0, & 0 \le x < \pi,\ t > 0 \\ u(0,t) = 0, & t > 0 \\ u(\pi,t) - u_x(\pi,t) = 0, & t > 0 \\ u(x,0) = \sin x, & 0 \le x < \pi \\ u_t(x,0) = 0, & 0 \le x < \pi \end{cases}$$

Exercise 6.31 *Solve the wave propagation problem with Robin boundary condition*

$$\begin{cases} u_{tt} = c^2 u_{xx}, & 0 < x < \pi,\ t > 0 \\ u(0,t) = 0, & t > 0 \\ u(\pi,t) - u_x(\pi,t) = 0, & t > 0 \\ u(x,0) = 0, & 0 < x < \pi \\ u_t(x,0) = f(x), & 0 < x < \pi \end{cases}$$

where

$$f(x) = \begin{cases} 0, & 0 < x < \frac{\pi}{2} \\ 1, & \frac{\pi}{2} < x < \pi \end{cases}$$

Exercise 6.32 *Find a bounded solution for the wave problem with Robin boundary condition*

$$\begin{cases} u_{tt} = c^2 u_{xx}, & 0 < x, \quad t > 0 \\ u_t(0,t) + a u_x(0,t) = h(t), & t > 0 \\ u(x,0) = f(x), & 0 < x \\ u_t(x,0) = g(x), & 0 < x \end{cases}$$

Exercise 6.33 *For constants T_0, d and $h > 0$, consider*

$$u_t = k u_{xx}, \quad 0 < x < L,$$

$$u(0,t) = T_0, \quad u_x(L,t) + h u(L,t) - hd = 0, \quad u(x,0) = f(x).$$

(a) Solve for $V(x)$, where

$$kV_{xx}(x) = 0, \quad V(0) = T_0, \quad V_x + hV(L) - hd = 0,$$

is the corresponding steady-state problem.

(b) Let $w(x,t) = u(x,t) - V(x)$. Show that $w(x,t)$ satisfies a heat flow problem with homogeneous boundary conditions.

(c) Find the solution $u(x,t)$ to the original problem.

Exercise 6.34 *Solve the boundary value problem*

$$u_t = k u_{xx}, \quad 0 < x < L, \quad t > 0,$$

with boundary and initial conditions

$$u(0,t) = 0, \quad u_t(L,t) + hk\,u_x(L,t) = 0, \quad u(x,0) = 1,$$

that describes the diffusion of gas through a membrane.

Ans: $u(x,t) = 4\sum_{n=1}^{\infty}\left(\frac{1-\cos\beta_n}{2\beta_n\sin 2\beta_n}\right)\exp\left(-\frac{\beta_n^2 kt}{L^2}\right)\sin\frac{\beta_n x}{L}$, *where* $\{\beta_n\}_{n=1}^{\infty}$ *are the positive roots of the equation* $\tan x = \frac{hL}{x}$.

Exercise 6.35 *Solve the heat problem with a source and Robin's boundary conditions*

$$u_t = ku_{xx} + Q(x,t), \quad 0 < x < L,$$

$$u(0,t) = 0, \quad u_x(L,t) + hu(L,t) - B(t) = 0, \quad u(x,0) = f(x)$$

for a continuous function $B(t)$ *on the interval* $[0,L]$.

6.7 Robin's Boundary Conditions in Laplacian

Consider the Laplace equation for a function $u(x,y)$ that is harmonic in a rectangular region with boundary conditions

$$u_{xx}(x,y) + u_{yy}(x,y) = 0, \quad 0 < x < a, \quad 0 < y < b, \tag{6.39}$$
$$u(x,0) = f_1(x), \quad u(x,b) = f_2(x), \tag{6.40}$$
$$u(0,y) = 0, \quad u_x(a,y) + hu(a,y) = 0. \tag{6.41}$$

The Robin condition at $x = a$ (6.41) combines both the temperature and the flux at this boundary. The rate of change of the temperature $u_x(a,y)$ is related to the temperature itself by a proportionality constant h. This type of boundary condition can be interpreted as a situation where there is heat exchange with the surrounding medium at a rate proportional to the temperature on the boundary. In this case, h represents the heat transfer coefficient, indicating the intensity of heat exchange. We assume a nonzero solution and by substituting the product $u = X(x)Y(y)$ into (6.39)-(6.41), we obtain

$$X''(x) + \lambda X(x) = 0, \quad X(0) = 0, \quad hX(a) + X'(a) = 0, \tag{6.42}$$

$$Y''(y) - \lambda Y(y) = 0, \quad Y(0) = 0. \tag{6.43}$$

We begin solving equation (6.42), where we know that $\lambda \geq 0$. We obtain the trivial solution for $\lambda = 0$. If $\lambda > 0$, we set $\lambda = \alpha^2$ with $\alpha > 0$. Then, (6.42) becomes

$$X''(x) + \alpha^2 X(x) = 0,$$

whose general solution is

$$X(x) = A\cos\alpha x + B\sin\alpha x.$$

Applying $X(0) = 0$ gives $A = 0$. Substituting $A = 0$ into $X(x)$, we obtain

$$X(x) = B\sin\alpha x.$$

Using the Robin boundary condition,

$$X'(b) + hX(b) = \alpha B\cos\alpha b + Bh\sin\alpha b = B\big(\alpha\cos\alpha b + h\sin\alpha b\big) = 0.$$

Since $B \neq 0$, we must have

$$\alpha\cos\alpha b + h\sin\alpha b = 0$$

which implies

$$\tan\alpha b = -\frac{\alpha}{h}.$$

So we obtain the eigenvalues

$$\lambda_n = \alpha_n^2, \quad \text{where} \quad \tan\alpha_n b = -\frac{\alpha_n}{h}, \quad n = 1, 2\ldots$$

and α_n are the intersection points of $\tan\alpha b$ and $-\frac{\alpha}{h}$. The eigenfunctions are

$$X_n(x) = \sin\alpha_n x, \quad n = 1, 2\ldots$$

Consequently, the normalized eigenfunctions are

$$\psi_n(x) = \frac{X_n(x)}{\|X_n\|} = \sqrt{\frac{2h}{hb + \cos^2\alpha_n b}}\sin\alpha_n x, \quad n = 1, 2\ldots$$

Additionally,

$$Y_n(y) = a_n\cosh\alpha_n y + b_n\sinh\alpha_n y, \quad n = 1, 2\ldots$$

where a_n and b_n are to be determined using (6.40). Thus, the general solution is

$$u(x, y) = \sum_{n=1}^{\infty}\Big[a_n\cosh\alpha_n y + b_n\sinh\alpha_n y\Big]\psi_n(x). \tag{6.44}$$

Using the boundary condition $u(x, 0) = f_1(x)$ in (6.44) gives

$$f_1(x) = \sum_{n=1}^{\infty} a_n\psi_n(x).$$

This implies that the coefficients a_n are the Fourier sine series coefficients of $f_1(x)$, and therefore

$$a_n = \int_0^a f_1(x)\psi_n(x)\,dx, \quad n = 1, 2, \ldots. \tag{6.45}$$

The second boundary condition $u(x,b) = f_2(x)$ leads to

$$f_2(x) = \sum_{n=1}^{\infty} \left[a_n \cosh \alpha_n b + b_n \sinh \alpha_n b\right] \psi_n(x).$$

This implies that

$$B_n = a_n \cosh \alpha_n b + b_n \sinh \alpha_n b,$$

where B_n must be the Fourier sine coefficients of $f_2(x)$. Thus,

$$B_n = \int_0^a f_2(x)\psi_n(x)dx, \quad n = 1,2,\ldots. \tag{6.46}$$

Solving for b_n in (6.46) yields

$$b_n = \frac{B_n - a_n \cosh \alpha_n b}{\sinh \alpha_n b}, \quad n = 1,2,\ldots. \tag{6.47}$$

Finally, the general solution of (6.39)–(6.41) is obtained by substituting a_n and b_n into (6.44), where a_n, B_n, and b_n are given by the equations (6.45), (6.46), and (6.47), respectively.

6.7.1 Exercises

Exercise 6.36 *Solve*

$$\begin{aligned} &u_{xx}(x,y) + u_{yy}(x,y) = 0, \quad 0 < x < 1, \quad 0 < y < 1, \\ &u(x,0) = 0, \quad u_y(x,1) + hu(x,1) = 0, \\ &u(0,y) = 0, \quad u(1,y) = f(y). \end{aligned}$$

Ans: $u(x,y) = \sum_{n=1}^{\infty} B_n \frac{\sinh(\alpha_n x)}{\sinh(\alpha_n)} \sin(\alpha_n y)$, *where* $B_n = \frac{2h}{h+\cos^2\alpha_n} \int_0^1 f(y)\sin(\alpha_n y)\,dy$ $(n = 1,2,\ldots)$.

Exercise 6.37 *Solve*

$$\begin{aligned} &u_{xx}(x,y) + u_{yy}(x,y) = 0, \quad 0 < x < a, \quad 0 < y < b, \\ &u_x(0,y) = 0, \quad u_x(a,y) + hu(a,y) = 0, \\ &u(x,0) = 0, \quad u(x,b) = f(x). \end{aligned}$$

Ans: $u(x,y) = 2h \sum_{n=1}^{\infty} \frac{\cos(\alpha_n x)}{ha + \sin^2(\alpha_n a)} \cdot \frac{\sinh(\alpha_n y)}{\sinh(\alpha_n b)} \int_0^a f(s)\cos(\alpha_n s)\,ds$, $\tan \alpha_n a = h/\alpha_n$.

Exercise 6.38 *Find a* **bounded solution** $u(x,y)$ *for*

$$u_{xx}(x,y) + u_{yy}(x,y) = 0, \quad 0 < x < a, \quad y > 0,$$

$$u_x(0,y) = 0, \quad u_x(a,y) + hu(a,y) = 0,$$
$$u(x,0) = f(x),$$

where h is a positive constant.

Ans: $u(x,y) = \sum_{n=1}^{\infty} A_n e^{-\alpha_n y} \cos(\alpha_n x)$, *where*

$$\tan(\alpha_n) = \frac{h}{\alpha_n}, \quad A_n = \frac{2h}{h + \sin^2(\alpha_n)} \int_0^1 f(x) \cos(\alpha_n x)\, dx.$$

6.8 Bessel's Function

Bessel's equation arises in problems with cylindrical or spherical symmetry, such as heat conduction, wave propagation, or vibrations in circular membranes. Despite the singularity at $x = 0$, its solutions (Bessel functions) are well-behaved for physical problems, often subject to boundary conditions at a specific finite radius or infinity.

We make the following definition.

Definition 6.5 *Consider the second-order differential equation*

$$y''(x) + P(x)y'(x) + Q(x)y(x) = 0. \tag{6.48}$$

(a) *A function is analytic at x_0 if there exists a Taylor series expansion of the function about x_0 with a nonzero radius of convergence.*

(b) *A point x_0 is said to be an **ordinary point** of* (6.48) *if both $P(x)$ and $Q(x)$ are **analytic** at x_0. A point that is not an ordinary point is said to be a **singular point** of* (6.48).

(c) *A singular point x_0 of* (6.48) *is said to be a **regular singular point** if both $(x - x_0)P(x)$ and $(x - x_0)^2 Q(x)$ are analytic at x_0. A singular point that is **not regular** is said to be an **irregular singular point** of* (6.48).

There are many special functions which arise as solutions to differential equations (Hermite, Legendre, Chebyshev, etc.). Here we will look at how one important class of functions, **Bessel Functions**, arise through a series solution to a differential equation.

Bessel Functions of the First Kind are particularly important in the study of partial differential equations, and arise in the study of vibrating circular drumheads, heat equations, and many other areas where cylindrical symmetry is present.

Bessel's equation is a differential equation of the form

$$x^2 y'' + xy' + (x^2 - v^2)y = 0 \tag{6.49}$$

where v is a real number. Dividing by x^2, we see that the equation is not analytic at $x=0$. Further, since $xP(x)=1$ and $x^2Q(x)=x^2-v^2$ are analytic at $x=0$, the point $x=0$ is a regular singular point. Therefore, we can try to determine at least one series solution of Bessel's equation which has the form

$$y=\sum_{n=0}^{\infty} a_n x^{r+n}.$$

Differentiating $y=\sum_{n=0}^{\infty} a_n x^{r+n}$ and substituting into Bessel's equation (6.49), we obtain

$$y=\sum_{n=0}^{\infty} a_n x^{r+n}, \quad y'=\sum_{n=0}^{\infty}(r+n)a_n x^{r+n-1}, \quad y''=\sum_{n=0}^{\infty}(r+n)(r+n-1)a_n x^{r+n-2}.$$

Substituting into the equation

$$x^2\sum_{n=0}^{\infty}(r+n)(r+n-1)a_n x^{r+n-2}+x\sum_{n=0}^{\infty}(r+n)a_n x^{r+n-1}+(x^2-v^2)\sum_{n=0}^{\infty}a_n x^{r+n}=0.$$

Simplifying

$$\sum_{n=0}^{\infty}(r+n)(r+n-1)a_n x^{r+n}+\sum_{n=0}^{\infty}(r+n)a_n x^{r+n}+\sum_{n=0}^{\infty}a_n x^{r+n+2}-\sum_{n=0}^{\infty}v^2 a_n x^{r+n}=0.$$

Reindexing the third series gives

$$\sum_{n=0}^{\infty}(r+n)(r+n-1)a_n x^{r+n}+\sum_{n=0}^{\infty}(r+n)a_n x^{r+n}+\sum_{n=2}^{\infty}a_{n-2} x^{r+n}-\sum_{n=0}^{\infty}v^2 a_n x^{r+n}=0.$$

Factoring out x^r and simplifying leads to

$$\sum_{n=0}^{\infty}\left[(r+n)(r+n-1)a_n+(r+n)a_n+a_{n-2}-v^2a_n\right]x^n=0.$$

Thus, the coefficients of x^n must individually vanish to satisfy the equation.

$$r(r-1)a_0+ra_0-v^2a_0x^0+(r+1)ra_1+(r+1)a_1-v^2a_1x^1+$$

$$\sum_{n=2}^{\infty}\left[(r+n)^2a_n+a_{n-2}-v^2a_n\right]x^n=0.$$

We want this to be true for all values of x. Therefore, the coefficients of powers of x must be zero. This leads to the equations

$$r(r-1)a_0+ra_0-v^2a_0=0, \tag{6.50}$$

$$(r+1)ra_1+(r+1)a_1-v^2a_1=0, \tag{6.51}$$

$$(r+n)^2 a_n + a_{n-2} - \nu^2 a_n = 0, \quad n = 2,3,\ldots \tag{6.52}$$

These equations are the recursion and indicial equations. We can choose any one of them as the indicial equation; let's choose the indicial equation to be (6.50)

$$r(r-1)a_0 + ra_0 - \nu^2 a_0 = 0.$$

If we assume $a_0 \neq 0$, the solutions to the indicial equation are

$$r = \pm\nu.$$

We can determine one solution based on our choice of whether $r_1 = \nu$ or $r_2 = -\nu$.

- $r_1 = \nu > 0$.

Then the recursion equations (6.51) and (6.52) become

$$(2\nu+1)a_1 = 0 \tag{6.53}$$

$$a_n = -\frac{a_{n-2}}{n(n+2\nu)}, \quad n = 2,3,\ldots \tag{6.54}$$

Equation (6.53) tells us that $a_1 = 0$, unless $\nu = -1/2$, in which case the equation is satisfied regardless of the value of a_1. Since we have assumed that $\nu > 0$, this situation does not arise, and we may safely assume $a_1 = 0$.

We can now use (6.54) along with $a_1 = 0$ to generate the first few a_n, and then try to determine the pattern. Thus,

$$a_0 = \text{unspecified, does not equal zero,}$$

$$a_1 = 0,$$

$$a_2 = -\frac{a_0}{2(2+2\nu)},$$

$$a_3 = 0,$$

$$a_4 = -\frac{a_2}{4(4+2\nu)} = \frac{a_0}{2\cdot 4(2+2\nu)(4+2\nu)} = \frac{a_0}{2^2\cdot 2!(1+\nu)(2+\nu)},$$

$$a_5 = 0,$$

$$a_6 = -\frac{a_4}{6(6+2\nu)} = -\frac{a_0}{2\cdot 4\cdot 6(2+2\nu)(4+2\nu)(6+2\nu)} = -\frac{a_0}{2^3\cdot 3!(1+\nu)(2+\nu)(3+\nu)}.$$

This is enough for us to recognize the pattern. First, let's work with $(1+\nu)(2+\nu)\cdots(m+\nu)$. If ν is an integer, we observe that:

$$\frac{1}{(1+\nu)(2+\nu)\cdots(m+\nu)} = \frac{1\cdot 2\cdot 3\cdots(\nu-1)(\nu)}{1\cdot 2\cdot 3\cdots(\nu-1)(\nu)(1+\nu)(2+\nu)\cdots(m+\nu)} = \frac{\nu!}{(\nu+m)!}.$$

If ν is not an integer, we can replace the factorial with a gamma function using the relation $w! = \Gamma(1+w)$. Therefore,

$$\frac{1}{(1+\nu)(2+\nu)\cdots(m+\nu)} = \frac{\Gamma(1+\nu)}{\Gamma(1+\nu+m)}.$$

The general term in our expansion is therefore

$$a_{2m} = \frac{(-1)^m a_0 \Gamma(1+\nu)}{2^{2m} m! \Gamma(1+\nu+m)}, \qquad m = 0,1,2,\ldots.$$

Hence, a solution to Bessel's equation is

$$y = \sum_{n=0}^{\infty} a_n x^{r+n} = \sum_{m=0}^{\infty} a_{2m} x^{\nu+2m} = \sum_{m=0}^{\infty} \frac{(-1)^m a_0 \Gamma(1+\nu)}{2^{2m} m! \Gamma(1+\nu+m)} x^{\nu+2m}.$$

By factoring out terms, this becomes

$$y = a_0 x^{\nu} \Gamma(1+\nu) \sum_{m=0}^{\infty} \frac{(-1)^m x^{2m}}{2^{2m} m! \Gamma(1+\nu+m)}.$$

Bessel functions of the first kind of order ν are obtained by choosing $a_0 = \frac{1}{2^{\nu}\Gamma(1+\nu)}$, and they are commonly denoted $J_{\nu}(x)$,

$$J_{\nu}(x) = \left(\frac{x}{2}\right)^{\nu} \sum_{m=0}^{\infty} \frac{(-1)^m x^{2m}}{2^{2m} m! \Gamma(1+\nu+m)} = \sum_{m=0}^{\infty} \frac{(-1)^m}{m! \Gamma(1+\nu+m)} \left(\frac{x}{2}\right)^{2m+\nu}. \tag{6.55}$$

Although our analysis assumed $\nu > 0$, it is evident that $J_{\nu}(x)$ is well-defined for all $\nu \in \mathbb{R}$. An important consideration in solving partial differential equations is that $J_{\nu}(x)$ remains finite at $x = 0$ when ν is an integer. Following a similar line of reasoning, for the second component $r_2 = -\nu$, we find

$$J_{-\nu}(x) = \left(\frac{x}{2}\right)^{-\nu} \sum_{m=0}^{\infty} \frac{(-1)^m x^{2m}}{2^{2m} m! \Gamma(1-\nu+m)} = \sum_{m=0}^{\infty} \frac{(-1)^m}{m! \Gamma(1-\nu+m)} \left(\frac{x}{2}\right)^{2m-\nu}. \tag{6.56}$$

The functions $J_{\nu}(x)$ and $J_{-\nu}(x)$ are known as the Bessel functions of orders ν and $-\nu$, respectively. Both (6.55) and (6.56) are convergent on the interval $(0,\infty)$. It is evident that $J_{\nu}(x) = J_{-\nu}(x)$ when $\nu = 0$. For non-integer values of ν, the functions $J_{\nu}(x)$ and $J_{-\nu}(x)$ converge for all x and are linearly independent. Consequently, the general solution of the Bessel equation can be expressed as

$$y(x) = c_1 J_{\nu}(x) + c_2 J_{-\nu}(x). \tag{6.57}$$

If ν is an integer, say $\nu = m$, then $J_{\nu}(x) + c_2 J_{-\nu}(x)$ are **not** linearly independent. To see this, we let $\nu = m$, then

$$\begin{aligned} J_{-m}(x) &= \sum_{k=n}^{\infty} \frac{(-1)^k x^{2k-m}}{2^{2k-m} k! \Gamma(-m+k+1)} = (-1)^m \sum_{k=0}^{\infty} \frac{(-1)^k x^{2k+m}}{2^{2k+m} k! \Gamma(m+k+1)} \\ &= (-1)^m J_m(x). \end{aligned} \tag{6.58}$$

This demonstrates that J_{-m} is dependent on J_m, and as a result, a second linearly independent solution becomes necessary. While various irregular solutions are explored in the literature, the most commonly utilized one, as described by Watson (1966), is

$$Y_\nu(x) = \frac{(\cos \nu\pi)J_\nu(x) - J_{-\nu}(x)}{\sin \nu\pi}. \tag{6.59}$$

For non-integer ν, it is obvious that $Y_\nu(x)$, being a linear combination of $J_\nu(x)$ and $J_{-\nu}(x)$, is linearly independent of $J_\nu(x)$. When ν is a nonnegative integer m, $Y_\nu(x)$ is indeterminate. But

$$Y_m(x) = \lim_{\nu \to \infty} Y_\nu(x)$$

exists and is a solution of the Bessel equation. Moreover, it is linearly independent of $J_\nu(x)$. (For an extended treatment, see Watson (1966)). The function $Y_\nu(x)$ is called the Bessel function of the second kind of order ν. We have the following theorem.

Theorem 6.4 *The general solution of Bessel's function of order ν given by* (6.49) *is*

$$y(x) = c_1 J_\nu(x) + c_2 Y_\nu(x), \quad \textit{for } \nu \geq 0.$$

Example 6.6 The solution to

$$x^2 y''(x) + xy'(x) + (x^2 - 16) = 0,$$

is

$$y(x) = c_1 J_4(x) + c_2 Y_4(x).$$

□

Corollary 6 *The solution to the modified Bessel's equation containing a parameter λ, namely,*

$$x^2 y''(x) + xy'(x) + (\lambda^2 x^2 - \nu^2)y(x) = 0 \tag{6.60}$$

is given by

$$y(x) = c_1 J_\nu(\lambda x) + c_2 Y_\nu(\lambda x). \tag{6.61}$$

Proof *To introduce λ^2, we perform a substitution for x of the form*

$$u = \frac{x}{\lambda}.$$

Then, apply the chain rule to the derivatives of $y(x)$

$$\frac{dy}{dx} = y'(x) = \frac{dy}{du} \cdot \frac{du}{dx} = \frac{1}{\lambda} y'(u),$$

where $u = \frac{x}{\lambda}$. Similarly,

$$y''(x) = \frac{d}{dx}\left(\frac{1}{\lambda} y'(u)\right) = \frac{1}{\lambda} \cdot \frac{d}{dx}\left(y'(u)\right),$$

and since $\frac{d}{dx} = \frac{1}{\lambda}\frac{d}{du}$, *this becomes*

$$y''(x) = \frac{1}{\lambda^2} y''(u).$$

Substitute $x = \lambda u$, *we obtain*

$$x^2 = \lambda^2 u^2, \quad xy'(x) = \lambda u \cdot \frac{1}{\lambda} y'(u) = uy'(u), \quad y''(x) = \frac{1}{\lambda^2} y''(u).$$

Substituting these into (6.60) *gives*

$$\lambda^2 u^2 \cdot \frac{1}{\lambda^2} y''(u) + \lambda u \cdot \frac{1}{\lambda} y'(u) + (\lambda^2 u^2 - \nu^2) y(u) = 0.$$

Simplify to get

$$u^2 y''(u) + uy'(u) + (\lambda^2 u^2 - \nu^2) y(u) = 0.$$

This is the desired equation after exchanging x *with* u

$$x^2 y''(x) + xy'(x) + (\lambda^2 x^2 - \nu^2) y(x) = 0.$$

Since the solution of (6.49) *is* $y(x) = c_1 J_\nu(x) + c_2 Y_\nu(x)$ *and* $x = \lambda u$, *we have*

$$y(\lambda u) = c_1 J_\nu(\lambda u) + c_2 Y_\nu(\lambda u).$$

We have the following lemma.

Lemma 12 *Let* $J_\nu(x)$ *and* $J_{-\nu}(x)$ *be given by* (6.55) *and* (6.56), *respectively. Then*

(a) $J_{1/2}(x) = \sqrt{\frac{2}{\pi x}} \sin x,$

(b) $J_{-1/2}(x) = \sqrt{\frac{2}{\pi x}} \cos x.$

Proof *We only prove (a) and leave (b) as an exercise. Simplify the argument of the Gamma function*

$$\Gamma\left(1 + \frac{1}{2} + n\right) = \Gamma\left(\frac{3}{2} + n\right).$$

The Gamma function satisfies the recurrence relation

$$\Gamma(z+1) = z\Gamma(z).$$

Applying this repeatedly to $\Gamma\left(\frac{3}{2} + n\right)$, *we get*

$$\Gamma\left(\frac{3}{2} + n\right) = \left(\frac{3}{2} + n - 1\right)\left(\frac{3}{2} + n - 2\right) \cdots \left(\frac{3}{2}\right) \Gamma\left(\frac{3}{2}\right) = \prod_{k=0}^{n-1} \left(\frac{3}{2} + k\right) \Gamma\left(\frac{3}{2}\right).$$

From the properties of the Gamma function

$$\Gamma\left(\frac{3}{2}\right) = \frac{1}{2}\sqrt{\pi}.$$

Substituting $\Gamma\left(\frac{3}{2}\right) = \frac{1}{2}\sqrt{\pi}$, *we obtain*

$$\Gamma\left(1+\frac{1}{2}+n\right) = \prod_{k=0}^{n-1}\left(\frac{3}{2}+k\right)\cdot\frac{1}{2}\sqrt{\pi}.$$

The product can be written as

$$\prod_{k=0}^{n-1}\left(\frac{3}{2}+k\right) = \frac{3}{2}\cdot\frac{5}{2}\cdot\frac{7}{2}\cdots\frac{2n+1}{2}.$$

Factoring out 2^n *from each term in the denominator gives*

$$\prod_{k=0}^{n-1}\left(\frac{3}{2}+k\right) = \frac{1}{2^n}\cdot(3\cdot 5\cdot 7\cdots(2n+1)).$$

The product $3\cdot 5\cdot 7\cdots(2n+1)$ *is the double factorial, denoted as* $(2n+1)!!$. *A property of double factorials is*

$$(2n+1)!! = \frac{(2n+1)!}{2^n\cdot n!}.$$

Substituting $(2n+1)!!$ *into the expression, we have that*

$$\prod_{k=0}^{n-1}\left(\frac{3}{2}+k\right) = \frac{1}{2^n}\cdot\frac{(2n+1)!}{2^n\cdot n!} = \frac{(2n+1)!}{2^{2n}\cdot n!}.$$

The full expression for $\Gamma\left(1+\frac{1}{2}+n\right)$ *takes the form*

$$\Gamma\left(1+\frac{1}{2}+n\right) = \prod_{k=0}^{n-1}\left(\frac{3}{2}+k\right)\cdot\frac{1}{2}\sqrt{\pi}.$$

Substituting the product

$$\Gamma\left(1+\frac{1}{2}+n\right) = \frac{(2n+1)!}{2^{2n}\cdot n!}\cdot\frac{1}{2}\sqrt{\pi} = \frac{(2n+1)!\cdot\sqrt{\pi}}{2^{2n+1}\cdot n!}.$$

Let $\nu = \frac{1}{2}$. *Then from (6.55), we obtain*

$$J_{1/2}(x) = \left(\frac{x}{2}\right)^{1/2}\sum_{m=0}^{\infty}\frac{(-1)^m x^{2m}}{2^{2m}m!\frac{(2m+1)!\cdot\sqrt{\pi}}{2^{2m+1}\cdot m!}} = \sqrt{\frac{2}{\pi x}}\sum_{m=0}^{\infty}\frac{(-1)^m x^{2m+1}}{(2m+1)!}.$$

Since the series in the last line is the Maclaurin series for $\sin x$, *we conclude that*

$$J_{1/2}(x) = \sqrt{\frac{2}{\pi x}}\sin x.$$

This completes the proof of (a). □

Example 6.7 Consider the second-order differential equation

$$y''(x) + k^2 xy(x) = 0. \tag{6.62}$$

We use the transformations

$$u = \frac{y}{\sqrt{x}}, \quad w = \frac{2kx^{3/2}}{3}$$

to transform (6.62) into a Bessel's equation. Then,

$$\frac{du}{dx} = \frac{\frac{y'}{\sqrt{x}} - \frac{1}{2}x^{-1/2}y}{x} = \frac{y'}{\sqrt{x}} - \frac{y}{2x^{3/2}}.$$

Hence,

$$\frac{du}{dw} = \frac{\frac{du}{dx}}{\frac{dw}{dx}} = \frac{\frac{y'}{\sqrt{x}} - \frac{y}{2x^{3/2}}}{kx^{1/2}} = \frac{y'}{kx} - \frac{y}{2kx^2},$$

and

$$\frac{d^2u}{dw^2} = \frac{\frac{d}{dx}\left(\frac{du}{dw}\right)}{\frac{dw}{dx}}.$$

We do a bit more calculations.

$$\frac{d}{dx}\left(\frac{du}{dw}\right) = \frac{d}{dx}\left(\frac{y'}{kx} - \frac{y}{2kx^2}\right) = \frac{y''x - y'}{kx^2} - \frac{y'x - 2y}{2kx^3}.$$

Thus,

$$\frac{d^2u}{dw^2} = \frac{\frac{y''x-y'}{kx^2} - \frac{y'x-2y}{2kx^3}}{kx^{\frac{1}{2}}} = \frac{y''}{k^2x^{3/2}} - \frac{3y'}{2k^2x^{5/2}} + \frac{y}{k^2x^{7/2}}.$$

Consequently,

$$w^2\frac{d^2u}{dw^2} + w\frac{du}{dw} = \frac{4}{9}x^{\frac{3}{2}}y'' + \frac{1}{9}\frac{y}{\sqrt{x}}.$$

Substituting $y'' = -k^2xy$, leads us to

$$w^2\frac{d^2u}{dw^2} + w\frac{du}{dw} = -\frac{4}{9}k^2x^3\left(\frac{y}{\sqrt{x}}\right) + \frac{1}{9}\left(\frac{y}{\sqrt{x}}\right) = -w^2u + \frac{u}{9}.$$

We arrive at the Bessel's equation of order $\nu = \frac{1}{3}$,

$$w^2\frac{d^2u}{dw^2} + w\frac{du}{dw} + (w^2 - \frac{1}{9})u = 0,$$

which has the solution

$$u(w) = C_1J_{1/3}(w) + C_2J_{-1/3}(w).$$

Consequently, the original equation (6.62) has the general solution

$$y(x) = \sqrt{x}\Big[C_1J_{1/3}(2kx^{3/2}/3) + C_2J_{-1/3}(2kx^{3/2}/3)\Big].$$

□

The next lemma contains some important properties involving Bessel's functions and their derivatives.

Lemma 13 *Let $J_\nu(x)$ be given by (6.55). Then*

(a)

$$\frac{d}{dx}\left[x^{-\nu}J_\nu(x)\right] = -x^{-\nu}J_{\nu+1}(x), \tag{6.63}$$

(b)

$$\frac{d}{dx}\left[x^{\nu}J_\nu(x)\right] = x^{\nu}J_{\nu-1}(x). \tag{6.64}$$

Proof *We prove (a) and leave (b) as an exercise. The key idea of the proof is to multiply both sides of the derivative of* (6.55) *with x and carry out some calculations to derive a first-order differential equation in terms of J. Thus,*

$$\begin{aligned} xJ_\nu'(x) &= \sum_{m=0}^{\infty} \frac{(-1)^m(2m+\nu)}{m!\Gamma(1+\nu+m)}\left(\frac{x}{2}\right)^{2m+\nu} \\ &= \nu\sum_{m=0}^{\infty} \frac{(-1)^m}{m!\Gamma(1+\nu+m)}\left(\frac{x}{2}\right)^{2m+\nu} + 2\sum_{m=0}^{\infty} \frac{(-1)^m m}{m!\Gamma(1+\nu+m)}\left(\frac{x}{2}\right)^{2m+\nu}. \end{aligned}$$

The first sum is $\nu J_\nu(x)$, and in the second summation term, we substitute $k = n-1$ to obtain

$$\begin{aligned} xJ_\nu'(x) &= \nu J_\nu(x) - x\sum_{m=0}^{\infty} \frac{(-1)^k}{k!\Gamma(1+\nu+k)}\left(\frac{x}{2}\right)^{2k+\nu+1} \\ &= \nu J_\nu(x) - xJ_{\nu+1}(x). \end{aligned}$$

This relation is significant in its own right, so we label it for future reference. That is

$$xJ_\nu'(x) = \nu J_\nu(x) - xJ_{\nu+1}(x). \tag{6.65}$$

Since $x \in (0,\infty)$ we divide both sides of (6.65) *by x and arrive at the linear differential equation*

$$J_\nu'(x) - \frac{\nu}{x}J_\nu(x) = -J_{\nu+1}(x),$$

that has the integrating factor

$$\mu = e^{-\int \frac{\nu}{x}dx} = x^{-\nu}.$$

Consequently, (6.65) *is equivalent to*

$$\frac{d}{dx}\left[x^{-\nu}J_\nu(x)\right] = -x^{-\nu}J_{\nu+1}(x).$$

This completes the proof. □

Many important properties can be derived from equations (6.63) and (6.64). To demonstrate this, we expand the left-hand sides of both equations using the product rule and obtain

$$x^{\nu} J'_{\nu}(x) + \nu x^{\nu-1} J_{\nu}(x) = x^{\nu} J_{\nu-1}(x)$$

and

$$x^{-\nu} J'_{\nu}(x) - \nu x^{\nu} J_{\nu}(x) = -x^{-\nu} J_{\nu+1}(x).$$

Simplifying both sides of the above two equations we arrive at the relations

$$xJ'_{\nu}(x) = xJ_{\nu-1}(x) - \nu J_{\nu}(x), \tag{6.66}$$

and

$$xJ'_{\nu}(x) = \nu J_{\nu}(x) - xJ_{\nu+1}(x). \tag{6.67}$$

Subtracting (6.66) from (6.66) gives the relation

$$xJ_{\nu+1}(x) - 2\nu J_{\nu}(x) + xJ_{\nu-1}(x) = 0. \tag{6.68}$$

On the other hand, if we add both we obtain

$$2J'_{\nu}(x) = J_{\nu-1}(x) - J_{\nu+1}(x). \tag{6.69}$$

All of the above relations play a crucial role in solving Bessel functions, as they enable us to express higher-order Bessel functions in terms of lower-order ones.

Example 6.8 In this example we try to express $J_3(x)$ in terms of $J_1(x)$ and $J_0(x)$. Let $\nu = 2, 1$ in (6.68) and obtain the two equations

$$xJ_3(x) - 4J_2(x) + xJ_1(x) = 0, \quad xJ_2(x) - 2J_1(x) + xJ_0(x) = 0.$$

Solving the two equations gives

$$J_3(x) = \left(\frac{8}{x^3} - 1\right) J_1(x) - 2\frac{4}{x} J_0(x).$$

□

Example 6.9 In this example we want to evaluate

$$\int x^4 J_1(x)\,dx.$$

We proceed by performing integration by parts by letting $u = x^2$ and $dV = x^2 J_1(x)\,dx$. Note that using (6.64) we have

$$V = \int \frac{dV}{dx}\,dx = \int x^2 J_1(x)\,dx = x^2 J_2(x).$$

Thus,

$$\int x^4 J_1(x)\,dx = x^4 J_2(x) - 2\int x^3 J_2(x)\,dx.$$

Again, by invoking (6.64), we get $\int x^3 J_2(x)\,dx = x^3 J_3(x)$. Substituting into the previous integral, we obtain

$$\int x^4 J_1(x)\,dx = x^4 J_2(x) - 2x^3 J_3(x) + C.$$

□

We end this section with the following application regarding buckling of a vertical column.

Example 6.10 *(Buckling of a Vertical Column)* Buckling refers to the sudden lateral deformation of a structural member (such as a column) under compressive load. When the load on a column exceeds its critical load (Euler's buckling load), it can lose its stability, resulting in a bending or lateral displacement. The critical load at which this occurs depends on the column's length, cross-sectional properties, material properties, and boundary conditions. One end is fully embedded (fixed), while the other end is free to move or rotate. We take $x = 0$ at the free top end of the column and $x = L$ as its bottom. Let $y(x)$ be the deflection of the column at point x. From the theory of elasticity the beam is governed by the second-order differential equation

$$y''(x) + k^2 xy(x) = 0, \quad y'(0) = 0, \quad y(L) = 0, \tag{6.70}$$

where $k^2 = \frac{P}{EI}$. Here, P is the axial compressive load, E is Young's modulus of the column material, and I is the moment of inertia of the column's cross-section. This specific formulation accounts for a non-standard variation in the equation due to the x-dependence in $k^2 xy(x)$, which may represent a variation in the load or geometry. From Example 6.7 the solution is the general solution

$$y(x) = \sqrt{x}\left[C_1 J_{1/3}(2kx^{3/2}/3) + C_2 J_{-1/3}(2kx^{3/2}/3)\right].$$

Now we apply the boundary conditions in order to find the constants C_1 and C_2. Let $\nu = \frac{1}{3}$. Then using (6.55) and (6.56) we arrive at, after computing few terms in the series,

$$y(x) = \frac{C_1 k^{1/3}}{3^{1/3}\Gamma\left(\frac{4}{3}\right)}\left(x - \frac{k^2x^4}{12} + \frac{k^4x^7}{504} - \dots\right) + \frac{C_2 k^{-1/3}}{3^{-1/3}\Gamma\left(\frac{2}{3}\right)}\left(1 - \frac{k^2x^3}{6} + \frac{k^4x^6}{180} - \dots\right).$$

Differentiating and setting $x = 0$ yields $C_1 = 0$, and so we are left with

$$y(x) = C_2 J_{-1/3}(2kx^{3/2}/3).$$

Substituting $x = L$ gives

$$J_{-1/3}(2kx^{3/2}/3) = 0.$$

Thus, the beam will buckle at $z = 2kx^{3/2}/3$, where z is a root of the Bessel equation $J_{-1/3}(z) = 0$. □

6.8.1 Exercises

Exercise 6.39 *Prove (b) of Lemma 12.*

Exercise 6.40 *Give all details for obtaining* (6.56).

Exercise 6.41 *Prove part (b) of Lemma 13.*

Exercise 6.42 *Express $J_5(x)$ in terms of $J_1(x)$ and $J_0(x)$.*

Exercise 6.43 *Use the substitution $y(x) = x^{-1/2}w(x)$ to find the solution of*

$$x^2y''(x) + 2xy'(x) + \lambda^2x^2y(x) = 0, \quad x > 0.$$

Exercise 6.44 *Show that $y(x) = x^{\nu}J_{\nu}(x)$ is a solution of*

$$x^2y''(x) + (1-2\nu)y'(x) + xy(x) = 0, \quad x > 0.$$

Exercise 6.45 *Show that*

(a) $\displaystyle\int x^3J_0(x)\,dx = x^3J_1(x) + 2x^2J_0(x) - 4xJ_1(x) + C.$

(b) $\displaystyle\int J_0(x)\cos x\,dx = xJ_0(x)\cos x + xJ_1(x)\sin x + C.$

(c) $\displaystyle\int J_0(x)\sin x\,dx = xJ_0(x)\sin x - xJ_1(x)\cos x + C.$

Exercise 6.46 *Use* (6.68) *to express $J_{3/2}(x)$ in terms of* $\sin x$ *and* $\cos x$.

Exercise 6.47 *Express $J_{5/2}(x)$ in terms of* $\sin x$ *and* $\cos x$.

Exercise 6.48 *Solve*

(a) $x^2y'' + xy' + (b^2x^2 - \nu^2)y = 0.$

(b) $xy'' - y' + xy = 0.$

(c) $x^2y'' + (x^2 + \frac{1}{4})y = 0.$

6.9 Orthogonality of Bessel's Functions

In this section, we delve into the concept of orthogonality of Bessel functions, a property that plays a crucial role in various applications across mathematics, physics, and engineering. Bessel functions, which are solutions to Bessel's differential equation, exhibit orthogonality over a specific interval with respect to a weighted inner product. This orthogonality property is particularly important in problems involving cylindrical or spherical symmetry, such as heat conduction in a cylindrical rod, wave propagation, and vibration analysis.

By leveraging the orthogonality of Bessel functions, we can expand functions into series of Bessel functions, analogous to how Fourier series are used to expand periodic functions. This allows for efficient analysis and solution of partial differential equations, especially in boundary value problems where the geometry naturally leads to Bessel functions as part of the solution.

In Corollary 6 we converted the standard Bessel equation

$$x^2y'' + xy' + (x^2 - \nu^2)y = 0$$

to the parametric Bessel equation

$$x^2y''(x) + xy'(x) + (\lambda^2x^2 - \nu^2) = 0, \tag{6.71}$$

whose solution is given by

$$y(x) = c_1J_\nu(\lambda x) + c_2Y_\nu(\lambda x). \tag{6.72}$$

The self-adjoint form of (6.71)

$$\frac{d}{dx}\left(x\frac{dy}{dx}\right) + (\lambda^2x - \frac{\nu^2}{x})y = 0, \tag{6.73}$$

has a bounded solution

$$\phi(x) = J_\nu(\lambda x).$$

We prove the next theorem regrading orthogonality of solutions of Bessel function.

Theorem 6.5 *Let μ_n be positive roots of some transcendent equation involving Bessel functions of the first kind. The sequence of solutions given by*

$$\phi_n(x) = J_\nu\left(\frac{\mu_n x}{\ell}\right), \quad n = 1, 2, \ldots$$

are orthogonal with respect to the weight function $p(x) = x$. That is,

$$(\phi_n(x), \phi_k(x)) = \int_0^\ell J_\nu\left(\frac{\mu_n x}{\ell}\right) J_\nu\left(\frac{\mu_k x}{\ell}\right) x\,dx = \begin{cases} 0, & \text{if } n \neq k \\ \|J_\nu\|^2, & \text{when } n = k \end{cases}$$

where the value of the norm squared, $\|J_\nu\|^2$, depends on the boundary condition at the right endpoint $x = \ell$.

Proof *For two distinct positive numbers k_1 and k_2, we consider the two functions*

$$\phi_1(x) = J_\nu(k_1x) \quad \text{and} \quad \phi_2(x) = J_\nu(k_2x). \tag{6.74}$$

These functions are solutions of the equations

$$\frac{d}{dx}\left(x\frac{d\phi_1(x)}{dx}\right) + (k_1^2x - \nu x)\phi_1(x) = 0$$

and

$$\frac{d}{dx}\left(x\frac{d\phi_2(x)}{dx}\right)+(k_2^2x-\nu x)\phi_2(x)=0,$$

respectively. Multiplying the former by $\phi_2(x)$ *and the latter by* $\phi_1(x)$*, and subtracting the results, we obtain*

$$(k_1^2-k_2^2)\phi_1(x)\phi_2(x)x=-\frac{d}{dx}\left(x\frac{d\phi_1(x)}{dx}\right)\phi_2(x)+\frac{d}{dx}\left(x\frac{d\phi_2(x)}{dx}\right)\phi_1(x).$$

Integrating both sides of the latter with respect to $x\in[0,\ell]$*, we get*

$$(k_1^2-k_2^2)\int_0^\ell \phi_1(x)\phi_2(x)x\,dx=-\int_0^\ell\frac{d}{dx}\left(x\frac{d\phi_1(x)}{dx}\right)\phi_2(x)\,dx+\int_0^\ell\frac{d}{dx}\left(x\frac{d\phi_2(x)}{dx}\right)\phi_1(x)\,dx.$$

Performing integration by parts shows

$$\begin{aligned}-\int_0^\ell\frac{d}{dx}\left(x\frac{d\phi_1(x)}{dx}\right)\phi_2(x)\,dx &+ \int_0^\ell\frac{d}{dx}\left(x\frac{d\phi_2(x)}{dx}\right)\phi_1(x)\,dx\\ &= \left[x\frac{d\phi_2(x)}{dx}\phi_1(x)-x\frac{d\phi_1(x)}{dx}\phi_2(x)\right]_{x=0}^{x=\ell}\\ &+ \int_0^\ell x\frac{d\phi_1}{dx}\frac{d\phi_2}{dx}\,dx-\int_0^\ell x\frac{d\phi_1}{dx}\frac{d\phi_2}{dx}\,dx\\ &= \left[x\frac{d\phi_2(x)}{dx}\phi_1(x)-x\frac{d\phi_1(x)}{dx}\phi_2(x)\right]_{x=0}^{x=\ell}.\end{aligned}$$

If $\nu>1$*, the lower limit becomes zero, and we get*

$$(k_1^2-k_2^2)\int_0^\ell \phi_1(x)\phi_2(x)x\,dx=\left[x\frac{d\phi_2(x)}{dx}\phi_1(x)-x\frac{d\phi_1(x)}{dx}\phi_2(x)\right]\Bigg|_{x=\ell}. \tag{6.75}$$

In view of (6.74) *relation* (6.75) *give the new integral relation in terms of Bessel's function*

$$(k_1^2-k_2^2)\int_0^\ell xJ_\nu(k_1x)J_\nu(k_2x)\,dx=\left[k_2xJ_\nu'(k_2x)J_\nu(k_1x)-k_1xJ_\nu'(k_1x)J_\nu(k_2x)\right]\Bigg|_{x=\ell}. \tag{6.76}$$

Setting $k_1=\frac{\mu_n}{\ell}$*,* $k_2=\frac{\mu_k}{\ell}$ *in* (6.76)*, we obtain the new integral relation*

$$\begin{aligned}\frac{(\mu_n^2-\mu_k^2)}{\ell^2}\int_0^\ell xJ_\nu\left(\frac{\mu_nx}{\ell}\right)J_\nu\left(\frac{\mu_kx}{\ell}\right)dx &= k_2lJ_\nu(\mu_n)J_\nu'(\mu_k)-k_1lJ_\nu(\mu_k)J_\nu'(\mu_n)\\ &= \mu_kJ_\nu(\mu_n)J_\nu'(\mu_k)-\mu_nJ_\nu(\mu_k)J_\nu'(\mu_n).\end{aligned} \tag{6.77}$$

If parameters μ_n *and* μ_k *are chosen in a way to annihilate the right-hand side of* (6.77)*, we get orthogonality of Bessel's functions. This is expected since* μ_n *(*$n=1,2,3,\ldots$*) is a sequence of positive roots of the equation*

$$J_\nu(\mu)=0.$$

On the other hand, if $n = k$ we have

$$(\phi_n(x), \phi_n(x)) = \int_0^\ell J_\nu\left(\frac{\mu_n x}{\ell}\right) J_\nu\left(\frac{\mu_n x}{\ell}\right) x\,dx = \int_0^\ell J_\nu^2\left(\frac{\mu_n x}{\ell}\right) dx = \|J_\nu\|^2.$$

We consider three important cases of boundary conditions for which Bessel's functions are orthogonal.

- **Dirichlet Boundary Conditions**

Let μ_n $(n = 1, 2, 3, \ldots)$ be a sequence of positive roots of the equation

$$J_\nu(\mu) = 0.$$

Then the right-hand side of (6.77) *will be zero for $n \neq k$. So we need to determine*

$$\|J_\nu(\mu_n x/\ell)\|^2 = (J_\nu(\mu_n x/\ell), J_\nu(\mu_n x/\ell)) = \int_0^\ell J_\nu^2\left(\frac{\mu_n x}{\ell}\right) x\,dx.$$

From (6.77) *, one has*

$$\|J_\nu\|^2 = \frac{\ell^2}{\mu_k^2 - \mu_n^2}\left[\mu_n J_\nu(\mu_k) J_\nu'(\mu_n) - \mu_k J_\nu(\mu_n) J_\nu'(\mu_k)\right]. \tag{6.78}$$

We will arrive at our objective by taking limit in (6.78) *as $\mu_k \to \mu_n$. Thus, applying L'Hôpital's rule in* (6.78) *and then applying the Dirichlet condition $J_\nu(\mu_n) = 0$ we arrive at*

$$\begin{aligned}
\|J_\nu\|^2 &= \lim_{\mu_k \to \mu_n} \frac{\ell^2}{2\mu_k} \frac{d}{d\mu_k}\left[\mu_n J_\nu(\mu_k) J_\nu'(\mu_n) - \mu_k J_\nu(\mu_n) J_\nu'(\mu_k)\right] \\
&= \lim_{\mu_k \to \mu_n} \frac{\ell^2}{2\mu_k}\left[\mu_n J_\nu'(\mu_k) J_\nu'(\mu_n) - J_\nu(\mu_n) J_\nu'(\mu_k) - \mu_k J_\nu(\mu_n) J_\nu''(\mu_k)\right] \\
&= \lim_{\mu_k \to \mu_n} \frac{\ell^2}{2\mu_k}\left[\mu_n J_\nu'(\mu_k) J_\nu'(\mu_n)\right] = \frac{\ell^2}{2}[J_\nu'(\mu_n)]^2. \qquad (6.79)
\end{aligned}$$

Since μ_n is a zero of $J_\nu(\mu)$ we have $J_\nu(x) = 0$. Hence from (6.68) *we have that*

$$J_{\nu-1}(x) = -J_{\nu+1}(\mu_n).$$

Therefore, from (6.69), *we obtain*

$$2J_\nu'(\mu_n) = J_{\nu-1}(\mu_n) - J_{\nu+1}(\mu_n) = -2J_{\nu+1}(\mu_n)$$

or,

$$J_\nu'(\mu_n) = -J_{\nu+1}(\mu_n).$$

Substituting this into (6.79) *yields,*

$$\|J_\nu\|^2 = \frac{\ell^2}{2}[J_\nu'(\mu_n)]^2 = \frac{\ell^2}{2}[J_{\nu+1}(\mu_n)]^2.$$

Finally, we arrive at

$$\left(J_\nu\left(\frac{\mu_n x}{\ell}\right), J_\nu\left(\frac{\mu_k x}{\ell}\right)\right) = \begin{cases} 0, & \textit{if } n \neq k \\ \frac{\ell^2}{2}[J_{\nu+1}(\mu_n)]^2, & \textit{when } n = k \end{cases} \tag{6.80}$$

• **Neumann Boundary Conditions**
Let μ_n $(n = 1,2,3,\dots)$ be a sequence of positive roots of the equation

$$J_\nu'(\mu) = 0.$$

Applying L'Hôpital's rule to (6.78) *and using the Neuman condition,* $J_\nu'(\mu_n) = 0$, *results into*

$$\begin{aligned} \|J_\nu\|^2 &= \lim_{\mu_k \to \mu_n} \frac{\ell^2}{2\mu_k} \frac{d}{d\mu_k}\left[\mu_n J_\nu(\mu_k) J_\nu'(\mu_n) - \mu_k J_\nu(\mu_n) J_\nu'(\mu_k)\right] \\ &= \lim_{\mu_k \to \mu_n} \frac{\ell^2}{2\mu_k}\left[\mu_n J_\nu'(\mu_k) J_\nu'(\mu_n) - J_\nu(\mu_n) J_\nu'(\mu_k) - \mu_k J_\nu(\mu_n) J_\nu''(\mu_k)\right] \\ &= \frac{\ell^2}{2\mu_n}\left[-J_\nu(\mu_n) J_\nu'(\mu_n) - \mu_n J_\nu(\mu_n) J_\nu''(\mu_n)\right] = \frac{\ell^2}{2\mu_n}\left[-\mu_n J_\nu(\mu_n) J_\nu''(\mu_n)\right]. \end{aligned} \tag{6.81}$$

Making use of Bessel's function along with $J_\nu'(\mu_n) = 0$, we get

$$\mu_n^2 J_\nu''(\mu_n) = (\nu^2 - \mu_n^2) J_\nu(\mu_n).$$

Substituting this into (6.81) *gives*

$$\|J_\nu\|^2 = \frac{\ell^2}{2}\left(1 - \frac{\nu^2}{\mu_n^2}\right) J^2(\mu_n),$$

which is the desired result. Thus,

$$\left(J_\nu\left(\frac{\mu_n x}{\ell}\right), J_\nu\left(\frac{\mu_k x}{\ell}\right)\right) = \begin{cases} 0, & \textit{if } n \neq k \\ \frac{\ell^2}{2}\left(1 - \frac{\nu^2}{\mu_n^2}\right) J^2(\mu_n), & \textit{when } n = k \end{cases}$$

• **Robin Boundary Conditions**
Let μ_n $(n = 1,2,3,\dots)$ be a sequence of positive roots of the equation

$$a\ell J_\nu(\mu) + b\mu J_\nu'(\mu) = 0,$$

where a and b are some real numbers. It is not hard to verify that the right-hand side of (6.77) *will be zero for $n \neq k$. To determine the value of the square norm when $n = k$, we take the limit*

$$\|J_\nu\|^2 = \int_0^\ell J_\nu^2\left(\frac{\mu_n x}{\ell}\right) x\,dx = \lim_{\mu_k \to \mu_n} \frac{\ell^2}{\mu_k^2 - \mu_n^2}\left[\mu_n J_\nu(\mu_k) J_\nu'(\mu_n) - \mu_k J_\nu(\mu_n) J_\nu'(k)\right].$$

We again apply L'Hôpital's rule and obtain

$$\begin{aligned}\|J_\nu\|^2 &= \frac{\ell^2}{2\mu_n}\lim_{\mu_k\to\mu_n}\frac{d}{d\mu_k}\left[\mu_n J_\nu(\mu_k)J_\nu'(\mu_n)-\mu_k J_\nu(\mu_n)J_\nu'(\mu_k)\right]\\ &= \frac{\ell^2}{2\mu_n}\left[\mu_n(J_\nu'(\mu_n))^2-J_\nu(\mu_n)J_\nu'(\mu_n)-\mu_n J_\nu(\mu_n)J_\nu''(\mu_n)\right].\end{aligned}$$

From Bessel's equation, we have

$$-\mu J_\nu''(\mu)=J_\nu'(\mu)+\left(\frac{\mu^2-\nu^2}{\mu}\right)J_\nu(\mu).$$

So

$$-\mu J_\nu(\mu)J_\nu''(\mu)=J_\nu(\mu)J_\nu'(\mu)+\left(\frac{\mu^2-\nu^2}{\mu}\right)J_\nu^2(\mu).$$

Substituting into the expression of $\|J_\nu\|^2$ *we have*

$$\|J_\nu\|^2=\frac{\ell^2}{2}\left[\left(J_\nu'(\mu_n)\right)^2+\left(1-\frac{\nu^2}{\mu_n^2}\right)J_\nu^2(\mu_n)\right].$$

□

6.10 Vibrations of Circular Membrane; Bessel's Equation

• **Circular Symmetry;** $\frac{\partial^2 u}{\partial \phi^2}=0$.

For an application, we consider the wave equation

$$u_{tt}=c^2\nabla^2 u=c^2\left(u_{\rho\rho}+\frac{1}{\rho}u_\rho+\frac{1}{\rho^2}u_{\phi\phi}\right), \tag{6.82}$$

on a circular membrane with radius a. To simplify the problem, we assume that the displacement $u(\rho,t)$ is radially symmetric, meaning u is independent of the angular variable ϕ. Under this assumption, the wave equation reduces to

$$u_{tt}=c^2\left(u_{\rho\rho}+\frac{1}{\rho}u_\rho\right),\quad 0<\rho<a, \tag{6.83}$$

as there is no dependence on ϕ. The boundary condition for this problem is that the membrane is fixed along the edge, implying

$$u(a,t)=0. \tag{6.84}$$

Additionally, we impose initial conditions for the displacement and velocity of the membrane:

$$u(\rho,0)=f(\rho),\quad u_t(\rho,0)=g(\rho). \tag{6.85}$$

This setup captures the dynamics of a radially symmetric wave propagating on a circular membrane constrained by fixed boundary conditions.

We seek a nontrivial solution of the form

$$u(\rho,t) = R(\rho)T(t).$$

Differentiating and then substituting into (6.83), we obtain, by separating the variables, that

$$\frac{R'' + \frac{R'}{\rho}}{R} = \frac{T''}{c^2 T} = -\lambda.$$

Equating both sides we arrive the second-order differential equations

$$T''(t) + \lambda c^2 T(t) = 0, \tag{6.86}$$

and

$$\rho^2 R''(\rho) + \rho R'(\rho) + \lambda \rho^2 R(\rho) = 0, \quad 0 < \rho < a, \tag{6.87}$$
$$R(a) = 0.$$

Equation (6.87) is a parametric Bessel equation of order zero (see Corollary 6).

• $\lambda = 0$. In this case (6.87) has the solution

$$R(\rho) = A \ln \rho + B.$$

As $\rho \to 0^+$, we have $\ln \rho \to -\infty$ and so we must take $A = 0$; otherwise, the solution becomes unbounded. Hence, we are left with

$$R(\rho) = B.$$

Applying the boundary condition $R(a) = 0$ implies $B = 0$.

• $\lambda > 0$. To avoid square-roots, we let $\lambda = \alpha^2,\ \ \alpha > 0$. Then, equation (6.87) becomes

$$\rho^2 R''(\rho) + \rho R'(\rho) + \alpha^2 \rho^2 R(\rho) = 0, \quad 0 < \rho < a,$$

which is a Bessel equation with order zero, or $\nu = 0$. Then from Corollary 6 the solution is

$$R(\rho) = C_1 J_0(\alpha\rho) + C_2 Y_0(\alpha\rho).$$

Since the deflection of the membrane is always finite while Y_0 becomes infinite as $\rho \to 0^+$, we must take $C_2 = 0$. So we are left with

$$R(\rho) = C_1 J_0(\alpha\rho).$$

Applying $R(a) = 0$ we infer that

$$C_1 J_0(\alpha a) = 0.$$

Denoting the positive zeros of $J_0(\alpha r)$ by $\mu_1, \mu_2, \ldots$ we thus obtain

$$\alpha a = \mu_n \quad \text{or} \quad \alpha = \alpha_m = \frac{\mu_m}{a}, \quad m = 1, 2, \ldots$$

Hence the functions

$$R_m(\rho) = J_0(\alpha\rho) = J_0\left(\frac{\mu_m \rho}{a}\right), \quad m = 1, 2. \ldots$$

are solutions of (6.87). At the same time, the corresponding general solution to (6.86) is

$$T_m(t) = a_m \cos \alpha_m ct + b_m \sin \alpha_m ct.$$

Finally, the general solution to (6.85) is given by

$$u(\rho, t) = \sum_{m=1}^{\infty} (a_m \cos \alpha_m t + b_m \sin \alpha_m t) J_0\left(\frac{\mu_m \rho}{a}\right). \tag{6.88}$$

Applying the first initial condition to (6.88) gives

$$f(\rho) = \sum_{m=1}^{\infty} a_m J_0\left(\frac{\mu_m \rho}{a}\right).$$

Due to the orthogonality of the Bessel functions with respect to the weight function ρ we have

$$\left(J_0\left(\frac{\mu_m \rho}{a}\right), f(\rho)\right) = a_m \left(J_0\left(\frac{\mu_m \rho}{a}\right), J_0\left(\frac{\mu_m \rho}{a}\right)\right).$$

This implies that

$$a_m = \frac{\left(J_0\left(\frac{\mu_m \rho}{a}\right), f(\rho)\right)}{\left(J_0\left(\frac{\mu_m \rho}{a}\right), J_0\left(\frac{\mu_m \rho}{a}\right)\right)} = \frac{\int_0^a \rho f(\rho) J_0\left(\frac{\mu_m \rho}{a}\right) d\rho}{\int_0^a J_0^2\left(\frac{\mu_m \rho}{a}\right) \rho \, d\rho}.$$

Here we have Dirichlet boundary conditions and therefore from (6.80), we have that

$$\int_0^\rho J_0^2\left(\frac{\mu_m \rho}{a}\right) \rho \, d\rho = \frac{\rho^2}{2} J_1^2(\mu_m).$$

Similar to the case above, if we apply the initial velocity condition, we obtain

$$g(\rho) = \sum_{m=1}^{\infty} \alpha_m b_m c J_0\left(\frac{\mu_m \rho}{a}\right), \text{ from which we have}$$

$$b_m = \frac{\int_0^\rho \rho g(\rho) J_0\left(\frac{\mu_m \rho}{a}\right) d\rho}{\alpha_m c \int_0^\rho J_0^2\left(\frac{\mu_m \rho}{a}\right) \rho \, d\rho},$$

where

$$\int_0^\rho J_0^2\left(\frac{\mu_m \rho}{a}\right) \rho \, d\rho = \frac{\rho^2}{2} J_1^2(\mu_m).$$

- **Dependence on ϕ; $\frac{\partial^2 u}{\partial \phi^2} \neq 0$.**

Now we consider the full blown (6.82) and seek a solution where $u = u(\rho, \phi, t)$ of

$$u_{tt} = c^2 \nabla^2 u = c^2 \left(u_{\rho\rho} + \frac{1}{\rho} u_\rho + \frac{1}{\rho^2} u_{\phi\phi} \right), \quad -\pi < \phi < \pi, \quad 0 < \rho < a. \tag{6.89}$$

We impose the homogeneous Dirichlet boundary condition

$$u(a, \phi, t) = 0, \tag{6.90}$$

and the implicit periodic boundary conditions in ϕ

$$u(\rho, -\pi, t) = u(\rho, \pi, t), \quad \frac{\partial u}{\partial \phi}(\rho, -\pi, t) = \frac{\partial u}{\partial \phi}(\rho, \pi, t). \tag{6.91}$$

We require a bounded solution in ρ, that is

$$\lim_{\rho \to 0^+} |u(\rho, \phi, t)| < \infty. \tag{6.92}$$

Finally, we specify initial position and initial velocity by

$$u(\rho, \phi, 0) = f(\rho, \phi), \quad u_t(\rho, \phi, 0) = 0. \tag{6.93}$$

We assume a separable solution of the form

$$u(\rho, \phi, t) = h(t)\Psi(\rho)g(\phi).$$

Substituting into the governing wave equation, we obtain two Sturm–Liouville problems

$$g''(\phi) + \mu g(\phi) = 0,$$

with periodic boundary conditions

$$g(-\pi) = g(\pi), \quad g'(-\pi) = g'(\pi).$$

The eigenvalues and associated eigenfunctions are

$$\mu_m = m^2, \quad g_0(\phi) = a_0, \quad g_m(\phi) = a_m \cos(m\phi) + b_m \sin(m\phi), \quad m = 0, 1, 2, \ldots$$

The second Sturm–Liouville Problem in ρ is

$$\frac{d}{d\rho}\left(\rho \frac{d\Psi}{d\rho}\right) + \left(\lambda\rho - \frac{m^2}{\rho}\right)\Psi = 0,$$

with boundary conditions

$$\Psi(a) = 0, \quad |\Psi(0)| < \infty.$$

The eigenvalues and associated eigenfunctions are

$$\lambda_{mn} = \left(\frac{z_{mn}}{a}\right)^2, \quad \Psi_{mn}(\rho) = J_m\left(\frac{z_{mn}\rho}{a}\right),$$

where z_{mn} is the n-th zero of the Bessel function $J_m(z)$, satisfying

$$J_m(z_{mn}) = 0.$$

Given that $\lambda_{mn} > 0$, the general solution of

$$h''(t) + c^2\lambda_{mn}h(t) = 0,$$

is

$$h(t) = c_{mn}\cos\left(c\sqrt{\lambda_{mn}}\,t\right) + d_{mn}\sin\left(c\sqrt{\lambda_{mn}}\,t\right).$$

Using the simplifying assumption that $u_t(\rho,\phi,0) = 0$, we can omit the term involving $\sin\left(c\sqrt{\lambda_{mn}}\,t\right)$, giving

$$h(t) = c_{mn}\cos\left(c\sqrt{\lambda_{mn}}\,t\right).$$

The superposition principle, applied to the product solution, yields the full solution

$$u\rho,\phi,t) = \sum_{n=1}^{\infty} A_{0n}J_0\left(\sqrt{\lambda_{0n}}\,\rho\right)\cos\left(c\sqrt{\lambda_{0n}}\,t\right)$$

$$+\sum_{m=1}^{\infty}\sum_{n=1}^{\infty}\Big(A_{mn}\cos(m\phi) + B_{mn}\sin(m\phi)\Big)J_m\left(\sqrt{\lambda_{mn}}\,\rho\right)\cos\left(c\sqrt{\lambda_{mn}}\,t\right).$$

From the initial condition (6.93) we have

$$f(\rho,\phi) = \sum_{n=1}^{\infty} A_{0n}J_0\left(\sqrt{\lambda_{0n}}\,\rho\right) + \sum_{m=1}^{\infty}\sum_{n=1}^{\infty}\Big(A_{mn}\cos(m\phi) + B_{mn}\sin(m\phi)\Big)J_m\left(\sqrt{\lambda_{mn}}\,\rho\right).$$

This represents a standard double Fourier series in ϕ and a Fourier-Bessel series in ρ. Using orthogonality, the coefficients are given by

$$A_{0n} = \frac{\int_{-\pi}^{\pi}\int_0^a f(\rho,\phi)J_0\left(\sqrt{\lambda_{0n}}\,\rho\right)\rho\,d\rho\,d\phi}{2\pi\int_0^a J_0^2\left(\sqrt{\lambda_{0n}}\,\rho\right)\rho\,d\rho},$$

and

$$A_{mn} = \frac{\int_{-\pi}^{\pi}\int_0^a f(\rho,\phi)\cos(m\phi)J_m\left(\sqrt{\lambda_{mn}}\,\rho\right)\rho\,d\rho\,d\phi}{\pi\int_0^a J_m^2\left(\sqrt{\lambda_{mn}}\,\rho\right)\rho\,d\rho}.$$

Additionally,

$$B_{mn} = \frac{\int_{-\pi}^{\pi} \int_0^a f(\rho,\phi)\sin(m\phi)J_m\left(\sqrt{\lambda_{mn}}\,\rho\right)\rho\,d\rho\,d\phi}{\pi \int_0^a J_m^2\left(\sqrt{\lambda_{mn}}\,\rho\right)\rho\,d\rho}.$$

- **Heat Flow in a Long Circular Cylinder**

The work of this section can be easily extended to the study of heat flow in a long circular cylinder, subject to boundary conditions of the Dirichlet, Neumann, or Robin type. In other words, the temperatures $u = u(\rho,t)$ in the cylinder satisfy the special case of the heat equation in cylindrical coordinates,

$$u_t = k\left(u_{\rho\rho} + \frac{1}{\rho}u_\rho\right), \tag{6.94}$$

for $0 < \rho < c$ and $t > 0$, where $u(\rho,t)$ represents the temperature, k is the thermal diffusivity, and ρ is the radial coordinate with and initial condition

$$u(\rho,0) = f(\rho), \quad 0 < \rho < c. \tag{6.95}$$

The heat flow problem (6.94) and (6.95) can be considered in each of the following cases.

(1) $u(c,t) = 0$, for $t > 0$, (Dirichlet boundary condition),

(2) $u_\rho(c,t) = 0$, for $t > 0$, (Neuman boundary condition),

(3) $hu(c,t) + du_\rho(c,t) = 0$, where $h > 0, d > 0$, (Robin boundary conditions).

6.10.1 Exercises

Exercise 6.49 *Consider the wave equation*

$$u_{tt} = c^2\left(u_{rr} + \frac{1}{r}u_r + \frac{1}{r^2}u_{\theta\theta}\right)$$

Assume a solution of the form $u(r,\theta,t) = F(r,\theta)T(t)$. Use separation of variables to obtain the two differential equations

$$T''(t) + \lambda c^2 T(t) = 0,$$

$$F_{rr} + \frac{1}{r}F_r + \frac{1}{r^2}F_{\theta\theta} + \lambda^2 F = 0.$$

Exercise 6.50 *In the circular membrane problem, let $a = 1$ and $c = 1$. Find $u(\rho,t)$ when:*

(a) $f(\rho) = \lambda(1-\rho^2)$,

(b) $f(\rho) = \lambda(1-\rho^4)$.

Exercise 6.51 *Solve*

$$u_{tt} = c^2\left(u_{\rho\rho} + \frac{1}{\rho}u_\rho + \frac{1}{\rho^2}u_{\phi\phi}\right), \quad -\pi < \phi < \pi, \quad 0 < \rho < a,$$
$$u(a,\phi,t) = 0, \quad \lim_{\rho\to 0^+} |u(\rho,\phi,t)| < \infty,$$
$$u(\rho,\phi,0) = f(\rho,\phi), \quad u_t(\rho,\phi,0) = 0.$$

Exercise 6.52 *Solve the heat flow problem in a long circular cylinder, governed by (6.94) and (6.95), for each of the three cases: Dirichlet, Neumann, and Robin boundary conditions.*

Exercise 6.53 *Solve*

$$u_{tt} = c^2\left(u_{\rho\rho} + \frac{1}{\rho}u_\rho\right), \quad 0 < \rho < a,$$
$$u(a,t) = 0, \quad \lim_{\rho\to 0^+} |u(\rho,t)| < \infty,$$
$$u(\rho,0) = f(\rho), \quad u_t(\rho,0) = 0.$$

Ans:

$$u(\rho,t) = \sum_{n=1}^{\infty} b_n J_0\left(\frac{\gamma_n \rho}{a}\right)\cos\left(\frac{\gamma_n ct}{a}\right), \text{ where } \gamma_n \text{ are the positive roots of } J_0(x) = 0,$$

$$b_n = \frac{2}{a^2 J_1^2(\gamma_n)} \int_0^a \rho f(\rho) J_0\left(\frac{\gamma_n \rho}{a}\right) d\rho.$$

Exercise 6.54 *Solve*

$$u_{tt} = c^2\left(u_{\rho\rho} + \frac{1}{\rho}u_\rho\right), \quad 0 < \rho < a,$$
$$u(a,t) = 0, \quad \lim_{\rho\to 0^+} |u(\rho,t)| < \infty,$$
$$u(\rho,0) = 0, \quad u_t(\rho,0) = v_0 \text{ (constant)}.$$

Exercise 6.55 *Solve the wave equation in each of the following cases.*

$$u_{tt} = c^2\left(u_{rr} + \frac{1}{r}u_r + \frac{1}{r^2}u_{\theta\theta}\right), \quad \text{subject to} \quad u_\rho(a,\theta,t) = 0.$$

with initial conditions

(a) $u(\rho,\theta,0) = 0, \quad u_t(\rho,\theta,0) = f(\rho)\cos 5\theta,$

(b) $u(\rho,\theta,0) = 0, \quad u_t(\rho,\theta,0) = f(\rho),$

(c) $u(\rho,\theta,0) = 0, \quad u_t(\rho,\theta,0) = f(\rho,\theta).$

Exercise 6.56 *Consider the heat flow problem*

$$u_t = k\left(u_{rr} + \frac{1}{r}u_r + u_{zz}\right), \quad 0 \le r \le c, \quad 0 \le z \le h,$$

where $u(r,z,t)$ is the temperature at position (r,z) and time t. This equation represents heat conduction in a 3D cylindrical domain, assuming there is no dependence on the angular coordinate θ. The boundary conditions are

$$u(r,0,t) = u(r,h,t) = 0, \quad \textit{(axial boundaries at } z = 0 \textit{ and } z = h),$$

$$u(c,z,t) = 0, \quad \textit{(radial boundary at } r = c).$$

The initial temperature distribution is

$$u(r,z,0) = f(r,z).$$

The equation models the diffusion of heat (or a similar scalar quantity, like the concentration of a diffusing substance) in a finite cylindrical domain with radius c and height h. The temperature evolves over time under the influence of thermal diffusion, starting from an initial distribution $f(r,z)$. The boundaries are maintained at a fixed temperature (zero in this case), enforcing heat loss or insulation at the edges. This setup is commonly encountered in problems such as cooling of a cylindrical rod, temperature distribution in a cylindrical reactor, or diffusion of substances in a cylindrical medium.

(a) Assume a solution of the form $u(r,z,t) = R(r)Z(z)T(t)$ and obtain the following differential equations.

$$r^2R''(r) + rR'(r) + \alpha^2 r^2 R(r) = 0, \quad R(c) = 0,$$
$$Z''(z) + \beta^2 Z(z) = 0, \quad Z(0) = Z(h) = 0,$$
$$T'(t) + (\alpha^2 + \beta^2)kT(t) = 0,$$

where α and β are the parameters of separations.

(b) Derive the series solution

$$u(r,z,t) = \sum_{m=1}^{\infty}\sum_{n=1}^{\infty} b_{mn} e^{-\lambda_{mn}kt} J_0\left(\frac{\gamma_m r}{c}\right) \sin\frac{n\pi z}{h},$$

where γ_m are the positive roots of $J_0(x) = 0$, $\lambda mn = \dfrac{\gamma_m^2}{c^2} + \dfrac{n^2\pi^2}{h^2}$ and the coefficients are given by

$$b_{mn} = \frac{4}{hc^2 J_1^2(\gamma_m)} \int_0^c \int_0^h r f(r,z) J_0\left(\frac{\gamma_m r}{c}\right) \sin\frac{n\pi z}{h}\, dz\, dr.$$

(c) Let f_0 be a constant so that $f(r,z) = f_0$. Then show that

$$u(r,z,t) = \frac{8f_0}{\pi} \sum_{m=1}^{\infty}\sum_{n=1}^{\infty} \frac{e^{-\lambda_{m(2n-1)}kt}}{(2n-1)\gamma_m J_1(\gamma_m)} J_0\left(\frac{\gamma_m r}{c}\right) \sin\left(\frac{(2n-1)\pi z}{h}\right).$$

6.11 Legendre Polynomials

Legendre polynomials are a crucial set of orthogonal polynomials widely used in mathematical physics, engineering, and numerical analysis. They arise naturally in solving problems involving spherical symmetry, such as the Laplace equation in spherical coordinates, making them essential in fields like quantum mechanics, electromagnetics, and gravitational studies. Their orthogonality property simplifies the representation of functions on finite intervals, particularly on $[-1,1]$, enabling efficient approximation and integration techniques, such as Gaussian quadrature. Furthermore, Legendre polynomials are central to expanding solutions in terms of series in boundary value problems, significantly contributing to the mathematical modeling of physical systems.

We consider the parametric Legendre second-order differential equation

$$(1-x^2)y'' - 2xy' + \gamma y = 0, \tag{6.96}$$

that has the self-adjoin form

$$\left[(1-x^2)y'\right]' + \gamma y = 0. \tag{6.97}$$

The Legendre equation is of the type of singular Sturm–Liouville since it has regular singular points at $x=-1$ and $x=1$, while $x=0$ is an ordinary point, and therefore its coefficients are analytic at the point $x=0$. Consequently, we may expand these coefficients near zero. That is

$$\frac{-2x}{1-x^2} = -2x\sum_{m=0}^{\infty} x^{2m} = \sum_{m=0}^{\infty}(-2)x^{2m+1},$$

and

$$\frac{\gamma}{1-x^2} = \gamma\sum_{m=0}^{\infty} x^{2m} = \sum_{m=0}^{\infty}\gamma x^{2m}$$

converge for $|x|<1$. Thus, the Legendre equation on $|x|<1$ has a convergent power series solution at $x=0$. As a matter of fact, the only case in which the Legendre equation has a bounded solution on $[-1,1]$ is when the parameter γ has the form $\gamma = n(n+1)$ with $n=0$ or $n\in\mathbb{Z}^+$. We find power series solutions centered at $x=0$ (the radius of convergence will be at least 1) by assuming that

$$y = \sum_{j=0}^{\infty} c_j x^j$$

is a solution of (6.96). A substitution into (6.96) gives

$$(1-x^2)\sum_{j=2}^{\infty} j(j-1)c_j x^{j-2} - 2x\sum_{j=1}^{\infty} jc_j x^{j-1} + \gamma\sum_{j=0}^{\infty} c_j x^j = 0.$$

Expanding the terms, this becomes

$$\sum_{j=2}^{\infty} j(j-1)c_j x^{j-2} - \sum_{j=2}^{\infty} j(j-1)c_j x^j - \sum_{j=1}^{\infty} 2jc_j x^j + \gamma \sum_{j=0}^{\infty} c_j x^j = 0.$$

After re-indexing the first series and simplifying we arrive at

$$\sum_{j=0}^{\infty} (j+2)(j+1)c_{j+2}x^j - \sum_{j=0}^{\infty} (j^2+j-\gamma)c_j x^j = 0.$$

Factoring x^j and putting both series under one summation and then setting the coefficients of the sum to zero, we obtain the recurrence relations

$$c_{j+2} = \frac{(j+1)j-\gamma}{(j+2)(j+1)}c_j, \quad j = 0,1,2,\ldots \tag{6.98}$$

Our goal is to obtain two linearly independent solutions. As for obtaining the first solution, we set $c_0 \neq 0$ and $c_1 = 0$. In this case, the recurrence relation gives

$$c_3 = \frac{2-\gamma}{6}c_1 = 0, \quad c_5 = \frac{12-\gamma}{20}c_3 = 0, \quad \ldots, \quad c_{\text{odd}} = 0.$$

The coefficients with even indices can be written in terms of c_0 and hence

$$c_2 = \frac{-\gamma}{2}c_0, \quad c_4 = \frac{(2\cdot 3-\gamma)}{4\cdot 3}c_2 = \frac{(2\cdot 3-\gamma)(-\gamma)}{4!}c_0.$$

By an induction argument, it can be shown that

$$\begin{aligned} c_{2k} &= \frac{[(2k-1)(2k-2)-\gamma][(2k-3)(2k-4)-\gamma]\cdots[3\cdot 2-\gamma][-\gamma]}{(2k)!}c_0 \\ &= \frac{c_0}{(2k)!}\prod_{i=1}^{k-1}[(2i+1)(2i)-\gamma]. \end{aligned}$$

This gives a solution

$$y_1(x) = c_0 \sum_{k=0}^{\infty} \frac{\left(\prod_{i=0}^{k-1}[(2i+1)(2i)-\gamma]\right)x^{2k}}{(2k)!}.$$

A second series solution, independent of the first, can be obtained by setting $c_0 = 0$ and $c_1 \neq 0$. In this case, the coefficients for even terms are zero, i.e., $c_{\text{even}} = 0$, and the coefficients for odd terms are given by

$$c_{2k+1} = \frac{c_1}{(2k+1)!}\prod_{i=1}^{k}[2i(2i-1)-\lambda].$$

The corresponding solution is

$$y_2(x) = c_1 \sum_{k=0}^{\infty} \left(\prod_{i=1}^{k}[2i(2i-1)-\lambda]\right)\frac{x^{2k+1}}{(2k+1)!}.$$

Now we set $\gamma = n(n+1)$ and consider (6.98) from which, we obtain

$$c_{n+2} = \frac{(n+1)n - n(n+1)}{(n+2)(n+1)} c_n = 0, \quad c_{n+4} = \frac{(n+3)(n+2) - (n+2)(n+3)}{(n+4)(n+3)} c_{n+2} = 0,$$

and in general

$$c_{n+2k} = 0, \quad \text{for} \quad k \geq 1.$$

Consequently, for even n, or $n = 2p$, the series for y_1 terminates at c_{2p}, making y_1 a polynomial of degree $2p$. In contrast, the series for y_2 is infinite, with a radius of convergence equal to 1, and y_2 is unbounded. Similarly, for odd n, or $n = 2p+1$, the series for y_2 terminates at c_{2p+1}, making y_2 is a polynomial of degree $2p+1$. In contrast, the series of y_1 is unbounded.

From (6.98) we have

$$c_j = \frac{(j+2)(j+1)}{(j+1)j - \gamma} c_{j+2} = -\frac{(j+2)(j+1)}{(n-j)(n+j+1)} c_{j+2},$$

for $j = n-2, n-4, \ldots 1$, or zero. By induction one shows that

$$c_{n-2k} = -\frac{(n-2k+2)(n-2k+1)}{(2k)(2n-2k+1)} c_{n-2k+2}, \quad k = 1, 2, \ldots, \left\lfloor \frac{n}{2} \right\rfloor.$$

After some tweaking, the above expression becomes

$$c_{n-2k} = \frac{(-1)^k}{2^k k!} \frac{n(n-1)\cdots(n-2k+1)}{(2n-1)(2n-3)\cdots(2n-2k+1)} c_n.$$

Then our polynomial solution is therefore

$$y(x) = \sum_{k=0}^{\lfloor \frac{n}{2} \rfloor} (-1)^k \frac{2^k}{k!} \frac{n(n-1)\cdots(n-2k+1)}{(2n-1)(2n-3)\cdots(2n-2k+1)} c_n x^{n-2k},$$

where c_n is an arbitrary constant. To find the polynomial solution $P_n(x)$ from $y(x)$ it is convenient to choose c_n so that $P_n(1) = 1$. For that reason, we set

$$c_n = \frac{(2n)!}{2^n (n!)^2}.$$

Then after some manipulation of the right-hand side of $y(x)$, we have

$$P_n(x) = \frac{1}{2^n} \sum_{k=0}^{\lfloor \frac{n}{2} \rfloor} (-1)^k \frac{(2n-2k)!}{k!(n-k)!(n-2k)!} x^{n-2k}, \tag{6.99}$$

where

$$\left\lfloor \frac{n}{2} \right\rfloor = \begin{cases} \frac{n}{2}, & \text{if } n \text{ is even} \\ \frac{n-1}{2}, & \text{if } n \text{ is odd} \end{cases}$$

Observe that when n is even (or odd), the powers of x present in P_n are exclusively even (or odd), implying that P_n is an even (or odd) function. We will make a note in the next corollary that P_n given by (6.99) is a solution of (6.96) with $\gamma = n(n+1)$.

Corollary 7 *Let $y(x)$ be a bounded solution on the interval $(-1,1)$ of the Legendre equation* (6.96) *with $\lambda = n(n+1)$. If P_n is given by* (6.99)*, then there exists a constant C such that*

$$y(x) = CP_n(x).$$

The first six Legendre polynomials are given by

$$P_0(x) = 1, \quad P_1(x) = x,$$

$$P_2(x) = \frac{1}{2}\left(3x^2 - 1\right), \quad P_3(x) = \frac{1}{2}\left(5x^3 - 3x\right),$$

$$P_4(x) = \frac{1}{8}\left(35x^4 - 30x^2 + 3\right), \quad P_5(x) = \frac{1}{8}\left(63x^5 - 70x^3 + 15x\right).$$

Recall that for a given nonnegative integer n, only one of the two solutions $y_1(x)$ and $y_2(x)$ of Legendre's equation is a polynomial, while the other is an infinite series. This infinite series, when appropriately normalized, is called the **Legendre function of the second kind.** It is defined for $|x| < 1$ by

$$Q_n(x) = \begin{cases} y_1(1)y_2(x) & \text{for even } n \\ -y_2(1)y_1(x) & \text{for odd } n \end{cases}$$

Thus, when n is a nonnegative integer, the general solution of the Legendre equation is given by

$$y(x) = c_1 P_n(x) + c_2 Q_n(x).$$

Lemma 14 *(Rodrigues' Formula) The Legendre polynomial $P_n(x)$ given by* (6.99) *is equivalent to*

$$P_n(x) = \frac{1}{2^n n!}\frac{d^n}{dx^n}\left[(x^2-1)^n\right]. \tag{6.100}$$

Thus, expression (6.100) *solves* (6.96) *for $\gamma = n(n+1)$.*

Proof *There are many ways to prove the lemma, but we chose to use the Binomial theorem that states*

$$(x+y)^n = \sum_{i=0}^{n} \frac{n!}{i!(n-i)!} x^{n-i} y^i.$$

Then

$$(x^2-1)^n = \sum_{i=0}^{n} \frac{n!}{i!(n-i)!}(x^2)^{n-i}(-1)^i$$

Taking the n-th derivative in the above expression and then multiplying it by the coefficient $\frac{1}{2^n n!}$, we obtain

$$\frac{1}{2^n n!}\frac{d^n}{dx^n}\left[(x^2-1)^n\right] = \frac{1}{2^n n!}\frac{d^n}{dx^n}\left[\sum_{i=0}^{n} \frac{n!}{i!(n-i)!}(x^2)^{n-i}(-1)^i\right]$$

$$= \frac{1}{2^n n!}\sum_{i=0}^{n}\frac{n!}{i!(n-i)!}\frac{d^n}{dx^n}\left(x^2\right)^{n-i}(-1)^i. \qquad (6.101)$$

Now we compute $\frac{d^n}{dx^n}\left(x^2\right)^{n-i}$. *We make use of the general formula for derivatives, that says the n-th derivative of* x^m *is given by*

$$\frac{d^n}{dx^n}x^m = \begin{cases} m(m-1)(m-2)\cdots(m-n+1)x^{m-n}, & \text{if } n \le m \\ 0, & \text{if } n > m \end{cases}$$

As

$$m(m-1)(m-2)\cdots(m-n+1) = \frac{m!}{(m-n)!},$$

we obtain for $m = 2(n-i)$ *that*

$$m(m-1)(m-2)\cdots(m-n+1) = \frac{(2(n-i))!}{(2(n-i)-n)!}.$$

Consequently,

$$\frac{d^n}{dx^n}\left(x^2\right)^{n-i} = \frac{(2(n-i))!}{(2(n-i)-n)!}x^{2(n-i)-n} = \frac{(2n-2i)!}{(n-2i)!}x^{n-2i}. \qquad (6.102)$$

Substituting (6.102) *into* (6.101) *gives the desired result. Thus,*

$$P_n(x) = \frac{1}{2^n}\sum_{i=0}^{n}\frac{(-1)^i(2n-2i)!}{i!(n-i)!(n-2i)!}x^{n-2i},$$

and by changing the dummy variable i to k, we get the same general formula of Legendre Polynomials

$$P_n(x) = \frac{1}{2^n}\sum_{k=0}^{\lfloor \frac{n}{2} \rfloor}\frac{(-1)^k(2n-2k)!}{k!(n-k)!(n-2k)!}x^{n-2k}.$$

This completes the proof. □

Example 6.11 Express $f(x) = x^4 + 2x^3 + 2x^2 - x - 3$ in terms of Legendre's polynomials. We already know that

$$P_0(x) = 1, \quad P_1(x) = x, \quad P_2(x) = \frac{1}{2}\left(3x^2-1\right),$$

and

$$P_3(x) = \frac{1}{2}\left(5x^3-3x\right), \quad P_4(x) = \frac{1}{8}\left(35x^4-30x^2+3\right).$$

Using the above polynomials, we obtain

$$x^4 = \frac{8}{35}P_4(x) + \frac{6}{7}x^2 - \frac{3}{35}, \quad x^3 = \frac{2}{5}P_3(x) + \frac{3}{5}x,$$

$$x^2 = \frac{2}{3}P_2(x) + \frac{1}{3}, \quad x = P_1(x), \quad 1 = P_0(x).$$

Substituting successively the values of x^4, x^3, and so on in the given polynomial, we can express $f(x)$ in terms of the Legendre polynomials, and obtain after some calculations that

$$f(x) = \frac{8}{35}P_4(x) + \frac{4}{5}P_3(x) + \frac{40}{21}P_2(x) + \frac{1}{5}P_1(x) - \frac{224}{105}P_0(x).$$

□

We have the following property between three consecutive Legendre polynomials.

Lemma 15 *If $P_n(x)$ is given by* (6.100)*, then*

$$(2n+1)P_n(x) = P'_{n+1}(x) - P'_{n-1}(x). \tag{6.103}$$

Proof *From Lemma 14 we have that*

$$\begin{aligned} P'_k(x) &= \frac{d}{dx}\left(\frac{1}{2^k k!}\frac{d^k}{dx^k}\left[(x^2-1)^k\right]\right) = \frac{2k}{2^k k!}\frac{d^k}{dx^k}\left[x(x^2-1)^{k-1}\right] \\ &= \frac{1}{2^{k-1}(k-1)!}\frac{d^{k-1}}{dx^{k-1}}\left[\left((2k-1)x^2-1\right)(x^2-1)^{k-2}\right]. \end{aligned}$$

Letting $k = n+1$ implies that

$$P'_{n+1}(x) = \frac{1}{2^n n!}\frac{d^n}{dx^n}\left[\left((2n+1)x^2-1\right)(x^2-1)^{n-1}\right].$$

By adjusting Rodrigues' formula at $n-1$, we obtain

$$P'_{n-1}(x) = \frac{d}{dx}\left(\frac{1}{2^{n-1}(n-1)!}\frac{d^{n-1}}{dx^{n-1}}\left[(x^2-1)^{n-1}\right]\right) = \frac{2n}{2^n n!}\frac{d^n}{dx^n}\left[(x^2-1)^{n-1}\right].$$

Consequently,

$$\begin{aligned} P'_{n+1}(x) - P'_{n-1}(x) &= \frac{1}{2^n n!}\frac{d^n}{dx^n}\left[(x^2-1)^{n-1}\left(x^2(2n+1)-1\right) - 2n(x^2-1)^{n-1}\right] \\ &= \frac{1}{2^n n!}\frac{d^n}{dx^n}\left[(x^2-1)^{n-1}\left(x^2(2n+1)-1-2n\right)\right] \\ &= \frac{1}{2^n n!}\frac{d^n}{dx^n}\left[(x^2-1)^{n-1}\left(2n(x^2-1)+(x^2-1)\right)\right] \\ &= \frac{1}{2^n n!}\frac{d^n}{dx^n}\left[(x^2-1)^n\right](2n+1) \\ &= (2n+1)P_n(x). \end{aligned}$$

This completes the proof. □

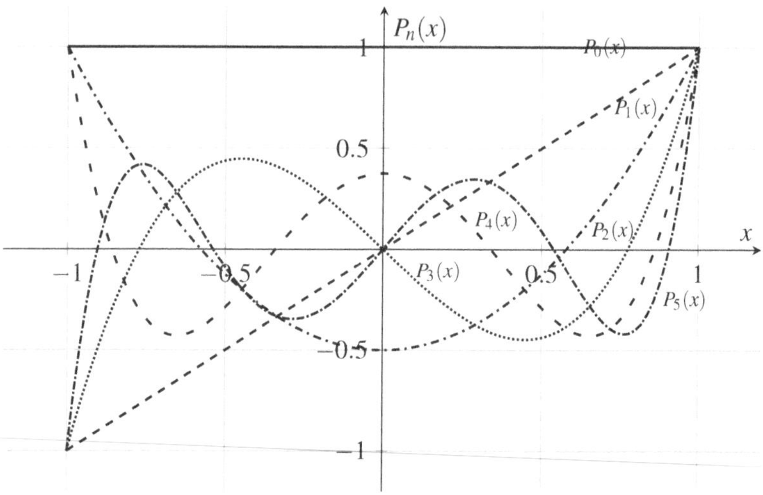

FIGURE 6.2: First six Legendre polynomials.

6.11.1 Exercises

Exercise 6.57 *Use the recurrence relation that gives the coefficients of the Legendre polynomials to show that*

$$P_{2n}(0) = \frac{(-1)^n (2n)!}{2^{2n}(n!)^2}.$$

Exercise 6.58 *Prove that*

$$P_n'(x) - P_{n-2}'(x) = (2n-1)P_{n-1}(x).$$

Exercise 6.59 *Prove that*

$$xP_9'(x) - P_8'(x) = 9P_9(x).$$

Exercise 6.60 *Prove that*

$$x^4 = \frac{1}{35}\left[8P_4(x) + 20P_2(x) + 7P_0(x)\right].$$

Exercise 6.61 *Prove the following properties of Legendre polynomials.*

(a) $(n+1)P_{n+1}(x) - (2n+1)xP_n(x) + nP_{n-1}(x) = 0, \quad n \geq 1.$

(b) $(x^2-1)P_n'(x) = nxP_n(x) - nP_{n-1}(x), \quad n \geq 1.$

(c) $nP_n(x) + P_{n-1}'(x) - xP_n'(x) = 0, \quad n \geq 1.$

(d) $P_{n+1}'(x) = xP_n'(x) + (n+1)P_n(x), \quad n \geq 0.$

(e) $P_{2n}(-x) = P_{2n}(x).$

(f) $P_{2n+1}(-x) = -P_{2n+1}(x).$

6.12 The Generating Function and Eigenfunctions Expansion

The generating function

$$g(x,t) = \frac{1}{\sqrt{1-2xt+t^2}}, \tag{6.104}$$

provides a single formula from which all the Legendre polynomials $P_n(x)$ can be derived. This compact representation avoids the need to compute each polynomial individually. Expanding the generating function in a Taylor series with respect to t

$$\frac{1}{\sqrt{1-2xt+t^2}} = \sum_{n=0}^{\infty} P_n(x)t^n, \tag{6.105}$$

gives the coefficients $P_n(x)$, which are the Legendre polynomials. This makes it possible to calculate $P_n(x)$ directly from the series. Expression (6.105) can be used to derive recurrence relations, such as

$$(n+1)P_{n+1}(x) - (2n+1)xP_n(x) + nP_{n-1}(x) = 0,$$

which is a fundamental property of Legendre polynomials. These recurrence relations are essential in numerical computation and theoretical analysis. In what to follow, we highlight how (6.105) can be used to obtain many of the recurrence relations. We first differentiate (6.105) with respect to t and obtain

$$\frac{\partial g(t,x)}{\partial t} = \frac{x-t}{(1-2xt+t^2)^{3/2}} = \sum_{n=0}^{\infty} nP_n(x)t^{n-1}.$$

If we multiply by $1-2xt+t^2$, we get

$$(1-2xt+t^2)\sum_{n=0}^{\infty} nP_n(x)t^{n-1} + \frac{t-x}{\sqrt{1-2xt+t^2}} = 0,$$

which, using the generating function (6.105), becomes

$$(1-2xt+t^2)\sum_{n=0}^{\infty} nP_n(x)t^{n-1} + (t-x)\sum_{n=0}^{\infty} P_n(x)t^n = 0.$$

Consequently, expanding further, yields

$$\sum_{m=1}^{\infty} mP_m(x)t^{m-1} - 2x\sum_{n=0}^{\infty} nP_n(x)t^n + \sum_{s=0}^{\infty} sP_s(x)t^{s+1} + \sum_{s=0}^{\infty} P_s(x)t^{s+1} - \sum_{n=0}^{\infty} xP_n(x)t^n = 0.$$

Letting $m = n+1$ and $s = n-1$, the above expression simplifies to

$$[P_1(x) - xP_0(x)]t^0 + \sum_{n=1}^{\infty}\Big[(n+1)P_{n+1}(x) - (2n+1)xP_n(x) + nP_{n-1}(x)\Big]t^n = 0.$$

The coefficient of t^0 vanishes since $P_0(x) = 1$ and $P_1(x) = x$. As

$$\sum_{n=1}^{\infty} \Big[(n+1)P_{n+1}(x) - (2n+1)xP_n(x) + nP_{n-1}(x)\Big]t^n = 0,$$

must hold for all power of t, the coefficient of each power of t must therefore, and hence we obtain

$$(n+1)P_{n+1}(x) - (2n+1)xP_n(x) + nP_{n-1}(x) = 0.$$

This relation can be used to generate Legendre polynomials of higher order if one knows them of lower order. For example, knowing that

$$P_0(x) = 1, \quad P_1(x) = x,$$

we can compute higher-order polynomials. For $n = 1$, the recurrence relation gives

$$2P_2(x) = 3xP_1(x) - P_0(x),$$

which implies

$$P_2(x) = \frac{1}{2}(3x^2 - 1).$$

- **Orthogonality of Legendre Polynomials**

As previously stated, the Legendre equation is a singular Sturm–Liouville problem and can be written in its self-adjoint form as

$$\left[(1-x^2)y'\right]' + \gamma y = 0, \quad -1 < x < 1, \tag{6.106}$$

where the parameter γ is given by

$$\gamma = n(n+1). \tag{6.107}$$

The next theorem is regrading orthogonality of Legendre polynomials.

Theorem 6.6 *Let γ_m and γ_n be distinct eigenvalues of* (6.106) *with corresponding eigenfunctions $P_m(x)$ and $P_n(x)$, respectively. Then the sequence of polynomials $P_n(x)$ are orthogonal on $(-1,1)$ with respect to the weight function $p(x) = 1$. Moreover,*

$$(P_n(x), P_m(x)) = \int_{-1}^{1} P_n(x)P_m(x)\,dx = \begin{cases} 0, & \text{if } n \neq m \\ \|P_n\|^2 = \frac{2}{2n+1}, & \text{when } n = m \end{cases}$$

Proof *Define the operator*

$$\mathscr{L} := \frac{d}{dx}\Big((x^2-1)\frac{d}{dx}\Big). \tag{6.108}$$

Then the self-adjoint (6.106) *can be written as*

$$\mathscr{L}P_n(x) = \gamma_n P_n(x). \tag{6.109}$$

where

$$\mathscr{L}P_n(x) = \frac{d}{dx}\left((x^2-1)\frac{dP_n(x)}{dx}\right).$$

For two distinct positive numbers γ_m *and* γ_n*, and corresponding eigenfunctions* $P_m(x)$ *and* $P_n(x)$*, we have*

$$\mathscr{L}P_n(x) = \gamma_n P_n(x),$$

and

$$\mathscr{L}P_m(x) = \gamma_m P_m(x).$$

respectively. Multiplying the former by $P_m(x)$ *and the latter by* $P_n(x)$*, and subtracting the results, we obtain using* (6.7) *of Lemma 11 that*

$$\big(P_n(x)\mathscr{L}P_m(x) - P_m(x)\mathscr{L}P_n(x)\big)dx = d\left\{(x^2-1)[P_m(x)P_n'(x) - P_n(x)P_m'(x)]\right\}. \tag{6.110}$$

Noticing that the left-hand side of (6.110) *is* $(\gamma_n - \gamma_m)P_m(x)P_n(x)$*, we deduce that*

$$(\gamma_n - \gamma_m)P_m(x)P_n(x) = d\left\{(x^2-1)[P_m(x)P_n'(x) - P_n(x)P_m'(x)]\right\}$$

Integrating both sides of the latter with respect to $x \in (-1,1)$*, we obtain*

$$(\gamma_n - \gamma_m)\int_{-1}^{1} P_m(x)P_n(x)\,dx = \left[(x^2-1)\Big(P_m(x)P_n'(x) - P_n(x)P_m'(x)\Big)\right]_{-1}^{1} = 0. \tag{6.111}$$

Consequently, we infer that

$$\int_{-1}^{1} P_m(x)P_n(x)\,dx = 0,$$

for $\gamma_m \neq \gamma_n$*. Now we tackle the case* $n = m$*. Assume* $n = m$*, then*

$$\|P_n\|^2 = (P_n(x), P_n(x)) = \int_{-1}^{1} P_n^2(x)\,dx.$$

To obtain our result we will make use of the generating function and in the process we will need the following.

Recall that for $|z| < 1$*, the Taylor series of* $\ln(1+z)$ *is*

$$\ln(1+z) = \sum_{j=1}^{\infty}\frac{(-1)^{j-1}}{j}z^j.$$

Hence, for $|t| < 1$*, we have*

$$\begin{aligned}\frac{1}{t}\ln\left(\frac{1+t}{1-t}\right) &= \frac{1}{t}(\ln(1+t) - \ln(1-t)) \\ &= \frac{1}{t}\left(\sum_{j=1}^{\infty}\frac{(-1)^{j-1}}{j}t^j - \sum_{j=1}^{\infty}\frac{(-1)^{j-1}}{j}(-t)^j\right)\end{aligned}$$

$$= \frac{1}{t}\sum_{j=1}^{\infty}\frac{(-1)^{j-1}+1}{j}t^j = \sum_{n=0}^{\infty}\frac{2}{2n+1}t^{2n}. \tag{6.112}$$

Squaring both sides of (6.105) *we see that*

$$\frac{1}{1-2xt+t^2} = \left[\sum_{n=0}^{\infty}P_n(x)t^n\right]^2 = \sum_{n=0}^{\infty}\sum_{m=0}^{\infty}P_n(x)P_m(x)t^{n+m}.$$

Integrating from -1 *to* 1 *and using the orthogonality of the Legendre polynomials, we have*

$$\int_{-1}^{1}\frac{dx}{1-2xt+t^2} = \sum_{n=0}^{\infty}\sum_{m=0}^{\infty}t^{n+m}\int_{-1}^{1}P_n(x)P_m(x)\,dx = \sum_{n=0}^{\infty}t^{2n}\int_{-1}^{1}\left[P_n^2(x)\right]dx. \tag{6.113}$$

However,

$$\int_{-1}^{1}\frac{dx}{1-2xt+t^2} = -\frac{1}{2t}\left[\ln\left|1-2xt+t^2\right|\right]_{x=-1}^{x=1} = \frac{1}{t}\ln\left(\frac{1+t}{1-t}\right).$$

Comparing the above expression with (6.112), *yields*

$$\int_{-1}^{1}\frac{dx}{1-2xt+t^2} = \sum_{n=0}^{\infty}\frac{2}{2n+1}t^{2n}. \tag{6.114}$$

Comparing (6.114) *with* (6.113) *we arrive at the expression*

$$\sum_{n=0}^{\infty}\left[\frac{2}{2n+1} - \int_{-1}^{1}\left[P_n^2(x)\right]dx\right]t^{2n} = 0,$$

from which we conclude that

$$\int_{-1}^{1}\left[P_n^2(x)\right]dx = \frac{2}{2n+1}.$$

This completes the proof. □

We have the following lemma regarding the normalization of Legendre polynomials.

Lemma 16 *(a) The set*

$$\varphi_n(x) = \sqrt{\frac{2n+1}{2}}P_n(x), \quad n = 0,1,2,\ldots$$

is orthonormal on the interval $-1 < x < 1$.

(b) The set

$$\varphi_n(x) = \sqrt{4n+1}P_{2n}(x), \quad n = 0,1,2,\ldots$$

is orthonormal on the interval $0 < x < 1$.

(c) The set

$$\varphi_n(x) = \sqrt{4n+3}P_{2n+1}(x), \quad n = 0, 1, 2, \ldots$$

is orthonormal on the interval $0 < x < 1$.

Proof *Part (a) is proved by setting* $\varphi_n(x) = \frac{P_n(x)}{\|P_n\|}$. *Now we use Theorem 6.6 to prove the other two parts. For (b), we notice that*

$$\|P_{2n}\|^2 = \int_0^1 [P_{2n}(x)]^2\, dx = \frac{1}{2}\int_{-1}^1 [P_{2n}(x)]^2\, dx = \frac{1}{2}\cdot\frac{2}{2(2n)+1} = \frac{1}{4n+1}.$$

Therefore, the orthonormal set is

$$\phi_n(x) = \frac{P_{2n}(x)}{\|P_{2n}\|} = \sqrt{4n+1}\,P_{2n}(x).$$

The proof of (c) is similar by noting that

$$\|P_{2n+1}\|^2 = \int_0^1 [P_{2n+1}(x)]^2\, dx = \frac{1}{2}\int_{-1}^1 [P_{2n+1}(x)]^2\, dx = \frac{1}{2}\cdot\frac{2}{2(2n+1)+1} = \frac{1}{4n+3}.$$

So

$$\phi_n(x) = \frac{P_{2n+1}(x)}{\|P_{2n+1}\|} = \sqrt{4n+3}\,P_{2n+1}(x).$$

This completes the proof of the Lemma.

□

- **Eigenfunctions Expansions**

The Fourier-Legendre series expansion is important for representing a function on the interval from negative one to one as a sum of Legendre polynomials. This expansion is particularly useful because Legendre polynomials form an orthogonal basis on this interval, ensuring that the series converges efficiently under suitable conditions. It allows for the approximation of complex functions in terms of simpler, well-understood components, which is valuable in solving differential equations, modeling physical systems, and performing numerical computations in fields like physics and engineering.

The Legendre polynomials form a complete orthogonal set in the space of piecewise smooth functions defined on the interval from negative one to one. This means that any piecewise smooth function on this interval can be expressed as an infinite series of Legendre polynomials, which serve as the "building blocks" for the function. The completeness of this family guarantees that no information is lost in the representation, provided enough terms are included in the series.

This representation, known as the Legendre series, is a type of generalized Fourier series where the basis functions are Legendre polynomials rather than sines and cosines. Each coefficient in the series is determined by projecting the function onto the corresponding Legendre polynomial, leveraging their orthogonality property. This orthogonality ensures that the series converges efficiently, often requiring fewer terms to approximate the function compared to other basis sets.

Let $f : [-1,1] \to \mathbb{R}$ be piecewise smooth function. Then its associated Legendre series is

$$f(x) \sim \sum_{n=0}^{\infty} c_n P_n(x), \quad -1 < x < 1, \tag{6.115}$$

where

$$c_n = \frac{(f(x), P_n(x))}{\|P_n\|^2} = \frac{2n+1}{2} \int_{-1}^{1} f(x) P_n(x)\, dx. \tag{6.116}$$

We have a parallel theorem to Theorem 2.1.

Theorem 6.7 *Let f be a piecewise smooth function on $[-1,1]$. Then, at the point of discontinuities $x \in (-1,1)$ the Legendre series given by (6.115) converges to*

$$\frac{f(x^+) + f(x^-)}{2},$$

and to $f(x)$ at the points x where f is continuous. That is

$$f(x) = \sum_{n=0}^{\infty} c_n P_n(x).$$

Remark 11 *For each m, the Legendre polynomials $P_0, P_1, \ldots, P_m$ form a basis in the space of polynomials of degree m. This means that any polynomial $D(x)$ of degree m can be expressed as a linear combination of these Legendre polynomials. Consequently, the Legendre series of $D(x)$ terminates at the order m (i.e., $c_k = 0$ for $k > m$). Specifically, for x^m, we have*

$$x^m = c_0 P_0(x) + c_1 P_1(x) + \cdots + c_m P_m(x) = c_0 + c_1 x + c_2 \frac{1}{2}(3x^2 - 1) + \ldots$$

We have the following corollary.

Corollary 8 *Let f be a piecewise smooth function on $[-1,1]$.*

(a) If $f(x)$ is odd, that is $f(-x) = -f(x)$, then the odd Legendre series extension of f is given by

$$f(x) = \sum_{j=0}^{\infty} c_{2j+1} P_{2j+1}(x),$$

where

$$c_{2j+1} = (2(2j+1)+1) \int_0^1 f(x) P_{2j+1}(x)\, dx = (4j+3) \int_0^1 f(x) P_{2j+1}(x)\, dx.$$

(b) *If $f(x)$ is even, that is $f(-x) = f(x)$, then the even Legendre series extension of f is given by*

$$f(x) = \sum_{j=0}^{\infty} c_{2j} P_{2j}(x),$$

where

$$c_{2j} = (2(2j)+1)\int_0^1 f(x)P_{2j}(x)\,dx = (4j+1)\int_0^1 f(x)P_{2j}(x)\,dx.$$

As a consequence of Lemma 16 and Corollary 8, we have the following results that are handy when it comes to applications. It plays a crucial role in the integration of a polynomial multiplied by a Legendre polynomial on the interval $(-1,1)$.

Lemma 17 *Let $P_n(x)$ be the Legendre polynomial. Then*

$$\int_{-1}^{1} x^m P_n(x)\,dx = 0 \quad \text{for} \quad m < n.$$

Proof *Using Rodrigues' formula we have*

$$\int_{-1}^{1} x^m P_n(x)\,dx = \frac{1}{2^n n!}\int_{-1}^{1} x^m \frac{d^n}{dx^n}\left((x^2-1)^n\right)dx. \tag{6.117}$$

Since $m < n$, we can integrate by parts m-times to show the result, using the properties $P_n(1) = 1$ and $P_n(-1) = (-1)^n$. This completes the proof. □

Example 6.12 In this example we expand $f(x) = x^4$, in Legend series. First of all, Lemma 17 tells us that

$$\int_{-1}^{1} x^4 P_n(x)\,dx = 0, \quad n = 5, 6, \ldots.$$

We will make use of (6.116) to compute c_n. Since f is even and P_n is even when n is even we automatically have $c_1 = c_3 = 0$. Thus, we compute c_0, c_2 and c_4, where

$$c_n = \frac{2n+1}{2}\int_{-1}^{1} f(x)P_n(x)\,dx.$$

Hence,

$$c_0 = \frac{1}{2}\int_{-1}^{1} x^4 P_0(x)\,dx = \frac{1}{2}.2\int_0^1 x^4\,dx = \frac{1}{5},$$

$$c_2 = \frac{5}{2}.2\int_0^1 x^4 P_2(x)\,dx = 5\int_0^1 x^4\frac{1}{2}(3x^2-1)\,dx = \frac{20}{35},$$

$$c_4 = \frac{9}{2}.2\int_0^1 x^4 P_4(x)\,dx = 9\int_0^1 x^4\frac{1}{8}(35x^4-30x^2+3)\,dx = \frac{8}{35}.$$

Then (6.115) yields

$$x^4 = \frac{1}{5}P_0(x) + \frac{20}{35}P_2(x) + \frac{8}{35}P_4(x) = \frac{1}{35}\left[8P_4(x) + 20P_2(x) + 7P_0(x)\right].$$

□

Another example.

Example 6.13 We want to expand the Heaviside function

$$H(x) = \begin{cases} 1, & x > 0 \\ 0, & x < 0 \end{cases} \tag{7.2.19}$$

in Legendre series. Since f is odd, its Legendre series contains only odd indexed polynomials. Using (a) of Corollary 8, we have

$$c_{2n+1} = (4n+3)\int_0^1 P_{2n+1}(x)\,dx.$$

Replacing n with $2n+1$ in (6.103) of Lemma 15, we obtain the recurrence relation

$$(4n+3)P_{2n+1} = P'_{2n+2} - P'_{2n}.$$

Consequently,

$$c_{2n+1} = \int_0^1 \left(P'_{2n+2}(x) - P'_{2n}(x)\right)\,dx = P_{2n+2}(1) - P_{2n+2}(0) - P_{2n}(1) + P_{2n}(0).$$

From Exercise 6.57 we have $P_{2j}(1) = 1$ and $P_{2j}(0) = \frac{(-1)^j(2j)!}{2^{2j}(j!)^2}$, and so it follows that

$$c_{2n+1} = P_{2n}(0) - P_{2n+2}(0) = \frac{(-1)^n(4n+3)}{2^{n+1}(n+1)}\frac{(2n)!}{(n!)^2}.$$

We then have the expansion

$$1 = \sum_{n=0}^{\infty} \frac{(-1)^n(4n+3)}{2^{n+1}(n+1)}\frac{(2n)!}{(n!)^2}P_{2n+1}(x), \quad 0 < x < 1.$$

□

6.12.1 Exercises

Exercise 6.62 *Show the following.*

(a) $\int_{-1}^{1} xP_n(x)P_{n-1}(x)\,dx = \frac{2n}{4n^2-1}$.

(b) $\int_{-1}^{1}(1-x^2)\left[P'_n(x)\right]^2 dx = \frac{2n(n+1)}{2n+1}$.

(c) $\int_{-1}^{1}\left[P'_n(x)\right]^2 dx = n(n+1)$.

Exercise 6.63 *Use the substitution* $x = \cos\theta$ *and transform*

$$\sin\theta\, y''(\theta) + \cos\theta\, y'(\theta) + n(n+1) = 0,$$

into Legendre's equation.

Exercise 6.64 *Use expression* (6.103) *and a property (even/odd) of the Legendre polynomials to verify that*

(a) $\int_{-1}^{h} P_n(x)\,dx = \frac{1}{2n+1}\left[P_{n+1}(h) - P_{n-1}(h)\right],$

(b) $\int_{h}^{1} P_n(x)\,dx = \frac{1}{2n+1}\left[P_{n-1}(h) - P_{n+1}(h)\right].$

Exercise 6.65 *Find the Legendre series of the functions*

(a)

$$f(x) = \begin{cases} 0, & -1 < x < 0 \\ x, & 0 < x < 1 \end{cases}$$

(b)

$$f(x) = \begin{cases} 0, & -1 < x < h \\ 1, & h < x < 1 \end{cases}$$

6.13 Applications of Legendre's Equation

When working with the Laplacian operator in spherical coordinates, the partial differential equations often separate into angular and radial components. The angular part frequently reduces to a singular Sturm–Liouville problem, which takes the form of Legendre's equation or its generalized forms. These equations arise naturally when solving problems with spherical symmetry, such as in quantum mechanics, electromagnetism, or gravitational fields. We have the following lemma.

Lemma 18 *The eigenvalues and eigenfunctions of equation*

$$\Phi''(\phi) + \frac{\cos\phi}{\sin\phi}\Phi'(\phi) + \gamma\Phi(\phi) = 0, \quad 0 < \phi < \pi, \tag{6.118}$$

are given by

$$\gamma = n(n+1), \quad \Phi_n(\phi) = P_n(\cos\phi), \quad n = 0, 1, \ldots$$

Proof *For* $\phi \in (0,\pi)$, *consider the change of variable*

$$x = \cos\phi, \quad x \in (-1,1),$$

and let

$$y(x) = \Phi(\phi) = \Phi(\arccos x) \quad \textit{if and only if} \quad \Phi(\phi) = y(\cos\phi).$$

Then

$$\Phi'(\phi) = -\sin\phi\, y'(\cos\phi) = -\sqrt{1-x^2}\, y'(x),$$
$$\Phi''(\phi) = \sin^2\phi\, y''(\cos\phi) - \cos\phi\, y'(\cos\phi) = (1-x^2)y''(x) - xy'(x).$$

By replacing these expressions of Φ' *and* Φ'' *in* (6.118), *we get*

$$(1-x^2)y''(x) - 2xy'(x) + \gamma y(x) = 0,$$

which is the Legendre equation. Since $\gamma = n(n+1)$, *we know the solution* y *is bounded. In this case,* y *is a multiple of the Legendre polynomial* $P_n(x)$. *By going back to the function* Φ *and variable* ϕ, *we have*

$$\Phi_n(\phi) = P_n(\cos\phi), \quad n = 0, 1, \ldots$$

This completes the proof. □

The next corollary is a special case of Lemma 18.

Corollary 9 *Assume* (6.118) *of Lemma 18.*

(a) If $0 < \varphi < \frac{\pi}{2}$ *and there is a Neuman boundary condition of the form* $\Phi'(\frac{\pi}{2}) = 0$, *then the eigenvalues and corresponding eigenfunctions of* (6.118) *are*

$$\gamma_n = 2n(2n+1), \quad \Phi_n(\varphi) = P_{2n}(\cos\varphi), \quad n = 0, 1, \ldots$$

where the associated Fourier Legendre series coefficients c_{2n} *are given by (b) of Lemma 8.*

(b) If $0 < \varphi < \frac{\pi}{2}$ *and there is a Dirichlet boundary condition of the form* $\Phi(\frac{\pi}{2}) = 0$, *then the eigenvalues and corresponding eigenfunctions of* (6.118) *are*

$$\gamma_n = (2n+1)(2n+2), \quad \Phi_n(\varphi) = P_{2n+1}(\cos\varphi), \quad n = 0, 1, \ldots$$

where the associated Fourier Legendre series coefficients c_{2n+1} *are given by (a) of Lemma 8.*

• Laplacian In Spherical Coordinates

To simplify the notation, we exchange θ with ϕ in (4.107). Specifically, we use the coordinates

$$x = \rho\cos\theta\sin\phi, \quad y = \rho\sin\theta\sin\phi, \quad z = \rho\cos\phi,$$

instead. We are interested in finding a bounded solutions $u = u(\rho, \theta, \phi)$ of the Laplace equation $\Delta u = 0$. That is, $u(\rho, \theta, \phi)$ is bounded inside the sphere $\rho < a$ and satisfies

$$\frac{\partial^2 u}{\partial \rho^2} + \frac{2}{\rho}\frac{\partial u}{\partial \rho} + \frac{1}{\rho^2\sin^2\phi}\frac{\partial^2 u}{\partial \theta^2} + \frac{1}{\rho^2}\frac{\partial^2 u}{\partial \phi^2} + \frac{\cot\phi}{\rho^2}\frac{\partial u}{\partial \phi} = 0, \tag{6.119}$$

$$\rho > 0, \quad 0 < \phi < \pi, \quad 0 < \theta < 2\pi.$$

We assume a non-zero solution of the form

$$u(\rho,\theta,\phi) = R(\rho)\Theta(\theta)\Phi(\phi),$$

where Θ and Θ' are 2π-periodic. After replacing u and its derivatives in terms of R, Θ, Φ and their derivatives, we can rewrite (6.119) as

$$\rho^2\frac{R''(\rho)}{R(\rho)} + 2\rho\frac{R'(\rho)}{R(\rho)} + \frac{\Theta''(\theta)}{\sin^2\phi\,\Theta(\theta)} + \frac{\Phi''(\phi)}{\Phi(\phi)} + \frac{\cot\phi\,\Phi'(\phi)}{\Phi(\phi)} = 0. \tag{6.120}$$

Separating the variable ρ from θ and ϕ leads to

$$\rho^2 R''(\rho) + 2\rho R'(\rho) - \gamma R(\rho) = 0,$$

$$\frac{\Theta''(\theta)}{\sin^2\phi\,\Theta(\theta)} + \frac{\Phi''(\phi)}{\Phi(\phi)} + \frac{\cot\phi\,\Phi'(\phi)}{\Phi(\phi)} = -\gamma,$$

where *gamma* is the separation constant. A further separation of the second equation with a separation constant α leads to the three ODEs,

$$\rho^2 R''(\rho) + 2\rho R'(\rho) - \gamma R(\rho) = 0, \tag{6.121}$$

$$\Theta''(\theta) + \alpha\Theta(\theta) = 0, \tag{6.122}$$

$$\Phi''(\phi) + \frac{\cos\phi}{\sin\phi}\Phi'(\phi) + \left(\gamma - \frac{\alpha}{\sin^2\phi}\right)\Phi(\phi) = 0. \tag{6.123}$$

We begin by solving (6.122). For $\alpha = 0$, then the general solution of (6.122) is

$$\Theta(\theta) = A\theta + B.$$

Using the fact that Θ and Θ' are 2π-periodic, we obtain $A = 0$ and B is arbitrary that we take to be one. Thus $\Theta(\theta) = 1$. Consequently, for $\alpha = 0$, (6.123) becomes (6.118), and therefore, by Lemma 18 it has he eigenvalues and corresponding eigenfunctions

$$\gamma_n = n(n+1), \quad \Phi_n(\phi) = P_n(\cos\phi), \quad n = 0, 1, \ldots$$

Equation (6.121) is a Cauchy-Euler equation. Under the transformation $\rho = e^t$ the equation is transformed into

$$R''(t) + R'(t) - n(n+1) = 0,$$

with its auxiliary equation having the two roots

$$r_1 = n, \quad r_2 = -(n+1).$$

Consequently, the general solution with respect to t is

$$R(t) = C_1 e^{nt} + C_2 e^{-(n+1)t}.$$

In terms of ρ, the original equation has the solution

$$R(\rho) = C_1\rho^n + C_2\rho^{-(n+1)}.$$

Since we are seeking a bounded solution, we must set $C_2 = 0$; otherwise, the solution becomes unbounded because

$$\rho^{-(n+1)} \to \infty \quad \text{as} \quad \rho \to 0^+.$$

Thus the solution is

$$u(\rho,\phi) = \sum_{n=0}^{\infty} c_n\rho^n P_n(\cos\phi). \tag{6.124}$$

Now assume the case with longitudinal symmetry means that the problem is independent of the angle θ. Thus, $f = f(\phi)$, and the temperature u depends only on the radius ρ and the altitude ϕ ($u = u(\rho,\phi)$). Problem (6.119) can be written as

$$\begin{cases} u_{\rho\rho} + \frac{2}{\rho}u_\rho + \frac{1}{\rho^2}u_{\phi\phi} + \frac{\cos\phi}{\rho^2\sin\phi}u_\phi = 0 \\ u(a,\phi) - f(\phi) \end{cases} \tag{6.125}$$

Since u is assumed to be a bounded function, the method of separation of variables can be applied and the solution is given by (6.124). Applying the boundary condition, we obtain

$$u(a,\phi) = f(\phi) = \sum_{n=0}^{\infty} C_n a^n P_n(\cos\phi).$$

The last series is a Legendre series but is expressed in terms of $\cos\phi$. To find the coefficients C_n, we need to express the series as a Legendre series in standard form. We resort to the substitution $x = \cos\phi$ with $-1 < x < 1$. We rewrite f in terms of the variable x (for instance, $g(x) = f(\phi)$). The series becomes

$$g(x) = \sum_{n=0}^{\infty} C_n a^n P_n(x).$$

Now this is the usual Legendre series of the function $g(x)$. Its n-th Legendre coefficient $C_n a^n$ is given by

$$C_n a^n = \frac{2n+1}{2}\int_{-1}^{1} g(x)P_n(x)\,dx.$$

We can rewrite C_n in terms of the variable ϕ in the sense that

$$C_n = \frac{2n+1}{2a^n}\int_0^{\pi} \sin\phi\, f(\phi)P_n(\cos\phi)\,d\phi. \tag{6.126}$$

The solution to problem (6.125) is therefore

$$u(\rho,\phi) = \sum_{n=0}^{\infty} C_n\rho^n P_n(\cos\phi).$$

Suppose that

$$u(10,\phi) = f(\phi) = \begin{cases} 100, & \text{if } 0 < \phi < \frac{\pi}{2} \\ 0, & \text{if } \frac{\pi}{2} < \phi < \pi \end{cases}$$

Then,

$$\begin{aligned} C_n = \frac{2n+1}{2(10^n)} \int_0^{\pi} f(\phi) \sin\phi \, P_n(\cos\phi)\, d\phi &= \frac{50(2n+1)}{10^n} \int_0^{\pi/2} \sin\phi \, P_n(\cos\phi)\, d\phi \\ &= \frac{50(2n+1)}{10^n} \int_0^1 P_n(x)\, dx \qquad (6.127) \end{aligned}$$

We try to compute $\int_0^1 P_n(x)\,dx$. Replacing n in the expression of Exercise 6.58 with $n+1$ and then integrating the resulting expression we have

$$\int_0^1 P_n(x)\, dx = \frac{1}{2n+1}\left[P_{n-1}(0) - P_{n+1}(0)\right]. \qquad (6.128)$$

Note that $n-1$ and $n+1$ are odds, only when n is even. Since $P_{odd}(0) = 0$, we have that

$$C_{even} = \frac{50(2n+1)}{10^n} \int_0^1 [P_{n-1}(0) - P_{n+1}(0)]\, dx = \frac{50(2n+1)}{10^n} \int_0^1 [P_{odd}(0) - P_{odd}(0)]\, dx = 0.$$

Next we compute C_{odd}. From Exercise 6.57 we have that

$$P_{2n}(0) = \frac{(-1)^n (2n)!}{2^{2n}(n!)^2}.$$

Using a combination of (6.127) and (6.128) we arrive at

$$\begin{aligned} C_{2n+1} &= \frac{50(2n+3)}{10^{2n+1}} \int_0^1 P_n(x)\, dx = \frac{50(4n+3)}{10^{2n+1}} \frac{1}{4n+3}\left[P_{2n}(0) - P_{2n+2}(0)\right] \\ &= \frac{50}{10^{2n+1}}\left[P_{2n}(0) - P_{2n+2}(0)\right] = \frac{25(-1)^n (2n)!}{10^{2n+1} 2^{2n} (n+1)(n!)^2}. \end{aligned}$$

Finally, the solution to the problem is given by

$$u(\rho,\phi) = 25 \sum_{n=0}^{\infty} \frac{(-1)^n (2n)!}{2^{2n}(n+1)(n!)^2} \left(\frac{\rho}{10}\right)^{2n+1} P_{2n+1}(\cos\phi).$$

- **Spherical Shell** In the following example, we demonstrate several techniques for computing Fourier-Legendre coefficients, particularly in scenarios where boundary conditions and initial data are involved.

Example 6.14 In this example we consider the Dirichlet problem that models the steady-state temperature distribution within a thin spherical shell subjected to specific thermal boundary conditions. Physically, the governing partial differential equation accounts for heat conduction in spherical coordinates, where ρ represents the radial distance and ϕ represents the angular position within the shell. The boundary conditions reflect real-world constraints: the inner surface ($\rho = 1$) is maintained at 50 degrees, the outer surface ($\rho = 2$) at 100 degrees, and the base of the shell ($\phi = \pi/2$) at 0 degrees. The angular variation in temperature is captured by Legendre polynomials, reflecting how temperature distribution depends on the geometry of the shell. The radial dependence indicates how heat flows between the inner and outer hemispheres. This formulation is crucial in understanding heat distribution in spherical geometries, which appear in various applications, such as in geophysics, astrophysics, or engineering systems involving thermal insulation or cooling in spherical designs. Such model is given by the boundary value problem

$$u_{\rho\rho} + \frac{2}{\rho}u_\rho + \frac{1}{\rho^2}u_{\phi\phi} + \frac{\cos\phi}{\rho^2\sin\phi}u_\phi = 0, \quad 1 < \rho < 2, \quad 0 < \phi < \frac{\pi}{2},$$

with boundary conditions

$$u(1,\phi) = 50, \quad u(2,\phi) = 100, \quad 0 < \phi < \frac{\pi}{2},$$

$$u(\rho,\pi/2) = 0, \quad 1 < \rho < 2.$$

Such model is depicted in Fig. 6.3. Assuming a nonzero and bounded solution of the form $u(\rho,\phi) = R(\rho)\Phi(\phi)$. Then separation of variables for the homogeneous part of the equation leads to the following ordinary differential equations

$$\rho^2 R'' + 2\rho R' - \lambda R = 0,$$

$$\sin\phi\,\Phi'' + \cos\phi\,\Phi' + \lambda\sin\phi\,\Phi = 0, \quad \Phi\left(\frac{\pi}{2}\right) = 0.$$

As previously analyzed, the eigenvalues of the Φ-equation are

$$\lambda = n(n+1),$$

with the corresponding eigenfunctions being the Legendre polynomials $P_n(\cos\phi)$. However, we also require that

$$\Phi_n\left(\frac{\pi}{2}\right) = P_n(0) = 0.$$

This implies that n must be odd. From previous work, the solutions with separated variables for the homogeneous part are:

$$\rho^n P_n(\cos\phi) \quad \text{and} \quad \rho^{-(n+1)}P_n(\cos\phi),$$

where n is odd. Before, we disposed of the term $\rho^{-(n+1)}P_n(\cos\phi)$ since ρ could approach zero from the right, which in turns, will make $\rho^{-(n+1)}P_n(\cos\phi)$ unbounded. Thus, since $\rho > 1$, the second solution $\rho^{-(n+1)}$ of the R-equation must be considered. The series solution for $u(\rho,\phi)$ is given by

$$u(\rho,\phi) = \sum_{n=0}^{\infty}\left[A_{2n+1}\rho^{2n+1} + \frac{B_{2n+1}}{\rho^{2n+2}}\right]P_{2n+1}(\cos\phi),$$

where $P_{2n+1}(\cos\phi)$ are the Legendre polynomials. Now, we use the non-homogeneous boundary conditions to find the coefficients. The boundary conditions lead to the following equations:

$$100 = \sum_{n=0}^{\infty}\left[A_{2n+1}2^{2n+1} + \frac{B_{2n+1}}{2^{2n+2}}\right]P_{2n+1}(\cos\phi),$$

$$50 = \sum_{n=0}^{\infty}[A_{2n+1} + B_{2n+1}]P_{2n+1}(\cos\phi).$$

These equations can be solved to determine the coefficients A_{2n+1} and B_{2n+1}. For $0 < \phi < \frac{\pi}{2}$, the above two equations are equivalently to

$$100 = \sum_{n=0}^{\infty}\left[A_{2n+1}2^{2n+1} + \frac{B_{2n+1}}{2^{2n+2}}\right]P_{2n+1}(x), \quad 0 < x < 1,$$

$$50 = \sum_{n=0}^{\infty}[A_{2n+1} + B_{2n+1}]P_{2n+1}(x), \quad 0 < x < 1.$$

From Example 6.13 we have that

$$1 = \sum_{n=0}^{\infty}\frac{(-1)^n(2n)!}{2^{2n+2}(n+1)(n!)^2}P_{2n+1}(x), \quad 0 < x < 1.$$

Thus, the above two expressions yield

$$100\sum_{n=0}^{\infty}\frac{(-1)^n(2n)!}{2^{2n+2}(n+1)(n!)^2}P_{2n+1}(x) = \sum_{n=0}^{\infty}\left[A_{2n+1}2^{2n+1} + \frac{B_{2n+1}}{2^{2n+2}}\right]P_{2n+1}(x), \quad 0 < x < 1,$$

$$50\sum_{n=0}^{\infty}\frac{(-1)^n(2n)!}{2^{2n+2}(n+1)(n!)^2}P_{2n+1}(x) = \sum_{n=0}^{\infty}[A_{2n+1} + B_{2n+1}]P_{2n+1}(x), \quad 0 < x < 1.$$

Consequently, we arrive at the relations

$$A_{2n+1} + B_{2n+1} = \frac{50(-1)^n(2n)!}{2^{2n+2}(n+1)(n!)^2},$$

$$2^{2n+1}A_{2n+1} + \frac{B_{2n+1}}{2^{2n+2}} = \frac{100(-1)^n(2n)!}{2^{2n+2}(n+1)(n!)^2},$$

from which, we can solve A_{2n+1} and B_{2n+1}. □

- **More On Laplacian In Spherical Coordinates**

When solving the Dirichlet problem in a spherical shell, one may encounter the **parametric Legendre equation**, given by

$$(1-x^2)\Phi''(x) - 2x\Phi'(x) + \left[n(n+1) - \frac{m^2}{1-x^2}\right]\Phi(x) = 0, \tag{6.129}$$

which can also be expressed in its self-adjoint form as

$$\left[(1-x^2)\Phi'(x)\right]' + \left[n(n+1) - \frac{m^2}{1-x^2}\right]\Phi(x) = 0. \tag{6.130}$$

We have the following lemma.

Lemma 19 *The solution to the parametrized Legendre equation* (6.129) *is given by*

$$\Phi(x) = C_1 P_n^m(x) + C_2 Q_n^m(x), \tag{6.131}$$

where

$$P_n^m(x) = (x^2-1)^{\frac{m}{2}} \frac{d^m P_n}{dx^m}, \tag{6.132}$$

and

$$Q_n^m(x) = (x^2-1)^{\frac{m}{2}} \frac{d^m Q_n}{dx^m}. \tag{6.133}$$

Proof *We let*

$$\Phi(x) = (x^2-1)^{\frac{m}{2}} u(x).$$

Differentiating, using the product rule one obtains,

$$\frac{d\Phi}{dx} = mx(x^2-1)^{\frac{m}{2}-1} u + (x^2-1)^{\frac{m}{2}} \frac{du}{dx},$$

and after simplifications,

$$\begin{aligned} \frac{d^2\Phi}{dx^2} &= (x^2-1)^{\frac{m}{2}} \frac{d^2u}{dx^2} + 2mx(x^2-1)^{\frac{m}{2}-1} \frac{du}{dx} \\ &+ m(x^2-1)^{\frac{m}{2}-1} u + m(\frac{m}{2}-1)x(x^2-1)^{\frac{m}{2}-2} 2xu. \end{aligned}$$

Substituting the last three expressions into (6.129) *we have that*

$$(1-x^2)\frac{d^2u}{dx^2} - 2(m+1)x\frac{du}{dx} + (n-m)(n+m+1)u = 0. \tag{6.134}$$

To obtain a solution of (6.134) *we differentiate m-times the Legendre equation*

$$(1-x^2)\frac{d^2y}{dx^2} - 2x\frac{dy}{dx} + n(n+1)y = 0.$$

A first differentiation leads to

$$(1-x^2)\frac{d^3y}{dx^3} - 4x\frac{d^2y}{dx^2} + \left[n(n+1)-2\right]\frac{dy}{dx} = 0.$$

Another differentiation yields

$$(1-x^2)\frac{d^4y}{dx^4} - 6x\frac{d^3y}{dx^3} + \left[n(n+1)-6\right]\frac{d^2y}{dx^2} = 0.$$

By continuing the differentiation process, we observe a distinct pattern emerging

$$(1-x^2)\frac{d^{m+2}y}{dx^{m+2}} - (2m+1)x\frac{d^{m+1}y}{dx^{m+1}} + (n-m)(n+m+1)\frac{d^m y}{dx^m} = 0. \tag{6.135}$$

Comparing the derivatives of (6.134) *and* (6.135) *we detect that*

$$\frac{d^2u}{dx^2} = \frac{d^{m+2}y}{dx^{m+2}}, \quad \frac{du}{dx} = \frac{d^{m+1}y}{dx^{m+1}}, \quad u = \frac{d^m y}{dx^m}.$$

Thus, since the solution of the regular Legendre equation is a combination of $P_n(x)$ *and* $Q_n(x)$*, it follows that*

$$\Phi(x) = (x^2-1)^{\frac{m}{2}}u(x) = (x^2-1)^{\frac{m}{2}}\left[C_1\frac{d^m P_n}{dx^m} + C_2\frac{d^m Q_n}{dx^m}\right] = C_1 P_n^m(x) + C_2 Q_n^m(x),$$

where $P_n^m(x)$ *and* $Q_n^m(x)$ *are given by* (6.132) *and* (6.133)*, respectively. The expressions* $P_n^m(x)$ *and* $Q_n^m(x)$ *are called the associated Legendre polynomials (of degree* n *and order* m*). Note that since* P_n *is a polynomial of degree* n*, then if* $m > n$*, the function* P_n^m *is identically zero (*$P_n^m(t) = 0$*). Note also that* $P_0^n(x) = P_n(x)$*.* □

Example 6.15 In this example, we consider the Laplacian in spherical coordinates, as previously discussed in equations (6.121)–(6.123). However, this time we assume that the solutions depend on the angle θ. Assume (6.121), (6.122), where the function Θ is 2π-periodic, and R and Φ are bounded. In the previous section, we assumed u to be independent of θ, or eigenvalue $\alpha = 0$ was only considered. The function Θ is 2π-periodic means that

$$\Theta(0) = \Theta(2\pi), \quad \Theta'(0) = \Theta'(2\pi).$$

Setting $\alpha = m^2$, $m = 1, 2, \ldots$, then the Θ equation or (6.122) has the corresponding eigenfunctions

$$\{\cos m\theta, \sin m\theta\}.$$

Substituting $\alpha = m^2$ into (6.123), we obtain (6.130) and by Lemma 19, it has the solution

$$\Phi(\phi) = P_n^m(\cos\phi), \quad m \le n.$$

Notice that we have neglected the other linearly independent solution $Q_n^m(x)$, since it is not well behaved at $x = \pm 1$ corresponding to the north and south poles of the original problem. So, we can throw out this solutions in many physical cases. As for the equation in R, we have from previous work that it has the two linearly independent solutions ρ^n and $\rho^{-(n+1)}$. Since we are seeking a bounded solution, we must throw away $\rho^{-(n+1)}$, otherwise, the solution becomes unbounded because

$$\rho^{-(n+1)} \to \infty \quad \text{as} \quad \rho \to 0^+.$$

Finally, the solution of (6.121)–(6.123) takes the general form

$$u(\rho,\theta,\phi) = \sum_{n=0}^{\infty}\sum_{m=0}^{n}\left[c_{nm}\cos m\theta + d_{nm}\sin m\theta\right]\rho^n P_n^m(\cos\phi). \tag{6.136}$$

□

6.13.1 Exercises

Exercise 6.66 *Solve* (6.125) *for*

$$u(a,\phi) = 1 - \cos 2\phi.$$

Ans: $u(\rho,\phi) = \frac{4}{3}P_0(\cos\phi) - \frac{4}{3}\left(\frac{\rho}{a}\right)^2 P_2(\cos\phi) = \frac{4}{3} - \frac{2}{3}\left(\frac{\rho}{a}\right)^2\left(3\cos^2\phi - 1\right).$

Exercise 6.67 *Solve* (6.125) *for* $\rho \le 1$, *and*

$$u(1,\phi) = f(\phi) = \begin{cases} 1, & 0 < \phi < \frac{\pi}{2} \\ 0, & \frac{\pi}{2} < \phi < \pi \end{cases}$$

Ans: $u(r,\phi) = \frac{1}{2} + \frac{1}{2}\sum_{n=0}^{\infty}\left[P_{2n}(0) - P_{2n+2}(0)\right]\rho^{2n+1}P_{2n+1}(\cos\phi).$

Exercise 6.68 *Solve the following Dirichlet problem inside the sphere*

$$\begin{cases} u_{\rho\rho} + \frac{2}{\rho}u_\rho + \frac{1}{\rho^2}u_{\phi\phi} + \frac{\cos\phi}{\rho^2\sin\phi}u_\phi = 0, & 0<\rho<a,\ 0<\phi<\pi \\ u(L,\phi) = f(\phi), & 0<\phi<\pi \end{cases}$$

where $u(\rho,\phi)$ *is bounded.*

(a) Given $a = 5$ *and*

$$f(\phi) = \begin{cases} 50 & \text{for } 0<\phi<\frac{\pi}{2} \\ 100 & \text{for } \frac{\pi}{2}<\phi<\pi \end{cases}$$

(b) Given $a = 1$ *and* $f(\phi) = \cos 2\phi$.

(c) Given $a = 5$ *and*

$$f(\phi) = \begin{cases} 50 & \text{for } 0<\phi<\frac{\pi}{4} \\ 0 & \text{for } \frac{\pi}{4}<\phi<\pi \end{cases}$$

Exercise 6.69 *Solve the following Dirichlet problem in a spherical shell*

$$\begin{cases} u_{\rho\rho} + \frac{2}{\rho}u_\rho + \frac{1}{\rho^2}u_{\phi\phi} + \frac{\cos\phi}{\rho^2\sin\phi}u_\phi = 0, & a<\rho<b,\ 0<\phi<\pi \\ u(a,\phi) = \cos\phi, & 0<\phi<\pi \\ u(b,\phi) = \sin^2\phi, & 0<\phi<\pi \end{cases}$$

Exercise 6.70 *Consider the radial vibrations of a spherical membrane of radius a. Let* $u(\theta,\phi,t)$ *denote the radial displacement at time t, from equilibrium, of the point on the a-sphere with coordinates* (θ,ϕ). *The function u satisfies the wave equation*

$$u_{tt} = c^2\left(u_{\phi\phi} + \cot\phi\, u_\phi + \frac{1}{\sin^2\phi}u_{\theta\theta}\right).$$

Use the method of separation of variables and show the solution is of the form

$$P_n^m(\cos\phi)\cos(m\theta)\cos(\omega_{mn}t),$$

where $m \le n$, $\omega_{mn} = c\sqrt{n(n+1)}$, *and* $P_n^m(\cos\phi)$ *denotes the associated Legendre polynomial. Moreover, the general solution is given by*

$$u(t,\theta,\phi) = a + \sum_{n=0}^{\infty}\sum_{m=0}^{n} c_{mn} P_n^m(\cos\phi)\cos(m\theta)\cos(\omega_{mn}t).$$

Exercise 6.71 *For a positive constant b, consider the heat flow in a bar lying along the x-axis from* -1 *to* 1 *given by the equation*

$$u_t = b(1-x^2)u_{xx} - 2bxu_x, \quad b > 0.$$

Assume both ends of the bar are insulated. That is

$$u_x(-1,t) = 0, \quad u_x(1,t) = 0.$$

Assume the bar has an initial temperature given by

$$u(x,0) = f(x), \quad -1 < x < 1.$$

Show that the solution can be expressed as

$$u(x,t) = \sum_{n=0}^{\infty} A_n \exp[-n(n+1)bt] P_n(x),$$

where

$$A_n = \frac{2n+1}{2}\int_{-1}^{1} f(x)P_n(x)\,dx, \quad n = 0,1,2,\ldots$$

and $P_n(x)$ *are the Legendre polynomials.*

6.14 Non-Homogeneous Sturm–Liouville Problem

In this last section of a long chapter, we are interested in solving the non-homogeneous Sturm–Liouville Problem

$$\left(r(x)y'\right)' + q(x)y + \mu p(x)y = f(x), \quad a \le x \le b \tag{6.137}$$

with the boundary conditions

$$\alpha_1 y(a) + \beta_1 y'(a) = 0, \quad \alpha_2 y(b) + \beta_2 y'(b) = 0. \tag{6.138}$$

Let λ_n, $n = 0,1,\ldots$ be the eigenvalues with corresponding eigenfunctions y_n, $n = 0,1,\ldots$ for the homogeneous S-L problem with boundary conditions given by (6.138). Then we have

$$\left(r(x)y_n'\right)' + q(x)y_n + \lambda_n p(x)y_n = 0, \tag{6.139}$$

and

$$\alpha_1 y_n(a) + \beta_1 y_n'(a) = 0, \quad \alpha_2 y_n(b) + \beta_2 y_n'(b) = 0.$$

Suppose that $y(x)$ is a continuous solution of (6.137) and (6.138). Then, we can write y using generalized eigenfunctions expansion and have

$$y(x) = \sum_{n=0}^{\infty} c_n y_n(x). \tag{6.140}$$

Substitute the expansion (6.140) into (6.137) and using the fact that the S-L eigenfunctions $y_n(x)$ satisfy (6.139) and (6.138), we obtain

$$\begin{aligned} f(x) &= \frac{d}{dx}\left(r(x)\frac{d}{dx}\sum_{n=0}^{\infty} c_n y_n(x)\right) + q(x)\sum_{n=0}^{\infty} c_n y_n(x) + \mu p(x)\sum_{n=0}^{\infty} c_n y_n(x) \\ &= \sum_{n=0}^{\infty} c_n\left(\left[\frac{d}{dx}\left(r(x)\frac{d}{dx}\right) + q(x)\right] y_n(x)\right) + \sum_{n=0}^{\infty} \mu c_n p(x) y_n(x) \\ &= \sum_{n=0}^{\infty} c_n(-\lambda_n p(x) y_n(x)) + \sum_{n=0}^{\infty} \mu c_n p(x) y_n(x) \\ &= \sum_{n=0}^{\infty} c_n(\mu - \lambda_n) p(x) y_n(x). \end{aligned}$$

Now, we multiply both sides of the last equation by $y_m(x)$ and integrate over the interval $[a,b]$ and have

$$\begin{aligned} \int_a^b f(x) y_m(x)\,dx &= \int_a^b \left(\sum_{n=0}^{\infty} c_n(\mu - \lambda_n) p(x) y_n(x)\right) y_m(x)\,dx \\ &= \sum_{n=0}^{\infty} c_n(\mu - \lambda_n)\int_a^b p(x) y_n(x) y_m(x)\,dx \\ &= c_m(\mu - \lambda_m)\int_a^b p(x)(y_m(x))^2\,dx. \end{aligned}$$

Thus, switching to n we have

$$\int_a^b f(x) y_n(x)\,dx = c_n(\mu - \lambda_n)\int_a^b p(x)(y_n(x))^2\,dx. \tag{6.141}$$

Case 1: $\mu \neq \lambda_n$. Then expression (6.141) permits us to solve for the coefficients c_n and obtain

$$c_n = \frac{1}{\mu - \lambda_n}\frac{\int_a^b f(x) y_n(x)\,dx}{\int_a^b p(x)(y_n(x))^2\,dx}. \tag{6.142}$$

Consequently, (6.137) has the unique solution given by (6.140) and the coefficient c_n is given by (6.142).

Case 2: $\mu = \lambda_n$, for some n. Then expression (6.141) implies that if

$$\int_a^b f(x)y_n(x)\,dx \neq 0,$$

there is no solution to (6.141), and consequently no solution to the original non-homogeneous problem (6.137). On the other hand, if $\mu = \lambda_n$ and

$$\int_a^b f(x)y_n(x)\,dx = 0,$$

then (6.141) becomes vacuous. In that case, if we define

$$c_m = \frac{1}{\mu - \lambda_m} \frac{\int_a^b f(x)y_m(x)\,dx}{\int_a^b p(x)(y_m(x))^2\,dx}, \qquad m \neq n$$

$$c_n = \text{arbitrary constant}$$

and consequently,

$$y(x) = \sum_{m=0}^{\infty} c_m y_m(x)$$

is a solution of the non-homogeneous differential equation (6.137) satisfying the boundary conditions (6.138). We have proved the following theorem.

Theorem 6.8 *Let* $\{\lambda_0, \lambda_1, \lambda_2, \ldots\}$ *be the set of eigenvalues of the homogeneous Sturm–Liouville problem*

$$\left(r(x)y'\right)' + q(x)y + \mu p(x)y = 0$$

and let $\{y_0, y_1, \ldots\}$ *be the corresponding eigenfunctions, that satisfy* (6.138) *too. Then the non-homogeneous boundary value problem given by* (6.137) *along with boundary conditions* (6.138) *has a unique solution whenever* $\mu \notin \{\lambda_0, \lambda_1, \lambda_2, \ldots\}$ *that is given by* (6.138) *where* c_n *is given by* (6.142). ***If, on the other hand,*** $\mu = \lambda_m$, *then the non-homogeneous problem* (6.137) *has no solution unless*

$$\int_0^1 f(x)y_m(x)\,dx = 0. \tag{6.143}$$

If, in fact, $\mu = \lambda_m$ *and* (6.143) *is true, then there is a one-parameter family of solutions given by*

$$y(x) = \sum_{n=0}^{\infty} c_n y_n(x)$$

where

$$c_n = \begin{cases} \frac{1}{\lambda_n - \mu} \frac{\int_a^b f(x)y_n(x)\,dx}{\int_a^b p(x)(y_n(x))^2\,dx}, & \text{if } n \neq m, \\ \text{an arbitrary constant}, & \text{if } n = m. \end{cases}$$

Legendre and Bessel equations commonly arise as solutions to non-homogeneous Sturm–Liouville problems, which involve second-order linear differential equations with variable coefficients. As we have seen, these special functions play a key role in solving physical problems with spherical and cylindrical symmetry, respectively. Using Corollary 6, the modified Bessel's equation containing a parameter λ,

$$x^2y''(x)+xy'(x)+(\lambda^2x^2-\nu^2)y(x)=F(x)$$

is an example of non-homogeneous S–L problem since it can be written as

$$\frac{d}{dx}\left(x\frac{dy}{dx}\right)+\left(-\frac{\nu^2}{x}+\lambda^2x^2\right)y=f(x),$$

where $xf(x)=F(x)$. As for the Legendre equation we have

$$\frac{d}{dx}\left((1-x^2)\frac{dy}{dx}\right)+n(n+1)y=f(x).$$

We provide the following example.

Example 6.16 Consider the non-homogeneous boundary value problem

$$y''+\alpha^2y=f(x),\quad y(0)=0,\quad y(b)=0.$$

Clearly, the homogeneous boundary value problem

$$y''+\lambda y=0,\quad y(0)=0,\quad y(b)=0$$

has the eigenvalues and corresponding eigenfunctions

$$\lambda_n=\left(\frac{n\pi}{b}\right)^2,\quad y_n(x)=\sin\frac{n\pi}{b}x,\quad n=1,2\ldots$$

In our case, $\mu=\alpha^2$, and hence the solution to the non-homogeneous problem is given by

$$\begin{aligned}y(x)=\sum_{n=1}^{\infty}c_ny_n(x) &= \sum_{n=1}^{\infty}\frac{1}{\alpha^2-\lambda_n}\frac{\int_0^b f(x)y_n(x)\,dx}{\int_0^b (y_n(x))^2\,dx}\sin\frac{n\pi}{b}x\\ &= \sum_{n=1}^{\infty}\frac{2\int_0^b f(s)\sin\frac{n\pi}{b}s\,ds}{b\left(\alpha^2-\left(\frac{n\pi}{b}\right)^2\right)}\sin\frac{n\pi}{b}x\end{aligned}$$

provided that

$$\alpha^2\neq\lambda_n=\left(\frac{n\pi}{b}\right)^2,\quad n=1,2\ldots$$

Next, we consider a specific cases. Consider the non-homogeneous boundary value problem

$$y''+4\pi^2y=x,\quad y(0)=0,\quad y(1)=0.$$

Then the homogeneous boundary value problem

$$y'' + \lambda y = 0, \quad y(0) = 0, \quad y(1) = 0$$

has the eigenvalues and corresponding eigenfunctions

$$\lambda_n = (n\pi)^2, \quad y_n(x) = \sin n\pi x, \quad n = 1, 2 \ldots$$

Since $\mu = 4\pi^2 = \lambda_2$, and $c_2 = \int_0^1 x \sin 2\pi x\, dx = \frac{1}{\pi} \neq 0$, there is no solution. As a matter of fact, we will directly solve the differential equation. The general solution is

$$y = \frac{x}{4\pi^2} + A\cos 2\pi x + B\sin 2\pi x.$$

From the boundary condition at $x = 0$, we get $A = 0$. Then at $x = 1$ we have

$$y(1) = \frac{1}{4\pi^2} \neq 0$$

which contradicts the boundary condition $y(1) = 0$, hence no solution exists.
For the sake of comparison, we now consider

$$y'' + 4\pi^2 y = 1, \quad y(0) = 0, \quad y(1) = 0.$$

Then

$$c_2 = 2\int_0^1 \sin 2\pi x\, dx = 0$$

so there should be solutions where the coefficient of $\sin 2\pi x$ is arbitrary. The general solution of

$$y'' + 4\pi^2 y = 1$$

is

$$y = \frac{1}{4\pi^2} + A\cos 2\pi x + B\sin 2\pi x$$

From $x = 0$ we get $A = -\frac{1}{4\pi^2}$, and then the boundary condition at $x = 1$ is also satisfied no matter what B is. □

6.14.1 Exercises

Exercise 6.72 *(See Example 6.2) Discuss the solution of the non-homogeneous Sturm–Liouville problem*

$$(xy')' + (2x + \mu x^{-1})y = f(x)$$

subject to a set of homogeneous boundary conditions

$$y'(1) = 0, \quad y'(2) = 0.$$

Find the solution when $f(x) = 1$.

Exercise 6.73 *Solve the non-homogeneous Sturm–Liouville problem*

$$y''(x)+y(x)=1, \quad y(0)=0, \quad y(\frac{\pi}{2})=0.$$

Exercise 6.74 *Solve the non-homogeneous Sturm–Liouville problem*

$$y''(x)+y(x)=\pi-2x, \quad y(0)=0, \quad y(\pi)=0.$$

Exercise 6.75 *Show the non-homogeneous Sturm–Liouville problem*

$$y''(x)+\pi^2y(x)=-x, \quad y(0)=0, \quad y(1)=0$$

has no solution.

Exercise 6.76 *Solve the non-homogeneous Sturm–Liouville problem*

$$y''(x)+4y(x)=x^2, \quad y'(0)=0, \quad y'(1)=0.$$

Exercise 6.77 *Solve the non homogeneous Sturm–Liouville problem*

$$y''(x)+\mu y(x)=x, \quad y(0)=0, \quad y'(\frac{\pi}{2})=0.$$

Exercise 6.78 *Solve the non-homogeneous Sturm–Liouville problem*

$$\left(e^x y'(x)\right)'+e^x\mu y(x)=1, \quad y(0)=0, \quad y(b)=0.$$

Exercise 6.79 *Discuss the solution of the non-homogeneous Sturm–Liouville problem*

$$y''(x)+y'(x)+(1+\mu)y(x)=f(x), \quad y(0)=0, \quad y(1)=0.$$

Exercise 6.80 *Discuss the solution of the non-homogeneous Sturm–Liouville problem*

$$X''(x)+\mu X(x)=3, \quad X'(1)=0, \quad hX(0)-X'(0)=0, \quad h>0.$$

Exercise 6.81 *(See Section 4.14) Find the solution of the non-homogeneous Sturm–Liouville problem*

$$\left(xX'(x)\right)'+\frac{\mu}{x}X(x)=a_0, \quad 1<x<e$$

subject to a set of homogeneous boundary conditions

$$X(1)=0, \quad X'(e)=0,$$

where a_0 is a positive constant.

7

Fourier Integrals and Applications

We now consider any periodic function of period $2L$ that can be represented by a Fourier series and find out what happens if we let $L \to \infty$.

For positive constant a, consider the periodic function

$$f_L(x) = \begin{cases} 0 & \text{if } -L \leq x \leq -1 \\ a & \text{if } -1 \leq x \leq 1 \\ 0 & \text{if } 1 \leq x \leq L \end{cases}$$

which is periodic for $2L = 4, 8, 16, \ldots$ Then as $L \to \infty$, we have

$$f(x) = \lim_{L\to\infty} f_L(x) = \begin{cases} a & \text{if } -1 \leq x \leq 1 \\ 0 & \text{otherwise} \end{cases}$$

which is non-periodic. Moreover,

$$\lim_{L\to\infty} a_0 = \lim_{L\to\infty} \frac{1}{2L} \int_{-L}^{L} 1\,dx = \lim_{L\to\infty} \frac{1}{L} = 0,$$

$$\begin{aligned} \lim_{L\to\infty} a_n &= \lim_{L\to\infty} \frac{1}{L} \int_{-L}^{L} \cos \frac{n\pi x}{L}\,dx \\ &= \lim_{L\to\infty} \frac{2}{L} \int_{0}^{L} \cos \frac{n\pi x}{L}\,dx = \lim_{L\to\infty} \frac{2}{L} \frac{\sin \frac{n\pi}{L}}{n\pi/L} = 0. \end{aligned}$$

This set of Fourier coefficients is referred to as the amplitude spectrum of $f(x)$, since a_n denotes the peak amplitude of the wave. As L increases, we notice that the amplitudes become progressively more densely packed along the positive a_n-axis, where $\alpha_n = \frac{n\pi}{L}$. The results of this example offer an intuitive sense of what we can anticipate when moving from this specific function to a more general one, as we will investigate further.

DOI: 10.1201/9781003652540-7

7.1 Transition from Fourier Series to Fourier Integral

We begin by stating the following definition, which is essential to our work.

Definition 7.1 *A function $f:\mathbb{R}\to\mathbb{C}$ is said to be Lebesgue integrable or belong to $L^1(\mathbb{R})$ if it satisfies*

$$\int_{-\infty}^{\infty} |f(x)|\,dx < \infty.$$

In other words, the integral of the absolute value of f over the entire real line is finite.

Example 7.1 The function

$$f(x) = \frac{1}{1+x^2}$$

belongs to $L^1(\mathbb{R})$, whereas

$$f(x) = \frac{1}{x}, \quad x \neq 0$$

does not. Also, the function

$$f(x) = e^{-|x|}$$

is in $L^1(\mathbb{R})$ because

$$\int_{-\infty}^{\infty} |e^{-|x|}|\,dx = 2\int_0^{\infty} e^{-x}\,dx = 2 < \infty.$$

On the other hand, the function

$$f(x) = \frac{\sin x}{x}$$

is not in $L^1(\mathbb{R})$, as

$$\int_{-\infty}^{\infty} \left|\frac{\sin x}{x}\right| dx,$$

diverges. □

We have seen that for any piecewise $2L$-periodic function f, its Fourier series is given by

$$f(x) = \frac{a_0}{2} + \sum_{n=1}^{\infty}\left(a_n \cos\frac{n\pi x}{L} + b_n \sin\frac{n\pi x}{L}\right), \tag{7.1}$$

where the Fourier coefficients of $f(x)$ are given by the formulas

$$a_0 = \frac{1}{L}\int_{-L}^{L} f(x)\,dx,$$

$$a_n = \frac{1}{L}\int_{-L}^{L} f(x)\cos\frac{n\pi x}{L}\,dx, \quad n = 1,2\ldots$$

and

$$b_n = \frac{1}{L}\int_{-L}^{L} f(x)\sin\frac{n\pi x}{L}\,dx, \quad n = 1,2\ldots$$

Setting

$$\alpha_n = \frac{n\pi}{L}, \qquad n = 1,2,\ldots$$

we have $\Delta\alpha = \alpha_{n+1} - \alpha_n = \frac{\pi}{L}$, and then we may write rewrite (7.1) as

$$\begin{aligned} f(x) &= \frac{1}{2L}\int_{-L}^{L} f(v)\,dv + \frac{1}{\pi}\sum_{n=1}^{\infty}\left[(\cos\alpha_n x)\Delta\alpha\int_{-L}^{L} f(v)\cos\alpha_n v\,dv\right] \\ &+ \left.(\sin\alpha_n x)\Delta\alpha\int_{-L}^{L} f(v)\sin\alpha_n v\,dv\right]. \end{aligned} \tag{7.2}$$

Assume $L^1(\mathbb{R})$ and take the limit in (7.2) as $L \to \infty$. This plausibly leads us to the Fourier integral representation

$$f(x) = \frac{1}{\pi}\int_0^{\infty}\left[\cos\,\alpha x\int_{-\infty}^{\infty} f(v)\cos\,\alpha v\,dv + \sin\,\alpha x\int_{-\infty}^{\infty} f(v)\sin\,\alpha v\,dv\right]d\alpha. \tag{7.3}$$

If we introduce the notations

$$A(\alpha) = \frac{1}{\pi}\int_{-\infty}^{\infty} f(v)\cos\,\alpha v\,dv, \quad B(\alpha) = \frac{1}{\pi}\int_{-\infty}^{\infty} f(v)\sin\,\alpha v\,dv, \tag{7.4}$$

we put (7.3) in the form

$$f(x) = \int_0^{\infty}[A(\alpha)\cos\,\alpha x + B(\alpha)\sin\,\alpha x]\,d\alpha, \tag{7.5}$$

which is the Fourier integral representation for the function $f(x)$. Its convergence to $f(x)$ is suggested, but by no means established by the preceding arguments.

Example 7.2 Express the function $f(x)$ in terms of Fourier integral representation, where

$$f(x) = \begin{cases} 0, & x < 0 \\ 1, & 0 < x < 2 \\ 0, & x > 2 \end{cases}$$

The function f is absolutely integrable; therefore, its Fourier integral can be computed as follows:

$$\begin{aligned} nA(\alpha) &= \int_{-\infty}^{\infty} f(v)\cos\alpha v\,dv \\ &= \int_{-\infty}^{0} f(v)\cos\alpha v\,dv + \int_0^2 f(v)\cos\alpha v\,dv + \int_2^{\infty} f(v)\cos\alpha v\,dv. \end{aligned}$$

Since $f(x) = 0$ for $x < 0$ and $x > 2$, only the integral over $(0,2)$ remains

$$\mathrm{n}A(\alpha) = \int_0^2 \cos \alpha v \, dv = \frac{\sin 2\alpha}{\alpha}.$$

Similarly, for $B(\alpha)$

$$\mathrm{n}B(\alpha) = \int_{-\infty}^{\infty} f(v) \sin \alpha v \, dv = \int_0^2 \sin \alpha v \, dv = \frac{1 - \cos 2\alpha}{\alpha}.$$

Substituting these coefficients into the Fourier integral yields

$$f(x) = \frac{1}{\pi} \int_0^{\infty} \left[\frac{\sin 2\alpha}{\alpha} \cos \alpha x + \frac{1 - \cos 2\alpha}{\alpha} \sin \alpha x \right] d\alpha.$$

Using trigonometric identities, the last integral simplifies to

$$f(x) = \frac{2}{\pi} \int_0^{\infty} \frac{\sin \alpha \cos \alpha (x-1)}{\alpha} d\alpha.$$

□

When a function $f(x)$ is defined solely in the interval $0 < x < \infty$, it can be extended to an even or odd function. This leads to a Fourier integral representation involving only cos ax or sin αx, commonly referred to as the Fourier cosine and sine integral representations of f, respectively. Thus, as in the case of representing functions using Fourier series, we have analogous results for the cosine and sine Fourier integral representations.

Theorem 7.1 *Suppose that f is piecewise continuous on every bounded interval of the x axis, and it is absolutely integrable on $(0,\infty)$.*

- *If f is even, then its Fourier integral representation has only cosine term. The Fourier coefficients are given by*

$$B(\alpha) = 0, \text{ and } A(\alpha) = \frac{2}{\pi} \int_0^{\infty} f(v) \cos \alpha v \, dv, \tag{7.6}$$

  and the corresponding Fourier integral representation is given by

$$f(x) = \int_0^{\infty} A(\alpha) \cos \alpha x \, d\alpha. \tag{7.7}$$

- *If f is odd, then its Fourier integral representation has only sine term. The Fourier coefficients are given by*

$$A(\alpha) = 0, \text{ and } B(\alpha) = \frac{2}{\pi} \int_0^{\infty} f(v) \sin \alpha v \, dv, \tag{7.8}$$

  and the corresponding Fourier integral representation is given by

$$f(x) = \int_0^{\infty} B(\alpha) \sin \alpha x \, d\alpha. \tag{7.9}$$

We have the following example.

Example 7.3 *We shall derive the Fourier cosine and Fourier sine integrals of $f(x) = e^{-kx}$, where $x > 0$ and $k > 0$. The result will be used to evaluate the so-called Laplace integrals. We perform integration by parts. Let*

$$u = \cos(\alpha v), \quad dv = e^{-kv}dv, \text{ then } du = -\alpha \sin(\alpha v)dv, \quad v = -\frac{e^{-kv}}{k}.$$

$$\begin{aligned} I = \int_0^\infty e^{-kv}\cos(\alpha v)dv &= \left[-\cos(\alpha v)\frac{e^{-kv}}{k}\right]_0^\infty - \frac{\alpha}{k}\int_0^\infty e^{-kv}\sin(\alpha v)dv \\ &= \frac{1}{k} - \frac{\alpha}{k}\int_0^\infty e^{-kv}\sin(\alpha v)dv. \end{aligned}$$

Let

$$J = \int_0^\infty e^{-kv}\sin(\alpha v)dv.$$

Apply integration by parts again with

$$u = \sin(\alpha v), \quad dv = e^{-kv}dv, \text{ then } du = \alpha \cos(\alpha v)dv, \quad v = -\frac{e^{-kv}}{k}.$$

This gives

$$J = \left[-\sin(\alpha v)\frac{e^{-kv}}{k}\right]_0^\infty + \frac{\alpha}{k}\int_0^\infty e^{-kv}\cos(\alpha v)dv.$$

Hence

$$J = \frac{\alpha}{k}\int_0^\infty e^{-kv}\cos(\alpha v)dv = \frac{\alpha}{k}I.$$

Substituting $J = \frac{\alpha}{k}I$ back into the expression for I, we obtain

$$I = \frac{1}{k} - \frac{\alpha^2}{k^2}I.$$

Solving for I

$$I\left(1 + \frac{\alpha^2}{k^2}\right) = \frac{1}{k}$$

$$I = \frac{k}{k^2 + \alpha^2}.$$

Now, by integration by parts,

$$\int_0^\infty e^{-kv}\cos\alpha v\, dv = \frac{k}{k^2 + \alpha^2}.$$

Hence,

$$A(\alpha) = \frac{2}{\pi}\int_0^\infty e^{-kv}\cos(\alpha v)dv = \frac{2}{\pi}\frac{k}{k^2 + \alpha^2}.$$

By substituting this into the first integral in (7.6), we obtain the Fourier cosine integral representation

$$e^{-kx} = \frac{2}{\pi}\int_0^\infty \frac{k\cos\alpha x}{k^2+\alpha^2}\,d\alpha, \quad (x>0, k>0). \tag{7.10}$$

As for Fourier sine integral,

$$B(\alpha) = \frac{2}{\pi}\int_0^\infty f(v)\sin\alpha v\,dv = \frac{2}{\pi}\int_0^\infty e^{-kv}\sin(\alpha v)dv = \frac{2\alpha}{\pi}\frac{1}{k^2+\alpha^2}. \tag{7.11}$$

Therefore, the Fourier sine integral representation of f is

$$e^{-kx} = \frac{2}{\pi}\int_0^\infty \frac{\alpha\sin\alpha x}{k^2+\alpha^2}\,d\alpha, \quad (x>0, k>0).$$

Remark 12 *Note that the results of Example 7.3 immediately imply that*

$$\frac{\pi}{2k}e^{-kx} = \int_0^\infty \frac{\cos\alpha x}{k^2+\alpha^2}\,d\alpha, \quad x>0,\ k>0,$$

and

$$\frac{\pi}{2}e^{-kx} = \int_0^\infty \frac{\alpha\sin\alpha x}{k^2+\alpha^2}\,d\alpha, \quad x>0,\ k>0,$$

that are known as **Laplace integrals**

7.1.1 Exercises

Exercise 7.1 *Derive* (7.11).

In Exercises 7.2-7.7, find the cosine Fourier integral representations of $f(x)$.

Exercise 7.2 $f(x) = \begin{cases} be^x, & 0 \le x \le a \\ 0, & x > a \end{cases}$

Exercise 7.3 $f(x) = \begin{cases} b\sin x, & 0 \le x \le \pi \\ 0, & x > \pi \end{cases}$

Exercise 7.4 $f(x) = \begin{cases} b(a^2-x^2), & 0 \le x \le a \\ 0, & x > a \end{cases}$

Exercise 7.5 $f(x) = \frac{1}{1+x^2}, \quad x \ge 0$

Exercise 7.6 $f(x) = \begin{cases} bx^2, & 0 \le x \le 1 \\ 0, & x > 1 \end{cases}$

Exercise 7.7 $f(x) = \begin{cases} b, & 0 \le x \le 1 \\ 0, & x > 1 \end{cases}$

Exercise 7.8 *Verify that the Fourier integral representation of the function* $f(x) = \begin{cases} x^2, & 0 < x < a \\ 0, & x > a \end{cases}$ *is*

$$f(x) = \frac{2}{\pi}\int_0^{\infty}\left((a^2 - \frac{2}{\alpha^2})\sin a\alpha + \frac{2a}{\alpha}\cos ak\right)\cos\alpha x\, d\alpha.$$

Exercise 7.9 *Use Example 7.3 to show that the Fourier integral representation of*

$$f(x) = \begin{cases} 0, & x < 0 \\ e^{-x}, & x > 0 \end{cases}$$

$$f(x) = \frac{1}{\pi}\int_0^{\infty}(\cos\alpha x + \alpha\sin\alpha x)\frac{1}{1+\alpha^2}\, d\alpha.$$

Exercise 7.10 *Show that the Fourier integral representation of* $f(x) = \exp(-|x|)$, $-\infty < x < \infty$,

$$\exp(-|x|) = \frac{2}{\pi}\int_0^{\infty}\frac{\cos\, \alpha x)}{1+\alpha^2}\, d\alpha, \quad -\infty < x < \infty.$$

Exercise 7.11 *Show that the Fourier integral representation of*

$$f(x) = \begin{cases} -1, & -a < x < 0 \\ 1, & 0 \le x < a \\ 0, & \textit{otherwise} \end{cases}$$

is given by

$$f(x) = \frac{2}{\pi}\int_0^{\infty}\frac{1}{\alpha}(1 - \cos\alpha a)\sin\alpha x\, d\alpha.$$

Exercise 7.12 *Find the Fourier sine and cosine integral representations of the function given by:*

$$f(x) = \begin{cases} \frac{a-x}{a}, & 0 < x < a \\ 0, & x > a \end{cases}$$

Exercise 7.13 *Find the Fourier integral representation of the function:*

$$f(x) = \begin{cases} \frac{1}{\varepsilon}, & a < x < a + \varepsilon \\ 0, & \textit{elsewhere} \end{cases}$$

In Exercises 7.14-7.19, find the Fourier sine integral representations of $f(x)$, and sketch the odd extension of each function.

Exercise 7.14 $f(x) = \begin{cases} \cos x, & 0 \le x \le \pi \\ 0, & x > \pi \end{cases}$

Exercise 7.15 $f(x) = \begin{cases} e^{-x}, & 0 \leq x \leq 1 \\ 0, & x > 1 \end{cases}$

Exercise 7.16 $f(x) = \begin{cases} x, & 0 \leq x \leq a \\ 0, & x > a \end{cases}$

Exercise 7.17 $f(x) = \begin{cases} b, & 0 \leq x \leq 1 \\ 0, & x > 1 \end{cases}$

Exercise 7.18 $f(x) = \begin{cases} 1-x, & 0 < x < 1 \\ 0, & 1 < x \end{cases}$

Exercise 7.19 $f(x) = \begin{cases} 1, & 0 < x < 1 \\ 2-x, & 1 < x < 2 \\ 0, & 2 < x \end{cases}$

Exercise 7.20 *Show that the Fourier cosine coefficient $A(\alpha)$ of the integral representation of*

$$f(x) = e^{-x}\cos x$$

can be written as

$$A(\alpha) = \frac{1}{\pi}\int_0^\infty e^{-v}\cos((\alpha+1)v)\,dv + \frac{1}{\pi}\int_0^\infty e^{-v}\cos((\alpha-1)v)\,dv.$$

Use the result of Example 7.3 to show that the Fourier integral representation of $f(x$ is given by

$$e^{-x}\cos x = \frac{2}{\pi}\int_0^\infty \frac{\alpha^2+2}{\alpha^4+4}\cos\alpha x\,d\alpha, \quad x \geq 0.$$

Exercise 7.21 *Assume the constant k in (7.10) of Example 7.3 is a variable. Differentiate both sides of (7.10) with respect to k and derive the expression*

$$(1+x)e^{-x} = \frac{4}{\pi}\int_0^\infty \frac{\cos\alpha x}{(\alpha^2+1)^2}\,d\alpha, \quad x \geq 0.$$

Exercise 7.22 *Use integration by parts to establish*

$$\int_0^\infty e^{-\alpha}\cos\alpha x\,d\alpha = \frac{1}{1+x^2}.$$

Exercise 7.23 *The expression in Exercise 7.22 is valid for all x. Explain why its validity implies that*

$$\frac{2}{\pi}\int_0^\infty \frac{\cos\alpha x}{1+x^2}\,dx = e^{-\alpha}, \quad \alpha > 0.$$

Exercise 7.24 *Integrate both sides of the equality in Exercise 7.22 from 0 to t to derive the expression*

$$\int_0^\infty \frac{e^{-\alpha}\sin\alpha t}{\alpha}\,d\alpha = \tan^{-1}(t).$$

Exercise 7.25 *Does the equality in Exercise 7.24 imply that*

$$\frac{2}{\pi}\int_0^\infty \tan^{-1}(t)\sin \alpha t\,dt = \frac{e^{-\alpha}}{\alpha}?$$

7.2 Fourier Integral Theorem

The Fourier Integral Theorem is crucial because it extends Fourier series to non-periodic functions, allowing them to be represented as an integral of sinusoidal components. This is fundamental in signal processing, physics, and engineering, enabling the analysis of wave propagation, heat conduction, and quantum mechanics by decomposing functions into continuous frequency components. The next two lemmas are crucial to the proof of Fourier Integral Theorem. We begin by stating and proving the Riemann–Lebesgue Lemma.

Lemma 20 *(Riemann–Lebesgue Lemma). If $g(x)$ is piecewise continuous on the interval $[a,b]$, then*

(a)

$$\lim_{\lambda\to\infty}\int_a^b g(x)\sin \lambda x\,dx = 0,$$

(b)

$$\lim_{\lambda\to\infty}\int_a^b g(x)\cos \lambda x\,dx = 0.$$

Proof *We will prove only part (a), as it is directly relevant to our work, while part (b) is left as an exercise. Let*

$$I(\lambda) = \int_a^b g(x)\sin(\lambda x)\,dx$$

with the change of variable $x = v + \frac{\pi}{\lambda}$. Then we have

$$\sin(\lambda x) = \sin\left(\lambda(v+\frac{\pi}{\lambda})\right) = -\sin \lambda v,$$

and therefore,

$$I(\lambda) = -\int_{a-\pi/\lambda}^{b-\pi/\lambda} g(v+\pi/\lambda)\sin \lambda v\,dv.$$

Switching back to x (since v is a dummy variable), we have

$$I(\lambda) = -\int_{a-\pi/\lambda}^{b-\pi/\lambda} g(x+\pi/\lambda)\sin \lambda x\,dx.$$

Then

$$2I(\lambda) = \int_a^b g(x)\sin\lambda x\,dx - \int_{a-\pi/\lambda}^{b-\pi/\lambda} g(x+\pi/\lambda)\sin\lambda x\,dx.$$

For a better manipulation we decompose the above integral to obtain

$$\begin{aligned} 2I(\lambda) &= -\int_{a-\pi/\lambda}^{a} g(x+\pi/\lambda)\sin\lambda x\,dx \\ &+ \int_{b-\pi/\lambda}^{b} g(x)\sin(\lambda x)\,dx + \int_a^{b-\pi/\lambda} [g(x)-g(x+\pi/\lambda)]\sin\lambda x\,dx. \end{aligned}$$

Since $g(x)$ is continuous function in $[a,b]$, there is a positive constant D such that $|g(x)| \leq D$. Thus,

$$\left|\int_{a-\pi/\lambda}^{a} g(x+\pi/\lambda)\sin\lambda x\,dx\right| = \left|\int_a^{a+\pi/\lambda} g(x)\sin\lambda x\,dx\right| \leq \frac{\pi D}{\lambda}.$$

Along the same lines, we have that

$$\left|\int_{b-\pi/\lambda}^{b} g(x)\sin\lambda x\,dx\right| \leq \frac{\pi D}{\lambda}.$$

Consequently,

$$|I(\lambda)| \leq \frac{\pi D}{\lambda} + \int_a^{b-\pi/\lambda} |g(x)-g(x+\pi/\lambda)|\,dx.$$

Since $g(x)$ is a continuous function on the closed interval $[a,b]$, it is uniformly continuous on $[a,b]$, meaning

$$|g(x)-g(x+\pi/\lambda)| < \frac{\varepsilon}{b-a},$$

for all $\lambda > 0$ and all x in $[a,b]$. By choosing λ such that $\frac{\pi D}{\lambda} < \frac{\varepsilon}{2}$ for $\lambda > 0$, we obtain

$$|I(\lambda)| < \frac{\varepsilon}{2} + \frac{\varepsilon}{2} = \varepsilon.$$

This completes the proof. □

Lemma 21 *If f is piecewise smooth in the interval $[0,b]$, then for $b > 0$,*

$$\lim_{\lambda\to\infty}\int_0^b f(x)\frac{\sin\lambda x}{x}\,dx = \frac{\pi}{2}f(0^+). \tag{7.12}$$

Proof *By writing*

$$f(x)\frac{\sin\lambda x}{x} = \frac{\sin\lambda x}{x}\left[f(x)-f(0^+)+f(0^+)\right],$$

we have

$$\begin{aligned}\int_0^b f(x)\frac{\sin\lambda x}{x}dx &= \int_0^b f(0^+)\frac{\sin\lambda x}{x}dx+\int_0^b \frac{f(x)-f(0^+)}{x}\sin\lambda x dx\\ &= f(0^+)\int_0^{\lambda b}\frac{\sin t}{t}dt+\int_0^b \frac{f(x)-f(0^+)}{x}\sin\lambda x dx,\end{aligned}$$

where we have made the substitution $t=\lambda x$ *in the first integral* $\int_0^b f(0^+)\frac{\sin\lambda x}{x}dx$. *Since* f *is piecewise smooth, or* f' *is piecewise continuous, it implies that*

$$\lim_{x\to 0^+}\frac{f(x)-f(0^+)}{x}=f'(0^+),$$

exists. Hence, for a positive constant M *we have that*

$$\left|\frac{f(x)-f(0^+)}{x}\sin\lambda x\right|\le M,$$

uniformly for all $\lambda>0$ *and* $x\in(0,b)$. *Thus, by the Riemann–Lebesgue lemma, Lemma 20*

$$\lim_{\lambda\to\infty}\int_0^b \frac{f(x)-f(0^+)}{x}\sin\lambda x dx=0.$$

Finally, taking limit as λ *approaches infinity, we obtain*

$$\lim_{\lambda\to\infty}\int_0^b f(x)\frac{\sin\lambda x}{x}dx=\lim_{\lambda\to\infty}\int_0^b f(0^+)\frac{\sin\lambda x}{x}dx=\frac{\pi}{2}f(0^+),$$

since

$$\int_0^\infty \frac{\sin k}{k}dk=\frac{\pi}{2}.$$

Theorem 7.2 *Let* f *and its derivative* f' *be piecewise continuous on every finite interval, and suppose that* f *is absolutely integrable on* $(-\infty,\infty)$. *Then, the Fourier integral of* f *converges to* $f(x)$ *at every point of continuity. At a point of discontinuity, the Fourier integral converges to the average of the left-hand and right-hand limits of* f. *Precisely, to*

$$\frac{f(x^+)+f(x^-)}{2},$$

where $f(x^+)$ *and* $f(x^-)$ *denote the limit of* f *at* x *from the right and from the left, respectively. In other words,*

$$\frac{1}{\pi}\int_0^\infty\int_{-\infty}^\infty f(v)\cos\alpha(v-x)\,dv\,d\alpha=\frac{1}{2}\left[f(x^+)+f(x^-)\right]. \tag{7.13}$$

Proof *The Fourier integral representation of* $f(x)$ *given by expression* (7.3) *can be written as*

$$\frac{1}{\pi}\int_0^\infty\int_{-\infty}^\infty f(v)\cos\alpha(v-x)\,dv\,d\alpha.$$

Notice that

$$\left|\int_{-\infty}^{\infty} f(v)\cos\alpha(v-x)\,dv\right| \leq \int_{-\infty}^{\infty} |f(v)|\,dv < \infty,$$

and we conclude that the integral converges independently of α and x. Let

$$I = \int_{0}^{\lambda}\int_{-\infty}^{\infty} f(v)\cos\alpha(v-x)\,dv\,d\alpha. \tag{7.14}$$

Then the order of integration may be interchanged. We then have

$$I = \int_{-\infty}^{\infty} f(v)\left(\int_{0}^{\lambda}\cos\alpha(v-x)\,d\alpha\right)dv = \int_{-\infty}^{\infty} f(v)\frac{\sin\lambda(v-x)}{v-x}\,dv.$$

We will decompose the above integral into the sum of four integrals. Hence, for real number H we have

$$\begin{aligned} I &= \int_{-\infty}^{-H} f(v)\frac{\sin\lambda(v-x)}{v-x}\,dv + \int_{-H}^{x} f(v)\frac{\sin\lambda(v-x)}{v-x}\,dv \\ &+ \int_{x}^{H} f(v)\frac{\sin\lambda(v-x)}{v-x}\,dv + \int_{H}^{\infty} f(v)\frac{\sin\lambda(v-x)}{v-x}\,dv. \end{aligned}$$

Introducing the substitution $u = v-x$ in the third integral on the right, we obtain

$$\int_{x}^{H} f(v)\frac{\sin\lambda(v-x)}{v-x}\,dv = \int_{0}^{H-x} f(u+x)\frac{\sin\lambda u}{u}\,du.$$

Taking limit as λ approaches infinity and applying Lemma 21, we obtain

$$\lim_{\lambda\to\infty}\int_{x}^{H} f(v)\frac{\sin\lambda(v-x)}{v-x}\,dv = \lim_{\lambda\to\infty}\int_{0}^{H-x} f(u+x)\frac{\sin\lambda u}{u}\,du = \frac{\pi}{2}f(x^{+}).$$

Following a similar approach, we set $u = -v+x$ in the second integral on the right. Then, taking the limit as λ approaches infinity and applying Lemma 21, we obtain

$$\lim_{\lambda\to\infty}\int_{-H}^{x} f(v)\frac{\sin\lambda(v-x)}{v-x}\,dv = \lim_{\lambda\to\infty}\int_{0}^{H+x} f(x-u)\frac{\sin\lambda u}{u}\,du = \frac{\pi}{2}f(x^{-}).$$

Next, we analyze the first and last integrals and verify that they vanish as $\lambda\to\infty$. Since f is absolutely integrable on $(-\infty,\infty)$ there is a number $\beta > H$ such that

$$\int_{\beta}^{\infty} |f(x)|dx < \frac{\varepsilon}{4},$$

and consequently,

$$\left|\int_{\beta}^{\infty} f(v)\frac{\sin\lambda(v-x)}{v-x}\,dv\right| < \frac{\varepsilon}{4}.$$

By Lemma 20 there is a number a_0

$$\left| \int_H^\beta f(v) \frac{\sin \lambda (v-x)}{v-x} dv \right| < \frac{\varepsilon}{4}, \quad \lambda > a_0.$$

Adding the last two inequalities gives

$$\left| \int_H^\infty f(v) \frac{\sin \lambda (v-x)}{v-x} dv \right| < \frac{\varepsilon}{2}, \quad \lambda > a_0.$$

Along the same lines, we can show that the absolute value of the first integral is less than $\varepsilon/2$. Thus,

$$\begin{aligned} \frac{1}{\pi} \int_0^\infty \int_{-\infty}^\infty f(v) \cos \alpha (v-x)\, dv\, d\alpha &= \frac{1}{\pi} \lim_{\lambda \to \infty} I = \frac{1}{\pi} \left[\frac{\pi}{2} f(x^+) + \frac{\pi}{2} f(x^-) \right] \\ &= \frac{f(x^+) + f(x^-)}{2}, \end{aligned}$$

at the point of discontinuity x. On the other hand, if f is continuous at x, then $f(x^+) = f(x^-) = f(x)$, and consequently

$$\frac{1}{\pi} \int_0^\infty \int_{-\infty}^\infty f(v) \cos \alpha (v-x)\, dv\, d\alpha = \frac{f(x^+) + f(x^-)}{2} = f(x).$$

This completes the proof. □

We have the following corollary.

Corollary 10 *(Dirichlet's integral)*

$$\int_0^\infty \frac{\sin \alpha}{\alpha} d\alpha = \frac{\pi}{2}.$$

Proof *Let*

$$f(x) = \begin{cases} 1, & 0 < x < 1 \\ 0, & -\infty < x < 0 \end{cases}$$

The function satisfies the assumptions of the Theorem 7.2 and hence we may compute its Fourier integral representation.

$$A(\alpha) = \frac{1}{\pi} \int_0^1 \cos \alpha x\, dx = \frac{\sin \alpha}{\pi \alpha}.$$

Similarly,

$$B(\alpha) = \frac{1}{\pi} \int_0^1 \sin \alpha x\, dx = \frac{-1 + \cos \alpha}{\pi \alpha}.$$

Substituting these coefficients into the Fourier integral yields

$$\frac{1}{\pi} \int_0^\infty \left[\frac{\sin \alpha}{\alpha} \cos \alpha x + \frac{-1 + \cos \alpha}{\alpha} \sin \alpha x \right] d\alpha. \tag{7.15}$$

Hence, we may apply Theorem 7.2, which states that due to the continuity of f on $(0,1)$ and $(-\infty,0)$, the Fourier integral representation given by (7.15) converges to f at each point in these intervals. However, the function f has a discontinuity at $x=0$, and therefore, by Theorem 7.2, its Fourier integral representation given by (7.15) converges to

$$\frac{f(0^+)+f(0^-)}{2}=\frac{1+0}{2}=\frac{1}{2}.$$

Thus, substituting $x=0$ in (7.15) gives

$$\frac{1}{2}=\frac{1}{\pi}\int_0^\infty \frac{\sin\alpha}{\alpha}d\alpha,$$

from which we obtain

$$\int_0^\infty \frac{\sin\alpha}{\alpha}d\alpha=\frac{\pi}{2}.$$

This completes the proof. □

7.2.1 Exercises

Exercise 7.26 *Prove part (b) of Lemma 20.*

Exercise 7.27 *Argue why you are able to use the result of Exercise 7.10 in order to show that the Fourier representation of*

$$g(x)=\begin{cases}-e^{-x}, & 0<x\\ e^{x}, & x<0\end{cases}$$

is given by

$$\int_0^\infty \frac{2}{\pi}\frac{-\alpha}{1+\alpha^2}\sin\alpha x\,d\alpha.$$

To what value does the Fourier integral representation converge at $x=0$?

Exercise 7.28 *Verify that all of the conditions in Theorem 7.2 are satisfied by the function f defined by the equations*

$$f(x)=\begin{cases}1, & |x|<1\\ 0, & |x|>1\end{cases}$$

and $f(\pm1)=\frac{1}{2}$.

Thus, show that for every x ($-\infty<x<\infty$),

$$f(x)=\frac{1}{\pi}\int_0^\infty \frac{\sin\alpha(1+x)+\sin\alpha(1-x)}{\alpha}d\alpha=\frac{2}{\pi}\int_0^\infty \frac{\sin\alpha\cos\alpha x}{\alpha}d\alpha.$$

Moreover, show that

$$\int_0^\infty \frac{\sin\alpha}{\alpha}d\alpha=\frac{\pi}{2}.$$

Exercise 7.29 *Explain why you are able to use the results of Exercise 7.21 in showing that*

$$\frac{\pi}{4} = \int_0^\infty \frac{d\alpha}{(\alpha^2+1)^2}.$$

Exercise 7.30 *Explain why* $f(x) = \frac{1}{2}$, $-\infty < x < \infty$, *fails to have a Fourier integral representation.*

Exercise 7.31 *Verify that the Fourier integral representation, given by the following integral, converges to the function* $f(x)$. *That is,*

$$\int_0^\infty \frac{\cos\alpha x + \alpha \sin\alpha x}{1+\alpha^2}\,d\alpha = f(x) = \begin{cases} 0, & x<0 \\ \frac{\pi}{2}, & x=0 \\ \pi e^{-x}, & x>0 \end{cases}$$

Exercise 7.32 *Verify that the Fourier integral representation, given by the following integral, converges to the function* $f(x)$. *That is,*

$$\int_0^\infty \frac{(\sin\alpha - \alpha\cos\alpha)\sin\alpha x}{\alpha^2}\,d\alpha = f(x) = \begin{cases} \frac{\pi x}{2}, & 0<x<1 \\ \frac{\pi}{4}, & x=1 \\ 0, & x>1 \end{cases}$$

Exercise 7.33 *An important application of the Fourier integral representation is in modeling heat flow in an unbounded domain. To illustrate this, differentiate directly to show that*

$$u(x,t) = \int_0^\infty B(\alpha)e^{-\alpha^2 kt}\sin\alpha x\,d\alpha$$

is a solution to the heat flow equation

$$u_t(x,t) = ku_{xx}(x,t), \quad -\infty < x < \infty, \quad t>0.$$

Exercise 7.34 *With the aid of Exercise 7.24 establish the equality*

$$\int_0^\infty \frac{1-e^{-\alpha}}{\alpha}\sin\alpha x\,d\alpha = \frac{\pi}{2} - \tan^{-1}(x), \quad x>0.$$

7.3 Complex Form of Fourier Integral Representation

Writing Fourier integrals in complex form is important because it simplifies mathematical analysis and computation by expressing oscillatory functions in terms of exponentials. The use of Euler's formula allows us to replace trigonometric functions with complex exponentials, making differentiation, integration, and algebraic manipulations more straightforward. In particular, complex Fourier integrals unify the Fourier sine and cosine representations into a single expression, facilitating a more elegant and general approach to solving problems in signal processing, quantum mechanics, and differential equations. This form also provides deeper insight into the frequency components of a function and is essential in the study of analytic functions and contour integration in the complex plane.

We have already established that the Fourier integral representation of a function $f(x)$ is

$$f(x)=\frac{1}{\pi}\int_0^{\infty}\int_{-\infty}^{\infty} f(v)\cos\alpha(v-x)\,dv\,d\alpha. \tag{7.16}$$

To express the Fourier integral representation (7.16) in complex form we make use of

$$\cos\alpha(v-x)=\cos\alpha(x-v)=\frac{1}{2}\left(e^{i\alpha(x-v)}+e^{-i\alpha(x-v)}\right),$$

and then substituting it into (7.16) to obtain the two separated integrals

$$f(x)=\frac{1}{2\pi}\int_0^{\infty}\int_{-\infty}^{\infty} f(v)e^{i\alpha(x-v)}dv\,d\alpha+\frac{1}{2\pi}\int_0^{\infty}\int_{-\infty}^{\infty} f(v)e^{-i\alpha(x-v)}dv\,d\alpha.$$

Changing the integration variable from α to $-\alpha$ in the second integral, we obtain

$$\begin{aligned}
f(x) &= \frac{1}{2\pi}\left[\int_0^{\infty}\int_{-\infty}^{\infty} f(v)e^{i\alpha(x-v)}dv\,d\alpha-\int_0^{-\infty}\int_{-\infty}^{\infty} f(v)e^{i\alpha(x-v)}dv\,d\alpha\right]\\
&= \frac{1}{2\pi}\left[\int_0^{\infty}\int_{-\infty}^{\infty} f(v)e^{i\alpha(x-v)}dv\,d\alpha+\int_{-\infty}^{0}\int_{-\infty}^{\infty} f(v)e^{i\alpha(x-v)}dv\,d\alpha\right]\\
&= \frac{1}{2\pi}\int_{-\infty}^{\infty}\int_{-\infty}^{\infty} f(v)e^{i\alpha(x-v)}dv\,d\alpha.
\end{aligned} \tag{7.17}$$

Equation (7.17) is called the **complex form of the Fourier integral representation** for $f(x)$. In some of the literature, expression (7.17) is written in the equivalent form

$$\begin{aligned}
f(x) &= \frac{1}{\sqrt{2\pi}}\int_{-\infty}^{\infty} e^{i\alpha x}d\alpha\left[\frac{1}{\sqrt{2\pi}}\int_{-\infty}^{\infty} e^{-i\alpha v}f(v)dv\right]\\
&= \frac{1}{\sqrt{2\pi}}\int_{-\infty}^{\infty} F(\alpha)e^{i\alpha x}d\alpha,
\end{aligned} \tag{7.18}$$

where

$$F(\alpha)=\frac{1}{\sqrt{2\pi}}\int_{-\infty}^{\infty} e^{-i\alpha v}f(v)dv. \tag{7.19}$$

Equation (7.19) is referred to as the **Fourier transform** of $f(x)$, which we will discuss further in Section 7.6.

Example 7.4 In this example we will find the complex Fourier integral representation of

$$f(x) = \begin{cases} b, & -a < x < a \\ 0, & x < |a| \end{cases}$$

We use the coefficient $F(\alpha)$ given by (7.19). Thus

$$F(\alpha) = \frac{1}{\sqrt{2\pi}} \int_{-a}^{a} be^{-i\alpha x} dx = \frac{1}{\sqrt{2\pi}} b \frac{e^{-i\alpha x}}{-i\alpha}\Big|_{-a}^{a}$$

$$= \frac{b}{\sqrt{2\pi}} \frac{e^{i\alpha a} - e^{-i\alpha a}}{i\alpha} = \sqrt{\frac{2}{\pi}} \frac{b \sin \alpha a}{\alpha}.$$

Substituting $F(\alpha)$ into (7.18) gives the representation of f

$$f(x) = \int_{-\infty}^{\infty} \frac{b \sin \alpha a}{\pi \alpha} e^{i\alpha x} d\alpha, \quad -\infty < x < \infty.$$

Note that at $x = \pm a$, the integral converges to $\frac{b}{2}$. □

Remark 13 *This example highlights an important property of Fourier transforms. If $f(x)$ is even, then $F(\alpha)$ is real, whereas if $f(x)$ is odd, then $F(\alpha)$ is purely imaginary.*

7.3.1 Exercises

Exercise 7.35 *Prove the assertion of Remark 13.*

Exercise 7.36 *Find the complex Fourier integral representation of*

$$f(x) = \begin{cases} \sin \pi x, & -a < x < a \\ 0, & x < |a| \end{cases}$$

Exercise 7.37 *Find the complex Fourier integral for the following functions.*

(a)

$$f(x) = \begin{cases} xe^{-x}, & x > 0 \\ 0, & otherwise \end{cases}$$

(b)

$$f(x) = e^{-a|x|} \sin x.$$

Exercise 7.38 *Use the identity*

$$\int_{-\infty}^{\infty} e^{-z^2} dz = \sqrt{\pi},$$

to show the complex Fourier coefficient of

$$f(x) = e^{-x^2}, \quad -\infty < x < \infty,$$

is

$$F(\alpha) = \frac{e^{-\alpha^2/4}}{\sqrt{2}}.$$

Exercise 7.39 *Compute the complex Fourier integral representation of the following functions.*

(a) $f(x) = e^{-|x|}, \quad -\infty < x < \infty,$ *(b)* $f(x) = \begin{cases} 1, & -a < x < a \\ 0, & \textit{otherwise} \end{cases}$

(c) $f(x) = e^{-\beta|x|}\cos x, \quad -\infty < x < \infty,$ *(d)* $f(x) = \frac{1}{1+x^2}, \quad -\infty < x < \infty$

(e) $f(x) = e^{-x^2}, \quad -\infty < x < \infty$

7.4 Heat Flow in an Infinite Rod

The Fourier integral representation plays a crucial role in solving heat flow problems on unbounded domains, where conventional Fourier series methods are not applicable due to the lack of periodic boundary conditions. In such cases, the Fourier integral provides a powerful tool for representing and analyzing the continuous spectrum of temperature distributions. In an infinite medium, heat spreads out symmetrically and dissipates over time. The Fourier integral formulation allows us to decompose the initial heat distribution into sinusoidal components, each evolving according to its frequency-dependent dissipation rate.

For a positive constant M, we consider the heat flow problem in an infinite rod that is governed by the equations

$$\begin{cases} u_t - ku_{xx} = 0, & -\infty < x < \infty, \ t > 0 \\ u(x,0) = \phi(x), & -\infty < x < \infty \\ \left|u(x,t)\right| \leq M, & \text{as } x \to \pm\infty, \ t > 0 \end{cases} \tag{7.20}$$

We seek separated solutions of the form

$$u(x,t) = X(x)T(t),$$

where X is a function of x alone, and T is a function of t alone. Note that both X and T must be nontrivial to ensure a nontrivial solution, i.e., $X \neq 0$ and $T \neq 0$. By differentiating with respect to t and x and substituting into the first equation of (7.20), we obtain the relation

$$\frac{X''(x)}{X(x)} = \frac{T'(t)}{kT(t)}.$$

Since the left-hand side is a function of x alone and the right-hand side is a function of t alone, the two sides must have a common constant value $-\alpha$ (we choose $-\alpha$ for convenience). Therefore, we arrive at

$$X''(x) + \alpha^2 X(x) = 0,$$

$$X(x), \quad \text{bounded} \quad \text{as } x \to \pm\infty,$$

which is a singular because of the infinite interval, and the first-order ordinary differential equation

$$T'(t) + \alpha k T(t) = 0, \quad t > 0.$$

The equation in $X(x)$ has the general solution

$$\phi(x) = c_1 \cos \alpha x + c_2 \sin \alpha x,$$

which is bounded for any choice of the constants and for any value of α. On the other hand, the solution of the equation in $T(t)$ is

$$T(t) = e^{-\alpha^2 kt}.$$

Thus, for any value of α^2, the function

$$u(x,t;\alpha) = \left(c_1 \cos \alpha x + c_2 \sin \alpha x\right) e^{-\alpha^2 kt}.$$

Since the parameter α may take on any value, we must use an integral, which is the continuous analogue of a sum or series, in order to include all possibilities. Thus the solution $u(x,t)$ should have the form

$$u(x,t) = \int_0^\infty [A(\alpha) \cos \alpha x + B(\alpha) \sin \alpha x] \, e^{-\alpha^2 kt} \, d\alpha. \tag{7.21}$$

At time $t = 0$, in (7.21) the exponential factor becomes 1, and the initial condition is

$$\int_0^\infty [A(\alpha) \cos \alpha x + B(\alpha) \sin \alpha x] \, d\alpha = \phi(x), \quad -\infty < x < \infty.$$

As this is clearly a Fourier integral problem, we must choose $A(\alpha)$ and $B(\alpha)$ to be the Fourier integral coefficient functions,

$$A(\alpha) = \frac{1}{\pi} \int_{-\infty}^\infty \phi(x) \cos \alpha x \, dx, \quad B(\alpha) = \frac{1}{\pi} \int_{-\infty}^\infty \phi(x) \sin \alpha x \, dx. \tag{7.22}$$

Then the function $u(x,t)$ in (7.21) satisfies the partial differential equation given in (7.20). The Fourier integral coefficients $A(\alpha)$ and $B(\alpha)$ are well defined provided that ϕ is sectionally smooth and $\int_{-\infty}^\infty |\phi(x)| \, dx < \infty$. It can be proved that the boundedness condition, $|u(x,t)| \le M$, is also satisfied, provided that the initial value $\phi(x)$ is bounded as $x \to \pm\infty$. Suppose that the initial function ϕ is given by

$$\phi(x) = \begin{cases} 1, & 0 < x < 1 \\ 0, & -\infty < x < 0 \end{cases}$$

Then ϕ satisfies the assumptions of the Theorem 7.2 and from Corollary 10 we have that

$$A(\alpha) = \frac{1}{\pi}\int_0^1 \cos\,\alpha x dx = \frac{\sin\,\alpha}{\pi\alpha}.$$

Similarly,

$$B(\alpha) = \frac{1}{\pi}\int_0^1 \sin\,\alpha x dx = \frac{-1+\cos\,\alpha}{\pi\alpha}.$$

Substituting these coefficients into (7.21) yields the solution

$$u(x,t) = \frac{1}{\pi}\int_0^\infty \left[\frac{\sin\,\alpha}{\alpha}\cos\alpha x + \frac{-1+\cos\,\alpha}{\alpha}\sin\alpha x\right] e^{-\alpha^2 kt}\,d\alpha. \tag{7.23}$$

Now, we explore alternative representations of the solution to (7.20). Specifically, when the initial function ϕ is constant, the solution can be expressed in terms of the error function, which we define below along with other necessary preliminaries.

Definition 7.2 *The error function is the following improper integral considered as a real function* $erf : \mathbb{R} \to \mathbb{R}$, *such that*

$$\text{erf}(x) = \frac{2}{\sqrt{\pi}}\int_0^x e^{-z^2}dz,$$

where exponential is the real exponential function. Moreover,

$$\int_a^b e^{-z^2}dz = \frac{\sqrt{\pi}}{2}\left[\text{erf}(b) - \text{erf}(a)\right].$$

In addition, the complementary error function,

$$\text{erfc}(x) = 1 - \text{erf}(x) = \frac{2}{\sqrt{\pi}}\int_x^0 e^{-z^2}dz.$$

Expressing the solution of heat flow in an infinite rod using the error function offers several advantages. The erf function naturally arises in the solution of the heat equation with a point or step-like initial condition, simplifying the mathematical representation of temperature distribution. It avoids the need for complex integral evaluations and provides a closed-form expression that is well-tabulated and efficiently computable. Additionally, the erf function's asymptotic properties help describe heat diffusion behavior at large and small times, making it particularly useful for analyzing transient heat conduction problems.

The following two lemmas are crucial for establishing the next result

Lemma 22 *For* $x \in \mathbb{R}$,

$$\int_0^\infty e^{-x^2}dx = \frac{\sqrt{\pi}}{2}.$$

Proof *Let*

$$I = \int_0^\infty e^{-x^2}\,dx.$$

Then, for $y \in \mathbb{R}$,

$$I^2 = I \cdot I = \int_0^\infty e^{-x^2}\,dx \int_0^\infty e^{-y^2}\,dy = \int_0^\infty \int_0^\infty e^{-(x^2+y^2)}\,dxdy.$$

We switch to polar coordinates by letting

$$x = r\cos(\theta), \quad y = r\sin(\theta).$$

Then,

$$\begin{aligned} I^2 &= \int_0^{\pi/2}\int_0^\infty e^{-r^2} r dr\, d\theta = \int_0^{\pi/2}\left(\int_0^\infty e^{-r^2} r dr\right) d\theta \\ &= \int_0^{\pi/2}\left(-\frac{1}{2}e^{-r^2}\Big|_0^\infty\right) d\theta = \int_0^{\pi/2}\frac{1}{2}\,d\theta = \frac{\pi}{4}. \end{aligned}$$

Taking the square root on both sides we arrive at

$$I = \int_0^\infty e^{-x^2}\,dx = \frac{\sqrt{\pi}}{2}.$$

This completes the proof. □

Lemma 23 *For a positive constant a, we have*

$$\int_0^\infty e^{-\alpha^2 a}\cos\alpha b\, d\alpha = \frac{1}{2}\sqrt{\frac{\pi}{a}}\, e^{-\frac{b^2}{4a}}.$$

Proof *Set*

$$y(x) = \int_0^\infty e^{-\alpha^2 a}\cos\alpha x\, d\alpha. \tag{7.24}$$

Then

$$y'(x) = -\int_0^\infty \alpha e^{-\alpha^2 a}\sin\alpha x\, d\alpha.$$

Performing an integration by parts with

$$u = \sin\alpha x, \quad dv = -\alpha e^{-\alpha^2 a} d\alpha$$

yields

$$\begin{aligned} y'(x) &= \frac{\sin\alpha x\ e^{-\alpha^2 a}}{2\alpha}\Big|_{\alpha=0}^\infty - \frac{x}{2a}\int_0^\infty e^{-\alpha^2 a}\cos\alpha x\, d\alpha \\ &= -\frac{x}{2a}\int_0^\infty e^{-\alpha^2 a}\cos\alpha x\, d\alpha = -\frac{x}{2a}y(x), \end{aligned}$$

which is a first-order differential equation that we can easily solve by separating the variables. That is,

$$\frac{y'(x)}{y(x)} = -\frac{x}{2a},$$

whose general solution is $\ln|y(x)| = -\frac{x^2}{4a} + K$, *for a constant* K. *Taking exponential on both sides leads to*

$$y(x) = Ce^{-\frac{x^2}{4a}}.$$

Our next task is to determine the constant C. *From* (7.24), *we have*

$$y(0) = \int_0^\infty e^{-\alpha^2 a}\, d\alpha,$$

which can be rewritten as

$$y(0) = \frac{1}{\sqrt{a}} \int_0^\infty e^{-u^2}\, du = \frac{1}{2}\sqrt{\frac{\pi}{a}},$$

where we made the substitution $u = \alpha\sqrt{a}$ *and then applied Lemma 22. Thus,*

$$C = \frac{1}{2}\sqrt{\frac{\pi}{a}},$$

and consequently, we arrive at

$$y(x) = \int_0^\infty e^{-\alpha^2 a} \cos\alpha x\, d\alpha = \frac{1}{2}\sqrt{\frac{\pi}{a}} e^{-\frac{x^2}{4a}}.$$

Replacing x *with* b *gives the desired result.* □

Substituting the coefficients $A(\alpha)$ and $B(\alpha)$, given by (7.22), into (7.21), and reversing the order of integration, we obtain

$$\begin{aligned} u(x,t) &= \frac{1}{\pi}\int_0^\infty e^{-\alpha^2 kt} \int_{-\infty}^\infty \phi(s)\cos\alpha(s-x)\, ds\, d\alpha \\ &= \frac{1}{\pi}\int_{-\infty}^\infty \phi(s)\left[\int_0^\infty e^{-\alpha^2 kt}\cos\alpha(s-x)\, d\alpha\right] ds. \end{aligned} \tag{7.25}$$

Utilizing Lemma 23 with $a = kt$ and $x \to s - x$ we arrive at

$$\int_0^\infty e^{-\alpha^2 kt}\cos\alpha(s-x)\, d\alpha = \frac{1}{2}\sqrt{\frac{\pi}{kt}} e^{-\frac{(s-x)^2}{4kt}}. \tag{7.26}$$

Finally, substituting (7.26) into (7.25) we have

$$u(x,t) = \frac{1}{2\sqrt{\pi kt}} \int_{-\infty}^\infty \phi(s) e^{-\frac{(s-x)^2}{4kt}}\, ds. \tag{7.27}$$

By setting $\sigma = \frac{(s-x)}{\sqrt{4kt}}$, (7.27) may be written in the alternative form

$$u(x,t) = \frac{1}{\sqrt{\pi}} \int_{-\infty}^{\infty} \phi(x + 2\sigma\sqrt{kt}) e^{-\sigma^2} \, d\sigma. \tag{7.28}$$

• **Constant initial function**

Now suppose that the initial function $\phi(x) = u_0$, where u_0 is a positive constant. That is,

$$u(x,0) = u_0.$$

Thus, substituting u_0 for ϕ into (7.27) and using the fact that the exponential term is even, we obtain

$$u(x,t) = \frac{u_0}{\sqrt{\pi kt}} \int_0^{\infty} e^{-\frac{(s-x)^2}{4kt}} \, ds. \tag{7.29}$$

Setting $z = \frac{(s-x)}{\sqrt{4kt}}$, then (7.29) becomes

$$u(x,t) = \frac{2u_0}{\sqrt{\pi}} \int_{\frac{-x}{\sqrt{4kt}}}^{\infty} e^{-z^2} \, dz. \tag{7.30}$$

Notice that (7.30) is equivalent to

$$\begin{aligned} u(x,t) &= \frac{2u_0}{\sqrt{\pi}} \left[\int_{\frac{-x}{\sqrt{4kt}}}^{0} e^{-z^2} dz + \int_0^{\infty} e^{-z^2} dz \right] \\ &= \frac{2u_0}{\sqrt{\pi}} \left[\int_0^{\frac{x}{\sqrt{4kt}}} e^{-z^2} dz + \frac{\sqrt{\pi}}{2} \right] \\ &= u_0 \,\mathrm{erf}\left(\frac{x}{2\sqrt{kt}} \right) + u_0. \end{aligned}$$

Notice that

$$\lim_{x \to \infty} u(x,t) = u_0 \lim_{x \to \infty} \mathrm{erf}\left(\frac{x}{2\sqrt{kt}} \right) + u_0 = 2u_0$$

since

$$\lim_{x \to \infty} \mathrm{erf}\left(\frac{x}{2\sqrt{kt}} \right) = 1.$$

7.4.1 Exercises

Exercise 7.40 *By differentiating (7.23) show it is a solution to the first equation in system (7.20). Additionally, show that the solution given by (7.23) is bounded as $x \to \pm\infty$.*

Exercise 7.41 *Verify that the solution*

$$u(x,t) = \mathrm{erf}\left(\frac{x}{\sqrt{4kt}} \right)$$

satisfies the initial condition

$$u(x,0)=\begin{cases}1, & x>0\\ -1, & x<0\end{cases}$$

Exercise 7.42 *Find the solution of system (7.20) by assuming the initial function*

$$\phi(x)=e^{-b|x|},\ \textit{for a positive constant } b.$$

Exercise 7.43 *Show the heat flow problem*

$$\begin{cases} u_t-4u_{xx}=0, & -\infty<x<\infty,\ t>0\\ u(x,0)=\phi(x), & -\infty<x<\infty\end{cases}$$

with initial data

$$\phi(x)=\begin{cases}1-|x|, & |x|<1\\ 0, & |x|\geq 1\end{cases}$$

has the solution

$$\begin{aligned} u(x,t) &= \frac{1}{2}(1+x)\Big(\operatorname{erf}\Big(\frac{l+x}{\sqrt{16t}}\Big)-\operatorname{erf}\Big(\frac{x}{\sqrt{16t}}\Big)\Big)\\ &+ \frac{1}{2}(1-x)\Big(\operatorname{erf}\Big(\frac{x}{\sqrt{16t}}\Big)-\operatorname{erf}\Big(\frac{x-1}{\sqrt{16}}\Big)\Big)\\ &+ \frac{2\sqrt{t}}{\sqrt{\pi}}\Big(e^{-\frac{(x+1)^2}{16t}}-e^{-\frac{x^2}{16t}}\Big)+\frac{2\sqrt{t}}{\sqrt{\pi}}\Big(e^{-\frac{(x-1)^2}{16t}}-e^{-\frac{x^2}{16t}}\Big).\end{aligned}$$

Exercise 7.44 *Show the heat flow problem*

$$\begin{cases} u_t-ku_{xx}=0, & -\infty<x<\infty,\ t>0\\ u(x,0)=\operatorname{sgn}(x), & -\infty<x<\infty\end{cases}$$

has the solution

$$u(x,t)=\operatorname{erf}\Big(\frac{x}{\sqrt{4kt}}\Big).$$

Exercise 7.45 *For positive constants u_0 and u_1, solve the heat flow problem*

$$\begin{cases} u_t-ku_{xx}=0, & -\infty<x<\infty,\ t>0\\ u(x,0)=\phi(x), & -\infty<x<\infty\end{cases}$$

with initial data

$$\phi(x)=\begin{cases}u_0, & x<0\\ u_1, & x>0\end{cases}$$

Exercise 7.46 *(Heat flow in semi-infinite rod) Consider the heat flow problem on semi-infinite rod*

$$\begin{cases} u_t-ku_{xx}=0, & 0<x<\infty,\ t>0\\ u(0,t)=0, & 0<x<\infty\\ u(x,0)=\phi(x), & 0<x<\infty\end{cases}$$

(a) Show that the solution is given by

$$u(x,t) = \frac{2}{\pi}\int_0^\infty e^{-\alpha^2 kt}\sin\alpha x\int_0^\infty \phi(s)\sin\alpha s\,ds\,d\alpha.$$

(b) Show that the solution in part (a) can be written in the form

$$u(x,t) = \frac{1}{\sqrt{\pi}}\int_{-\frac{x}{2\sqrt{kt}}}^{\infty}\phi(x+2\sigma\sqrt{kt})e^{-\sigma^2}\,d\sigma - \frac{1}{\sqrt{\pi}}\int_{\frac{x}{2\sqrt{kt}}}^{\infty}\phi(-x+2\sigma\sqrt{kt})e^{-\sigma^2}\,d\sigma.$$

(c) If $f(x) = u_0$, where u_0 is a constant, show that it follows from part (b) that

$$u(x,t) = u_0\frac{1}{\sqrt{\pi}}\int_{-\frac{x}{2\sqrt{kt}}}^{\infty} e^{-\sigma^2}\,d\sigma - \frac{1}{\sqrt{\pi}}\int_{\frac{x}{2\sqrt{kt}}}^{\infty} e^{-\sigma^2}\,d\sigma.$$

(d) In terms of the error function erf, *show that the solution in part (c) can be written as*

$$u(x,t) = u_0\,\mathrm{erf}\left(\frac{x}{2\sqrt{kt}}\right).$$

Exercise 7.47 *Consider the heat flow problem on semi-infinite rod*

$$\begin{cases} u_t - ku_{xx} = 0, & 0 < x < \infty,\ t > 0 \\ u(0,t) = 0, & 0 < x < \infty \\ u(x,0) = \phi(x), & 0 < x < \infty \end{cases}$$

where

$$\phi(x) = \begin{cases} u_0, & 0 < x < l \\ 0, & x > l \end{cases}$$

Show that the solution is given by

$$u(x,t) = \frac{2u_0}{\pi}\int_0^\infty \frac{1-\cos\alpha l}{\alpha}\sin\alpha x e^{-\alpha^2 kt}\,d\alpha.$$

Exercise 7.48 *Consider the heat problem with Dirichlet boundary condition*

$$\begin{cases} v_t - kv_{xx} = 0, & 0 < x < \infty,\ t > 0 \\ v(x,0) = \phi(x), & x > 0 \\ v(0,t) = 0, & t > 0 \end{cases}$$

and initial data

$$\phi(x) = \begin{cases} 0, & 0 < x < l \\ 1, & x > l \end{cases}$$

Show its solution $u(x,t)$ can be written as

$$v(x,t) = \frac{1}{2}\mathrm{erf}\left(\frac{l+x}{\sqrt{4kt}}\right) - \frac{1}{2}\mathrm{erf}\left(\frac{l-x}{\sqrt{4kt}}\right).$$

Exercise 7.49 *Consider the heat problem with Neumann boundary condition*

$$\begin{cases} v_t - kv_{xx} = 0, & 0 < x < \infty,\ t > 0 \\ v(x,0) = \phi(x), & x > 0 \\ v_x(0,t) = 0, & t > 0 \end{cases}$$

(a) Show that the solution is of the form

$$v(x,t) = \frac{1}{\sqrt{\pi}} \int_{-x/(2\sqrt{kt})}^{\infty} \phi(x + 2\sigma\sqrt{kt})e^{-\sigma^2}\, d\sigma + \frac{1}{\sqrt{\pi}} \int_{x/(2\sqrt{kt})}^{\infty} \phi(-x + 2\sigma\sqrt{kt})e^{-\sigma^2}\, d\sigma.$$

(b) Let the initial function be given by

$$\phi(x) = \begin{cases} 0, & 0 < x < l \\ 1, & x > l \end{cases}$$

Show that its solution $v(x,t)$ can be written as

$$v(x,t) = \frac{1}{2}\operatorname{erf}\left(\frac{l+x}{2\sqrt{kt}}\right) + \frac{1}{2}\operatorname{erf}\left(\frac{l-x}{2\sqrt{kt}}\right).$$

Exercise 7.50 *For a positive constant u_0, consider the heat flow problem on semi-infinite rod with a non-homogeneous boundary condition*

$$\begin{cases} u_t - ku_{xx} = 0, & 0 < x < \infty,\ t > 0 \\ u(0,t) = u_0, & 0 < x < \infty, \\ u(x,0) = 0, & 0 < x < \infty, \end{cases}$$

Show that its solution is

$$u(x,t) = \frac{2u_0}{\sqrt{\pi}} \int_{\frac{x}{2\sqrt{kt}}}^{\infty} e^{-\sigma^2}\, d\sigma.$$

Exercise 7.51 *(Navier–Stokes equations) The Navier–Stokes equations simplify to the one-dimensional unsteady diffusion equation with constant velocity v_0*

$$\frac{\partial u}{\partial t} = \nu \frac{\partial^2 u}{\partial y^2}, \quad 0 < y < \infty, \quad t > 0$$

where $u(y,t)$ is the velocity of the fluid in the x-direction, and ν is the kinematic viscosity. The initial condition

$$u(y,0) = 0$$

assumes the fluid is initially at rest and at the wall ($y = 0$), the fluid moves with the plate

$$u(0,t) = v_0.$$

At large distances from the plate ($y \to \infty$), the fluid remains undisturbed. That is

$$u(\infty,t) = 0.$$

Introduce the self-similar variables

$$\eta = \frac{y}{\sqrt{\nu t}}, \quad f(\eta) = \frac{u}{v_0},$$

and obtain the solution

$$u(y,t) = v_0 \operatorname{erfc}\left(\frac{y}{\sqrt{4\nu t}}\right).$$

7.5 Wave Motion for Infinite String

Consider the wave motion of infinite string that is stretched over a real line. The vibration is governed by

$$\begin{cases} u_{tt} - c^2 u_{xx} = 0, & -\infty < x < \infty, \ t > 0 \\ u(x,0) = f(x), & -\infty < x < \infty \\ \left|u(x,t)\right| \leq M, & -\infty < x < \infty, \ t > 0 \\ u_t(x,0) = g(x), & -\infty < x < \infty \end{cases} \tag{7.31}$$

We seek a nontrivial, bounded solution. Hence, we assume that the solution $u(x,t)$ can be expressed as the product of two functions, one depending on x and the other on t, such that

$$u(x,t) = X(x)T(t).$$

After differentiating and separating the variables, we obtain the two equations

$$X''(x) + \lambda X(x) = 0, \tag{7.32}$$

and

$$T''(t) + \lambda c^2 T(t) = 0. \tag{7.33}$$

If $\lambda = 0$, then

$$X''(x) = 0 \Rightarrow X(x) = A + Bx$$

which is an unbounded solution on the given domain, unless $A = 0$. Thus, the solution is

$$X(x) = B$$

for the eigenvalue.

Case 2: If $\lambda < 0$, let

$$\lambda^2 = -\alpha^2 \quad \text{with} \quad \alpha > 0.$$

Then,

$$X'' - \alpha^2 X = 0,$$

whose solution is

$$X(x) = Ae^{\alpha x} + Be^{-\alpha x}.$$

Since $\alpha > 0$, the first term on the right-hand side, $Ae^{\alpha x}$, is unbounded in the domain $[0,\infty)$, and the second term, $Be^{-\alpha x}$, is unbounded in the region $(-\infty,0)$. Therefore, for a bounded solution, we must assume that $A = 0$ and $B = 0$, leading to the trivial solution.

Case 3: If $\lambda > 0$, let

$$\lambda^2 = \alpha^2 \quad \text{with} \quad \alpha > 0.$$

Then,

$$X'' + \alpha^2 X = 0$$

with the solution

$$X(x) = a_1 \cos \alpha x + b_1 \sin \alpha x.$$

Notice that the function $X(x)$ is always bounded for every $\alpha > 0$. Now the problem for T is

$$T'' + c^2 \lambda T = 0.$$

When $\lambda^2 = \alpha^2 > 0$, we have

$$T'' + c^2 \alpha^2 T = 0,$$

that has the solution

$$T(t) = a_2 \cos c\alpha t + b_2 \sin c\alpha t.$$

To capture all the values of λ we integrate. Thus, after adjusting the coefficients, we have by using the superposition principle, that

$$u(x,t) = \int_{-\infty}^{\infty} [a_1 \cos \alpha x + b_1 \sin \alpha x]\,[a_2 \cos c\alpha t + b_2 \sin c\alpha t]\, d\alpha. \tag{7.34}$$

Note that the solution given by (7.34) incorporates the solution for $\lambda = 0$. Multiplying the two factors in the integrand of (7.34) and adjusting the constants, we arrive at

$$\begin{aligned} u(x,t) &= \int_{-\infty}^{\infty} \Big[A(\alpha) \cos \alpha x \cos c\alpha t + B(\alpha) \cos \alpha x \sin c\alpha t \\ &+ C(\alpha) \sin \alpha x \cos c\alpha t + D(\alpha) \sin \alpha x \sin c\alpha t \Big]\, d\alpha. \end{aligned}$$

Applying the initial displacement condition we have

$$u(x,0) = f(x) = \int_{-\infty}^{\infty} [A(\alpha) \cos \alpha x + C(\alpha) \sin \alpha x]\, d\alpha,$$

where

$$A(\alpha) = \frac{1}{\pi} \int_{-\infty}^{\infty} f(\xi) \cos \alpha \xi \, d\xi, \quad C(\alpha) = \frac{1}{\pi} \int_{-\infty}^{\infty} f(\xi) \sin \alpha \xi \, d\xi.$$

On the other hand, if we apply the initial velocity condition, we obtain

$$u_t(x,0) = g(x) = \int_{-\infty}^{\infty} [c\alpha B(\alpha)\cos\alpha x + c\alpha D(\alpha)\sin\alpha x]\,d\alpha,$$

where

$$B(\alpha) = \frac{1}{c\alpha\pi}\int_{-\infty}^{\infty} g(\xi)\cos\alpha\xi\,d\xi, \quad D(\alpha) = \frac{1}{c\alpha\pi}\int_{-\infty}^{\infty} g(\xi)\sin\alpha\xi\,d\xi.$$

7.5.1 Exercises

Exercise 7.52 *Find the solution of*

$$\begin{cases} u_{tt} - c^2 u_{xx} = 0, & -\infty < x < \infty, \ t > 0 \\ u(x,0) = f(x), & -\infty < x < \infty \\ |u(x,t)| \leq M, & \text{as } x \to \pm\infty, \ t > 0 \\ u_t(x,0) = 0, & -\infty < x < \infty \end{cases}$$

where

$$f(x) = \begin{cases} 0, & x < -0.5 \\ 1, & -0.5 \leq x \leq 0.5 \\ 0, & x > 0.5 \end{cases}$$

Exercise 7.53 *(Wave motion for a semi-infinite string) Find the solution of*

$$\begin{cases} u_{tt} - c^2 u_{xx} = 0, & x > 0, \ t > 0 \\ u(0,t) = 0, & x > 0, \ t > 0 \\ u(x,0) = f(x), & x > 0 \\ |u(x,t)| \leq M, & x > 0, \ t > 0 \\ u_t(x,0) = g(x), & x > 0 \end{cases}$$

Exercise 7.54 *Find the solution of*

$$\begin{cases} u_{tt} - c^2 u_{xx} = 0, & x > 0, \ t > 0 \\ u(0,t) = 0, & t > 0 \\ u(x,0) = f(x), & x > 0 \\ |u(x,t)| \leq M, & x > 0, \ t > 0 \\ u_t(x,0) = 0, & x > 0 \end{cases}$$

where

$$f(x) = \begin{cases} x(1-x), & 0 \leq x \leq 1 \\ 0, & x > 1 \end{cases}$$

Exercise 7.55 *Show that the solution $u(x,y)$ for the steady heat flow problem with insulated boundary conditions*

$$\begin{cases} u_{xx}(x,y)+u_{yy}(x,y)=0, & x>0, \quad 0<y<1 \\ u_x(0,y)=0, \quad u_y(x,0)=0, & \\ |u(x,y)| \le M, & x>0, \quad 0<y<1 \\ u(x,1)=e^{-x} & \end{cases}$$

is given by

$$u(x,y)=\frac{2}{\pi}\int_0^{\infty}\frac{\cos\alpha x\cosh\alpha y}{(\alpha^2+1)\cosh\alpha}\,d\alpha.$$

Exercise 7.56 *Show that the solution $u(x,y)$ for the steady heat flow problem with insulated boundary conditions*

$$\begin{cases} u_{xx}(x,y)+u_{yy}(x,y)=0, & x>0, \quad 0<y<1 \\ u_x(0,y)=0, \quad u_y(x,1)=-u(x,1) & \\ |u(x,y)| \le M, & x>0, \quad 0<y<1 \\ u(x,0)=\begin{cases} 1, & 0<x<1 \\ 0, & x>1 \end{cases} & \end{cases}$$

is given by

$$u(x,y)=\frac{2}{\pi}\int_0^{\infty}\frac{\alpha\cosh\alpha(1-y)+\sinh\alpha(1-y)}{\alpha^2\cosh\alpha+\alpha\sinh\alpha}\sin\alpha\cos\alpha x\,d\alpha.$$

7.6 Fourier Transforms

Fourier transforms are extremely useful in solving partial differential equations, especially those involving linear systems with constant coefficients. By transforming a PDE from the spatial or temporal domain into the frequency domain, the differential operators convert into algebraic expressions, significantly simplifying the problem. This method is particularly effective for solving the heat equation, wave equation, and Laplace's equation, where boundary conditions can be efficiently handled in the transformed space. After solving the equation in the frequency domain, the inverse Fourier transform is used to retrieve the solution in the original domain. This technique is widely applied in physics, engineering, and mathematical modeling for problems involving diffusion, vibrations, and electromagnetism. Fourier transform decomposes functions into their constituent frequencies, making it essential in fields like signal processing, physics, engineering, and image analysis. It allows us to analyze time-domain signals in the frequency domain, revealing hidden periodicities, filtering noise, and compressing data efficiently. In electrical engineering, it helps design filters and analyze circuits, while in quantum mechanics, it describes wave functions. In medical imaging techniques like MRI and CT scans, Fourier transforms reconstruct images from raw data. Their ability to convert complex signals into simpler sinusoidal components makes them indispensable in both theoretical and applied sciences.

We start by defining the Fourier transform using the integral expression (7.19), which was derived from the complex form of the Fourier integral.

Definition 7.3 (Fourier Transform) *Let $u(x,t)$ be a continuous, piecewise smooth, and absolutely integrable function. Then the* **Fourier transform** *of $u(x,t)$ with respect to $x \in \mathbb{R}$ is denoted by $U(k,t)$ and is defined as*

$$\mathscr{F}\{u(x,t)\} = U(k,t) = \frac{1}{\sqrt{2\pi}} \int_{-\infty}^{\infty} e^{-ikx} u(x,t)\, dx, \tag{7.35}$$

where k is called the Fourier transform variable, and e^{-ikx} is the kernel of the transform.

Then, for all $x \in \mathbb{R}$, the **inverse Fourier transform** *of $U(k,t)$ is given by*

$$\mathscr{F}^{-1}\{U(k,t)\} = u(x,t) = \frac{1}{\sqrt{2\pi}} \int_{-\infty}^{\infty} e^{ikx} U(k,t)\, dk. \tag{7.36}$$

The variable k is referred to as the frequency (or wave number) variable. It is important to note that there is no universally accepted convention for defining the Fourier transform. Some definitions include a prefactor in front of the integral, ours is $\frac{1}{\sqrt{2\pi}}$ while others use a positive exponent in the exponential term or even incorporate a factor of 2π. Physicists and engineers often denote this variable using ξ, ω, or p, representing wave number, frequency, or momentum, respectively. Therefore, when consulting different sources, one must be mindful of these variations, as the formulas may differ depending on the chosen convention. Before we embark on examples, we state and show that the Fourier transform is linear.

Theorem 7.3 *(Linearity) For constants a, b and functions $f(x)$ and $g(x)$, we have*

$$\mathscr{F}\{af(x) + bg(x)\} = a\mathscr{F}\{f(x)\} + b\mathscr{F}\{g(x)\}.$$

Proof *We know that*

$$\mathscr{F}\{f(x)\} = \frac{1}{\sqrt{2\pi}} \int_{-\infty}^{\infty} e^{-ikx} f(x)\, dx.$$

Then, for any constants a and b, we have

$$\begin{aligned}
\mathscr{F}\{af(x) + bg(x)\} &= \frac{1}{\sqrt{2\pi}} \int_{-\infty}^{\infty} [af(x) + bg(x)] e^{-ikx}\, dx, \\
&= a\frac{1}{\sqrt{2\pi}} \int_{-\infty}^{\infty} f(x) e^{-ikx}\, dx + b\frac{1}{\sqrt{2\pi}} \int_{-\infty}^{\infty} g(x) e^{-ikx}\, dx, \\
&= a\mathscr{F}\{f(x)\} + b\mathscr{F}\{g(x)\}.
\end{aligned}$$

□

Example 7.5 Let $f(x) = e^{-a|x|}$, $a > 0$. Then, using Definition 7.3 we have

$$\mathscr{F}\{e^{-a|x|}\} = \frac{1}{\sqrt{2\pi}} \int_{-\infty}^{\infty} e^{-a|x|-ikx}\, dx$$

$$= \frac{1}{\sqrt{2\pi}} \left(\int_{-\infty}^{0} e^{(a-ik)x}\, dx + \int_{0}^{\infty} e^{-(a+ik)x}\, dx \right)$$

$$= \frac{1}{\sqrt{2\pi}} \left(\frac{1}{a-ik} + \frac{1}{a+ik} \right) = \sqrt{\frac{2}{\pi}} \frac{a}{a^2+k^2}.$$

Example 7.6 Let $f(x) = e^{-a\frac{x^2}{2}}$, $a > 0$. By Definition 7.3 we have

$$\mathscr{F}\{e^{-a\frac{x^2}{2}}\} = \frac{1}{\sqrt{2\pi}} \int_{-\infty}^{\infty} e^{-ikx-a\frac{x^2}{2}}\, dx.$$

Now we complete the square.

$$\begin{aligned}
-\frac{a}{2}x^2 - ikx &= -\frac{a}{2}\left[x^2 + \frac{2ik}{a}x\right] \\
&= -\frac{a}{2}\left[x^2 + \frac{2ik}{a}x + \left(\frac{ik}{a}\right)^2 - \left(\frac{ik}{a}\right)^2\right] \\
&= -\frac{a}{2}\left(x + \frac{ik}{a}\right)^2 - \frac{k^2}{2a}.
\end{aligned}$$

We now put this expression into the integral and make the substitutions

$$y = x + \frac{ik}{a}, \quad \beta = \frac{a}{2}.$$

$$\begin{aligned}
\mathscr{F}\{e^{-a\frac{x^2}{2}}\} &= \frac{1}{\sqrt{2\pi}} \int_{-\infty}^{\infty} e^{-ax^2/2-ikx} dx \\
&= \frac{1}{\sqrt{2\pi}} e^{-\frac{k^2}{2a}} \int_{-\infty}^{\infty} e^{-\frac{a}{2}\left(x+\frac{ik}{a}\right)^2} dx \\
&= \frac{1}{\sqrt{2\pi}} e^{-\frac{k^2}{2a}} \int_{-\infty+\frac{ik}{a}}^{\infty+\frac{ik}{a}} e^{-\beta y^2} dy.
\end{aligned}$$

One might be tempted to absorb the term $\frac{ik}{a}$ into the limits of integration. However, since the integration occurs over a contour in the complex plane, we can deform this horizontal contour into one along the real axis without crossing any singularities of the integrand. Thus, we can safely write

$$\mathscr{F}\{e^{-a\frac{x^2}{2}}\} = \frac{1}{\sqrt{2\pi}} e^{-\frac{k^2}{4a}} \int_{-\infty}^{\infty} e^{-\beta y^2}\, dy$$

$$= \frac{1}{\sqrt{2\pi}} e^{-\frac{k^2}{4a}} \sqrt{\frac{\pi}{\beta}} = \frac{1}{\sqrt{a}} e^{-\frac{k^2}{2a}},$$

where we have used (see Lemma 23)

$$\int_{-\infty}^{\infty} e^{-\beta y^2}\, dy = \sqrt{\frac{\pi}{\beta}}.$$

Thus, for $a = 1$, we have, in particular, that

$$\mathscr{F}\{e^{-\frac{x^2}{2}}\} = e^{-\frac{k^2}{2}}.$$

We have the following definitions regarding Fourier cosine and sine transforms of $f(x)$.

Definition 7.4 *Let $f(x)$ be defined for $0 \le x < \infty$ and extended as an even function in $(-\infty, \infty)$, satisfying the conditions of Theorem 7.1. Then, at the points of continuity,*

(a) the Fourier cosine transform of $f(x)$ and its inverse transform are defined by

$$\mathscr{F}_c\{f(x)\} = F_c(k) = \sqrt{\frac{2}{\pi}} \int_0^{\infty} \cos(kx) f(x)\, dx,$$

$$\mathscr{F}_c^{-1}\{F_c(k)\} = f(x) = \sqrt{\frac{2}{\pi}} \int_0^{\infty} \cos(kx) F_c(k)\, dk,$$

where $\mathscr{F}_c$ is the Fourier cosine transformation and $\mathscr{F}_c^{-1}$ is its inverse transformation.

(b) Similarly, if $f(x)$ is defined for $0 \le x < \infty$ and extended as an odd function in $(-\infty, \infty)$ and satisfying the conditions of Theorem 7.1, then its Fourier sine transform and its inverse are defined by

$$\mathscr{F}_s\{f(x)\} = F_s(k) = \sqrt{\frac{2}{\pi}} \int_0^{\infty} \sin(kx) f(x)\, dx,$$

$$\mathscr{F}_s^{-1}\{F_s(k)\} = f(x) = \sqrt{\frac{2}{\pi}} \int_0^{\infty} \sin(kx) F_s(k)\, dk,$$

where $\mathscr{F}_s$ is called the Fourier sine transformation and $\mathscr{F}_s^{-1}$ is its inverse.

Example 7.7 We are interested in finding the Fourier cosine transform of the function

$$f(x) = \begin{cases} \cos x, & 0 < x < a \\ 0, & x > a \end{cases}$$

Using the definition, we have

$$F_c(k) = \sqrt{\frac{2}{\pi}} \int_0^{\infty} f(x) \cos kx\, dx$$

$$
\begin{aligned}
&= \sqrt{\frac{2}{\pi}}\int_0^a \cos x \cos kx\,dx + \sqrt{\frac{2}{\pi}}\int_a^\infty 0\cdot \cos kx\,dx \\
&= \sqrt{\frac{2}{\pi}}\frac{1}{2}\int_0^a [\cos(k+1)x + \cos(k-1)x]\,dx \\
&= \frac{1}{\sqrt{2\pi}}\left[\frac{\sin(k+1)a}{k+1} + \frac{\sin(k-1)a}{k-1}\right].
\end{aligned}
$$

□

In the following example, we apply the Fourier cosine transform to evaluate a specific integral.

Example 7.8 Consider the function

$$
f(x) = \begin{cases} 1-x^2, & |x| \le 1 \\ 0, & |x| > 1 \end{cases}
$$

Since $f(x)$ is even, its Fourier transform reduces to the Fourier cosine transform. Thus

$$
\begin{aligned}
\mathscr{F}_c\{f(x)\} = F_c(k) &= \sqrt{\frac{2}{\pi}}\int_0^\infty f(x)\cos kax\,dx = \sqrt{\frac{2}{\pi}}\int_0^1 (1-x^2)\cos kx\,dx \\
&= \sqrt{\frac{2}{\pi}}\left[(1-x^2)\frac{\sin kx}{k} - 2x\frac{\cos kx}{k^2} + 2\frac{\sin kx}{k^3}\right]_0^1 \\
&= \frac{2\sqrt{2}}{\sqrt{\pi}}\frac{\sin k - k\cos k}{k^3}.
\end{aligned}
$$

Then, by part (a) of Definition 7.4, we have that at any point x where f is continuous,

$$
\begin{aligned}
f(x) = \sqrt{\frac{2}{\pi}}\int_0^\infty \cos kx F_c(k)\,dk &= \sqrt{\frac{2}{\pi}}\int_0^\infty \frac{2\sqrt{2}}{\sqrt{\pi}}\frac{\sin k - k\cos k}{k^3}\cos(kx)\,dk \\
&= \frac{4}{\pi}\int_0^\infty \frac{\sin k - k\cos k}{k^3}\cos kx\,dk.
\end{aligned}
$$

Consequently,

$$
\int_0^\infty \frac{\sin k - k\cos k}{k^3}\cos kx\,dk = \frac{\pi}{4}f(x) = \frac{\pi}{4}\begin{cases} 1-x^2, & |x| \le 1 \\ 0, & |x| > 1 \end{cases} \tag{7.37}
$$

Setting $x = \frac{1}{2}$ in both sides of (7.37) and then replacing k with x, we obtain

$$
\int_0^\infty \frac{\sin x - x\cos x}{x^3}\cos\frac{x}{2}\,dx = \frac{\pi}{4}\left(1-\frac{1}{4}\right) = \frac{3\pi}{16}.
$$

Or,

$$
\int_0^\infty \frac{x\cos x - \sin x}{x^3}\cos\frac{x}{2}\,dx = -\frac{3\pi}{16}.
$$

□

Example 7.9 Show that, for $x > 0$,

$$\int_0^\infty \frac{\cos kx}{k^2+a^2}\,dk = \frac{\pi}{2a}e^{-ax}, \quad a > 0.$$

Let $f(x) = \frac{\pi}{2a}e^{-ax}$, where $a > 0$ and $x > 0$. Then, taking the Fourier cosine transform of $f(x)$, we have

$$\begin{aligned} F_c(k) = \sqrt{\frac{2}{\pi}}\int_0^\infty f(x)\cos kx\,dx &= \frac{\pi}{2a}\sqrt{\frac{2}{\pi}}\int_0^\infty e^{-ax}\cos kx\,dx \\ &= \frac{1}{a}\sqrt{\frac{\pi}{2}}\frac{a}{k^2+a^2}. \end{aligned}$$

Taking the inverse Fourier cosine transform, we obtain

$$\begin{aligned} f(x) = \sqrt{\frac{2}{\pi}}\int_0^\infty F_c(k)\cos kx\,dx &= \frac{1}{a}\sqrt{\frac{\pi}{2}}\sqrt{\frac{2}{\pi}}\int_0^\infty \frac{a}{k^2+a^2}\cos kx\,dx \\ &= \int_0^\infty \frac{\cos kx}{k^2+a^2}\,dx. \end{aligned}$$

Therefore,

$$f(x) = \int_0^\infty \frac{\cos kx}{k^2+a^2}\,dk = \frac{\pi}{2a}e^{-ax}, \text{ (see (a) of Exercise 7.60)}.$$

□

We end this section with the following example regarding inverse Fourier transform.

Example 7.10 Find $f(x)$ given that

$$F_c(k) = \sqrt{\frac{2}{\pi}}\frac{e^{-ak}}{k}, \quad a > 0.$$

We apply the rule for inverse sine Fourier transform. That is

$$f(x) = \sqrt{\frac{2}{\pi}}\int_0^\infty F_s(k)\sin kx\,dk = \frac{2}{\pi}\int_0^\infty \frac{e^{-ak}}{k}\sin kx\,dk.$$

Let

$$I(x) = \int_0^\infty \frac{e^{-ak}}{k}\sin kx\,dk.$$

Differentiating $I(x)$ with respect to x and then integrating, we obtain

$$I'(x) = \int_0^\infty e^{-ak}\cos kx\,dk$$

$$= \quad e^{-ak}\frac{-a\cos ax + x\sin ax}{a^2+x^2}\bigg|_0^\infty = \frac{a}{a^2+x^2}.$$

Integrating $I'(x)$ gives

$$I(x) = \int \frac{a}{a^2+x^2}\,dx = \tan^{-1}\frac{x}{a} + C.$$

Setting $x = 0$, and using

$$I(x) = \int_0^\infty \frac{e^{-ak}}{k}\sin(kx)\,dk,$$

we obtain $I(0) = 0$. Thus,

$$0 = \tan^{-1}\frac{0}{a} + C = C,$$

which gives

$$I(x) = \tan^{-1}\frac{x}{a}.$$

Consequently,

$$f(x) = \frac{2}{\pi}I(x) = \frac{2}{\pi}\tan^{-1}\frac{x}{a}.$$

□

7.6.1 Exercises

Exercise 7.57 *Explain why the following functions do not possess a well-defined Fourier transform.*

$$(a)\ f(x) = \sin\frac{1}{x}, \quad (b)\ f(x) = e^x, \quad (c)\ f(x) = \begin{cases} 1, & x \text{ is rational} \\ 0, & x \text{ is irrational} \end{cases}$$

Exercise 7.58 *Prove that*

$$(a)\ \mathscr{F}\{e^{-ax^2}\} = \frac{1}{\sqrt{2a}}e^{-\frac{k^2}{4a}}, \quad a > 0,$$

$$(b)\ \mathscr{F}\{\chi_{[-a,a]}(x)\} = \sqrt{\frac{2}{\pi}}\frac{\sin ak}{k}, \text{ where } \chi_{[-a,a]}(x) = H(a-|x|) = \begin{cases} 1, & |x| < a \\ 0, & |x| > a \end{cases}$$

$$(c)\ \mathscr{F}\{f(x)\} = \frac{1}{\sqrt{2\pi}}\frac{1}{a+ik}, \text{ where } f(x) = \begin{cases} e^{-ax}, & x \geq 0 \\ 0, & x < 0 \end{cases}, \quad a > 0$$

Exercise 7.59 *Find the Fourier transform of the following functions*

$$(a)\ f(x) = \begin{cases} a, & -l < x < 0, \quad a > 0 \\ 0, & \text{otherwise} \end{cases} \qquad (b)\ f(x) = \begin{cases} a, & -l < x < l, \quad a > 0 \\ 0, & \text{otherwise} \end{cases}$$

$$(c)\ f(x) = \begin{cases} a, & -c < x < 0 \\ b, & 0 < x < c, \quad a > 0,\ b > 0 \\ 0, & otherwise \end{cases}$$

Exercise 7.60 *Show that*

$$(a)\ \mathscr{F}_c\{e^{-ax}\} = \sqrt{\frac{2}{\pi}}\frac{a}{a^2+k^2}, \quad a > 0 \quad (b)\ \mathscr{F}_s\{e^{-ax}\} = \sqrt{\frac{2}{\pi}}\frac{k}{a^2+k^2}, \quad a > 0$$

Exercise 7.61 *Find $f(x)$, given that*

$$F_s(k) = \sqrt{\frac{2}{\pi}}\frac{\sin ak}{k}, \quad a \neq 0.$$

$$Ans{:}\,f(x) = \begin{cases} 1, & 0 < x < a, \quad a > 0 \\ 0, & x > a \end{cases}$$

Exercise 7.62 *Find the function $f(x)$ satisfying the integral equation*

$$\int_0^\infty f(x) \sin ax\, dx = \frac{a}{a^2+k^2}.$$

Exercise 7.63 *Show that, for $x > 0$,*

$$\int_0^\infty \frac{k \sin kx}{k^2+a^2}\, dk = \frac{\pi}{2a}e^{ax}, \quad a \leq 0.$$

Exercise 7.64 *Solve the integral equation*

$$\int_0^\infty f(x) \cos kx\, dx = \begin{cases} 1-k, & 0 \leq k \leq 1 \\ 0, & x > 1 \end{cases}$$

and then deduce the identity

$$\int_0^\infty \frac{\sin^2 x}{x^2}\, dx = \frac{\pi}{2}.$$

Exercise 7.65 *Show that*

$$\mathscr{F}_s\left\{\frac{e^{-ax}}{x}\right\} = \sqrt{\frac{2}{\pi}} \tan^{-1}\left(\frac{k}{a}\right),$$

and then evaluate

$$\int_0^\infty \tan^{-1}\left(\frac{k}{a}\right) \sin x\, dx.$$

7.7 Properties of Fourier Transform

The properties of the Fourier Transform describe how it behaves under various mathematical operations. Key properties include linearity, which allows the transform of a sum to be the sum of transforms, and time shifting, which results in a phase shift in the frequency domain. The frequency shifting property states that multiplying by an exponential shifts the spectrum. Scaling affects the spread of the function in both time and frequency domains, while convolution in one domain corresponds to multiplication in the other. Additional properties include Parseval's theorem, which preserves energy, and differentiation, where derivatives in time correspond to polynomial multiplication in frequency.

We begin with a shifting property of Fourier transform.

Theorem 7.4 *(Shifting). Let $\mathscr{F}\{f(x)\}$ be the Fourier transform of $f(x)$. Then, for a real constant c, we have*

$$\mathscr{F}\{f(x-c)\} = e^{-ikc}\mathscr{F}\{f(x)\}.$$

Proof *From Definition 7.3 we have that*

$$\mathscr{F}\{f(x-c)\} = \frac{1}{\sqrt{2\pi}}\int_{-\infty}^{\infty} f(x-c)e^{-ikx}\,dx.$$

Using the substitution $\xi = x-c$, we get

$$\begin{aligned}\mathscr{F}\{f(x-c)\} &= \frac{1}{\sqrt{2\pi}}\int_{-\infty}^{\infty} f(\xi)e^{-ik(\xi+c)}\,d\xi \\ &= e^{-ikc}\left[\frac{1}{\sqrt{2\pi}}\int_{-\infty}^{\infty} f(x)e^{-ik\xi}\,d\xi\right] = e^{-ikc}\mathscr{F}\{f(x)\}.\end{aligned}$$

The proof of the next theorem is left an exercise.

Theorem 7.5 *(Scaling). If $\mathscr{F}$ is the Fourier transform of f, then for nonzero constant c we have that*

$$\mathscr{F}\{f(cx)\} = \frac{1}{|c|}\mathscr{F}\left(\frac{k}{c}\right).$$

We have another theorem that describes how the Fourier transform of a function changes when the function is multiplied by a sinusoidal function. This process is known as modulation.

Theorem 7.6 *(Modulation) Let $F(k)$ be the Fourier transform of $f(x)$. We have the following results.*

(a) $\mathscr{F}\{f(x)\cos ax\} = \frac{1}{2}\left[F(k+a)+F(k-a)\right].$

(b) $\mathscr{F}_s\{f(x)\cos ax\} = \frac{1}{2}\left[F_s(k+a)+F_s(k-a)\right].$

(c) $\mathscr{F}_c\{f(x)\sin ax\} = \frac{1}{2}\left[F_s(k+a)-F_s(k-a)\right].$

(d) $\mathscr{F}_c\{f(x)\cos ax\} = \frac{1}{2}\left[F_c(k+a)+F_c(k-a)\right].$

(e) $\mathscr{F}_s\{f(x)\sin ax\} = \frac{1}{2}\left[F_c(k-a)-F_c(k+a)\right].$

Proof *We will prove parts (a) and (b) and leave the remaining parts as an exercise.*

$$\begin{aligned}
\mathscr{F}\{f(x)\cos ax\} &= \frac{1}{\sqrt{2\pi}}\int_{-\infty}^{\infty} f(x)\cos ax\, e^{-ikx}dx \\
&= \frac{1}{\sqrt{2\pi}}\int_{-\infty}^{\infty} f(x)\frac{e^{iax}+e^{-iax}}{2}e^{-ikx}dx \\
&= \frac{1}{\sqrt{2\pi}}\int_{-\infty}^{\infty} f(x)\frac{e^{-i(k+a)x}+e^{-i(k-a)x}}{2}dx \\
&= \frac{1}{2}\left[F(k+a)+F(k-a)\right].
\end{aligned}$$

This completes the proof of part (a). As for part (b) we have that

$$\begin{aligned}
\mathscr{F}s\{f(x)\cos ax\} &= \sqrt{\frac{2}{\pi}}\int_0^{\infty} f(x)\cos ax \sin kx dx \\
&= \frac{1}{2}\left[\sqrt{\frac{2}{\pi}}\int_0^{\infty} f(x)\left(\sin(k+a)x+\sin(k-a)x\right)dx\right] \\
&= \frac{1}{2}\left[\sqrt{\frac{2}{\pi}}\int_0^{\infty} f(x)\sin(k+a)x dx+\sqrt{\frac{2}{\pi}}\int_0^{\infty} f(x)\sin(k-a)x dx\right] \\
&= \frac{1}{2}\left[F_s(k+a)+F_s(k-a)\right].
\end{aligned}$$

This completes the proof of part (b).

Theorem 7.7 *(Differentiation)*

Let f be continuous and piecewise smooth in $(-\infty,\infty)$. If

$$\lim_{|x|\to\infty} f(x) = 0,$$

and f and f' are absolutely integrable, then

$$\mathscr{F}\{f'(x)\} = ik\mathscr{F}\{f(x)\} = ikF(k).$$

Proof *Using the definition of Fourier transform, we have*

$$\mathscr{F}\{f'(x)\} = \frac{1}{\sqrt{2\pi}} \int_{-\infty}^{\infty} f'(x) e^{-ikx}\, dx.$$

We perform an integration by parts on the previous integral by letting $u = e^{-ikx}$ *and* $dv = f'(x)$. *Then*

$$\mathscr{F}\{f'(x)\} = \frac{1}{\sqrt{2\pi}} \left[f(x)e^{-ikx}\Big|_{-\infty}^{\infty} + ik \int_{-\infty}^{\infty} f(x)e^{-ikx}\, dx \right].$$

Since $f(x) \to 0$ *as* $|x| \to \infty$, *the boundary terms vanish, leaving*

$$\mathscr{F}\{f'(x)\} = ik\mathscr{F}\{f(x)\} = ikF(k).$$

□

This result can be generalized further. If the function f and its first $(n-1)$ derivatives are continuous, and if its nth derivative is piecewise continuous, then the Fourier transform satisfies

$$\mathscr{F}\{f^{(n)}(x)\} = (ik)^n \mathscr{F}\{f(x)\} = (ik)^n F(k), \quad n = 0, 1, 2, \ldots$$

provided that f and all its derivatives up to order n are absolutely integrable. Furthermore, we assume that f and its first $(n-1)$ derivatives vanish as $|x| \to \infty$.

We extend Theorem 7.7 to functions of two variables, as these are most commonly encountered in applications.

Theorem 7.8 *(Differentiation)*

Let $u(x,t)$ *be continuous in both arguments on their respective domains. If*

$$\lim_{|x| \to \infty} u(x,t) = 0,$$

then

$$\mathscr{F}\{\frac{\partial u}{\partial x}\} = ik\mathscr{F}\{u(x,t)\} = ikU(k,t),$$

where $U(k,t)$ *is the Fourier transform of* $u(x,t)$.

Proof *Using the definition of Fourier transform, we have*

$$\mathscr{F}\{\frac{\partial u}{\partial x}\} = \frac{1}{\sqrt{2\pi}} \int_{-\infty}^{\infty} \frac{\partial u}{\partial x} e^{-ikx}\, dx.$$

By applying integration by parts to the above integral, we obtain

$$\mathscr{F}\{\frac{\partial u}{\partial x}\} = \frac{1}{\sqrt{2\pi}} \left[u(x,t)e^{-ikx}\Big|_{-\infty}^{\infty} + ik \int_{-\infty}^{\infty} u(x,t)e^{-ikx}\, dx \right].$$

Since $u(x,t) \to 0$ as $|x| \to \infty$, the boundary terms vanish, leaving

$$\mathscr{F}\{\frac{\partial u}{\partial x}\} = ik\mathscr{F}\{u(x,t)\} = ikU(k,t).$$

□

Similarly, if $u(x,t)$ is continuously differentiable up to order n, and if

$$\frac{\partial^m u}{\partial x^m} \to 0 \quad \text{as} \quad |x| \to \infty, \quad \text{for} \quad m = 1,2,3,\ldots,(n-1),$$

then the Fourier transform satisfies

$$\mathscr{F}\left\{\frac{\partial^n u}{\partial x^n}\right\} = (ik)^n \mathscr{F}\{u(x,t)\} = (ik)^n U(k,t).$$

Additionally, from Definition 7.3, it follows that

$$\begin{aligned} \mathscr{F}\left\{\frac{\partial u}{\partial t}\right\} &= \frac{1}{\sqrt{2\pi}} \int_{-\infty}^{\infty} \frac{\partial u}{\partial t} e^{-ikx}\,dx \\ &= \frac{d}{dt}\left[\frac{1}{\sqrt{2\pi}} \int_{-\infty}^{\infty} ue^{-ikx}\,dx\right] = \frac{dU}{dt}. \end{aligned}$$

In a similar way, it can be shown that

$$\mathscr{F}\left\{\frac{\partial^2 u}{\partial t^2}\right\} = \frac{d^2 U}{dt^2}, \quad \ldots, \quad \mathscr{F}\left\{\frac{\partial^n u}{\partial t^n}\right\} = \frac{d^n U}{dt^n}.$$

The Fourier cosine and sine derivatives of functions are useful for solving heat flow problems in semi-unbounded domains.

Theorem 7.9 (Differentiation Theorem) *Let $f(x)$ be absolutely integrable on $(0,\infty)$. If*

$$\lim_{x\to\infty} f(x) = \lim_{x\to\infty} f'(x) = 0,$$

then

(a)

$$\mathscr{F}_c\{f''(x)\} = -k^2 F_c(k) - \sqrt{\frac{2}{\pi}} f'(0),$$

where $F_c(k)$ is the Fourier cosine transform of $f(x)$.

(b)

$$\mathscr{F}_s\{f''(x)\} = -k^2 F_s(k) + \sqrt{\frac{2}{\pi}} k f(0),$$

where $F_s(k)$ is the Fourier sine transform of $f(x)$.

Proof *We prove part (b) and leave part (a) as an exercise. We integrate by parts twice. Thus*

$$\begin{aligned}
\mathscr{F}_s\{f''(x)\} &= \frac{2}{\pi}\int_0^\infty f''(x)\sin kx\,dx \\
&= \sqrt{\frac{2}{\pi}}\left[f'(x)\sin kx\right]_0^\infty - \sqrt{\frac{2}{\pi}}k\int_0^\infty f'(x)\cos kx\,dx \\
&= -\sqrt{\frac{2}{\pi}}k\int_0^\infty f'(x)\cos kx\,dx.
\end{aligned}$$

Another round of integration by parts yields

$$\begin{aligned}
\mathscr{F}_s\{f''(x)\} &= \sqrt{\frac{2}{\pi}}k\left[-f(x)\cos kx\right]_0^\infty - \sqrt{\frac{2}{\pi}}k^2\int_0^\infty f(x)\sin kx\,dx \\
&= \sqrt{\frac{2}{\pi}}kf(0) - k^2F_s(k).
\end{aligned}$$

The convolution of two functions in the time domain corresponds to the pointwise multiplication of their Fourier transforms in the frequency domain. This property, known as the Convolution Theorem, is essential for efficiently analyzing and processing signals, particularly in filtering and system response calculations and is widely used in solving differential equations on unbounded domains.

Definition 7.5 *The* **convolution** *of the functions f and g over the interval $(-\infty,\infty)$ is defined by*

$$(f*g)(x) = \frac{1}{\sqrt{2\pi}}\int_{-\infty}^{\infty} f(x-\xi)g(\xi)\,d\xi. \tag{7.38}$$

Theorem 7.10 (Convolution) *Suppose that $F(k)$ and $G(k)$ are the Fourier transforms of $f(x)$ and $g(x)$, respectively. Then the Fourier transform of the convolution $(f*g)$ is the product $F(k)G(k)$. That is,*

$$\mathscr{F}\{f(x)*g(x)\} = F(k)G(k).$$

That is

$$\mathscr{F}^{-1}\{F(k)G(k)\} = f(x)*g(x).$$

More explicitly,

$$\frac{1}{\sqrt{2\pi}}\int_{-\infty}^{\infty} F(k)G(k)e^{ikx}\,dk = (f*g)(x) = \frac{1}{\sqrt{2\pi}}\int_{-\infty}^{\infty} f(x-\xi)g(\xi)\,d\xi.$$

Proof *By definition 7.5, we have that*

$$\mathscr{F}[(f*g)(x)] = \frac{1}{2\pi}\int_{-\infty}^{\infty} e^{-ikx}dx\int_{-\infty}^{\infty} f(x-\xi)g(\xi)\,d\xi.$$

Rearranging the integrals, we obtain

$$\mathscr{F}[(f*g)(x)] = \frac{1}{2\pi}\int_{-\infty}^{\infty} g(\xi)e^{-ik\xi}d\xi \int_{-\infty}^{\infty} f(x-\xi)e^{-ik(x-\xi)}dx.$$

With the change of variable $\eta = x-\xi$, *this simplifies to*

$$\mathscr{F}[(f*g)(x)] = \frac{1}{\sqrt{2\pi}}\int_{-\infty}^{\infty} g(\xi)e^{-ik\xi}d\xi \cdot \frac{1}{\sqrt{2\pi}}\int_{-\infty}^{\infty} f(\eta)e^{-ik\eta}d\eta.$$

Thus, we conclude that

$$\mathscr{F}[(f*g)(x)] = F(k)G(k).$$

This completes the proof.

We have the following example.

Example 7.11 Let $g(x) = e^{-x^2}\cos 2x$. Then using by part (a) of Theorem 7.6 we have that

$$\mathscr{F}\{f(x)\cos ax\} = \frac{1}{2}\big[F(k+a)+F(k-a)\big].$$

Thus, if we let $f(x) = e^{-x^2}$, then by part (a) of Exercise 7.58, we obtain

$$F(k) = \mathscr{F}\{e^{-x^2}\} = \frac{1}{\sqrt{2}}e^{-\frac{k^2}{4}}.$$

Therefore,

$$\mathscr{F}\{g(x)\} = \mathscr{F}\{e^{-x^2}\cos 2x\} = \frac{1}{2}\Big[\frac{1}{\sqrt{2}}e^{-\frac{(k+2)^2}{4}} + \frac{1}{\sqrt{2}}e^{-\frac{(k-2)^2}{4}}\Big].$$

□

Example 7.12 In this example we are interested in finding

$$\mathscr{F}\{e^{-2(x-5)^2}\}.$$

To achieve this, we resort to the Shift Theorem. Let $f(x) = e^{-2x^2}$, then by part (a) of Exercise 7.58 we have

$$F(k) = \mathscr{F}\{e^{-2x^2}\} = \frac{1}{2}e^{-\frac{k^2}{8}}.$$

Therefore, by Theorem 7.4 we arrive at

$$\mathscr{F}\{e^{-2(x-5)^2}\} = e^{5ik}\frac{1}{2}e^{-\frac{k^2}{8}}.$$

□

Next, we shift our focus to examining the Fourier transform of step and impulse functions, which play a crucial role in signal processing and systems analysis. The Heaviside step function helps model abrupt changes in signals, while the Dirac delta function represents an idealized impulse with infinite amplitude and zero duration. Understanding their Fourier transforms provides insight into how these fundamental functions influence frequency-domain representations of signals.

Definition 7.6 *The* **Heaviside unit step function** *is defined by*

$$H(x-a)=\begin{cases}0, & x<a\\ 1, & x\geq a\end{cases}, \quad a\geq 0, \tag{7.39}$$

as shown in Fig. 7.1.

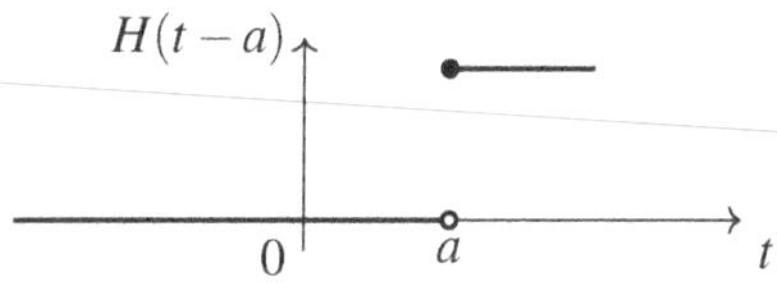

FIGURE 7.1: The Heaviside step function $H(t)$.

The Fourier transform of the Heaviside unit step function can be easily determined. We consider first

$$\mathscr{F}[H(x-a)]=\frac{1}{\sqrt{2\pi}}\int_{-\infty}^{\infty}H(x-a)e^{-ikx}dx=\frac{1}{\sqrt{2\pi}}\int_{a}^{\infty}e^{-ikx}dx.$$

This improper integral does not exist or converge in the conventional sense. However, by multiplying the Heaviside function $H(x-a)$ with an exponential damping factor $e^{-\beta x}$, $\beta\geq 0$, we obtain a modified function that ensures convergence. This approach allows us to rigorously define the integral and analyze its Fourier transform. Thus,

$$H(x-a)e^{-\beta x}=\begin{cases}0, & x<a\\ e^{-\beta x}, & x\geq a\end{cases} \tag{7.40}$$

Note that

$$\lim_{\beta\to 0}H(x-a)e^{-\beta x}=H(x-a).$$

Consequently, by utilizing this setup, we can determine the Fourier transform of the Heaviside unit step function. That is,

$$\begin{aligned}\mathscr{F}\{H(x-a)\} &= \lim_{\alpha\to 0}\mathscr{F}\left\{H(x-a)e^{-\beta x}\right\}\\ &= \lim_{\beta\to 0}\frac{1}{\sqrt{2\pi}}\int_{-\infty}^{\infty}H(x-a)e^{-\beta x}e^{-ikx}dx\\ &= \lim_{\beta\to 0}\frac{1}{\sqrt{2\pi}}\int_{a}^{\infty}e^{-(\beta+ik)x}dx=\frac{e^{-iak}}{\sqrt{2\pi}(ik)}.\end{aligned} \tag{7.41}$$

In particular, for $a = 0$, we obtain

$$\mathscr{F}\{H(x)\} = \frac{1}{\sqrt{2\pi}(ik)}.$$

Additionally, using the Shift Theorem, we have

$$\mathscr{F}\{H(x-a)e^{-\beta x}\} = \frac{1}{\sqrt{2\pi}} \frac{e^{-(ik+\beta)a}}{\beta + ik}, \tag{7.42}$$

and in particular, for $a = 0$, we obtain

$$\mathscr{F}\{H(x)e^{-\beta x}\} = \frac{1}{\sqrt{2\pi}} \frac{1}{\beta + ik}. \tag{7.43}$$

Notes that, it follows from (7.43) that

$$\mathscr{F}\{H(-x)e^{\beta x}\} = \frac{1}{\sqrt{2\pi}} \frac{1}{\beta - ik}. \tag{7.44}$$

Next, we define the impulse function of height h.

Definition 7.7 *(Impulse function) The* **impulse function** *of height h is defined by*

$$p(x) = \begin{cases} h, & a-\varepsilon < x < a+\varepsilon \\ 0, & x \leq a-\varepsilon \ or \ x \geq a+\varepsilon \end{cases} \tag{7.45}$$

where h is large and positive, $a > 0$, and ε is a small positive constant, as shown in Fig. 7.2.

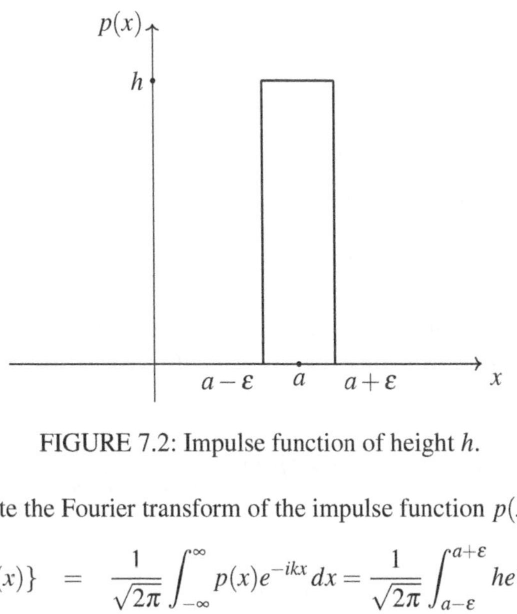

FIGURE 7.2: Impulse function of height h.

Next, we compute the Fourier transform of the impulse function $p(x)$.

$$\mathscr{F}\{p(x)\} = \frac{1}{\sqrt{2\pi}} \int_{-\infty}^{\infty} p(x)e^{-ikx}\,dx = \frac{1}{\sqrt{2\pi}} \int_{a-\varepsilon}^{a+\varepsilon} he^{-ikx}\,dx$$

$$= \frac{h}{\sqrt{2\pi}}e^{-iak}\frac{e^{ik\varepsilon}-e^{-ik\varepsilon}}{ik} = \frac{2h\varepsilon}{\sqrt{2\pi}}e^{-iak}\frac{\sin k\varepsilon}{k\varepsilon}. \tag{7.46}$$

Recall that the Dirac delta function $\delta(x-a)$ is not a function in the usual sense but rather a distribution. It has the following properties.

$$\delta(x-a) = \begin{cases} 0, & x \neq a \\ \infty, & x = a \end{cases}$$

with the additional feature that

$$\int_{-\infty}^{\infty} \delta(x-a)\,dx = 1. \tag{7.47}$$

The Dirac delta function is an example of a *distribution*-distributions are defined in terms of their integration properties. For any function $f(x)$ that is continuous at $x=a$, the delta distribution satisfies

$$\int_{-\infty}^{\infty} f(x)\delta(x-a)\,dx = f(a). \tag{7.48}$$

Next, we investigate the relationship between the Dirac delta function $\delta(x)$ and the impulse function $p(x)$, examining how the delta function can be derived as a limiting case of $p(x)$. The total area under $p(x)$ is

$$\int_{-\infty}^{\infty} p(x)\,dx = \int_{a-\varepsilon}^{a+\varepsilon} h\,dx = 2\varepsilon h. \tag{7.49}$$

Setting $h = \frac{1}{2\varepsilon}$, and substituting this into $p(x)$, we obtain

$$p(x) = \begin{cases} \frac{1}{2\varepsilon}, & a-\varepsilon < x < a+\varepsilon, \\ 0, & \text{otherwise.} \end{cases} \tag{7.50}$$

Now, taking the limit $\varepsilon \to 0$, we obtain

$$\delta(x-a) = \lim_{\varepsilon\to 0} p(x). \tag{7.51}$$

Additionally, with the choice $h = \frac{1}{2\varepsilon}$, (7.47) is satisfied. Now, we obtain Fourier Transform of the Dirac delta function. Using (7.46) we have, using L'Hôpital's rule, that

$$\mathscr{F}\{\delta(x-a)\} = \lim_{\varepsilon\to 0}\mathscr{F}\{p(x)\} = \lim_{\varepsilon\to 0}\frac{e^{-iak}}{\sqrt{2\pi}}\frac{\sin k\varepsilon}{k\varepsilon}.$$

Since

$$\lim_{\varepsilon\to 0}\frac{\sin k\varepsilon}{k\varepsilon} = 1,$$

by applying L'Hôpital's rule, we conclude

$$\mathscr{F}\{\delta(x-a)\} = \frac{e^{-iak}}{\sqrt{2\pi}}. \tag{7.52}$$

In particular, for $a = 0$, we obtain

$$\mathscr{F}\{\delta(x)\} = \frac{1}{\sqrt{2\pi}}. \tag{7.53}$$

In the next lemma, we state and prove an important property of the Dirac delta function.

Lemma 24 *For nonzero constant a we have the following scaling property.*

$$\delta(ax) = \frac{1}{|a|}\delta(x). \tag{7.54}$$

Proof *The only logical thing to do is to place $\delta(ax)$ inside an integral and make a substitution $y = ax$. Thus*

$$\int_{-\infty}^{\infty} \delta(ax)\,dx \quad = \quad \lim_{L\to\infty}\int_{-L}^{L} \delta(ax)\,dx = \lim_{L\to\infty}\frac{1}{a}\int_{-aL}^{aL} \delta(y)\,dy.$$

If $a > 0$, then

$$\int_{-\infty}^{\infty} \delta(ax)\,dx = \frac{1}{a}\int_{-\infty}^{\infty} \delta(y)\,dy.$$

However, if $a < 0$, then

$$\int_{-\infty}^{\infty} \delta(ax)\,dx = \frac{1}{a}\int_{\infty}^{-\infty} \delta(y)\,dy = -\frac{1}{a}\int_{-\infty}^{\infty} \delta(y)\,dy.$$

The multiplicative sign difference can be unified by replacing the factor $\frac{1}{a}$ with $\frac{1}{|a|}$, ensuring a consistent expression. Thus,

$$\int_{-\infty}^{\infty} \delta(ax)\,dx = \frac{1}{|a|}\int_{-\infty}^{\infty} \delta(y)\,dy.$$

Thus, the interpretation of the above discussion is only valid in the context of an integral and hence

$$\int_{-\infty}^{\infty} g(x)\delta(a(x-x_1))\,dx = \frac{1}{|a|}g(x_1). \tag{7.55}$$

Example 7.13 We want to evaluate $\int_{-\infty}^{\infty}(x^2+1)\delta(3x-6)\,dx$. Using (7.54) we have

$$\delta(3x-6) = \frac{1}{3}\delta(x-2).$$

Moreover, applying (7.55), we obtain

$$\begin{aligned}\int_{-\infty}^{\infty}(x^2+1)\delta(3x-6)\,dx &= \frac{1}{3}\int_{-\infty}^{\infty}(x^2+1)\delta(x-2)\,dx\\ &= \frac{1}{3}\left((2)^2+1\right)=\frac{5}{3},\end{aligned}$$

where we took $g(x)=x^2+1$.

Remark 14 *A broader generalization of $\delta(ax)$ is the delta function $\delta(f(x))$. The evaluation of the integral of $\delta(f(x))$ depends on the number of zeros of $f(x)$. If there is a single zero, where $f(x_1)=0$, then we obtain*

$$\int_{-\infty}^{\infty}\delta(f(x))\,dx=\int_{-\infty}^{\infty}\frac{1}{|f'(x_1)|}\delta(x-x_1)\,dx.$$

This can be proven using the substitution $y=f(x)$, and is left as an exercise for the reader. This result is often written as

$$\delta(f(x))=\frac{1}{|f'(x_1)|}\delta(x-x_1),$$

while noting that this expression is only valid in the context of an integral. As pointed out before, we conclude that

$$\int_{-\infty}^{\infty}g(x)\delta(f(x))\,dx=\frac{1}{|f'(x_1)|}g(x_1).$$

More generally, one can show that when $f(x_j)=0$ and $f'(x_j)\neq 0$ for $j=1,2,\ldots,n$, (i.e., when one has n simple zeros), then

$$\delta(f(x))=\sum_{j=1}^{n}\frac{1}{|f'(x_j)|}\delta(x-x_j).$$

In other words,

$$\int_{-\infty}^{\infty}g(x)\delta(f(x))\,dx=\sum_{j=1}^{n}\frac{1}{|f'(x_j)|}g(x_j). \tag{7.56}$$

Example 7.14 Consider

$$I=\int_0^{2\pi}\cos x\,\delta(x(x^2-\pi^2))\,dx.$$

Then

$$f(x)=x(x^2-\pi^2)=0 \quad \text{when} \quad x=0,\pm\pi.$$

Thus, f has three simple roots; however, but $x=-\pi$ lies outside the interval of integration. Furthermore, the derivative is $f'(x)=3x^2-\pi^2$. If we let $g(x)=\cos x$, then using (7.56) we have that

$$\begin{aligned}\int_0^{2\pi}\cos x\,\delta(x(x^2-\pi^2))\,dx &= \frac{1}{|f'(0)|}g(0)+\frac{1}{|f'(\pi)|}g(\pi)\\ &= \frac{1}{\pi^2}(1)+\frac{1}{2\pi^2}(-1)=\frac{1}{2\pi^2}.\end{aligned}$$

In the next example we use the convolution theorem to recover the original function.

Example 7.15 Find

$$\mathscr{F}^{-1}\left\{\frac{1}{6+5ik-k^2}\right\}.$$

Factoring the denominator and using Theorem 7.10 regarding convolution we arrive at

$$\begin{aligned}
\mathscr{F}^{-1}\left\{\frac{1}{(2+ik)}\cdot\frac{1}{(3+ik)}\right\} &= \mathscr{F}^{-1}\left\{\frac{1}{(2+ik)}\right\} * \mathscr{F}^{-1}\left\{\frac{1}{(3+ik)}\right\} \\
&= \left(\sqrt{2\pi}e^{-2x}H(x)\right) * \left(\sqrt{2\pi}e^{-3x}H(x)\right)) \\
&= \frac{1}{\sqrt{2\pi}}\int_{-\infty}^{\infty} 2\pi e^{-2\tau}H(\tau)e^{-3(x-\tau)}H(x-\tau)\,d\tau \\
&= \sqrt{2\pi}e^{-3x}\int_{-\infty}^{\infty} e^{\tau}H(\tau)H(x-\tau)\,d\tau,
\end{aligned}$$

where we have made use of (7.43). To be able to integrate the above expression, we must know the function value of $H(\tau)H(x-\tau)$.

$$H(\tau)H(x-\tau) = \begin{cases} 1, & \tau \geq 0,\ x-\tau \geq 0 \\ 0, & \tau < 0,\ x-\tau < 0 \end{cases} = \begin{cases} 1, & 0 \leq \tau \leq x \\ 0, & \tau < 0 \text{ and } \tau > x \end{cases}$$

Hence,

$$\sqrt{2\pi}e^{-3x}\int_{-\infty}^{\infty} e^{\tau}H(\tau)H(x-\tau)\,d\tau = \sqrt{2\pi}e^{-3x}\int_{0}^{x} e^{\tau}\,d\tau = \sqrt{2\pi}\left[e^{-2x}-e^{-3x}\right].$$

Finally,

$$\mathscr{F}^{-1}\left\{\frac{1}{6+5ik-k^2}\right\} = \sqrt{2\pi}\begin{cases} e^{-2x}-e^{-3x}, & x \geq 0 \\ 0, & x < 0 \end{cases}$$

7.8 Parseval's Formula

Parseval's formula is crucial in signal processing and physics as it states that the total energy of a function in the time domain is equal to its total energy in the frequency domain. Mathematically, it ensures that the integral of the square of a function is preserved under the Fourier transform. This principle is essential in analyzing energy distribution across different frequencies, making it invaluable in electrical engineering, acoustics, and quantum mechanics. It helps in filtering, signal compression, and understanding how energy propagates in various systems.

where $R(0)$ is the auto-correlation at $t = 0$, given by

$$R(0) = \frac{1}{\sqrt{2\pi}} \int_{-\infty}^{\infty} f(u)\overline{f(u)}\, du = \frac{1}{\sqrt{2\pi}} \int_{-\infty}^{\infty} |f(u)|^2\, du.$$

Substitute $R(t)$ and $R(0)$ into $\gamma(t)$ to get

$$\gamma(t) = \frac{\int_{-\infty}^{\infty} f(u) f^*(t-u)\, du}{\int_{-\infty}^{\infty} |f(u)|^2\, du},$$

where $*$ denotes complex conjugate.

Remark 15
- *The numerator $\int_{-\infty}^{\infty} f(u) f^*(t-u)\, du$ measures the overlap between $f(u)$ and its time-shifted version $f(t-u)$.*
- *The denominator $\int_{-\infty}^{\infty} |f(u)|^2\, du$ normalizes the auto-correlation by the total energy of $f(u)$, ensuring $\gamma(t)$ is dimensionless and satisfies $|\gamma(t)| \le 1$.*
- *At $t = 0$, $\gamma(0) = 1$, since $R(0) = \int_{-\infty}^{\infty} |f(u)|^2\, du$.*
- *For $t \neq 0$, $\gamma(t)$ measures the similarity between $f(u)$ and its time-shifted version $f(t-u)$.*

We provide the following example in which we compute $\gamma(t)$ for the Gaussian function.

Example 7.17 In this example, we compute the normalized auto-correlation function of the Gaussian function

$$f(t) = e^{-at^2},$$

where a is a positive constant. The Gaussian function is real-valued, so its complex conjugate satisfies

$$f^*(t) = f(t).$$

The denominator of $\gamma(t)$ is given by

$$R(0) = \frac{1}{\sqrt{2\pi}} \int_{-\infty}^{\infty} |f(u)|^2\, du = \frac{1}{\sqrt{2\pi}} \int_{-\infty}^{\infty} e^{-2au^2}\, du.$$

Using Lemma 23 with $(b = 0)$ where the Gaussian integral

$$\int_{-\infty}^{\infty} e^{-\ell u^2}\, du = \sqrt{\frac{\pi}{\ell}},$$

we obtain, for $\ell = 2a$ that

$$R(0) = \frac{1}{\sqrt{2\pi}} \sqrt{\frac{\pi}{2a}}.$$

The numerator of $\gamma(t)$ is

$$R(t) = \frac{1}{\sqrt{2\pi}} \int_{-\infty}^{\infty} f(u) f(t-u)\, du = \frac{1}{\sqrt{2\pi}} \int_{-\infty}^{\infty} e^{-au^2} e^{-a(t-u)^2}\, du.$$

Expanding the exponent, we get

$$u^2 + (t-u)^2 = 2u^2 - 2tu + t^2.$$

Consequently, we rewrite

$$R(t) = \frac{1}{\sqrt{2\pi}} \int_{-\infty}^{\infty} e^{-a(2u^2-2tu+t^2)}\, du.$$

By factoring and completing the square, we arrive at

$$-2a(u^2 - tu) = -2a\left((u - \frac{t}{2})^2 - \frac{t^2}{4}\right).$$

Thus,

$$R(t) = \frac{1}{\sqrt{2\pi}} e^{-at^2/2} \int_{-\infty}^{\infty} e^{-2a(u-\frac{t}{2})^2}\, du.$$

Since shifting the variable in a Gaussian integral does not change the total area under the curve, then by Lemma 23 we have

$$\int_{-\infty}^{\infty} e^{-2a(u-\frac{t}{2})^2}\, du = \sqrt{\frac{\pi}{2a}}.$$

Consequently,

$$R(t) = \frac{1}{\sqrt{2\pi}} e^{-at^2/2} \sqrt{\frac{\pi}{2a}}.$$

Finally, the normalized auto-correlation function is

$$\gamma(t) = \frac{R(t)}{R(0)} = \frac{e^{-at^2/2}\sqrt{\frac{\pi}{2a}}}{\sqrt{\frac{\pi}{2a}}} = e^{-at^2/2},$$

which is also Gaussian. The result shows that the correlation decays exponentially with t^2, reflecting the smoothness and localization of the Gaussian function. □

7.9.1 Exercises

Exercise 7.66 *Prove Theorem 7.5.*

Exercise 7.67 *Prove parts (c)–(e) of Theorem 7.6.*

Exercise 7.68 *For a positive constant b, evaluate the following.*

(a) $\mathscr{F}_s\{e^{-b|x|}\cos ax\}$, *(b)* $\mathscr{F}_c\{e^{-b|x|}\sin ax\}$.

(c) $\mathscr{F}_c\{e^{-b|x|}\cos ax\}$, *(d)* $\mathscr{F}_s\{e^{-b|x|}\sin ax\}$.

Exercise 7.69 *Prove part (a) of Theorem 7.9.*

Exercise 7.70 *Show that*

(a) $\mathscr{F}_s\left\{\frac{\partial u}{\partial t}\right\} = \frac{d}{dt}U_s$, (b) $\mathscr{F}_c\left\{\frac{\partial u}{\partial t}\right\} = \frac{d}{dt}U_c$.

Exercise 7.71 *Use convolution to find*

(a) $\mathscr{F}^{-1}\left\{\frac{1}{8+6ik-k^2}\right\}$, (b) $\mathscr{F}^{-1}\left\{\frac{1}{k^2+1}\right\}$, (c) $\mathscr{F}^{-1}\left\{\frac{1}{13+4k+k^2}\right\}$,

(d) $\mathscr{F}^{-1}\left\{\frac{1}{13+4k+k^2}\right\}$, (e) $\mathscr{F}^{-1}\left\{\frac{1}{(ik+4)^2}\right\}$.

Exercise 7.72 *For* $-\infty < x < \infty$, *use Fourier transform to solve*

(a) $y' - 4y = H(x)e^{-4x}$, (b) $y' + 3y = H(x)e^{-2x}$, (c) $y'' + 3y' + 2y = \delta(x-3)$.

Exercise 7.73 *Show that*

$$\int_0^{\infty} e^{-a^2x^2}\cos(bx)\,dx = \frac{\sqrt{\pi}}{2a}e^{-b^2/4a^2}, \quad a > 0.$$

Exercise 7.74 *Prove that*

$$\mathscr{F}[f(ax-b)] = \frac{1}{|a|}e^{ikb/a}F\left(\frac{k}{a}\right).$$

Exercise 7.75 *Prove the following properties of the Fourier convolution.*

(a) $\frac{d}{dx}\{f(x)*g(x)\} = f'(x)*g(x) = f(x)*g'(x)$.

(b) $\frac{d^2}{dx^2}[(f*g)(x)] = (f'*g')(x) = (f''*g)(x)$.

Exercise 7.76 *Prove the following properties of the Fourier convolution.*

(a) $\sin x * e^{-a|x|} = \sqrt{\frac{2}{\pi}}\frac{a\sin x}{1+a^2}$, (b) $e^{ax} * \chi_{[0,\infty)}(x) = \frac{1}{a}\frac{e^{ax}}{\sqrt{2\pi}}$, $a > 0$.

(c) $\frac{1}{\sqrt{2a}}e^{-\frac{x^2}{4a}} * \frac{1}{\sqrt{2b}}e^{-\frac{x^2}{4b}} = \frac{1}{\sqrt{2(a+b)}}e^{-\frac{x^2}{4(a+b)}}$, $a,\, b > 0$.

Exercise 7.77 *Use convolution to prove Theorem 7.11.*

Exercise 7.78 *Prove the result of Remark 14.*

Exercise 7.79 *Find the convolution of the box and triangular functions given by*

(a) $f(x) = \begin{cases} 1, & |x| \le 1 \\ 0, & |x| > 1 \end{cases}$, (b) $g(x) = \begin{cases} x, & 0 \le x \le 1 \\ 0, & \textit{otherwise} \end{cases}$

Exercise 7.80 *Evaluate the following integrals.*

(a) $\int_0^{\pi}(x^2+\sin x)\,\delta(2x-\pi)\,dx$, (b) $\int_{-\infty}^{\infty}\delta(3x-2)(x^3+1)\,dx$.

(c) $\int_0^{2\pi}\cos x\,\delta(x^2-\pi^2)\,dx$, (d) $\int_{-\infty}^{\infty}\delta(x^2-5x-6)(3x^2-7x+2)\,dx$.

Exercise 7.81 *Use Parseval's formula or Theorem 7.11 to prove*

$$\int_{-\infty}^{\infty} \frac{dx}{(a+x^2)^2} = \frac{\pi}{2}.$$

Exercise 7.82 *Show that the normalized auto-correlation function of the rectangular pulse function for*

$$\chi_{[-a,a]}(t) = \begin{cases} 1, & |t| \leq \frac{a}{2} \\ 0, & |t| > \frac{a}{2} \end{cases}$$

is given by

$$\gamma(t) = \begin{cases} 1 - \frac{|t|}{a}, & |t| \leq a \\ 0, & |t| > a \end{cases}$$

7.10 Application of Fourier Transform

In this section we use Fourier transform to solve heat flow and vibration on unbounded domains with Dirichlet conditions and Neumann conditions.

The Fourier Transform is a powerful mathematical tool that provides significant advantages over other methods, particularly when solving partial differential equations related to heat flow and vibration in unbounded domains. It allows us to convert differential equations into algebraic equations, making them easier to solve. Moreover, unlike traditional separation of variables or finite difference methods, the Fourier Transform naturally accommodates problems on infinite or semi-infinite domains. By leveraging the Fourier Transform, we obtain solutions that are not only computationally efficient but also provide deep insights into the frequency components of physical systems. This makes it a superior method compared to classical approaches in many practical scenarios.

7.10.1 Heat Equation

• Semi-Infinite Rod

Consider the heat flow problem that models the heat distribution in a semi-infinite rod

$$\begin{aligned} & u_t - \rho u_{xx} = 0, \quad x > 0, \quad t > 0, \\ & u(x,0) = 0, \quad x > 0, \quad u_x(0,t) = f(t), \quad t > 0 \\ & \lim_{x \to \infty} u(x,t) = \lim_{x \to \infty} u_x(x,t) = 0, \end{aligned} \tag{7.59}$$

where $u(x,t)$ is a bounded heat displacement. Let $U(k,t)$ be the Fourier cosine transform of $u(x,t)$. Then from Theorem 7.9 we have that

$$\mathscr{F}_c\{u_{xx}(x,t)\} = -k^2 U(k,t) - \sqrt{\frac{2}{\pi}} f(t).$$

Applying the Fourier cosine transform to heat flow equation yields

$$U_t + \rho k^2 U = -\sqrt{\frac{2}{\pi}} f(t)\rho. \tag{7.60}$$

Notice that we use ρ for conductivity to avoid confusing the reader with κ and k. Since,

$$U(k,t) = \sqrt{\frac{2}{\pi}} \int_0^\infty \cos kx\, u(x,t)\, dx,$$

we have that

$$U(k,0) = \sqrt{\frac{2}{\pi}} \int_0^\infty \cos kx\, u(x,0)\, dx = 0,$$

since $u(x,0) = 0$. Now we solve the linear equation (7.60) for $U(k,t)$ with initial condition $U(k,0) = 0$. The integrating factor of (7.60) is $e^{\rho k^2 t}$. Hence, (7.60) is equivalent to

$$\frac{d}{dt}\left(U(k,t)e^{\rho k^2 t}\right) = -\rho\sqrt{\frac{2}{\pi}} f(t) e^{\rho k^2 t}.$$

Integrating both sides and then dividing by $e^{\rho k^2 t}$ we arrive at

$$U(k,t) = Ce^{-\rho k^2 t} - \rho\sqrt{\frac{2}{\pi}} \int_0^t f(\tau) e^{-\rho k^2 (t-\tau)}\, d\tau.$$

Applying the boundary condition $U(\mathrm{k},0) = 0$, we see that $C = 0$. Substituting $U(k,t)$ into

$$u(x,t) = \sqrt{\frac{2}{\pi}} \int_0^\infty \cos kx\, U(k,t)\, dk,$$

we arrive at

$$u(x,t) = -\frac{2\rho}{\pi} \int_0^t f(\tau) d\tau \int_0^\infty e^{-\rho k^2 (t-\tau)} \cos kx\, dk.$$

With the aid of Lemma 23 and Exercise 7.73, the inside integral can be written as

$$\int_0^\infty e^{-k^2 \rho (t-\tau)} \cos kx\, dk = \frac{1}{2}\sqrt{\frac{\pi}{\rho(t-\tau)}} \exp\left(-\frac{x^2}{4\rho(t-\tau)}\right).$$

Therefore, the solution takes the final form

$$u(x,t) = -\sqrt{\frac{\rho}{\pi}} \int_0^t f(\tau) \frac{1}{\sqrt{t-\tau}} e^{-x^2/4\rho(t-\tau)}\, d\tau.$$

- **Infinite Rod**

Consider the heat flow problem that models the heat distribution in a semi-infinite rod

$$u_t - \rho u_{xx} = 0, \quad -\infty < x < \infty, \quad t > 0,$$

$$u(x,0) = f(x), \quad t > 0, \tag{7.61}$$
$$\lim_{|x|\to\infty} u(x,t) = \lim_{|x|\to\infty} u_x(x,t) = 0.$$

Let $U(k,t)$ be the Fourier transform of $u(x,t)$, where $u(x,t)$ is bounded in x and t. Taking the Fourier transform of the system (7.61) gives

$$U_t + \rho k^2 U = 0, \quad U(k,0) = F(k),$$

where we have used

$$\mathscr{F}\{u_{xx}(x,t)\} = (ik)^2 U(k,t) = -k^2 U(k,t),$$

and

$$U(k,t) = \frac{1}{\sqrt{2\pi}} \int_{-\infty}^{\infty} e^{-ikx} u(x,t)\, dx,$$

from which we obtain

$$U(k,0) = \frac{1}{\sqrt{2\pi}} \int_{-\infty}^{\infty} e^{-ikx} u(x,0)\, dx = \frac{1}{\sqrt{2\pi}} \int_{-\infty}^{\infty} e^{-ikx} f(x)\, dx = F(k).$$

The equation $U_t + \rho k^2 U = 0$ is a separable first-order ordinary differential equation, which has the general solution

$$U(k,t) = Ce^{-\rho k^2 t}.$$

Applying the initial condition $U(k,0) = F(k)$, we obtain $C = F(k)$. Thus, the solution takes the final form

$$U(k,t) = F(k)e^{-\rho k^2 t}.$$

Substituting $U(k,t)$ into

$$u(x,t) = \frac{2}{\sqrt{\pi}} \int_{-\infty}^{\infty} e^{ikx} U(k,t)\, dk$$

we arrive at

$$u(x,t) = \frac{1}{\sqrt{2\pi}} \int_{-\infty}^{\infty} F(k)e^{-\rho k^2 t} e^{ikx}\, dk. \tag{7.62}$$

Next, we use the Convolution Theorem to rewrite solution (7.62). For simplicity, we assume $\rho = 1$. Let $D(k) = e^{-k^2 t}$. Then, by a slight modification of Example 7.6, we find that its Fourier inverse $g(x)$ is given by

$$g(x) = \frac{1}{\sqrt{2\pi}} \int_{-\infty}^{\infty} e^{-k^2 t + ikx}\, dk = \frac{1}{\sqrt{2t}} e^{-\frac{x^2}{4t}}.$$

Using convolution, (7.62) can be written in the form

$$u(x,t) = \frac{1}{2\sqrt{\pi t}} \int_{-\infty}^{\infty} f(\xi) g(x-\xi)\, d\xi$$

$$= \frac{1}{\sqrt{4\pi t}} \int_{-\infty}^{\infty} f(\xi) e^{-\frac{(x-\xi)^2}{4t}} \, d\xi \tag{7.63}$$
$$= \int_{-\infty}^{\infty} f(\xi) G(x-\xi, t) \, d\xi,$$

where

$$G(x-\xi, t) = \frac{1}{\sqrt{4\pi t}} e^{-\frac{(x-\xi)^2}{4t}},$$

is called the **Green's function** or the **fundamental solution** of the diffusion equation.

7.10.2 Schrödinger Equation

The Schrödinger equation is a fundamental equation in quantum mechanics that describes how the quantum state of a system evolves over time.

We consider the Schrödinger equation with initial condition

$$\begin{aligned} &u_t - i\rho u_{xx} = 0, \quad -\infty < x < \infty, \quad t > 0, \\ &u(x,0) = f(x), \quad t > 0, \\ &\lim_{|x|\to\infty} u(x,t) = \lim_{|x|\to\infty} u_x(x,t) = 0. \end{aligned} \tag{7.64}$$

Let $U(k,t)$ be the Fourier transform of $u(x,t)$ where $u(x,t)$ is bounded in x and t. Taking the Fourier transform of the system (7.64) gives

$$U_t - i\rho k^2 U = 0, \quad U(k,0) = F(k).$$

Along the same lines of the previous section, we find the solution to be

$$u(x,t) = \frac{1}{\sqrt{2\pi}} \int_{-\infty}^{\infty} F(k) e^{-\rho i k^2 t} e^{ikx} \, dk. \tag{7.65}$$

We need to adjust (7.63) to accommodate the inclusion of i. That is, we need to substitute $-i\rho$ instead of ρ (as $t > 0$) into the formulae obtained earlier. Thus, instead of

$$\frac{1}{\sqrt{4\pi\rho t}} e^{-x^2/4kt},$$

we have to work with

$$\frac{1}{\sqrt{4\pi\rho i t}} e^{-x^2/(-4\rho i t)} = \frac{1}{\sqrt{4\pi\rho t}} e^{\frac{\pi i}{4} - ix^2/4\rho t},$$

since we need to take the correct branch of $\sqrt{-i} = e^{-i\pi/4}$. For $t < 0$, we replace t by $-t$ and i by $-i$, resulting in

$$\frac{1}{\sqrt{4\pi\rho|t|}} e^{-\frac{\pi i}{4} - ix^2/4\rho t}.$$

Therefore, with the aid of (7.63) the solution takes the form

$$u(x,t) = \frac{1}{\sqrt{4\pi\rho|t|}} \int_{-\infty}^{\infty} f(\xi) e^{\pm i\pi/4 - i(x-\xi)^2/4\rho t} \, d\xi.$$

7.10.3 Laplace Equation

We have dealt before with Laplacian in open domain. We revisit the equation on the upper-half plane applying Fourier transform. We want to find the solution $u(x,y)$ of the Dirichlet problem in the half-plane $y > 0$

$$u_{xx} + u_{yy} = 0, \quad -\infty < x < \infty, \quad y > 0,$$

$$u(x,0) = f(x), \quad -\infty < x < \infty,$$

where u and u_x vanish as $|x| \to \infty$, and u is bounded as $y \to \infty$. Note that if we do not require boundedness on u then $u(x,y) = y$ is a solution. Let $U(k,y)$ be the Fourier transform of $u(x,y)$ with respect to x

$$U(k,y) = \frac{1}{\sqrt{2\pi}} \int_{-\infty}^{\infty} u(x,y) e^{-ikx} dx.$$

An application of the Fourier transform with respect to x gives

$$U_{yy} - k^2 U = 0,$$

and the transformed boundary condition

$$U(k,0) = F(k), \quad \text{and} \quad U(k,y) \to 0 \quad \text{as} \quad y \to \infty.$$

The solution of this transformed system is

$$U(k,y) = Ae^{-|k|y} + Be^{|k|y}.$$

Since, we want a bounded solution, we take $B = 0$ and we are left with

$$U(k,y) = Ae^{-|k|y}.$$

Applying the boundary condition, gives $A = F(k)$, and consequently, the solution is

$$U(k,y) = F(k)e^{-|k|y}.$$

Since

$$F(k) = \frac{1}{\sqrt{2\pi}} \int_{-\infty}^{\infty} f(\xi) e^{-|k|y} e^{-ik\xi} d\xi,$$

we obtain the solution

$$u(x,y) = \frac{1}{\sqrt{2\pi}} \int_{-\infty}^{\infty} \left(\frac{1}{\sqrt{2\pi}} \int_{-\infty}^{\infty} f(\xi) e^{-|k|y} e^{-ik\xi} d\xi \right) e^{ikx} dk,$$

$$= \frac{1}{2\pi} \int_{-\infty}^{\infty} f(\xi) d\xi \int_{-\infty}^{\infty} e^{-k[i(\xi - x)] - |k|y} dk.$$

Adjusting the results of Example 7.5, we obtain

$$\int_{-\infty}^{\infty} e^{-k[i(\xi - x)] - |k|y} dk = \frac{2y}{(\xi - x)^2 + y^2}.$$

Hence, the solution of the Dirichlet problem in the half-plane $y > 0$ is

$$u(x,y) = \frac{y}{\pi} \int_{-\infty}^{\infty} \frac{f(\xi)}{(\xi - x)^2 + y^2} d\xi.$$

From this solution, we can readily deduce a solution of the Neumann problem in the half-plane $y > 0$, which we leave as an exercise.

• Laplace Equation with Robin Boundary Condition

We want to find the solution $u(x,y)$ of the Laplacian in the half-plane $y > 0$ with Robin boundary condition. That is, we consider the system

$$u_{xx} + u_{yy} = 0, \quad -\infty < x < \infty, \quad y > 0,$$

$$u_y(x,0) - \alpha u(x,0) = f(x), \quad -\infty < x < \infty,$$

where u and u_x vanish as $|x| \to \infty$, and u is bounded as $y \to \infty$. Let $U(k,y)$ be the Fourier transform of $u(x,y)$ with respect to x

$$U(k,y) = \frac{1}{\sqrt{2\pi}} \int_{-\infty}^{\infty} u(x,y) e^{-ikx} dx.$$

An applying of the Fourier transform with respect to x gives the second-order ordinary differential equation in y,

$$U_{yy} - k^2 U = 0.$$

As for transforming the Robin's boundary condition, we have

$$U_y(k,y) - \alpha U(k,y) = \frac{1}{\sqrt{2\pi}} \int_{-\infty}^{\infty} \left[u_y(x,y) - \alpha u(x,y) \right] e^{-ikx} dx.$$

Setting $y = 0$ gives

$$\begin{aligned} U_y(k,0) - \alpha U(k,0) &= \frac{1}{\sqrt{2\pi}} \int_{-\infty}^{\infty} \left[u_y(x,0) - \alpha u(x,0) \right] e^{-ikx} dx \\ &= \frac{1}{\sqrt{2\pi}} \int_{-\infty}^{\infty} f(x) e^{-ikx} dx = F(k), \end{aligned}$$

as expected, since Fourier transform is linear. The solution of $U_{yy} - k^2 U = 0$ is

$$U(k,y) = Ae^{-|k|y} + Be^{|k|y}.$$

Since, we want a bounded solution, we take $B = 0$ and we are left with

$$U(k,y) = Ae^{-|k|y}.$$

Applying the boundary condition gives

$$U_y(k,y) - \alpha U(k,y) = -A|k|e^{-|k|y} - \alpha Ae^{-|k|y}.$$

Setting $y = 0$ implies that

$$-A|k| - \alpha A = F(k).$$

Solving for the constant A yields

$$A = -\frac{F(k)}{|k| + \alpha},$$

and consequently, the solution is

$$U(k,y) = -\frac{F(k)}{|k| + \alpha} e^{-|k|y}.$$

Invoking the inverse Fourier transform, we arrive at the solution

$$u(x,y) = -\frac{1}{\sqrt{2\pi}} \int_{-\infty}^{\infty} \frac{F(k)}{|k| + \alpha} e^{-|k|y} e^{ikx} dk. \tag{7.66}$$

Let us further analyze (7.66). The integrand becomes singular when a Neumann boundary condition is imposed. Specifically, for $\alpha = 0$, a singularity arises at $k = 0$. To prevent this, we impose the condition $F(0) = 0$, which implies that

$$\int_{-\infty}^{\infty} f(x)\, dx = 0. \tag{7.67}$$

Condition (7.67) is essential and is justified from a physical perspective. For instance, when seeking a stationary heat distribution with a defined heat flow, we must assume that the total flow is zero; otherwise, a stationary distribution cannot exist. Let

$$G(k) = -\frac{1}{|k|} e^{-|k|y}.$$

Then, with the aid of Example 7.5, $G_y = e^{-|k|y}$ has the inverse Fourier transform,

$$g_y = \frac{1}{\sqrt{2\pi}} \int_{-\infty}^{\infty} e^{-|k|y} e^{ikx} dk = \frac{1}{\sqrt{2\pi}} \frac{2y}{x^2 + y^2} = \sqrt{\frac{2}{\pi}} \frac{y}{x^2 + y^2}.$$

An integration of both sides with respect to y, gives

$$g = \frac{1}{\sqrt{2\pi}} \log(x^2 + y^2).$$

Finally, we may write the solution $u(x,y)$ in the form

$$\begin{aligned} u(x,y) &= \frac{1}{\sqrt{2\pi}} \int_{-\infty}^{\infty} \left(\frac{1}{\sqrt{2\pi}} \int_{-\infty}^{\infty} f(\xi) e^{-|k|y} e^{-ik\xi} d\xi \right) e^{ikx} dk \\ &= \frac{1}{\sqrt{2\pi}} \int_{-\infty}^{\infty} f(\xi) \left(\frac{1}{\sqrt{2\pi}} \int_{-\infty}^{\infty} e^{-k[i(\xi - x)] - |k|y} dk \right) d\xi \\ &= \frac{1}{2\pi} \int_{-\infty}^{\infty} f(\xi) \log\left((x - \xi)^2 + y^2)\right)\, d\xi + C, \end{aligned} \tag{7.68}$$

where C is the constant of integration.

Remark 16 *(a) Given the heat flow, the solution is determined up to an additive constant, as the total heat energy can take any value.*

(b) Equation (7.68) *provides a solution that may grow as $|x| \to \infty$, even if f decays rapidly in that region (or even if $f(x) = 0$ for $|x| \geq c$). However, for $|x| \gg 1$ and rapidly decaying f, the term $(x-\xi)^2 + y^2$ can be approximated by $x^2 + y^2$ with a small error. Consequently, the growing component of u is given by*

$$\frac{1}{2\pi} \log(x^2+y^2) \int_{-\infty}^{\infty} f(\xi)\, d\xi,$$

which vanishes precisely due to condition (7.67).

7.10.4 Vibrating String

Consider the wave equation on the real line

$$u_{tt} - c^2 u_{xx} = 0, \quad -\infty < x < \infty, \quad t > 0,$$

with initial conditions

$$u(x,0) = f(x), \quad u_t(x,0) = g(x).$$

Once again, we have the flexibility to take the Fourier transform in either variable. Defining $U(k,t)$ as the Fourier transform in the x-variable, the problem transforms into

$$U_{tt} + (ck)^2 U = 0, \quad U(k,0) = F(k), \quad U_t(k,0) = G(k).$$

This corresponds to the initial value problem for a harmonic oscillator whose solution is given by

$$U(k,t) = F(k)\cos(ckt) + \frac{G(k)}{ck}\sin(ckt).$$

Since sines and cosines can be rewritten in terms of complex exponentials, we have

$$U(k,t) = \frac{1}{2}F(k)(e^{ickt} + e^{-ickt}) + \frac{1}{2ick}G(k)(e^{ickt} - e^{-ickt}).$$

Taking Fourier inverse, we write the solution $u(x,t)$ as

$$\begin{aligned} u(x,t) &= \frac{1}{\sqrt{2\pi}} \int_{-\infty}^{\infty} \left(\frac{1}{2}F(k)(e^{ickt} + e^{-ickt}) + \frac{1}{2ick}G(k)(e^{ickt} - e^{-ickt}) \right) e^{ikx}\, dk \\ &= \frac{1}{\sqrt{2\pi}} \int_{-\infty}^{\infty} \frac{1}{2}F(k) \left(e^{ik(x+ct)} + e^{ik(x-ct)} \right) dk \qquad (7.69) \\ &+ \frac{1}{\sqrt{2\pi}} \int_{-\infty}^{\infty} \frac{1}{2ick}G(k) \left(e^{ik(x+ct)} - e^{ik(x-ct)} \right) dk. \qquad (7.70) \end{aligned}$$

By careful examination of the formulae of inverse Fourier transform, we see that the integral (7.69) can be safely written as

$$\frac{1}{2}\left[f(x-ct) + f(x+ct)\right].$$

Along the same lines of thoughts, integral (7.70) can be written as

$$\frac{1}{2c}\int_{-\infty}^{x}(g(\tau+ct)-g(\tau-ct))\,d\tau.$$

Consequently, the solution may be written in the form

$$u(x,t)=\frac{1}{2}\left[f(x-ct)+f(x+ct)\right]+\frac{1}{2c}\int_{-\infty}^{x}(g(\tau+ct)-g(\tau-ct))\,d\tau$$

By applying the change of variables $\xi=\tau+ct$ for the first term and $\xi=\tau-ct$ for the second, we simplify the integral

$$\begin{aligned}\int_{-\infty}^{x}[g(\tau+ct)-g(\tau-ct)]\,d\tau &= \int_{-\infty}^{x+ct}g(\xi)\,d\xi-\int_{-\infty}^{x-ct}g(\xi)d\xi\\ &= \int_{x-ct}^{x+ct}g(\xi)d\xi.\end{aligned}$$

Thus, we obtain d'Alembert's formula for the wave equation

$$u(x,t)=\frac{1}{2}\left[f(x-ct)+f(x+ct)\right]+\frac{1}{2c}\int_{x-ct}^{x+ct}g(\xi)d\xi. \tag{7.71}$$

7.10.5 Exercises

Exercise 7.83 *Suppose that the heat flow on a semi-infinite rod is governed by the equation* (7.59). *We modify the boundary condition* $u_x(0,t)=f(t)$ *to* $u(0,t)=T_0$, *where* T_0 *is a constant temperature. We seek to show that the solution* $u(x,t)$ *of* (7.59) *is given by*

$$u(x,t)=\frac{2T_0}{\pi}\left[\frac{\pi}{2}-\frac{\pi}{2}\operatorname{erf}\left(\frac{x}{2\sqrt{\rho t}}\right)\right]=T_0\operatorname{erfc}\left(\frac{x}{2\sqrt{\rho t}}\right).$$

Exercise 7.84 *Consider* (7.63) *with*

$$f(x)=\begin{cases}a, & x>0\\ 0, & x<0\end{cases}$$

By letting

$$\eta=\frac{\xi-x}{2\sqrt{t}},$$

show that the solution given by (7.63) *can be written in the form*

$$u(x,t)=\frac{a}{2}\left[1+\operatorname{erf}\left(\frac{x}{2\sqrt{t}}\right)\right].$$

Exercise 7.85 *(Slowing-down of Neutrons (see Sneddon (1951), p. 215)[132] Consider the following physical problem*

$$u_t = u_{xx} + \delta(x)\delta(t),$$

with initial and boundary conditions

$$u(x,0) = \delta(x), \quad \lim_{|x|\to\infty} u(x,t) = 0.$$

This describes the problem of an infinite medium that slows neutrons, where a neutron source is located at the origin. Here, $u(x,t)$ represents the number of neutrons per unit volume per unit time, and $\delta(x)\delta(t)$ represents the source function. Show the solution to the problem is

$$u(x,t) = \frac{1}{\sqrt{2\pi}} \int_{-\infty}^{\infty} e^{-k^2t+ikx}\,dk = \frac{1}{\sqrt{4\pi t}} e^{-x^2/4t}.$$

Exercise 7.86 *Show that the solution of Neumann's problem in the half-plane $y > 0$*

$$u_{xx} + u_{yy} = 0, \quad -\infty < x < \infty, \quad y > 0,$$

$$u_y(x,0) = g(x), \quad -\infty < x < \infty,$$

where u is bounded as $y \to \infty$ and u and u_x vanish as $|x| \to \infty$ is

$$u(x,y) = \frac{1}{2\pi} \int_{-\infty}^{\infty} g(\xi) \log\left((\xi - x)^2 + y^2\right) d\xi.$$

Exercise 7.87 *Show that the solution of Neumann's problem in the half-plane $y > 0$*

$$u_{xx} + u_{yy} = 0, \quad -\infty < x < \infty, \quad 0 < y < b,$$

$$u(x,0) = g(x), \quad u(x,b) = h(x), \quad -\infty < x < \infty,$$

where u is bounded as $y \to \infty$ and u and u_x vanish as $|x| \to \infty$ is

$$u(x,y) = \frac{1}{2\pi} \int_{-\infty}^{\infty} \left\{ \left[G(k) \frac{\sinh(|k|(b-y))}{\sinh(|k|b)} + H(k) \frac{\sinh(|k|y)}{\sinh(|k|b)} \right] d\xi \right\} e^{ikx}\,dk,$$

where $G(k)$ and $H(k)$ denote the Fourier transforms of $g(x)$ and $h(x)$, respectively.

Exercise 7.88 *Use the Fourier transform to solve*

$$u_{xx} + u_{yy} = e^{-(x-1)^2-(y-2)^2}, \quad -\infty < x < \infty, \quad -\infty < y < \infty,$$

and discuss the decay of the solution $u(x,y)$ at large distance from the source.

Exercise 7.89 *Solve the heat flow problem*

$$u_t = u_{xx}, \quad x > 0, \quad t > 0,$$

with the initial and boundary conditions

$$u(x,0) = f(x), \quad u(0,t) = 0.$$

Exercise 7.90 *Use the Fourier transform with respect to the variable t given by*

$$\mathscr{F}\{u(x,t)\} = U(x,k) = \frac{1}{\sqrt{2\pi}} \int_{-\infty}^{\infty} e^{-ikt} u(x,t)\, dt,$$

to show that the solution $u(x,t)$ of the wave equation

$$u_{tt} = c^2 u_{xx}, \quad -\infty < x < \infty, \quad t > 0,$$

with the initial conditions

$$u(x,0) = f(x), \quad u_t(x,0) = g(x), \quad -\infty < x < \infty,$$

is identical to the solution given by (7.71).

Exercise 7.91 *Solve the wave equation with fourth-order spatial derivative*

$$u_{tt} = c^2 u_{xxxx}, \quad -\infty < x < \infty, \quad t > 0,$$

with the initial conditions

$$u(x,0) = f(x), \quad u_t(x,0) = 0, \quad -\infty < x < \infty.$$

Exercise 7.92 *Solve the wave equation with fourth-order spatial derivative and boundary conditions*

$$u_{tt} + c^2 u_{xxxx} = 0, \quad x > 0, \quad t > 0,$$

with the initial and boundary conditions

$$u(x,0) = 0, \quad u_t(x,0) = 0, \quad x > 0,$$

$$u(0,t) = g(t), \quad u_{xx}(0,t) = 0, \quad t > 0.$$

Exercise 7.93 *Solve the Laplace equation in the half-plane*

$$u_{xx} + u_{yy} = 0, \quad 0 < x < \infty, \quad 0 < y < \infty,$$

with the boundary conditions

$$u(x,0) = f(x), \quad 0 \le x < \infty,$$

$$u_x(0,y) = g(y), \quad 0 \le y < \infty,$$

and the conditions

$$u(x,y) \to 0 \text{ uniformly in } x \text{ as } x \to \infty \text{ and uniformly in } y \text{ as } x \to \infty.$$

Exercise 7.94 *Solve the heat equation with a source term*

$$u_t = u_{xx} + tu, \quad -\infty < x < \infty, \quad t > 0,$$

with the initial condition

$$u(x,0) = f(x), \quad u(x,t) \quad \text{is bounded} \quad -\infty < x < \infty.$$

Exercise 7.95 *Solve the heat equation with a source term and a damping term*

$$u_t - u_{xx} + hu = \delta(x)\delta(t), \quad -\infty < x < \infty, \quad t > 0,$$

with the initial condition

$$u(x,0) = 0, \quad u(x,t) \to 0 \text{ uniformly in } t \text{ as } |x| \to \infty.$$

Exercise 7.96 *Solve the heat equation with a source term, damping term, and a convective term*

$$u_t - u_{xx} + h(t)u_x = \delta(x)\delta(t), \quad 0 < x < \infty, \quad t > 0,$$

with the initial conditions

$$u(x,0) = 0, \quad u_x(0,t) = 0, \quad u(x,t) \to 0 \text{ uniformly in } t \text{ as } x \to \infty.$$

8

Classic PDEs

Classical Partial Differential Equations on unbounded domains often arise in problems involving infinite or semi-infinite physical regions, such as wave propagation in open space, heat diffusion over an infinite rod, or potential flow in an unbounded fluid. Solving PDEs on unbounded domains introduces unique challenges due to the lack of natural boundaries, requiring special techniques to handle boundary conditions at infinity or asymptotic behavior. For instance, solutions must typically decay or remain bounded as the spatial variables approach infinity to ensure physical relevance.

The method of characteristics is a powerful analytical tool used primarily for first-order PDEs, enabling the transformation of PDEs into simpler ordinary differential equations (ODEs). In the context of unbounded domains, this method identifies characteristic curves along which the PDE reduces to an ODE, providing insight into how information propagates through the domain. For hyperbolic PDEs, such as the wave equation, the method of characteristics can describe wavefronts traveling through unbounded regions. The approach is particularly effective for linear and nonlinear PDEs where the characteristics provide a direct path to constructing the solution.

8.1 Linear Equations

Now is the perfect time to master solving first-order linear partial differential equations, whether they have constant or variable coefficients, especially when the method of separation of variables is not applicable. We will start by exploring linear equations with constant coefficients.

8.1.1 Linear Equations with Constant Coefficients

We consider the first-order non-homogeneous general partial differential equation

$$Au_x(x,y) + Bu_y(x,y) + Cu(x,y) = G(x,y), \tag{8.1}$$

where A, B, and C are constants such that $A^2 + B^2 \neq 0$, and G is a given continuously differentiable function in x and y. We begin by examining the homogeneous solution

DOI: 10.1201/9781003652540-8

of the corresponding homogeneous equation

$$Au_x + Bu_y + Cu = 0, \tag{8.2}$$

where we have suppressed the independent variables x and y. Note that the requirement $A^2 + B^2 \neq 0$, implies that A and B are not both zero at the same time; otherwise, we would not have a differential equation to solve. Our aim is to use a linear transformation and transform (8.2) into an ordinary differential equation in a single independent variable, say, x, and the dependent variable u. We begin by introducing the transformations $\xi = \xi(x,y)$ and $\eta = \eta(x,y)$. Specifically, we define

$$\xi = c_{11}x + c_{12}y, \quad \eta = c_{21}x + c_{22}y,$$

where the constants c_{11}, c_{12}, c_{21}, and c_{22} are chosen appropriately to transform the given partial differential equation into an ordinary differential equation.

Using the chain rule, we have

$$u_x = u_\xi \xi_x + u_\eta \eta_x = c_{11}u_\xi + c_{21}u_\eta, \tag{8.3}$$

and

$$u_y = u_\xi \xi_y + u_\eta \eta_y = c_{12}u_\xi + c_{22}u_\eta. \tag{8.4}$$

Now substituting (8.3) and (8.4) into (8.2) and rearranging the terms we arrive at

$$(Ac_{11} + Bc_{12})u_\xi + (Ac_{21} + Bc_{22})u_\eta + Cu = 0. \tag{8.5}$$

Assume $A \neq 0$ and choose $c_{11} = 1$, $c_{12} = 0$, $c_{21} = B$, and $c_{22} = -A$. Then

$$\xi = x \quad \text{and} \quad \eta = Bx - Ay,$$

and consequently, (8.5) becomes

$$Au_\xi + Cu = 0,$$

which is a separable ODE. Separating the variables while fixing η we arrive at

$$\frac{u_\xi}{u} = -\frac{C}{A},$$

from which we get $\ln|u| = -\frac{C}{A}\xi + g(\eta)$, for some function g. Taking exponential on both sides, we obtain the homogeneous solution

$$u(\xi,\eta) = e^{-\frac{C}{A}\xi} f(\eta),$$

where f is an arbitrary continuously differentiable function. Thus, the general solution of the homogeneous equation (8.2) is

$$u_h(x,y) = e^{-\frac{C}{A}x} f(Bx - Ay). \tag{8.6}$$

If u_p is a particular solution of (8.1), then the general solution of (8.1) is

$$u(x,y) = u_h(x,y) + u_p(x,y),$$

where $u_h(x,y)$ is given by (8.6). Now if $B \neq 0$ then u_h of (8.1) can be found to be

$$u_h(x,y) = e^{-\frac{C}{B}y} f(Bx - Ay). \tag{8.7}$$

Remark 17 *If both A and B are nonzero, either* (8.6) *or* (8.7) *may be used.*

Example 8.1 Solve

$$u_x + u_y + u = x + y, \tag{8.8}$$

subject to the initial condition

$$u(0,y) = y^2.$$

First we find u_h and since neither A nor B is zero we have the luxury to use either (8.6) or (8.7). Using (8.7) with $A = B = C = 1$, we obtain

$$u_h(x,y) = e^{-y} f(x-y).$$

Using the concept of Chapter 1 of obtaining the particular solution, we try

$$u_p = a + bx + cy.$$

Then substituting into (8.8), we get

$$a + b + c + bx + cy = x + y.$$

It follows from the above expression that $a = -2$, $b = c = 1$. Then the general solution is

$$u(x,y) = e^{-y} f(x-y) + x + y - 2.$$

Our next task is to use the initial data to uniquely determine the function f. The initial data implies that when $x = 0$ and $y = y$ we have $u = y^2$. Thus, it follows from the general solution that

$$y^2 = e^{-y} f(-y) + y - 2.$$

This gives $f(-y) = e^y(y^2 - y + 2)$. Let $k = -y$. Then $f(k) = e^{-k}(k^2 + k + 2)$, from which we get

$$f(x-y) = e^{y-x}[(x-y)^2 + (x-y) + 2].$$

It follows from

$$u(x,y) = e^{-y} f(x-y) + x + y - 2$$

that

$$u(x,y) = e^{-x}[(x-y)^2 + (x-y) + 2] + x + y - 2.$$

□

Remark 18 *Note that if $C = 0$ in (8.2), then the homogeneous solution is reduced to*

$$u_h(x,y) = f(Bx - Ay),$$

for an arbitrary continuously differentiable function f.

Example 8.2 Solve the transport equation

$$u_t + cu_x = 0, \quad -\infty < x < \infty, \quad t > 0, \tag{8.9}$$

subject to

$$u(x,0) = f(x). \tag{8.10}$$

Using the above remark, we immediately obtain the solution to be

$$u(x,t) = f(x - ct).$$

Now suppose that we are given

$$f(x) = \begin{cases} 1, & x \leq 0 \\ 1 - x, & 0 < x \leq 1 \\ 0, & x > 1 \end{cases}$$

The easiest way is to replace x in f with $x - ct$ and rearrange the domains. Thus,

$$u(x,t) = f(x - ct) = \begin{cases} 1, & x - ct \leq 0 \\ 1 - x + ct, & 0 < x - ct \leq 1 \\ 0, & x - ct > 1 \end{cases}$$

It follows that

$$u(x,t) = \begin{cases} 1, & x \leq ct \\ 1 - x + ct, & ct < x \leq 1 + ct \\ 0, & x > 1 + ct \end{cases}$$

□

Next, we give a geometrical interpretation of the characteristic lines and solutions. Consider the simpler form of (8.2) with $C = 0$. That is,

$$Au_x + Bu_y = 0, \tag{8.11}$$

where $A^2 + B^2 \neq 0$. That is A and B cannot both be zero. Let $\cdot$ denote the inner product in $\mathbb{R}^2$. Then (8.11) is equivalent to

$$\langle A, B \rangle \cdot \langle u_x, u_y \rangle = 0.$$

If we let $\overrightarrow{v} = \langle A, B \rangle$, then the left side of the equation is the directional derivative of u in the direction of the vector $\overrightarrow{v}$. That is, the solution u of (8.11) must be constant in the direction of the vector $\overrightarrow{v} = Ai + Bj$. The lines parallel to the vector $\overrightarrow{v}$ have the equation for an arbitrary constant K

$$Bx - Ay = K, \tag{8.12}$$

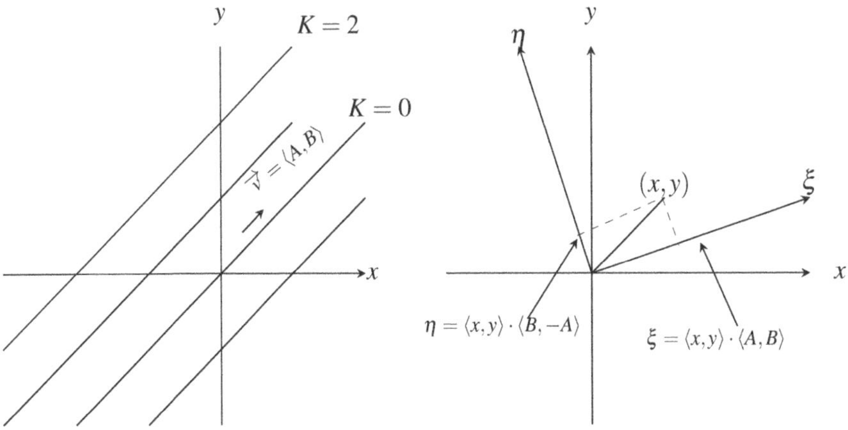

FIGURE 8.1: Characteristic lines; change of coordinates.

since the vector $\langle B, -A \rangle$ is orthogonal to $\vec{v}$, and as such is a normal vector to all lines that are parallel to $\vec{v}$. As shown in Fig. 8.1, (8.12) provides a family of lines and each one of them is uniquely determined by the specific value of K. The family of lines given by (8.12) are called *characteristic lines* for the equation (8.11). We conclude that the solution $u(x,y)$ is constant in the direction of $\vec{v}$, which is true also along the family of lines given by (8.12). Such any line is determined by $K = Bx - Ay$ and hence u will depend only on $Bx - Ay$. It follows that

$$u(x,y) = f(Bx - Ay),$$

as we have obtained in Remark 18, for an arbitrary function f.

8.1.2 Exercises

Exercise 8.1 *Give all the details in arriving at Equation* (8.7).

Exercise 8.2 *Redo Example 8.1 using* (8.6).

Exercise 8.3 *For constants A and B use the transformation*

$$\xi = Ax + By, \quad \eta = Bx - Ay,$$

and obtain the general solution

$$u(\xi, \eta) = e^{-\frac{C}{A^2+B^2}\xi}\Big(f(\eta) + \int g(\xi, \eta) e^{\frac{C}{A^2+B^2}\xi} d\xi\Big),$$

of the non-homogeneous equation

$$Au_x(x,y) + Bu_y(x,y) + Cu(x,y) = g(x,y).$$

Exercise 8.4 *Use the results of Exercise 8.3 to solve*

$$-2u_x + 4u_y + 5u = e^{x+3y},$$

subject to

$$u(0,y) = 2 + y.$$

Exercise 8.5 *Solve each of the given PDEs.*
(a) $u_x + u_y + u = \sin(x)$.
(b) $u_x + u = x, \quad u(0,y) = y^2$.
(c) $2u_x + u_y = x + e^y, \quad u(0,y) = y^2$.

Exercise 8.6 *Solve*
(a) $u_x - 3u_y = \sin(x) + \cos(y), \quad u(x,0) = x$.
(b) $2u_x + u_y = x + e^y, \quad u(0,y) = y^2$.

Exercise 8.7 *Solve*
(a) $u_x - 5u_y = 0, \quad u(x,0) = \dfrac{1}{1+x^2}$.
(b) $u_x + u_t = -3, \quad u(x,0) = e^{3x}$.

Exercise 8.8 *Solve for $u(x,t)$ and sketch $u(x,t)$ at $t = 1,2$.*

$$u_t - u_x = 0, \quad -\infty < x < \infty, \quad t > 0,$$

subject to

$$u(x,0) = f(x), \tag{8.13}$$

where

$$f(x) = \begin{cases} 1, & x \le 0 \\ 1 - x, & 0 < x \le 1 \\ 0, & x > 1 \end{cases}$$

8.2 Equations with Variable Coefficients

We turn our attention to the general first-order linear partial differential equation with variable coefficients

$$A(x,y)u_x(x,y) + B(x,y)u_y(x,y) + C(x,y)u(x,y) = G(x,y), \tag{8.14}$$

where A and B are continuously differentiable functions in the variables in the open set $D \subset \mathbb{R}^2$ and that A and B do not simultaneously identically vanish in D. In addition, the functions C and G are continuous in D. We begin by examining the homogeneous equation

$$Au_x + Bu_y + Cu = 0, \tag{8.15}$$

where we have suppressed the independent variables x and y. Let

$$\xi = \xi(x,y), \quad \eta = \xi(x,y),$$

be a transformation on D with

$$J = \begin{vmatrix} \xi_x & \eta_x \\ \xi_y & \eta_y \end{vmatrix} = \xi_x \eta_y - \xi_y \eta_x \neq 0. \tag{8.16}$$

Note that condition (8.16) is necessary so that the transformation is invertible. Then by (8.3) and (8.4) we have

$$u_x = u_\xi \xi_x + u_\eta \eta_x, \quad u_y = u_\xi \xi_y + u_\eta \eta_y.$$

Substituting into (8.15) and rearranging the terms, we obtain

$$(A\xi_x + B\xi_y)u_\xi + (A\eta_x + B\eta_y)u_\eta + Cu = 0, \tag{8.17}$$

where A, B, and C are functions of ξ and η. As before, our ultimate goal is to reduce (8.17) into an ODE and so we ask that

$$A\eta_x + B\eta_y = 0. \tag{8.18}$$

With this particular choice, the PDE given by (8.17) is reduced into an ODE in the variables ξ, and u. Assume that $A \neq 0$, and let $y(x)$ be a curve with slope $\frac{B}{A}$ that forms the surface, which is a solution to the PDE. Then,

$$\frac{d}{dx}\eta(x, y(x)) = \eta_x + \eta_y \frac{dy}{dx} = \eta_x + \frac{B}{A}\eta_y.$$

This implies that the characteristic lines given by $\eta(x,y) = k$, for constant k is a solution to the characteristic equation $\frac{dy}{dx} = \frac{B}{A}$, with $\eta_y \neq 0$, otherwise $\eta_x = 0$ as well, and this will not be a solution. Thus, $\eta(x,y) = k$ determines η. We may choose the other variable $\xi(x,y) = x$. We observe that for this change of variables

$$J = \begin{vmatrix} \xi_x & \eta_x \\ \xi_y & \eta_y \end{vmatrix} = \eta_y \neq 0,$$

and hence the transformation is invertible. Finally, the PDE given (8.17) is reduced to

$$A(\xi,\eta)u_\xi(\xi,\eta) + C(\xi,\eta)u(\xi,\eta) = 0,$$

which is a separable ODE whose general solution is

$$u_h(x,y) = f(\eta)e^{-\int \frac{C(\xi,\eta)}{A(\xi,\eta)}d\xi} = f(\eta)\Psi(\xi,\eta), \tag{8.19}$$

where f is an arbitrary function and Ψ is known. If u_p is a particular solution of (8.14), then the general solution of (8.14) is

$$u(x,y) - u_h(x,y) + u_p(x,y),$$

where u_h is given by (8.19).

Remark 19 *In some cases, computing the particular solution u_p of* (8.14) *and then adding it to the homogeneous solution u_h to obtain the general solution can be challenging. However, an alternative approach is to transform the entire equation* (8.14) *into a partial differential equation in terms of ξ and η. This transformation allows the use of ordinary differential equation methods to solve it more effectively. As a result,* (8.14) *is converted into the new PDE*

$$a(\xi,\eta)u_\xi + b(\xi,\eta)u_\eta + c(\xi,\eta)u = g(\xi,\eta). \tag{8.20}$$

This can be achieved by making use of

$$u_x = u_\xi \xi_x + u_\eta \eta_x, \quad u_y = u_\xi \xi_y + u_\eta \eta_y.$$

Example 8.3 Consider

$$x^2 u_x - xyu_y + yu = 0, \quad x \neq 0, \tag{8.21}$$

subject to

$$u(1,y) = e^{2y}.$$

Based on the above discussion, we have $A = x^2$, $B = -xy$, and $C = y$. Then, the characteristic equation

$$\frac{dy}{dx} = \frac{-xy}{x^2},$$

has the characteristic lines as its solution given by $xy = k$. Thus, we define $\eta(x,y) = xy$. Let $\xi(x,y) = x$. Then the Jacobian

$$J = \begin{vmatrix} \xi_x & \eta_x \\ \xi_y & \eta_y \end{vmatrix} = x \neq 0.$$

Hence the transformation is invertible. Moreover,

$$u_x = u_\xi + yu_\eta, \quad u_y = xu_\eta.$$

Substituting into (8.21), we obtain

$$x^2 u_\xi + yu = 0.$$

In terms of ξ and η the equation is reduced to

$$u_\xi = \frac{-\eta}{\xi^3}u.$$

An integration with respect to ξ yields the solution

$$u(\xi,\eta) = f(\eta)e^{-\int \frac{\eta}{\xi^3}d\xi} = f(\eta)e^{\frac{\eta}{2\xi^2}}.$$

In terms of x and y we have

$$u(x,y) = f(xy)e^{\frac{y}{2x}}.$$

Using the initial data, we obtain

$$e^{2y} = f(y)e^{\frac{y}{2}},$$

from which we have $f(y) = e^{\frac{3y}{2}}$. Thus, $f(xy) = e^{\frac{3xy}{2}}$, and the solution is

$$u(x,y) = e^{\frac{3xy}{2}} e^{\frac{y}{2x}}.$$

□

8.2.1 Exercises

Exercise 8.9 *Solve.*
(a) $yu_x + xu_y = 0, \quad u(0,y) = e^{-y^2}.$
(b) $xu_x - u_y = 0, \quad u(x,0) = x.$

Exercise 8.10 *Solve.*
(a) $3u_x - 2u_y + u = x, \quad u(x,0) = \dfrac{1}{1+x^2}.$
(b) $u_x + u_t = -3, \quad u(x,0) = e^{3x}.$

Exercise 8.11 *Use Remark 19 to solve*
(a) $xu_x + u_y = 1, \quad u(x,0) = \cos(3x).$
(b) $xu_x - yu_y + y^2u = y^2, \quad x,\ y \neq 0.$

8.3 Quasi-Linear Equations

We now investigate the general first-order *quasi-linear* partial differential equation

$$A(x,y,u)u_x(x,y) + B(x,y,u)u_y(x,y) = C(x,y,u), \tag{8.22}$$

where A, B and C are continuously differentiable functions of the variables x, y and u in some open set $\mathbb{D} \subset \mathbb{R}^3$. We represent the function $u(x,y)$ by a surface $z = u(x,y)$, in the xyz-space, see Fig. 8.2. Surfaces corresponding to solutions of a PDE are called *integral surfaces* of the PDE. Suppressing the argument, and using inner product, the left side of (8.22) can be written as

$$\langle A,\ B,\ C\rangle \cdot \langle u_x,\ u_y,\ -1\rangle = 0.$$

Hence, the normal vector $\langle u_x,\ u_y,\ -1\rangle$ to the surface at a given point is orthogonal to the vector $\langle A,\ B,\ C\rangle$ at that point. It follows that the vector $\langle A,\ B,\ C\rangle$ must be tangent to the surface $u(x,y) - z = 0$ and therefore *integral surfaces* must be formed from the *integral curves* of the vector field $\langle A,\ B,\ C\rangle$. Let $\mathbf{r}(t)$ be the parameterized characteristic curve. Then

$$\mathbf{r}(t) = \langle x(t), y(t), z(t)\rangle,$$

where

$$z(t) = u(x(t), y(t)).$$

The derivative of $\mathbf{r}(t)$ with respect to t must align with the vector field $\langle A, B, C\rangle$ and consequently we have

$$\frac{d\mathbf{r}}{dt} = \left\langle \frac{dx}{dt}, \frac{dy}{dt}, \frac{dz}{dt} \right\rangle = \langle A, B, C\rangle.$$

Since $z(t) = u(x(t), y(t))$, we apply the chain rule and obtain

$$\frac{dz}{dt} = u_x \frac{dx}{dt} + u_y \frac{dy}{dt}.$$

Substituting $\frac{dx}{dt} = A$ and $\frac{dy}{dt} = B$ from the tangent condition, we get

$$\frac{dz}{dt} = Au_x + Bu_y.$$

But from the original PDE,

$$Au_x + Bu_y = C.$$

Therefore,

$$\frac{dz}{dt} = C.$$

Thus, the integral curves are given as solutions to the system of ODEs

$$\frac{dx}{dt} = A(x,y,z), \quad \frac{dy}{dt} = B(x,y,z), \quad \frac{dz}{dt} = C(x,y,z). \tag{8.23}$$

Note that the choice of parameter t in (8.23) is artificial and it can be suppressed to write (8.23) in the form

$$\frac{dx}{A(x,y,z)} = \frac{dy}{B(x,y,z)} = \frac{dz}{C(x,y,z)}. \tag{8.24}$$

Either of the systems (8.23) or (8.24) is called the characteristic system associated with (8.22). Characteristics curves are solutions of either system (8.23) or (8.24).

If a surface $S : z = u(x,y)$ is a union of characteristic curves, then S is an integral surface. We have the following theorem.

Theorem 8.1 *Assume the point* $P = (x_0, y_0, z_0)$ *is a point on the integral surface* $S : z = u(x,y)$. *Let* γ *be the characteristic curve through* P. *Then* γ *lies entirely on* S.

Proof *Assume* $\gamma = (x(t),\ y(t),\ z(t))$ *is a solution of* (8.24) *and satisfies* $\gamma = (x(t_0),\ y(t_0),\ z(t_0)) = (x_0,\ y_0,\ z_0)$ *at some initial time* t_0. *Let*

$$W(t) = z(t) - u(x(t), y(t)).$$

Then $W(t_0) = z(t_0) - u(x_0, y_0) = 0$, *since the point P lies on the surface S. It follows from the chain rule and* (8.23) *that*

$$\begin{aligned} \frac{dW}{dt} &= \frac{dz}{dt} - u_x(x,y)\frac{dx}{dt} - u_y(x,y)\frac{dy}{dt} \\ &= C(x,y,z) - u_x(x,y)A(x,y,z) - u_y(x,y)B(x,y,z), \end{aligned}$$

which can be written in the form

$$\begin{aligned} \frac{dW}{dt} &= C(x,y,W+u(x,y)) - u_x(x,y)A(x,y,W+u(x,y)) \\ &- u_y(x,y)B(x,y,W+u(x,y)). \end{aligned} \quad (8.25)$$

Since $u(x,y)$ *satisfies* (8.22) *we see that* $W = 0$ *is a particular solution of* (8.25). *By uniqueness of solutions, (uniqueness theorem from ODEs) we know that* $W = 0$ *only hold at* $t = t_0$. *Thus, the function* $W(t) = z(t) - u(x(t),\ y(t))$ *vanishes identically. This implies that the curve gamma lies entirely on S. This completes the proof.*

Next, we discuss the *Cauchy Problem for quasi-linear equation* (8.22), which says, find the *integral surface* $z = u(x,y)$ of (8.22) containing an initial curve Γ.

• **Method of Solution**

Let Γ be *non-characteristic* contained in the surface $u(x,y) - z = 0$; (that is the tangent to Γ is nowhere tangent to the characteristic vector $\langle A,\ B,\ C\rangle$ along Γ), be given initial curve. Parametrize Γ by

$$x = f(s), \quad y = g(s), \quad z = h(s), \quad \text{at initial time} \quad t = 0.$$

To construct an integral surface containing Γ, one can proceed as follow;

1) Solve either (8.23) or (8.24) and then use the data given by the initial curve Γ to obtain the constants of integrations. This should produce two independent integral curves,

$$u_1\Big(f(s), h(s), h(s)\Big) = c_1, \quad u_2\Big(f(s), h(s), h(s)\Big) = c_2.$$

2) Eliminate s from the two equations in 1) and obtain the functional relation $F(c_1, c_2) = 0$ between c_1 and c_2.
3) Then the solution to the Cauchy problem is

$$F\left[u_1(x,y,z),\ u_2(x,y,z)\right] = 0.$$

4) If we are given an initial curve, then utilize it to compute the arbitrary function in 3).

Remark 20 *(1) You may obtain different looking solutions, but this depends on whether you use* (8.23) *or* (8.24). *However, once you apply the initial data given by the initial curve* Γ, *then solutions should match.*
(2) The method can be easily extended to PDEs of multiple variables by adapting the relations (8.23) *and* (8.24). *We shall explain this in one of the examples below.*

Example 8.4 Find $u(x,y)$ satisfying

$$xu_x + yu_y = u + 1,$$

with the initial condition

$$u(x,1) = 3x.$$

Let $z = u(x,y)$. First, we parametrize the initial curve Γ given by the data $u(x,1) = 3x$. Let $x = s$ at $t = 0$. Then,

$$\Gamma : x = s, \quad y = 1, \quad z = 3s \quad \text{for} \quad t = 0.$$

Next, we show that Γ is non-characteristic. Comparing the PDE with (8.22), we see that

$$\langle A,B,C \rangle = \langle x,y,z+1 \rangle.$$

The initial curve is $\Gamma : y = 1$, and its tangent vector is

$$T = \left\langle \frac{dx}{ds}, \frac{dy}{ds}, \frac{dz}{ds} \right\rangle = \langle 1,0,3 \rangle.$$

To determine whether Γ is characteristic, we check whether the characteristic vector $\langle A,B,C \rangle = \langle x,y,z+1 \rangle$ is parallel to $\langle 1,0,3 \rangle$. We observe that no scalar λ satisfies

$$\langle 1,0,3 \rangle = \lambda \langle x,y,z+1 \rangle.$$

Hence, Γ is non-characteristic, meaning the initial condition provides a well-posed problem. Now we solve the problem. Using system (8.23) we have

$$\frac{dx}{dt} = x, \quad \frac{dy}{dt} = y, \quad \frac{dz}{dt} = z + 1,$$

with corresponding solutions

$$x = c_1 e^t, \quad y = c_2 e^t, \quad z = c_3 e^t - 1.$$

Applying the initial data given by the parametrized curve Γ, we obtain $c_1 = s$, $c_2 = 1$, and $c_3 = 3s + 1$. Thus, we have the solutions

$$x = se^t, \quad y = e^t, \quad z = (3s+1)e^t - 1.$$

Eliminating s in the two equations $x = se^t$, $y = e^t$ gives $s = \frac{x}{e^t} = \frac{x}{y}$. Finally, substituting $s = \frac{x}{y}$ in $z = (3s+1)e^t - 1$ yields,

$$z = (3\frac{x}{y} + 1)y - 1 = 3x + y - 1.$$

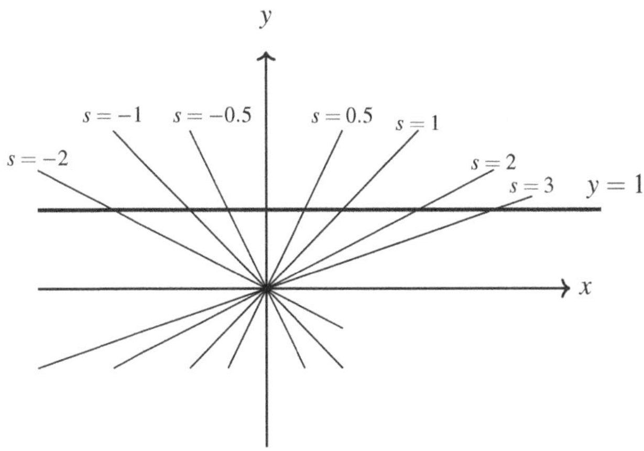

FIGURE 8.2: Characteristic lines $y = \frac{x}{s}$ intersecting initial curve $y = 1$.

So the solution is

$$u(x,y) = 3x + y - 1.$$

Fig. 8.2 shows the characteristic lines given by $y = \frac{x}{s}$, which intersect the initial curve $x = s, y = 1$ at exactly one point (since they are not in the same direction). As these curves cover the entire plane, the solution is therefore defined everywhere. □

In the next example, we revisit the transport equation with inconsistent data.

Example 8.5 Consider the transport equation

$$u_x + cu_y = 0, \quad -\infty < x < \infty, \quad y > 0, \tag{8.26}$$

subject to

$$u(x, cx) = f(x). \tag{8.27}$$

To parametrize the initial curve Γ we let $x = s$ at $t = 0$. Then

$$\Gamma : x = s, \quad y = cs, \quad z = f(s) \text{ for } t = 0.$$

Using system (8.23) we have

$$\frac{dx}{dt} = 1, \quad \frac{dy}{dt} = c, \quad \frac{dz}{dt} = 0,$$

and corresponding solutions

$$x = t + c_1, \quad y = ct + c_2, \quad z = c_3.$$

Applying the initial data given by the parametrized curve Γ, we obtain $c_1 = s$, $c_2 = cs$ and $c_3 = f(s)$. Thus, we have the solutions

$$x = s + t, \quad y = ct + cs, \quad z = f(s).$$

We see that it is not feasible to eliminate s and write the solution u in x and y. Notice that the characteristic line given by $y = c(s+t) = cx$, for a fixed value of c is the same as the equation of the initial curve $y = cx$. In other words, the direction of the characteristic lines are in the same direction as the initial curve. □

Example 8.6 Find $u(x,y)$ satisfying

$$u_x + u_y + u - 1 = 0,$$

subject to

$$u(x, x+x^2) = \sin(x), \quad x > 0.$$

Let $z = u(x,y)$. By parametrizing the initial curve Γ we have

$$\Gamma : x = s, \quad y = s + s^2, \quad z = \sin(s), \quad s > 0 \text{ for } t = 0.$$

Using system (8.23) we have

$$\frac{dx}{dt} = 1, \quad \frac{dy}{dt} = 1, \quad \frac{dz}{dt} = 1 - z,$$

and corresponding solutions evaluated at the initial curve

$$x = s + t, \quad y = s + s^2 + t, \quad z = 1 - \big(1 - \sin(s)\big)e^{-t}.$$

The two equations $x = s+t, \; y = s + s^2 + t$, give $y = x + s^2$. Then solving for s gives $s = \sqrt{y-x}$. Substituting s and $t = x - s$ into the equation $z = 1 - \big(1 - \sin(s)\big)e^{-t}$ and using $z = u(x,y)$, we obtain the solution

$$u(x,y) = 1 - \Big(1 - \sin(\sqrt{y-x})\Big)e^{(-x+\sqrt{y-x})},$$

which is defined for $y > x$. □

Example 8.7 Consider the transport equation

$$u_x + x u_y = 0, \quad -\infty < x, y < \infty, \tag{8.28}$$

subject to

$$u(x,0) = f(x). \tag{8.29}$$

To parametrize the initial curve Γ we let $x = s$ at $t = 0$. Then

$$\Gamma : x = s, \quad y = 0, \quad z = f(s), \quad \text{for} \quad t = 0.$$

Using system (8.23), we get $\dfrac{dx}{1} = \dfrac{dy}{x}$, which gives

$$y = \frac{x^2}{2} + \frac{s^2}{2}.$$

We also get $z = f(s)$ from the relation $\frac{dz}{dt} = 0$. Solving for s we arrive at $s = \pm\sqrt{x^2 - 2y}$. Thus, in order for a solution to exist we must require that f be an even function. Since the solution is constant on the characteristic curves $y = \frac{x^2}{2} + \frac{s^2}{2}$, and the curves intersect the line $y = 0$ (the initial curve), the solution exists in the region where the traces of the characteristic lines given by $y = \frac{x^2}{2}$, intersects the y-axis. Moreover, we must make sure that $s^2 = 2y - x^2 \geq 0$, which holds for $y \leq \frac{x^2}{2}$. Thus, the solution exists in the region $y \leq \frac{x^2}{2}$. □

Example 8.8 Find $u(x,y,z)$ satisfying

$$xu_x + yu_y + u_z = u,$$

with the initial condition

$$u(x,y,0) = h(x,y),$$

where the function h has first continuously partial derivatives. Let $w = u(x,y,z)$. First, we parametrize the initial curve Γ given by the data $u(x,y,0) = h(x,y)$. We set

$$x = s_1, \quad y = s_2, \quad z = 0 \quad \text{at} \quad t = 0.$$

Thus, the initial curve is given by

$$\Gamma: \quad x = s_1, \quad y = s_2, \quad z = 0, \quad w = h(s_1,s_2) \quad \text{for} \quad t = 0.$$

Using system (8.23), we obtain the characteristic equations

$$\frac{dx}{dt} = x, \quad \frac{dy}{dt} = y, \quad \frac{dz}{dt} = 1, \quad \frac{dw}{dt} = w,$$

with corresponding solutions

$$x = c_1e^t, \quad y = c_2e^t, \quad z = t + c_3, \quad w = c_4e^t.$$

Applying the initial conditions from the parametrized curve Γ, we get

$$c_1 = s_1, \quad c_2 = s_2, \quad c_3 = 0, \quad c_4 = h(s_1,s_2).$$

Thus, the solutions take the form

$$x = s_1e^t, \quad y = s_2e^t, \quad z = t, \quad w = h(s_1,s_2)e^t.$$

Solving for s_1, s_2 and using $z = t$, we obtain

$$s_1 = xe^{-t} = xe^{-z}, \quad s_2 = ye^{-t} = ye^{-z}.$$

Consequently, we have

$$w = h(s_1,s_2)e^t = h(xe^{-z}, ye^{-z})e^z.$$

Finally, the solution is given by

$$u(x,y,z) = w(x,y,z) = h(xe^{-z}, ye^{-z})e^z.$$

□

The next example demonstrates that, depending on the nature of the given equation, one must use either (8.23) or (8.24). Before proceeding, we make the following remark.

Remark 21 *A useful technique for integrating a system of first-order equations is the method of multipliers. Recall from algebra that if*

$$\frac{a}{b} = \frac{c}{d},$$

then, for arbitrary multipliers λ, μ, we have

$$\frac{a}{b} = \frac{c}{d} = \frac{\lambda a + \mu c}{\lambda b + \mu d}.$$

This can be further generalized to

$$\frac{a}{b} = \frac{c}{d} = \frac{e}{f} = \frac{\lambda a + \mu c + \nu e}{\lambda b + \mu d + \nu f} \tag{8.30}$$

for arbitrary multipliers λ, μ, and ν.

With appropriate choices of λ, μ, and ν, expression (8.30) *can lead to simpler systems of ODEs that are easier to solve. In particular, if*

$$\lambda A + \mu B + \nu C = 0,$$

then

$$\lambda dx + \mu dy + \nu dz = 0.$$

If there exists a function u such that

$$du = \lambda dx + \mu dy + \nu dz = 0,$$

then $u(x,y,z) = c_1$ defines an integral curve.

Example 8.9 Solve

$$xu_x + (x+u)u_y = y - x \tag{8.31}$$

containing the curve

$$u(x,2) = 1 + 2x.$$

A parametrization of the initial curve Γ gives

$$\Gamma : x = s, \quad y = 2, \quad z = 1 + 2s, \quad \text{for} \quad t = 0.$$

System (8.23) cannot be used since the obtained ODEs will not be separable. Hence we resort to using (8.23) in combination with (8.30). From (8.23) we see that

$$\frac{dx}{x} = \frac{dy}{x+z} = \frac{dz}{y-x}.$$

Then, it follows from (8.30)

$$\frac{\lambda dx + \mu dy + \nu dz}{\lambda x + \mu(x+z) + \nu(y-x)} = \frac{dx}{x} = \frac{dy}{x+z} = \frac{dz}{y-x}. \tag{8.32}$$

Setting in (8.32) $\lambda = 0, \mu = 1$ and $\nu = 1$ we arrive at

$$\frac{dy + dz}{y+z} = \frac{dx}{x}.$$

A direct integration yields

$$\ln(y+z) = \ln(x) + k,$$

or equivalently,

$$\ln\left(\frac{y+z}{x}\right) = k.$$

Taking the exponential on both sides, we obtain the first integral curve

$$\frac{y+z}{x} = e^k = c_1, \tag{8.33}$$

for some constant c_1. Applying the initial data given by the initial curve, we obtain $c_1 = \frac{3+2s}{s}$, or

$$s = \frac{3}{c_1 - 2}. \tag{8.34}$$

Similarly, if we set $\lambda = -1, \mu = 1$, and $\nu = 0$ in (8.32), we obtain $\dfrac{dy - dx}{z} = \dfrac{dz}{y-x}$, or

$$d(y-x)(y-x) = z dz.$$

An integration gives the second integral curve

$$(y-x)^2 - z^2 = c_2.$$

Applying the initial data, we get the relation

$$(2-s)^2 - (1+2s)^2 = c_2. \tag{8.35}$$

Substituting the value of s given by (8.34) into (8.35) yields the expression

$$\left(2 - \frac{3}{c_1 - 2}\right)^2 - \left(1 + 2\frac{3}{c_1 - 2}\right)^2 = c_2. \tag{8.36}$$

Finally, to obtain a functional relation of the solution substitute c_1 and c_2 where c_1 is given by (8.33) and $c_2 = (y-x)^2 - z^2$ into (8.36). Don't forget to replace z by u for the final answer. □

8.3.1 Exercises

Exercise 8.12 *Find $u(x,y)$ satisfying*

$$u_x + u_y = 1 - u, \quad u(x, x + x^2) = e^x, \quad x > 0$$

Exercise 8.13 *Find $u(x,y)$ satisfying*

$$xuu_x - u_y = 0, \quad u(x,0) = x.$$

Exercise 8.14 *Find $u(x,y)$ satisfying*

$$xu_x + yuu_y = -xy, \quad u(x, \frac{1}{x}) = 5. \ x > 0$$

Exercise 8.15 *Find $u(x,y)$ satisfying*

$$uu_x + u_y = 3, \quad u(2x^2, 2x) = 0.$$

Exercise 8.16 *Find $u(x,y)$ satisfying*

$$xuu_x + y^2 u_y = u^2, \quad x > 0, \quad y > 0,$$

and

$$u = s, \quad \text{when} \quad x = \frac{1}{s}, \quad y = 2s.$$

Exercise 8.17 *Find the general solution $u(x,y)$ of*

$$(x + u)u_x + (y + u)u_y = 0.$$

Exercise 8.18 *Find $u(x,y)$ satisfying*

$$xu_x + yu_y = xe^{-u}, \quad u(x, x^2) = 0.$$

Exercise 8.19 *Find $u(x,y)$ satisfying*

$$xu_x + yu_y = 1 + y^2, \quad u(x,1) = x + 1.$$

Answer: $u(x,y) = \ln(y) + \frac{y^2}{2} + \frac{1}{2} + \frac{x}{y}$.

Exercise 8.20 *Find $u(x,y)$ satisfying*

$$2xyu_x + (x^2 + y^2)u_y = 0, \quad x \neq y, \quad x > 0, \quad y > 0,$$

subject to

$$u(x, 1 - x) = \sin\left(\frac{x}{x - y}\right).$$

Exercise 8.21 *Find $u(x,y)$ satisfying*

$$x^2 u_x + uu_y = 1, \quad u(x, 1 - x) = 0, \quad x > 0.$$

Answer: $\dfrac{x}{1 + xu} = \dfrac{u^2}{2} + 1 - y.$

Exercise 8.22 *Find $u(x,y)$ satisfying*

$$3u_x - 2u_y = x - u,$$

subject to

(a) $u(x,x) = 2x,$ *(b)* $u\left(x, \frac{1-2x}{3}\right) = 0.$

Exercise 8.23 *Use the idea of Remark 21 to solve*

(a) $(x+y)u_x + yu_y = x - y;\ u(x,1) = 1 - 3x.$

(b) $(x+y)u_x + (x+yu)u_y = u^2 - 1;\ \ u(x,1) = x^2.$

(c) $(y-u)u_x + (u-x)u_y = x - y;\ \ u(x,2x) = 0.$

(d) $(y-x)u_x + (y+x)u_y = \frac{x^2+y^2}{u};\ \ u(x,0) = x.$

(e) $(y-u)u_x + (u-x)u_y = x - y;\ \ u = 0$ *on* $xy = 1.$

Exercise 8.24 *Find $u(x,y,z)$ satisfying*

(a) $xu_x + yu_y + u_z = u + 1;\ u(x,y,0) = h(x,y),$ *for a suitable function h.*

(b) $u_x + 3u_y - 2u_z = u;\ u(0,y,z) = h(y,z),$ *for a suitable function h.*

(c) $yu_x + xu_y + u_z = 0;\ u(x,y,0) = h(x,y),$ *for a suitable function h.*

Exercise 8.25 *For a constant α, consider the partial differential equation*

$$(x-y)\frac{\partial u}{\partial x} + (x+y)\frac{\partial u}{\partial y} = \alpha u.$$

(a) Show the general solution is given by

$$u(x,y) = (x^2+y^2)^{\alpha/2} F\left(\ln(x^2+y^2) - 2\tan^{-1}\left(\frac{y}{x}\right)\right),$$

where F is an arbitrary function.

(b) Assume the initial condition

$$u(x,0) = x^2, \quad x > 0.$$

Show that the solution satisfies this condition and is given by

$$u(x,y) = (x^2+y^2)\exp\left(\alpha\tan^{-1}\left(\frac{y}{x}\right)\right).$$

Exercise 8.26 *Use polar coordinates and show that the PDE in Exercise 8.25 can be transformed into*

$$\rho u_\rho + u_\theta = \alpha u,$$

and solve it.

Exercise 8.27 *Solve the given partial differential equation*

$$x^2 u_x + xyu_y = u^2, \quad u(y^2, y) = 1.$$

Exercise 8.28 *Solve the given partial differential equation*

$$u_x + x^2 yu_y = -u, \quad u(0,y) = y^2.$$

8.4 Burger's Equation

This section is devoted to the study of Burger's equation

$$u_y + a(u)u_x = 0, \quad -\infty < x < \infty, \quad y > 0, \tag{8.37}$$

subject to the initial data

$$u(x,0) = h(x), \tag{8.38}$$

for a suitable function h. We seek a solution $u = u(x,y)$ that satisfies (8.37) and (8.38). Burger's equation (8.37) can be thought of as nonlinear one-wave equation, where $a(u)$ is the wave speed. We begin by considering the wave speed $a(u)$ as a constant c. In particular, we are analyzing

$$u_y + cu_x = 0, \quad -\infty < x < \infty, \tag{8.39}$$

along the parametrized initial curve

$$\Gamma : x = s, \quad y = 0, \quad z = h(s), \quad \text{at} \quad t = 0.$$

From Example 8.5 we have the characteristic lines (Fig. 8.3)

$$s = x - cy, \quad \text{or} \quad y = \frac{1}{c}x - \frac{s}{c},$$

and the solution u is constant along those lines. Thus, the solution is given by

$$u(x,y) = h(x - cy).$$

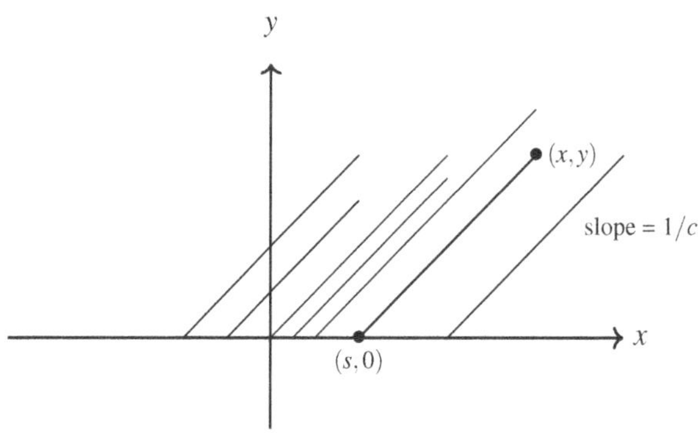

FIGURE 8.3: Characteristic lines $y = \frac{1}{c}x - \frac{s}{c}$ do not run into each others.

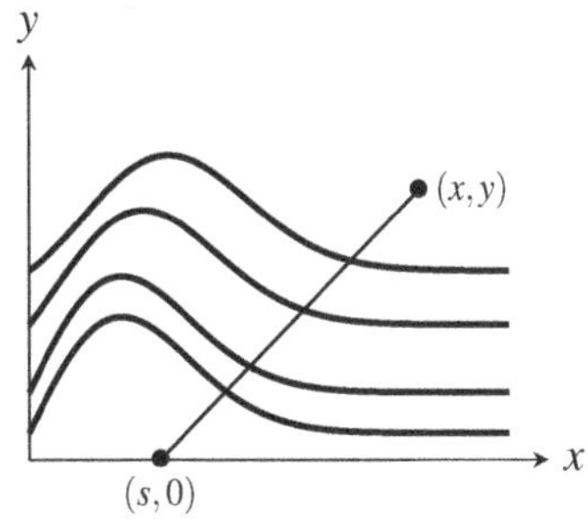

FIGURE 8.4: Wave propagation of the solutions $u(x+cy, y) = h(s-cy)$ by considering y as spatial time.

If we take the initial function $h(x) = e^{-x^2}$, then the solution becomes

$$u(x,y) = e^{-(x-cy)^2},$$

and are graphed in Fig. 8.5 for wave speed $c = 1$ and different values of y.

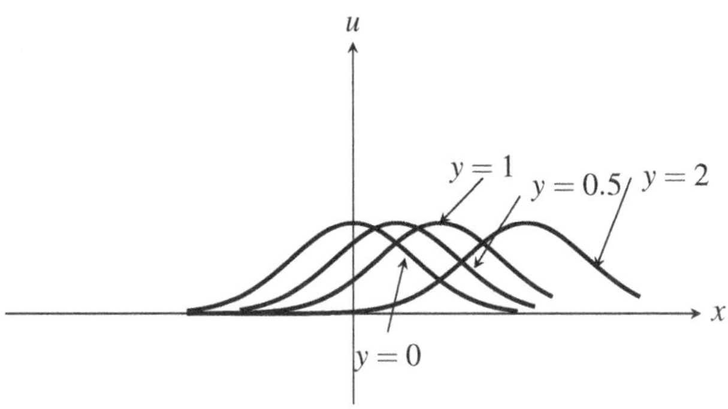

FIGURE 8.5: Wave propagation of the solutions $u(x,y) = e^{-(x-cy)^2}$ for wave speed $c = 1$.

Now we turn our attention to equation (8.37) subject to the initial data given by (8.38). One of the characteristic equation is

$$\frac{dx}{dy} = a(u).$$

If $x(t)$ is the corresponding characteristic curve, then

$$\frac{du(x,y)}{dy} = u_y + u_x \frac{dx}{dy} = u_y + a(u)u_x = 0.$$

This shows that the solution u is constant along the characteristics. Moreover, the characteristics are straight lines as it is obvious from the calculation below

$$\frac{d^2x}{dy^2} = \frac{d}{dy}(\frac{dx}{dy}) = a'\frac{du(x,y)}{dy} = 0.$$

Let

$$\Gamma : x = s, \quad y = 0, \quad z = h(s), \quad \text{at} \quad t = 0.$$

Using system (8.23) we have

$$\frac{dx}{dt} = a(z), \quad \frac{dy}{dt} = 1, \quad \frac{dz}{dt} = 0,$$

and corresponding solutions along Γ are

$$y = t, \quad x = s + a(h(s))t, \quad z = h(s).$$

Replacing t with y, we obtain the family of characteristic lines

$$x = s + a(h(s))y, \tag{8.40}$$

where s is the x-intercept of the characteristic curves. The characteristics are straight lines whose slopes are not constants but varies with s as Fig. 8.6 shows. Eliminating s and using $z = u$, we obtain the solution

$$u(x,y) = h\big(x - a(u)y\big). \tag{8.41}$$

We want a continuous solution with continuous u_x and u_y. Using (8.41) we have

$$u_x = h'\big(x - a(u)y\big)\big(1 - a'(u)u_x y\big),$$

or

$$u_x = \frac{h'(s)}{1 + a'(u)h'(s)y}. \tag{8.42}$$

Similarly, taking partial derivative with respect to y, we get

$$u_y = h'\big(x - a(u)y\big)\big(-a'(u)u_y y - a(u)\big).$$

Solving for u_y gives

$$u_y = \frac{h'(s)a(u)}{1 + a'(u)h'(s)y}. \tag{8.43}$$

Thus, along the characteristic $x = s + a(h(s))y$, we have u_x and u_y given by (8.42) and (8.43), respectively. Moreover, u_x and u_y become infinite at the positive time

$$y = -\frac{1}{a'(u)h'(s)} \quad \text{provided that} \quad a'(u)h'(s) < 0. \tag{8.44}$$

If $a'(u) > 0$, then in order for solutions to exist, expression (8.44) implies that $h'(s) > 0$. In other words, $h(s)$ is an increasing function. Otherwise, the solutions will experience a "blow-up". For example, if $a(u) = u$, then condition (8.44) takes the form

$$y = \min\{-\frac{1}{h'(s)}\},$$

and solutions will experience blow-up at and beyond the time $y = -\frac{1}{h'(s_0)}$, where $h(s_0)$ is the minimum of $h(s)$ at $s = s_0$, and $h'(s) < 0$; that is h is non-increasing.

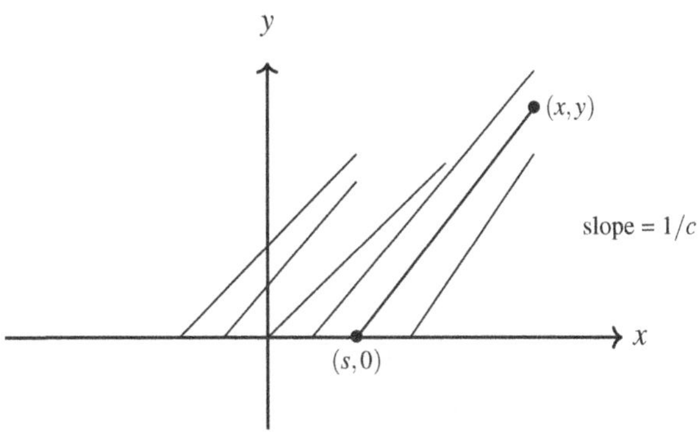

FIGURE 8.6: Characteristic lines running into each others in the nonlinear case, unlike Fig. 8.6.

Example 8.10 Solve

$$u_y + uu_x = 0, \quad -\infty < x < \infty, \quad y > 0 \tag{8.45}$$

subject to the initial data

$$u(x,0) = h(x) = \begin{cases} 1, & x \leq 0 \\ 1-x, & 0 < x \leq 1 \\ 0, & x > 1 \end{cases} \tag{8.46}$$

Parametrizing the initial curve Γ, we get

$$\Gamma : x = s, \quad y = 0, \quad z = \begin{cases} 1, & s \leq 0 \\ 1-s, & 0 < s \leq 1 \\ 0, & s > 1 \end{cases}$$

Using $a(u) = u$, it follows from the above discussion that

$$y = t, \quad x = zt + s, \quad z = h(s) = \begin{cases} 1, & s \leq 0 \\ 1-s, & 0 < s \leq 1 \\ 0, & s > 1 \end{cases}$$

Eliminating t from x and y, the characteristics satisfy

$$x = zy + s \quad \text{or} \quad s = x - zy. \tag{8.47}$$

We try to piece the solution together since our initial data is given piecewise.
1. For $s \leq 0$, $u = z = 1$. Moreover, we have from (8.47) that $s = x - y$. Since, $s \leq 0$, it follows that $x - y \leq 0$. We conclude that

$$u = 1, \quad \text{for} \quad x \leq y.$$

2. For $s > 1$, $u = z = 0$. Moreover, we have from (8.47) that $s = x$. We conclude that

$$u = 0, \quad \text{for} \quad x > 1.$$

3. For $0 < s \leq 1$, $u = z = 1 - s$. Moreover, we have from (8.47) that $s = x - zy = x - (1-s)y$. Solving for s yields $s = \dfrac{x-y}{1-y}$. Now substituting this value of s into $u = 1 - s$ gives

$$u = 1 - \frac{x-y}{1-y} = \frac{1-x}{1-y}.$$

As for the domain, we have $0 < s \leq 1$, which implies that $0 < \dfrac{x-y}{1-y} \leq 1$. Rearranging the terms we arrive at $y < x \leq 1$. We conclude that

$$u = \frac{1-x}{1-y}, \quad \text{for} \quad y < x \leq 1.$$

Finally, the solution is

$$u(x,y) = \begin{cases} 1, & x \leq y \\ \frac{1-x}{1-y}, & y < x \leq 1 \\ 0, & x > 1 \end{cases} \tag{8.48}$$

□

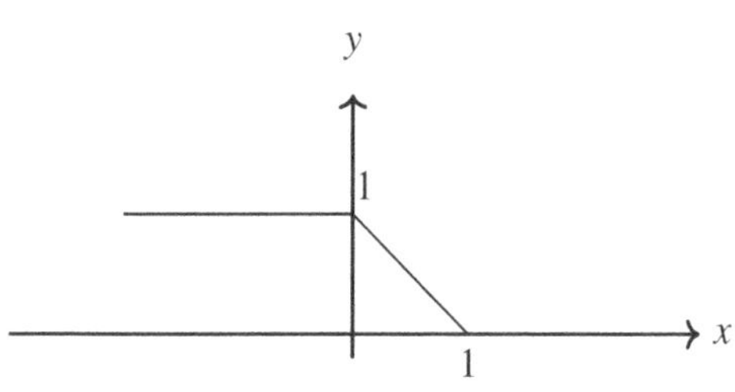

FIGURE 8.7: Initial wave profile.

The obtained solution in (8.48) is valid for $0 \leq y < 1$, and discontinuous at $y = 1$. The characteristics run into each others in the wedged region where $y > 1$ (See Fig. 8.8). Next, our goal is to extend the solution beyond $y \geq 1$. To do so, we introduce a curve starting at the discontinuity point $(1,1)$ and try to construct such curve (shock path) as shown in the next section.

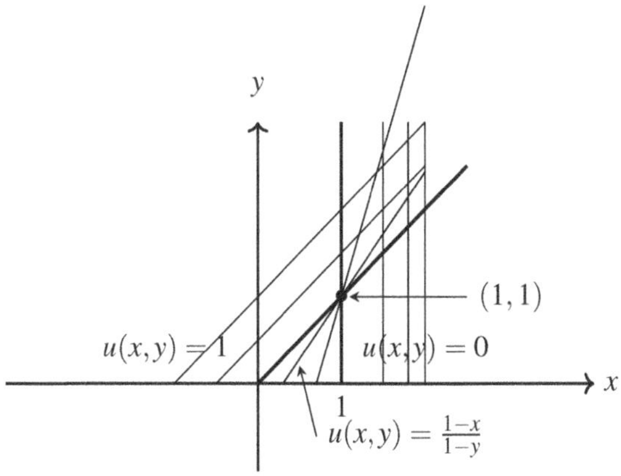

FIGURE 8.8: Characteristic lines running into each others in the nonlinear case, unlike Fig. 8.3.

Example 8.11 Find the blow-up time for

$$u_y + uu_x = 0, \quad -\infty < x < \infty, \quad y > 0,$$

subject to the initial data

$$u(x,0) = \frac{1}{1+x^2}.$$

According to (8.44) the blow-up time is $y = \min\{-\dfrac{1}{h'(\xi)}\}$ with $h(\xi) = \dfrac{1}{1+\xi^2}$. Wherefore, we have $h'(\xi) = -\dfrac{2\xi}{(1+\xi^2)^2}$. We need to find the minimum of the function g where

$$g(\xi) = -\frac{1}{h'(\xi)} = \frac{(1+\xi^2)^2}{2\xi}.$$

After some calculations, we find that

$$g'(\xi) = \frac{(1+\xi^2)(3\xi^2-1)}{2\xi^2}.$$

Setting $g'(\xi) = 0$, it follows that the only feasible solution (positive and real time) is $\xi = \frac{1}{\sqrt{3}}$, which minimizes the function g. Thus, the blow-up time is

$$y = g(\frac{1}{\sqrt{3}}) = \frac{8\sqrt{3}}{9}.$$

□

8.4.1 Shock Path

Consider the first-order PDE

$$u_t + \left(F(u)\right)_x = 0, \quad -\infty < x < \infty, \quad t > 0, \tag{8.49}$$

where F is continuously differentiable function. An equation of this form is called *conservation law* for the following reasons. Integrate (8.49) from $x = a$ to $x = b$ and get

$$\frac{d}{dt}\int_a^b u(x,t)dx + \int_a^b \left(F(u)\right)_x dx = 0.$$

But,

$$\int_a^b \left(F(u)\right)_x dx = F\left(u(b,t)\right) - F\left(u(a,t)\right),$$

and so we have

$$\frac{d}{dt}\int_a^b u(x,t)dx = F\left(u(a,t)\right) - F\left(u(b,t)\right). \tag{8.50}$$

If u is the amount of a quantity per unit length, then the left of (8.50) is the time rate of change of the total amount of the quantity inside the interval $[a,b]$. If $F\left(u(x,t)\right)$ is the flux through x, that is, the amount of the quantity per unit time positively flowing across x, then (8.50) implies that the rate of the quantity in $[a,b]$ equals the flux in at $x = a$ minus the flux out through $x = b$.

As depicted in Fig. 8.9, let $x = \gamma(t)$ be a smooth curve across which u is discontinuous. Assume u is smooth on each side of the curve γ. Let u_0 and u_1 denote the right and left limits of u at $\gamma(t)$, respectively. That is,

$$\lim_{x \to \gamma^+(t)} u(x,t) = u_0 \quad \text{and} \quad \lim_{x \to \gamma^-(t)} u(x,t) = u_1.$$

From (8.50) we have

$$\begin{aligned}
F\left(u(a,t)\right) - F\left(u(b,t)\right) &= \frac{d}{dt}\int_a^{\gamma^-(t)} u(x,t)dx + \frac{d}{dt}\int_{\gamma^+(t)}^b u(x,t)dx \\
&= \int_a^{\gamma^-(t)} u_t(x,t)dx + u(\gamma^-(t),t)\frac{d\gamma^-}{dt} \\
&+ \int_{\gamma^+(t)}^b u_t(x,t)dx - u(\gamma^+(t),t)\frac{d\gamma^+}{dt} \\
&= \int_a^{\gamma^-(t)} u_t(x,t)dx + u_1\frac{d\gamma^-}{dt} \\
&+ \int_{\gamma^+(t)}^b u_t(x,t)dx - u_0\frac{d\gamma^+}{dt}.
\end{aligned} \tag{8.51}$$

As

$$a \to \gamma^-(t) \quad \text{and} \quad b \to \gamma^+(t),$$

we have

$$\int_a^{\gamma^-(t)} u_t(x,t)dx \to 0 \quad \text{and} \quad \int_{\gamma^+(t)}^b u_t(x,t)dx \to 0.$$

As a consequence, expression (8.51) reduces to

$$F(u_1) - F(u_0) = (u_1 - u_0)\frac{d\gamma}{dt}. \tag{8.52}$$

Adopting the notation

$$[F(u)] = F(u_1) - F(u_0) \quad \text{and} \quad [u] = u_1 - u_0,$$

equation (8.52) takes the form

$$\frac{d\gamma}{dt} = \frac{[F(u)]}{[u]}. \tag{8.53}$$

Using (8.53), the weak solution that evolves will be a piecewise smooth function with a discontinuity or shock wave, that propagates with shock speed $\frac{d\gamma}{dt}$.

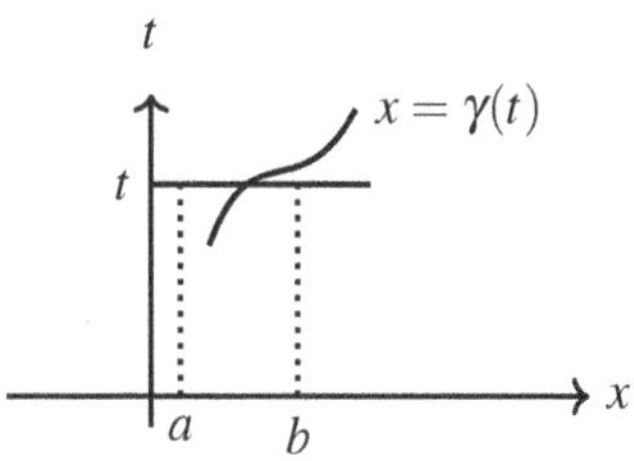

FIGURE 8.9: Shock path.

Example 8.12 Consider the problem of Example 8.10. By setting $t = y$, then $u_t + uu_x = 0$ is equivalent to

$$u_t + \left(\frac{1}{2}u^2\right)_x = 0, \quad x \in \mathbb{R}, \quad t > 0.$$

It follows that

$$F(u) = \frac{1}{2}u^2.$$

We know from Fig. 8.10 that the shock occurs at and beyond $(1,1)$. Also from Fig. 8.10, to the left of the shock we have $u_1 = 1$, and to the right of the shock, $u_0 = 0$. So, $[u] = u_1 - u_0 = 1$ and

$$[F(u)] = F(u_1) - F(u_0) = \frac{1}{2}u_1^2 - \frac{1}{2}u_0^2 = \frac{1}{2}.$$

Thus the path of shock, γ has the slope

$$\frac{d\gamma}{dt} = \frac{1}{2}.$$

Therefore, γ is of the form $2x = t + c$. Since this path passes through $(1,1)$, we see that $c = 1$. It follows that the shock line is given by

$$x(t) = \frac{t}{2} + \frac{1}{2}.$$

Therefore, for $t \geq 1$, the solution is given by

$$u(x,y) = \begin{cases} 1, & x(t) < \frac{t}{2} + \frac{1}{2} \\ 0, & x(t) > \frac{t}{2} + \frac{1}{2} \end{cases} \tag{8.54}$$

□

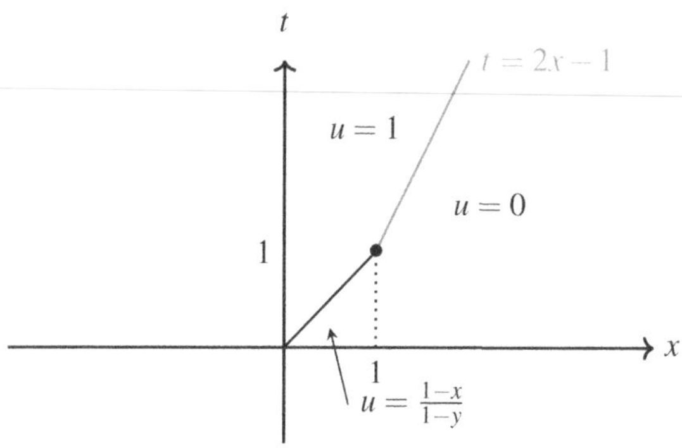

FIGURE 8.10: Characteristic lines intersecting the shock line $t = 2x - 1$.

8.4.2 Exercises

Exercise 8.29 *Find the breaking time for*

$$u_t + uu_x = 0, \quad u(x,0) = e^{-2x^2}, \quad x \in \mathbb{R}, \quad t \geq 0.$$

Exercise 8.30 *Consider the PDE*

$$u_t + u^2 u_x = 0, \quad u(x,0) = \frac{1}{x^4+1}, \quad x \in \mathbb{R}, \quad t \geq 0.$$

(a) Find and graph the characteristics.
(b) Determine the breaking time and find the shock line.
(c) Find the solution before the breaking time.

Exercise 8.31 *Find the breaking time for*

$$u_t + u^3 u_x = 0, \quad u(x,0) = x^{1/3}, \quad x \in \mathbb{R}, \quad t \geq 0,$$

and then find the solution.

Exercise 8.32 *Solve and find the shock line of the traffic flow problem and explain the physical meaning of the solution*

$$u_t + (1-2u)u_x = 0,$$

subject to

$$u(x,0) = \begin{cases} 1/2, & x < 0 \\ 1, & x > 0 \end{cases}$$

Exercise 8.33 *Solve and find the shock line.*

$$u_t + uu_x = 0, \quad u(x,0) = \begin{cases} 0, & x \leq 0 \\ x, & 0 \leq x \leq 2 \\ 2, & x > 2 \end{cases}$$

Exercise 8.34 *Solve and find the shock line.*

$$u_t + u^2u_x = 0, \quad u(x,0) = \begin{cases} 0, & x \leq 0 \\ 2, & 0 \leq x > 1 \\ 0, & x > 1 \end{cases}$$

Exercise 8.35 *Solve*

$$u_t + uu_x = 0, \quad u(x,0) = \begin{cases} 0, & x \leq -1 \\ 2, & -1 < x \leq 1 \\ 1, & x > 1 \end{cases}$$

Exercise 8.36 *Solve*

$$u_t + uu_x = 0, \quad u(x,0) = \begin{cases} 1, & x \leq 0 \\ 1 - \frac{x}{a}, & 0 \leq x < a \\ 1, & x \geq a \end{cases}$$

Exercise 8.37 *(a) Solve*

$$u_t + uu_x = 0, \quad u(x,0) = \begin{cases} 1, & x \leq 0 \\ 2, & x > 0 \end{cases}$$

(b) Draw the characteristics.

Exercise 8.38 *Find the solution of the Burgers' equation*

$$u_t + uu_x = -x, \quad t \geq 0$$

with the initial condition

$$u(x,0) = f(x), \quad -\infty < x < \infty.$$

Exercise 8.39 *(a) Consider the equation*

$$u_t + \left(\frac{u^3}{3}\right)_x = 0$$

for $t > 0$, $-\infty < x < \infty$ with initial data

$$u(x,0) = h(x) = \begin{cases} -a(1-e^{x}), & x < 0 \\ -a(1-e^{-x}), & x > 0 \end{cases}$$

where $a > 0$ is a constant. Solve until the first appearance of a discontinuous derivative and determine the critical time.

(b) Consider the equation

$$u_t + \left(\frac{u^3}{3}\right)_x = -cu.$$

How large does the constant $c > 0$ have to be so that a smooth solution (with no discontinuities) exists for all $t > 0$? Explain.

Exercise 8.40 *(a) Solve the equation*

$$u_t + u^3 u_x = 0$$

for $t > 0$, $-\infty < x < \infty$ with initial data

$$u(x,0) = \begin{cases} -a(1-e^{x}), & x < 0 \\ -a(1-e^{-x}), & x > 0 \end{cases}$$

where $a > 0$ is a constant. Determine the first time when a shock forms.

(b) Consider the equation

$$u_t + u^3 u_x = -cu.$$

With the same initial data and a positive constant c. How large does the constant $c > 0$ have to be in order to prevent shock formation?

8.5 Second-Order PDEs

In this section, we examine general second-order partial differential equations. We begin by looking into their classifications and reduction of orders. Thus, we consider the general second-order linear PDE given by

$$A(x,y)u_{xx} + 2B(x,y)u_{xy} + C(x,y)u_{yy} + D(x,y)u_x + E(x,y)u_y + F(x,y)u = G(x,y), \tag{8.55}$$

where the function u and the coefficients are twice continuously differentiable in some domain $\Omega \subset \mathbb{R}^2$. We shall consider (8.55) along with the Cauchy conditions imposed on some curve Γ that is defined by $y = f(x)$. Imposing Cauchy conditions implies that u_x and u_y are known on Γ. That is,

$$u_x(x,y(x)) = f(x), \quad u_y(x,y(x)) = g(x),$$

where f and g are known continuously differentiable functions. A differentiation of these relations with respect to x gives

$$u_{xx}(x,y(x)) + \frac{dy}{dx}u_{xy}(x,y(x)) = f_x(x) \tag{8.56}$$

$$u_{xy}(x,y(x)) + \frac{dy}{dx}u_{yy}(x,y(x)) = g_x(x). \tag{8.57}$$

In addition, along the curve Γ, equation (8.55) takes the form

$$Au_{xx}(x,y(x)) + 2Bu_{xy}(x,y(x)) + Cu_{yy}(x,y(x)) = H, \tag{8.58}$$

where H is a known function in x. Equations (8.56)–(8.58) determine u_{xx}, u_{yy}, and u_{xy} uniquely unless

$$\triangle = \begin{vmatrix} A & 2B & C \\ 0 & 1 & \frac{dy}{dx} \\ 1 & \frac{dy}{dx} & 0 \end{vmatrix} = -A\left(\frac{dy}{dx}\right)^2 + 2B\frac{dy}{dx} - C = 0.$$

Or,

$$A\left(\frac{dy}{dx}\right)^2 - 2B\frac{dy}{dx} + C = 0. \tag{8.59}$$

The above equation is quadratic in $\frac{dy}{dx}$ with solutions

$$\frac{dy}{dx} = \frac{B \pm \sqrt{B^2 - AC}}{A}. \tag{8.60}$$

When $B^2 - AC > 0$, there exists two families of curves such that no solution can be found when Cauchy conditions are imposed on them. The families of curves are known as the characteristics. On the other hand, there are no characteristics when $B^2 - AC < 0$, and one family of characteristics exists when $B^2 - AC = 0$. We call the initial curve Γ characteristic with respect to (8.55) and Cauchy conditions if $\triangle = 0$ along Γ, non-characteristic if $\triangle \neq 0$ along Γ. When Γ is non-characteristic, the Cauchy data uniquely determine the solution. However, in the case of a characteristic initial curve Γ, then (8.56)- (8.58) are inconsistent, unless more data is offered. Thus, when Cauchy data coincide with the initial curve Γ, the PDE (8.55) has no solution.

Definition 8.1 *The PDE in* (8.55) *has the following classifications. It is hyperbolic if* $B^2 - AC > 0$, *parabolic if* $B^2 - AC = 0$, *and elliptic if* $B^2 - AC < 0$.

Example 8.13

(1) The one-dimensional wave equation

$$u_{tt} - c^2 u_{xx} = 0 \quad \text{is hyperbolic.}$$

(2) The one-dimensional heat equation

$$u_t - k u_{xx} = 0 \quad \text{is parabolic.}$$

(3) The two-dimensional Laplace equation

$$u_{xx} + u_{yy} = 0 \quad \text{is elliptic.}$$

We conclude from Definition 8.1, that the classification of the PDE (8.55) depends on the *highest order terms*. Our next task is to use transformations that will reduce the complicated PDE (8.55) to a simpler one that we can easily solve using the knowledge of the previous sections of this chapter. We introduce the transformations

$$\xi = \xi(x,y), \quad \eta = \eta(x,y), \tag{8.61}$$

where ξ and η are twice continuously differentiable and that the Jacobian

$$J = \begin{vmatrix} \xi_x & \xi_y \\ \eta_x & \eta_y \end{vmatrix} \neq 0 \tag{8.62}$$

in the region of interest. Then, x and y are uniquely determined from the system (8.61). With this in mind, using the chain rule, we obtain

$$u_x = u_\xi \xi_x + u_\eta \eta_x, \quad u_y = u_\xi \xi_y + u_\eta \eta_y$$

$$\begin{aligned} u_{xx} &= (u_\xi)_x \xi_x + u_\xi \xi_{xx} + (u_\eta)_x \eta_x + u_\eta \eta_{xx} \\ &= u_{\xi\xi} \xi_x^2 + 2u_{\eta\xi} \xi_x \eta_x + u_{\eta\eta} \eta_x^2 + u_\xi \xi_{xx} + u_\eta \eta_{xx}. \end{aligned}$$

In a similar fashion, we obtain

$$u_{xy} = u_{\xi\xi} \xi_x \xi_y + u_{\xi\eta} (\xi_x \eta_y + \xi_y \eta_x) + u_{\eta\eta} \eta_x \eta_y + u_\xi \xi_{xy} + u_\eta \eta_{xy},$$

$$u_{yy} = u_{\xi\xi} \xi_y^2 + 2u_{\eta\xi} \xi_y \eta_y + u_{\eta\eta} \eta_y^2 + u_\xi \xi_{yy} + u_\eta \eta_{yy}.$$

Substituting into (8.55), we obtain the new and reduced PDE

$$\hat{A} u_{\xi\xi} + 2\hat{B} u_{\xi\eta} + \hat{C} u_{\eta\eta} + M(\xi, \eta, u, u_\xi, u_\eta) = 0, \tag{8.63}$$

where the new coefficients are known and we list the highest order terms. That is,

$$\hat{A} = A\xi_x^2 + 2B\xi_x \xi_y + C\xi_y^2,$$

$$\hat{B} = A\xi_x \eta_x + B(\xi_x \eta_y + \xi_y \eta_x) + C\xi_y \eta_y,$$

and

$$\hat{C} = A\eta_x^2 + 2B\eta_x \eta_y + C\eta_y^2.$$

Equation (8.63) is called the *canonical form* of (8.55). Thus, it can be easily shown that

$$\hat{B}^2 - \hat{A}\hat{C} = J^2(B^2 - AC),$$

which preserves the personification of the PDE (8.55) under the transformation (8.61). In the next discussion we explain how to find the transformations ξ and η.

Suppose that none of A, B, C is zero. Assume that under the transformations (8.61) $\hat{A}$ and $\hat{C}$ vanish. Let's consider $\hat{A} = 0$. Then it follows that

$$A\xi_x^2 + 2B\xi_x\xi_y + C\xi_y^2 = 0.$$

Divide by ξ_y^2 to obtain

$$A\Big(\frac{\xi_x}{\xi_y}\Big)^2 + 2B\Big(\frac{\xi_x}{\xi_y}\Big) + C = 0, \tag{8.64}$$

which is quadratic in $\frac{\xi_x}{\xi_y}$. Now along the curve ξ = constant we have

$$d\xi = \xi_x dx + \xi_y dy = 0,$$

or

$$\frac{dy}{dx} = -\frac{\xi_x}{\xi_y}. \tag{8.65}$$

Similarly, if we set $\hat{C} = 0$, then we obtain parallel equations to (8.64) and (8.65) in terms of η. A comparison of (8.60), (8.64) and (8.65) yields that ξ and η are the solutions to the ordinary differentials equations given by (8.60), that says

$$\frac{dy}{dx} = \frac{B \pm \sqrt{B^2 - AC}}{A}$$

along which ξ = constant and η = constant.

Remark 22 *(1) **Hyperbolic equations**: If* $B^2 - AC > 0$*, then* ξ *and* η *are uniquely obtained from* (8.60) *and satisfy* (8.62).

*(2) **Parabolic equations**: If* $B^2 - AC = 0$*, we can determine only one of* ξ *or* η *from* (8.60). *Suppose that* ξ *is a solution to*

$$\frac{dy}{dx} = \frac{B}{A}.$$

Then,

$$\xi = \frac{B}{A}x - y.$$

We may set $\eta = x$*, and under this transformation of* ξ *and* η, (8.62) *is satisfied. Furthermore, it follows that* $J = 1 \neq 0$.

*(3) **Elliptic equations**: If* $B^2 - AC < 0$*, we determine both* ξ *and* η *from* (8.60) *and obtain the transformation:*

$$\xi = \frac{B}{A}x - y + i\frac{\sqrt{B^2 - AC}}{A}x, \quad \eta = \frac{B}{A}x - y - i\frac{\sqrt{B^2 - AC}}{A}x.$$

One may show (see Exercise 8.44) that $J \neq 0$ *and that* $\hat{A} \neq 0$, $\hat{C} \neq 0$, *and* $\hat{B} = 0$.

Example 8.14 Solve

$$u_{xx} - 2u_{xy} - 3u_{yy} = 0,$$

subject to the Cauchy conditions

$$u(x,0) = x^2, \quad u_y(x,0) = \frac{1}{3}.$$

From (8.59) we have

$$\left(\frac{dy}{dx}\right)^2 + 2\frac{dy}{dx} - 3 = 0,$$

with

$$\frac{dy}{dx} = \frac{-1 \pm \sqrt{4}}{1} = 1, -3.$$

The two differential equations have the solutions

$$y = x + c_1, \quad y = -3x + c_2.$$

Solving for c_1 and c_2, we get $c_1 = y - x$, and $c_2 = y + 3x$. As a consequence, we may set

$$\xi = y + 3x, \quad \eta = y - x.$$

We note that the Jacobian $J = 4 \neq 0$. It follows that

$$u_x = 3u_\xi - u_\eta, \quad u_y = u_\xi + u_\eta$$

$$u_{xx} = 9u_{\xi\xi} - 6u_{\eta\xi} + u_{\eta\eta}$$

$$u_{xy} = 3u_{\xi\xi} + 2u_{\xi\eta} - u_{\eta\eta}$$

$$u_{yy} = u_{\xi\xi} + 2u_{\eta\xi} + u_{\eta\eta}.$$

Thus,

$$\begin{aligned} u_{xx} - 2u_{xy} - 3u_{yy} &= 9u_{\xi\xi} - 6u_{\eta\xi} + u_{\eta\eta} - 2(3u_{\xi\xi} + 2u_{\xi\eta} - u_{\eta\eta}) \\ &- 3(u_{\xi\xi} + 2u_{\eta\xi} + u_{\eta\eta}) = -16u_{\xi\eta} = 0. \end{aligned}$$

Thus, under the transformation $\xi = y + 3x$, $\eta = y - x$, the original PDE is transformed to the canonical form

$$u_{\xi\eta} = 0,$$

which has the solution

$$u(\xi, \eta) = F(\xi) + G(\eta),$$

for some functions F and G. In terms of x and y the general solution is

$$u(x,y) = F(y + 3x) + G(y - x).$$

Applying the Cauchy condition $x^2 = u(x,0)$, we arrive at

$$x^2 = F(3x) + G(-x). \tag{8.66}$$

To apply the second Cauchy condition, we notice that

$$u_y(x,y) = F'(y+3x) + G'(y-x).$$

Thus,

$$1/3 = F'(3x) + G'(-x).$$

Integrate both sides and then multiply the resulting equation with 3, to get

$$x = F(3x) - 3G(-x). \tag{8.67}$$

Solving the system of equations given by (8.66) and (8.67), we obtain

$$F(3x) = \frac{3x^2+x}{4}, \quad G(-x) = \frac{x^2-x}{4}.$$

Let $z = 3x$, then $F(z) = \frac{z^2}{12} + \frac{z}{12}$, and we conclude that

$$F(y+3x) = \frac{(y+3x)^2}{12} + \frac{(y+3x)}{12}.$$

Similarly, if we set $w = -x$, then $G(w) = \frac{w^2}{4} + \frac{w}{4}$ and as a consequence,

$$G(y-x) = \frac{(y-x)^2}{4} + \frac{(y-x)}{4}.$$

Finally, the solution is

$$\begin{aligned} u(x,y) &= F(y+3x) + G(y-x) \\ &= \frac{(y+3x)^2}{12} + \frac{(y+3x)}{12} + \frac{(y-x)^2}{4} + \frac{(y-x)}{4}. \end{aligned}$$

□

8.5.1 Exercises

Exercise 8.41 *Find the characteristics and reduce to canonical form and then solve.*

$$u_{xx} - 2\sin(x)u_{xy} - \cos^2(x)u_{yy} - \cos(x)u_y = 0.$$

Exercise 8.42 *Find the characteristics and reduce to canonical form and then solve*

$$4u_{xx} + 5u_{xy} + u_{yy} + u_x + u_y = 0.$$

Exercise 8.43 *Show that in the hyperbolic case when* $B^2 - AC > 0$, *we have* $\hat{A} = \hat{C} = 0$, *and* $\hat{B} \neq 0$.

Exercise 8.44 *Show that in the elliptic case when* $B^2 - AC < 0$, *we have* $\hat{A} \neq 0$, $\hat{C} \neq 0$, *and* $\hat{B} = 0$.

Exercise 8.45 *Solve*

$$u_{xx} - 4u_{xy} + 4u_{yy} = 0, \quad u(x,0) = e^{2x}, \quad u_y(x,0) = 5.$$

Exercise 8.46 *Solve*

$$3u_{xx} + 10u_{xy} + 3u_{yy} = 0, \quad u(x,0) = e^{-x}, \quad u_y(x,0) = x.$$

Exercise 8.47 *Find the characteristics and reduce to canonical form and then find the general solution*

$$4u_{xx} + 5u_{xy} + u_{yy} + u_x + u_y = 3.$$

Exercise 8.48 *Find the characteristics and reduce to canonical form and then find the general solution*

$$x^2u_{xx} + 2xyu_{xy} + y^2u_{yy} + xyu_x + y^2u_y = 0.$$

Exercise 8.49 *Find the characteristics and reduce to canonical form and then solve*

$$u_{xx} + 2u_{xy} + 2u_{yy} = 0.$$

Exercise 8.50 *For constants a and b use the transformation $v = ue^{-(a\xi + b\eta)}$, to transform the PDE*

$$u_{xx} - u_{yy} + 3u_x - 2u_y + u = 0,$$

into $v_{\xi\eta} = cv, \quad c =$ constant.

Exercise 8.51 *Solve*

$$x^2u_{xx} - y^2u_{yy} = xy, \quad u(x,1) = x, \quad u_y(x,1) = 6.$$

Exercise 8.52 *Solve*

$$(x-y)u_{xy} - u_x + u_y = 0.$$

Hint: set $v = (x-y)u_x + u$.

Exercise 8.53 *Solve*

$$xyu_{xx} + x^2u_{xy} - yu_x = 0, \quad u(x,0) = e^x, \quad u_y(x,0) = 5.$$

Exercise 8.54 *Consider the PDE*

$$Au_{xx} + Bu_{xy} + Cu_{yy} = 0,$$

where A, B, and C are constants.

(a) Use the transformation $r = x + My, \quad s = x + Ny$, where M and N are distinct constants, to write the PDE in terms of the new independent variables r and s.

(b) Assuming the PDE is of hyperbolic type, find values of M and N such that the transformed equation is

$$u_{rs} = 0.$$

(c) Write the general solution of the original PDE.

8.6 Inequalities and Energy Estimates

In this section, we employ various mathematical techniques that we have already studied to derive inequalities or upper bounds for solutions of certain higher-order partial differential equations, particularly those governing physical or geometric phenomena. The goal of these techniques is to estimate the behavior of solutions without necessarily solving the equations explicitly.

Recall that if $f : \mathbb{R} \to \mathbb{C}$ is a measurable function, then the L^2-norm of f, denoted $\|f\|$, is defined as

$$(f,f)^{\frac{1}{2}} = \|f\| = \left(\int_{-\infty}^{\infty} f(x)\overline{f(x)}\,dx\right)^{1/2} = \left(\int_{-\infty}^{\infty} |f(x)|^2\,dx\right)^{1/2}.$$

If f is defined on a bounded domain $\Omega \subset \mathbb{R}^n$, the L^2-norm is

$$\|f\|_{L^2(\Omega)} = \left(\int_{\Omega} |f(x)|^2\,dx\right)^{1/2}.$$

Theorem 8.2 *Consider the higher-order Cauchy problem*

$$u_t = -u_{xxxx} - 2u_{xx}, \quad -\infty < x < \infty, \quad 0 < t < \infty,$$

with initial condition

$$u(x,0) = \varphi(x),$$

and solution $u(x,t)$. Suppose that $u(x,t)$ and $\varphi(x)$ are C^∞ functions periodic in x with period 2π. That is,

$$u(x+2\pi,t) = u(x,t), \quad -\infty < x < \infty, \quad 0 < t < \infty.$$

Then, for some constant $K > 0$ the **energy estimate** *for u satisfies the estimate*

$$\|u(\cdot,t)\| \le Ke^t\|\varphi\|,$$

where

$$\|u(\cdot,t)\| = \left(\int_0^{2\pi} |u(x,t)|^2 dx\right)^{1/2}, \quad \|\varphi\| = \left(\int_0^{2\pi} |\varphi(x)|^2 dx\right)^{1/2}.$$

Proof *We will use the complex form of Fourier series (see Section 1.4), since u is 2π-periodic. Thus, we let*

$$u(x,t) = \sum_{n=-\infty}^{\infty} a_n(t)e^{inx},$$

where the sequence of functions $a_n(t)$ are to be found. By taking the appropriate derivatives and substituting into the PDE, we get

$$\sum_{n=-\infty}^{\infty} a_n'(t)e^{inx} = -\sum_{n=-\infty}^{\infty} n^4 a_n(t)e^{inx} + 2\sum_{n=-\infty}^{\infty} n^2 a_n(t)e^{inx}.$$

Combining all terms under one summation, we obtain the first-order differential equation in t given by

$$a_n'(t) = (-n^4 + 2n^2)a_n(t),$$

whose solution is easily found to be

$$a_n(t) = a_n(0)e^{(-n^4+2n^2)t}.$$

Applying the given initial condition yields

$$u(x,0) = \sum_{n=-\infty}^{\infty} a_n(0)e^{inx} = \varphi(x).$$

Consequently, from Exercise 1.48 we have

$$\sum_{n=-\infty}^{\infty} a_n^2(0) = \frac{1}{2\pi}\|\varphi\|^2.$$

Next, we compute the norm squared.

$$\|u(x,t)\|^2 = \int_0^{2\pi} u^2(x,t)\,dx = \int_0^{2\pi}\left(\sum_{n=-\infty}^{\infty} a_n(t)e^{inx}\right)\left(\sum_{m=-\infty}^{\infty} a_m(t)\overline{e^{imx}}\right)dx.$$

employing orthogonality of exponentials yields

$$\|u(x,t)\|^2 = \sum_{n=-\infty}^{\infty} a_n^2(t)\int_0^{2\pi} e^{inx}e^{-inx}dx = 2\pi\sum_{n=-\infty}^{\infty} a_n^2(t).$$

Substituting $a_n(t)$ into the expression for $u(x,t)$ results into

$$\begin{aligned}
\|u(x,t)\|^2 &= 2\pi\sum_{n=-\infty}^{\infty} a_n^2(0)e^{2(-n^4+2n^2)t} \\
&\le 2\pi\left|\sum_{n=-\infty}^{\infty} a_n^2(0)\right|\left|\sum_{n=-\infty}^{\infty} a_n^2(0)e^{2(-n^4+2n^2)t}\right| \\
&= 2\pi\left|\sum_{n=-\infty}^{\infty} a_n^2(0)\right|e^{2t}\sum_{n=-\infty}^{\infty} a_n^2(0)e^{-2(n^2-1)^2t}.
\end{aligned} \tag{8.68}$$

Since $\sum_{n=-\infty}^{\infty} e^{-2(n^2-1)^2t}$ is convergent, there exists a positive constant K, such that

$$\sum_{n=-\infty}^{\infty} a_n^2(0)e^{-2(n^2-1)^2t} \le K.$$

Thus, inequality (8.68) *reduces to*

$$\|u(x,t)\|^2 \le Ke^{2t}\|\varphi\|^2.$$

Taking the square root, we conclude that

$$\|u(x,t)\| \le Ke^{t}\|\varphi\|.$$

□

To facilitate learning new techniques, we provide the following theorem.

Theorem 8.3 *Let $u(x,t)$ be a solution of the higher-order Cauchy problem*

$$u_t + g(x,t)u_x + u_{xx} + u_{xxxx} = 0, \quad -\infty < x < \infty, \quad 0 < t < \infty,$$

with initial condition

$$u(x,0) = \varphi(x),$$

Suppose that $u(x,t), g(x,t)$ and $\varphi(x)$ are C^∞, where the functions g and φ are periodic in x with period 2π. That is,

$$g(x+2\pi,t) = g(x,t), \quad \varphi(x+2\pi) = \varphi(x), \quad -\infty < x < \infty, \quad 0 < t < \infty.$$

Let

$$M = \max_{0\le x\le 2\pi} |g(x,t), \quad t > 0.$$

Then, the **energy estimate** *for u satisfies*

$$\|u(\cdot,t)\|^2 \le \|\varphi\|^2 e^{Mt}.$$

Proof *First, we note that the periodicity conditions on u and φ imply*

$$g(2\pi,t) = g(0,t), \quad \varphi(2\pi) = \varphi(0), \quad \varphi'(2\pi) = \varphi'(0).$$

Furthermore, we observe that

$$uu_t = \frac{1}{2}\frac{d}{dt}(u^2).$$

Next, we multiply the PDE by u, integrate both sides with respect to x from $x = 0$ to $x = 2\pi$, and apply integration by parts to obtain

$$\begin{aligned}
\frac{1}{2}\frac{d}{dt}\int_0^{2\pi} u^2\,dx \quad &+ \quad \frac{1}{2}\int_0^{2\pi} g(x,t)\,(u_x)^2\,dx + \int_0^{2\pi} uu_{xx}\,dx + \int_0^{2\pi} uu_{xxxx}\,dx \\
&= \frac{1}{2}\frac{d}{dt}\|u\|^2 + \frac{1}{2}g(x,t)u^2\Big|_0^{2\pi} - \frac{1}{2}\int_0^{2\pi} g_x(x,t)u^2\,dx + uu_x\Big|_0^{2\pi} \\
&- \int_0^{2\pi} u_x^2\,dx + uu_{xxx}\Big|_0^{2\pi} - u_x u_{xx}\Big|_0^{2\pi} + \int_0^{2\pi} u_{xx}^2\,dx
\end{aligned}$$

$$= \frac{1}{2}\frac{d}{dt}\|u\|^2 - \frac{1}{2}\int_0^{2\pi} g_x(x,t)u^2\,dx - \int_0^{2\pi} u_x^2\,dx + \int_0^{2\pi} u_{xx}^2\,dx = 0,$$

where the integral $\int_0^{2\pi} uu_{xxxx}\,dx$, *was integrated by parts twice. Therefore, we arrive at the simplified expression*

$$\frac{1}{2}\frac{d}{dt}\|u\|^2 = \frac{1}{2}\int_0^{2\pi} g_x(x,t)u^2\,dx + \int_0^{2\pi} u_x^2\,dx - \int_0^{2\pi} u_{xx}^2\,dx. \tag{8.69}$$

Next, we need to find a way to control the size of the last two terms in (8.69). *Since* u *is* 2π*-periodic, we let*

$$u(x,t) = \sum_{n=-\infty}^{\infty} a_n(t)e^{inx}.$$

Then, $u_x = i\sum_{n=-\infty}^{\infty} na_n(t)e^{inx}$, *from which we have*

$$u_x^2 = -\left(\sum_{n=-\infty}^{\infty} na_n(t)e^{inx}\right)^2.$$

Similarly, $u_{xx} = -\sum_{n=-\infty}^{\infty} n^2a_n(t)e^{inx}$ *and hence,*

$$u_{xx}^2 = \left(\sum_{n=-\infty}^{\infty} n^2a_n(t)e^{inx}\right)^2.$$

Thus,

$$\int_0^{2\pi} u_x^2\,dx - \int_0^{2\pi} u_{xx}^2\,dx = -\int_0^{2\pi} \left(\sum na_n(t)e^{inx}\right)^2 dx - \int_0^{2\pi} \left(\sum n^2a_n(t)e^{inx}\right)^2 dx.$$

Now we evaluate each of the integrals. Expand the square of the sum we arrive at

$$\left(\sum_n a_n(t)e^{inx}\right)^2 = \sum_n\sum_m a_n(t)a_m(t)e^{i(n+m)x}.$$

Since $u \in C^\infty$, *we have uniform convergence, and therefore we may interchange the summation and integration. Thus,*

$$\int_0^{2\pi} \left(\sum_n na_n(t)e^{inx}\right)^2 dx = \sum_n\sum_m nma_n(t)a_m(t)\int_0^{2\pi} e^{i(n+m)x}\,dx.$$

We already know from complex form of Fourier series that $\int_0^{2\pi} e^{i(n+m)x}\,dx$ *is non-zero only when* $n+m=0$. *In particular,*

$$\int_0^{2\pi} e^{i(n+m)x}\,dx = \begin{cases} 2\pi, & n+m=0 \\ 0, & n+m\neq 0 \end{cases}$$

Thus, the double sum reduces to

$$\begin{aligned}\sum_n\sum_m n m a_n(t)a_m(t)\int_0^{2\pi} n^2 e^{i(n+m)x}\,dx &= 2\pi\sum_n n^2 a_n(t)a_{-n}(t)\\ &= 2\pi\sum_n n^2|a_n(t)|^2,\end{aligned}$$

where $a_n(t)$ and $a_{-n}(t)$ are complex conjugates. That is, $a_{-n}(t)=\overline{a_n(t)}$. Consequently,

$$\int_0^{2\pi}\left(\sum_n n a_n(t)e^{inx}\right)^2 dx = 2\pi\sum_n n^2|a_n(t)|^2.$$

Along the same lines, we obtain

$$\int_0^{2\pi}\left(\sum n^2 a_n(t)e^{inx}\right)^2 dx = 2\pi\sum_n n^4|a_n(t)|^2.$$

Thus, we have shown that

$$\int_0^{2\pi} u_x^2\,dx - \int_0^{2\pi} u_{xx}^2\,dx = -2\pi\sum a_n^2(t)(n^2+n^4)\le 0.$$

Therefore, expression (8.69) *simplifies to*

$$\frac{1}{2}\frac{d}{dt}\|u\|^2 \le \frac{1}{2}\int_0^{2\pi} g_x u^2\,dx \le \frac{1}{2}M\|u\|^2,$$

whose general solution is

$$\|u(x,t)\|^2 \le Ce^{Mt}, \quad t>0.$$

Applying the initial condition, we have $C=\|\varphi\|^2$, and with that we arrive at

$$\|u(\cdot,t)\|^2 \le \|\varphi\|^2 e^{Mt}, \quad t>0.$$

This completes the proof. □

8.6.1 Exercises

Exercise 8.55 *Let f be differentiable at x and periodic with period T. That is,*

$$f(x+T)=f(x) \quad \textit{for all} \quad x\in\mathbb{R}.$$

Then, the derivative $f'(x)$ is also periodic with period T.

Exercise 8.56 (Another way of proving Theorem 8.2) *Consider the PDE in Theorem 8.2. Multiply both sides of the PDE by u and then integrate by parts. To cancel out some of the resulting terms, use the periodicity condition on u and the result of Exercise 8.55. Finally, you may have to use the inequality $-2ab\le\frac{a^2}{2}+\frac{b^2}{2}$.*

Exercise 8.57 (Yet another way of proving Theorem 8.2) *Use Fourier transform to prove Theorem 8.2.*

Exercise 8.58 *Use Fourier transform to prove Theorem 8.3.*

Exercise 8.59 *Let $u(x,t)$ be the solution of the higher-order Cauchy problem*

$$u_t = -u_{xxxx} - 2u_{xx} + \alpha u, \quad -\infty < x < \infty, \quad 0 < t < \infty,$$

with initial condition

$$u(x,0) = \varphi(x).$$

Suppose that $u(x,t)$ and $\varphi(x)$ are C^∞ functions periodic in x with period 2π.

(a) Show that if $\alpha + 1 = 0$, then $\|u(\cdot,t)\|^2 \leq \|\varphi\|^2, \quad t > 0$.

(b) Show that if $\alpha + 1 > 0$, then $\|u(\cdot,t)\| \to \infty$, as $t \to \infty$.

(c) Show that if $\alpha + 1 \leq -\beta$ for a nonzero positive constant β then,

$$\|u(\cdot,t)\| \to 0, \text{ as } t \to \infty.$$

Exercise 8.60 *Let $u(x,t)$ be the solution of the higher-order Cauchy problem*

$$u_t = -uu_x - u_{xx} - u_{xxxx}, \quad -\infty < x < \infty, \quad 0 < t < \infty,$$

with initial condition $u(x,0) = \varphi(x)$. Suppose that $u(x,t)$ and $\varphi(x)$ are C^∞ functions periodic in x with period 2π.

(a) Find the solution $u(x,t)$ when $\varphi(x) = c$, where c is constant.

(b) For $\varepsilon << 1$ (very small) set $\varphi(x) = 1 + \varepsilon g(x)$, where $g(x + 2\pi) = g(x)$, and $\sup_{x\in\mathbb{R}} |g(x)| = 1$. For small ε, Set $u(x,t) = 1 + \varepsilon v(x,t)$ and show that $v(x,t)$ satisfies the PDE

$$v_t = -v_x - v_{xx} - v_{xxxx}, \quad -\infty < x < \infty, \quad 0 < t < \infty,$$

(c) Following the proof technique of Theorem 8.3, show that

$$\|v(\cdot,t)\|^2 \leq \|v(\cdot,0)\|^2, \quad t > 0.$$

9

Heat Equation on Unbounded Domains

In this chapter, we examine the solutions of the heat equation on unbounded domains. We begin by proving that the solution exists and is unique.

9.1 Uniqueness of Solution

In Section 4.2, we derived the heat equation. Here, we aim to establish the uniqueness of the solution to the non-homogeneous heat equation. To achieve this, we define an energy function V and demonstrate that, along the solutions of the heat equation, the energy function remains nonnegative and its derivative is less than or equal to zero. This approach ensures that the solution's behavior is consistent and unique. We begin by considering the non-homogeneous heat equation, along with the associated initial and boundary conditions.

$$\begin{cases} u_t - ku_{xx} = f(x,t), & 0 \leq x \leq l,\ t > 0 \\ u(x,0) = \phi(x), & 0 \leq x \leq l \\ u(0,t) = g(t),\ \ u(l,t) = h(t), & t > 0 \end{cases} \tag{9.1}$$

for given functions f, ϕ, g, h.

Theorem 9.1 *The heat equation given by* (9.1) *has at most one solution.*

Proof *Assume* (9.1) *has two solutions u and v. Set $w = u - v$. Then*

$$w(x,0) = u(x,0) - v(x,0) = \phi(x) - \phi(x) = 0,$$

$$w(0,t) = u(0,t) - v(0,t) = g(t) - g(t) = 0,$$

$$w(l,t) = u(l,t) - v(l,t) = h(t) - h(t) = 0.$$

Moreover,

$$\begin{aligned} w_t - kw_{xx} &= (u-v)_t - k(u-v)_{xx} = u_t - ku_{xx} - \big(v_t - kv_{xx}\big) \\ &= f(x,t) - f(x,t) = 0. \end{aligned}$$

DOI: 10.1201/9781003652540-9

Thus, we arrive at the homogeneous system w,

$$\begin{cases} w_t - kw_{xx} = 0, & 0 \le x \le l,\ t > 0 \\ w(x,0) = 0, & 0 \le x \le l \\ w(0,t) = 0,\ \ w(l,t) = 0, & t > 0 \end{cases} \tag{9.2}$$

Define the energy function

$$V[w](t) = \frac{1}{2}\int_0^l [w(x,t)]^2 dx. \tag{9.3}$$

Then it is clear that

$$V[w](0) = \frac{1}{2}\int_0^l [w(x,0)]^2 dx = \frac{1}{2}\int_0^l [0]^2 dx = 0,$$

and $V[w](t)$ *is positive for* $t > 0$. *To obtain any meaningful information from the energy function, we must show it is decreasing in time along the solutions of* (9.2). *Thus, using the first equation in* (9.2) *we arrive at*

$$\frac{d}{dt}V[w](t) = \int_0^l w(x,t)w_t(x,t)dx = k\int_0^l w(x,t)w_{xx}(x,t)dx.$$

An integration by parts yields

$$\begin{aligned} \frac{d}{dt}V[w](t) &= kw(x,t)w_x(x,t)\big|_0^l - k\int_0^l \big(w_x(x,t)\big)^2 dx \\ &= k\big[w(l,t)w_x(l,t) - w(0,t)w_x(0,t)\big] - k\int_0^l \big(w_x(x,t)\big)^2 dx \\ &= -k\int_0^l \big(w_x(x,t)\big)^2 dx \le 0. \end{aligned}$$

Since the energy function V is decreasing, we get

$$0 \le V[w](t) \le V[w](0) = 0.$$

Hence,

$$V[w](t) = \frac{1}{2}\int_0^l [w(x,t)]^2 dx = 0, \ \textit{for all}\ t \ge 0,$$

which implies $w \equiv 0$, *for all* $x \in [0,l]$, $t > 0$. *Wherefore,* $u - v = 0$ *for all* $x \in [0,l]$, $t > 0$. *This shows*

$$u = v \quad \textit{for all} \quad x \in [0,l], \quad t > 0.$$

This completes the proof.

Next we extend Theorem 9.1 to show uniqueness of solution for the heat equation on $\mathbb{R}$. Consider the heat equation

$$\begin{cases} u_t - ku_{xx} = f(x,t), & -\infty \le x \le \infty,\ t > 0 \\ u(x,0) = \phi(x), & -\infty \le x \le \infty \\ \lim\limits_{x\to\pm\infty} u = 0,\ \ \lim\limits_{x\to\pm\infty} u_x = 0, & t > 0 \end{cases} \tag{9.4}$$

for a given function ϕ.

Theorem 9.2 *The heat equation given by* (9.4) *has at most one solution.*

Proof *Assume* (9.4) *has two solutions* u *and* v. *Set* $w = u - v$. *Then by similar arguments as in the proof of Theorem 9.1* w *is a solution to the homogeneous heat equation*

$$\begin{cases} w_t - kw_{xx} = 0, & -\infty \le x \le \infty,\ t > 0 \\ w(x,0) = 0, & -\infty \le x \le \infty \\ \lim_{x\to\pm\infty} w = 0, \quad \lim_{x\to\pm\infty} w_x = 0, & t > 0 \end{cases} \tag{9.5}$$

Define the energy function for (9.5) *by*

$$V[w](t) = \frac{1}{2}\int_{-\infty}^{\infty} [w(x,t)]^2 dx. \tag{9.6}$$

Then it is clear that

$$V[w](0) = \frac{1}{2}\int_{-\infty}^{\infty} [w(x,0)]^2 dx = 0,$$

and $V[w](t) \ge 0$, *for* $t \ge 0$. *Moreover, using* (9.6), *we get*

$$\frac{d}{dt}V[w](t) = \int_{-\infty}^{\infty} w(x,t)w_t(x,t)dx = k\int_{-\infty}^{\infty} w(x,t)w_{xx}(x,t)dx.$$

An integration by parts yields

$$\begin{aligned} \frac{d}{dt}V[w](t) &= kw(x,t)w_x(x,t)\Big|_{x=-\infty}^{x=\infty} - k\int_{-\infty}^{\infty} \big(w_x(x,t)\big)^2 dx \\ &= -k\int_{-\infty}^{\infty} \big(w_x(x,t)\big)^2 dx \le 0. \end{aligned}$$

Since the energy function V *is decreasing, we get*

$$0 \le V[w](t) \le V[w](0) = 0.$$

Hence,

$$V[w](t) = \frac{1}{2}\int_{-\infty}^{\infty} [w(x,t)]^2 dx = 0, \ \textit{for all}\ t \ge 0,$$

which implies $w \equiv 0$, *for all* $x \in (-\infty,\infty)$, $t > 0$. *Consequently,* $u - v = 0$ *for all* $x \in (-\infty,\infty)$, $t > 0$. *This shows*

$$u = v \quad \textit{for all} \quad x \in (-\infty,\infty),\ t > 0.$$

This completes the proof.

The next theorem is about stability; it says that for a small change in the initial data, only produces a small change in the solution.

Theorem 9.3 (Stability) *Let $u^*(x,t)$ be another solution of (9.4) with initial data ϕ^*. Define the L^2 norm of a function h as*

$$||h||_2 = \left(\int_{-\infty}^{\infty} h^2(x)dx\right)^{\frac{1}{2}}.$$

Assume there is a small change in the initial data. That is, for small and positive ε we have that $||\phi - \phi^*||_2 < \varepsilon$. Then,

$$||u(x,t) - u^*(x,t)||_2 < \varepsilon.$$

Proof *The proof depends on the energy function. For simpler notation, we let*

$$w(x,t) = u(x,t) - u^*(x,t),$$

then w is a solution to the homogeneous heat problem

$$\begin{cases} w_t - kw_{xx} = 0, & -\infty \leq x \leq \infty,\ t > 0 \\ w(x,0) = \phi(x) - \phi^*(x), & -\infty \leq x \leq \infty \\ \lim\limits_{x \to \pm\infty} w = 0, \quad \lim\limits_{x \to \pm\infty} w_x = 0 & \end{cases} \tag{9.7}$$

Define the energy function V for (9.7) by (9.6). Then by Theorem 9.2 we have V is decreasing along the solutions of (9.7) with $V[w](t) \geq 0$, for $t \geq 0$. Thus

$$V[w](t) \leq V[w](0) = \frac{1}{2}\int_{-\infty}^{\infty} [w(x,0)]^2 dx.$$

Accordingly, we have

$$\frac{1}{2}||w||_2^2 = V[w] \leq \frac{1}{2}||w(x,0)||_2^2,$$

or

$$\frac{1}{2}||u - u^*||_2^2 \leq \frac{1}{2}||\phi - \phi^*||_2^2,$$

which implies that

$$||u - u^*||_2 \leq \varepsilon, \quad t \geq 0.$$

We have established stability for all $t \geq 0$ in terms of the square error. This completes the proof.

9.2 Solution of the Heat Equation

Finding the solution of the heat equation using the method of characteristics will not be of much success. To see this, we consider the heat equation over $\mathbb{R}$,

$$u_t - ku_{xx} = 0. \tag{9.8}$$

Then we have $A = k$, $B = C = 0$. Using (8.60), we obtain

$$\frac{dt}{dx} = \frac{B \pm \sqrt{B^2 - AC}}{A} = 0.$$

Wherefore, we only have one characteristic line given by

$$t = c,$$

for some constant c. Remember, we should be able to trace any point in the xt plane along the characteristic lines, which is not the case here since they are parallel to the x-axis.

Our aim then is to find another approach to establishing a bounded solution of the heat equation on an unbounded domain. We consider the heat problem with initial condition

$$\begin{cases} u_t - ku_{xx} = 0, & -\infty < x < \infty,\ t > 0 \\ u(x,0) = \phi(x), & -\infty < x < \infty \end{cases} \tag{9.9}$$

To arrive at the solution of (9.9), we begin by considering simple form of the initial condition. In particular, we first derive the solution of the heat problem with initial condition of the form

$$\begin{cases} V_t - kV_{xx} = 0, & -\infty < x < \infty,\ t > 0 \\ V(x,0) = H(x), \end{cases} \tag{9.10}$$

where $H(x)$ is the *Heavyside step function* defined by

$$H(x) = \begin{cases} 1, & x > 0 \\ 0, & x < 0 \end{cases} \tag{9.11}$$

In the next lemma we explore the invariance properties of the heat equation.

Lemma 25 *[Invariance properties of the heat equation] The heat equation* (9.8) *is invariant under these transformations.*
(a) If $u(x,t)$ is a solution of (9.8)*, then so is $u(x-z,t)$ for any fixed z. (Spatial translation)*
(b) If $u(x,t)$ is a solution of (9.8)*, then so are u_x, u_t, u_{xx}, and so on. (Differentiation)*
(c) If $u_1, u_2, \ldots, u_n$ are solutions of (9.8)*, then so is $\sum_{i=1}^{n} c_i u_i$ for any constants $c_1, c_2, \ldots, c_n$. (Linear combinations)*
(d) If $S(x,t)$ solves (9.8)*, then so is*

$$\int_{-\infty}^{\infty} S(x-y,t)g(y)dy$$

for function g as long as the integral converges.
(e) If $u(x,t)$ is a solution of (9.8)*, then so is $u(\sqrt{a}x, at)$ for any constant $a > 0$. (Dilation, or scaling)*

Proof *The proof of parts (a)–(c) are straightforward and we refer to Exercise 9.1. To prove (d), we assume a finite interval* $[-b,b]$ *partitioned by points* $\{y_i\}_{i=1}^n$ *such that* $-b = y_1 < y_2 < \cdots < y_n = b$ *with equal length* Δy. *Then using (c) combined with representing the integral with the Riemann sum we may write*

$$\int_{-\infty}^{\infty} S(x-y,t)g(y)dy = \lim_{b\to\infty}\int_{-b}^{b} S(x-y,t)g(y)dy = \lim_{n\to\infty}\sum_{i=1}^{n} S(x-y,t)g(y_i)\Delta y.$$

As for the proof of (d), we make use of the chain rule. Let $v(x,t) = u(\sqrt{a}x, at)$. *Then,* $v_t = au_t(\sqrt{a}x, at)$, $v_x = \sqrt{a}u_x(\sqrt{a}x, at)$, *and* $v_{xx} = au_{xx}(\sqrt{a}x, at)$. *Substituting into* (9.9) *we arrive at*

$$au_t(\sqrt{a}x, at) - kau_{xx}(\sqrt{a}x, at) = 0,$$

or

$$u_t(\sqrt{a}x, at) - ku_{xx}(\sqrt{a}x, at) = 0.$$

This completes the proof.

Now we are in a good position to solve (9.9). As we have mentioned before, we will find a particular solution of (9.10) and then make use of it to obtain the solution of the heat problem given in (9.9). First we note that $H(x)$ is invariant under the dilation, since

$$H(\sqrt{a}x) = \begin{cases} 1, & \sqrt{a}x > 0 \\ 0, & \sqrt{a}x < 0 \end{cases} = \begin{cases} 1, & x > 0 \\ 0, & x < 0 \end{cases} = H(x),$$

as $a > 0$. Due to (e) and the fact that H is invariant under the dilation, we know that $V(\sqrt{a}x, at)$ solves (9.10). Due to the uniqueness of solutions, we must have $V(\sqrt{a}x, at) = V(x,t)$, for all $x \in \mathbb{R}$ and $t > 0$. Thus, V is invariant under the dilation $(x,t) \to (\sqrt{a}x, at)$. Since our goal is to transfer (9.10) to an ODE, we want to eliminate one of the variables in V, which can be easily achieved by letting $a = \frac{1}{t}$. Then

$$V(x,t) = V\left(\frac{x}{\sqrt{t}}, \frac{1}{t}t\right) = V\left(\frac{x}{\sqrt{t}}, 1\right).$$

Let q be a function such that

$$V(x,t) = q\left(\frac{x}{\sqrt{t}}\right).$$

Thus, V is completely determined by the function of one variable q. Then,

$$q_t = -\frac{x}{2t^{\frac{3}{2}}}q'\left(\frac{x}{\sqrt{t}}\right), \quad q_{xx} = \frac{1}{t}q''\left(\frac{x}{\sqrt{t}}\right).$$

This reduces the PDE into,

$$V_t - kV_{xx} = -\frac{x}{2t^{\frac{3}{2}}}q'\left(\frac{x}{\sqrt{t}}\right) - \frac{k}{t}q''\left(\frac{x}{\sqrt{t}}\right) = 0,$$

which implies that

$$-\frac{x}{2t^{\frac{1}{2}}}q'\left(\frac{x}{\sqrt{t}}\right) - kq''\left(\frac{x}{\sqrt{t}}\right) = 0.$$

Setting, $z = \dfrac{x}{\sqrt{t}}$, the above second-order differential equation takes the form

$$q''(z) + \frac{z}{2k} q'(z) = 0.$$

Let $G(z) = q'(z)$, then the above equation reduces to the first-order differential equation

$$G'(z) + \frac{z}{2k} G(z) = 0,$$

which has the solution (see Appendix)

$$G(z) = q'(z) = Ce^{-\frac{z^2}{4k}}.$$

Integrating from 0 to z gives

$$q(z) = C \int_0^z e^{-\frac{s^2}{4k}} ds + D,$$

for unknown constants C and D. This yields

$$V(x,t) = q\left(\frac{x}{\sqrt{t}}\right) = C \int_0^{\frac{x}{\sqrt{t}}} e^{-\frac{s^2}{4k}} ds + D. \tag{9.12}$$

Note that by making a substitution in Lemma 22, one can show that

$$\int_0^\infty e^{-\frac{s^2}{4k}} ds = \sqrt{k\pi},$$

which we will need to compute the constants in (9.12). Since (9.12) is only valid for $t > 0$, so to check the initial condition in (9.10) we take the limit at $t \to 0^+$. Additionally, we observe that

$$\lim_{t \to 0+} \frac{x}{\sqrt{t}} = \begin{cases} \infty, & x > 0 \\ -\infty, & x < 0 \end{cases}$$

Thus, for $x > 0$, we have that

$$1 = \lim_{t \to 0^+} V(x,t) = C \int_0^\infty e^{-\frac{s^2}{4k}} ds + D = C\sqrt{k\pi} + D.$$

On the other hand, for $x < 0$, we obtain

$$0 = \lim_{t \to 0^+} V(x,t) = C \int_0^{-\infty} e^{-\frac{s^2}{4k}} ds + D = -C\sqrt{k\pi} + D.$$

By solving the system of equations

$$1 = C\sqrt{k\pi} + D, \quad 0 = -C\sqrt{k\pi} + D,$$

we obtain

$$C = \frac{1}{\sqrt{4k\pi}}, \quad D = \frac{1}{2}.$$

Plugging the constants into (9.12) we arrive at

$$V(x,t) = \frac{1}{\sqrt{4k\pi}} \int_0^{\frac{x}{\sqrt{t}}} e^{-\frac{s^2}{4k}} ds + \frac{1}{2}.$$

We try to put V in terms of the error function that we defined in Definition 7.2. Let $z = \frac{s}{\sqrt{4k}}$. Then

$$\frac{1}{\sqrt{4k\pi}} \int_0^{\frac{x}{\sqrt{t}}} e^{-\frac{s^2}{4k}} ds = \frac{1}{\sqrt{\pi}} \int_0^{\frac{x}{\sqrt{4kt}}} e^{-z^2} dz = \frac{1}{2} \operatorname{erf}\left(\frac{x}{\sqrt{4kt}}\right).$$

Hence the unique particular solution of (9.10) is given by

$$V(x,t) = \frac{1}{\sqrt{\pi}} \int_0^{\frac{x}{\sqrt{4kt}}} e^{-z^2} dz + \frac{1}{2}, \tag{9.13}$$

and in terms of the error function,

$$V(x,t) = \frac{1}{2} + \frac{1}{2} \operatorname{erf}\left(\frac{x}{\sqrt{4kt}}\right). \tag{9.14}$$

Solving (9.9).

Now we attempt to find the solution for the general heat problem given by (9.9). Define the function

$$S(x,t) = \frac{\partial V}{\partial x}(x,t). \tag{9.15}$$

By the invariance property (a), $S(x,t)$ solves (9.9). Therefore, by the invariance property (b),

$$u(x,t) = \int_{-\infty}^{\infty} S(x-y,t)\phi(y)dy, \quad \text{for } t > 0 \tag{9.16}$$

solves (9.9). We must show $u(x,t)$ given by (9.16) is the unique solution of (9.9). This can be accomplished by showing it satisfies the initial condition. Utilizing (9.15), we can write u as follows:

$$u(x,t) = \int_{-\infty}^{\infty} \frac{\partial V}{\partial x}(x-y,t)\phi(y)dy = -\int_{-\infty}^{\infty} \frac{\partial}{\partial y}[V(x-y,t)]\phi(y)dy.$$

We note that $S(x-y,t)$ decays exponentially as $y-x$ grows larger. For now, we assume

$$\phi(\pm\infty) = 0,$$

so we may perform an integration by parts on the above integral. That is

$$u(x,t) = -V(x-y,t)\phi(y)\Big|_{-\infty}^{\infty} + \int_{-\infty}^{\infty} V(x-y,t)\phi'(y)dy = \int_{-\infty}^{\infty} V(x-y,t)\phi'(y)dy.$$

Setting $t = 0$, and noticing that $V(x-y,0)$ is $H(x)$, we obtain

$$u(x,0) = \int_{-\infty}^{\infty} V(x-y,0)\phi'(y)dy = \int_{-\infty}^{x} \phi'(y)dy = \phi(y)\Big|_{-\infty}^{x} = \phi(x).$$

Thus, $u(x,t)$ does satisfy the initial condition of (9.9). Left to compute $S(x,t)$ which can be easily done from (9.15) with V given by (9.13). That is, using (9.13), we obtain

$$S(x,t) = \frac{\partial V}{\partial x}(x,t) = \frac{1}{\sqrt{4\pi kt}} e^{-\frac{x^2}{4kt}}. \tag{9.17}$$

Finally, substituting S given by (9.17) into (9.16) we have the explicit form of the solution of (9.9)

$$u(x,t) = \frac{1}{\sqrt{4\pi kt}} \int_{-\infty}^{\infty} e^{-\frac{(x-y)^2}{4kt}} \phi(y)dy, \quad \text{for } t > 0. \tag{9.18}$$

The function $S(x,t)$ is known as the *heat kernel.* Other resources may refer to it as *fundamental solution, source function, Green's function,* or *propagator* of the heat equation

We present the following example.

Example 9.1 Consider the heat problem

$$\begin{cases} u_t - 4u_{xx} = 0, & -\infty < x < \infty,\ t > 0 \\ u(x,0) = \phi(x), & -\infty < x < \infty \end{cases}$$

with initial data

$$\phi(x) = \begin{cases} 1 - |x|, & |x| < 1 \\ 0, & |x| \geq 1 \end{cases}$$

We are searching for a bounded solution $u(x,t)$. Substituting the initial data into (9.18) yields

$$\begin{aligned} u(x,t) &= \frac{1}{\sqrt{4\pi kt}} \int_{-\infty}^{\infty} e^{-\frac{(x-y)^2}{4kt}} \phi(y)dy \\ &= \frac{1}{\sqrt{16\pi t}} \Big[\int_{-1}^{0} e^{-\frac{(x-y)^2}{16t}} (1+y)dy + \int_{0}^{1} e^{-\frac{(x-y)^2}{16t}} (1-y)dy \Big]. \end{aligned}$$

We make the change of variables

$$s = \frac{y-x}{\sqrt{16t}},$$

and transform the above integrals to obtain

$$u(x,t) = \frac{1}{\sqrt{\pi}} \Big[\int_{-\frac{1+x}{\sqrt{16t}}}^{\frac{-x}{\sqrt{16t}}} e^{-s^2} (1 + x + 4s\sqrt{t})ds$$

$$+ \int_{\frac{-x}{\sqrt{16t}}}^{-\frac{x-1}{\sqrt{16t}}} e^{-s^2}(1 - x - 4s\sqrt{t})ds\Big].$$

The integrals with integrands in which e^{-s^2} is not multiplied by a term that has an s in it can be written in terms of the error function. The rest can be integrated out and at the end we end up with the solution

$$\begin{aligned} u(x,t) &= \frac{1}{2}(1+x)\left(\operatorname{erf}\left(\frac{l+x}{\sqrt{16t}}\right) - \operatorname{erf}\left(\frac{x}{\sqrt{16t}}\right)\right) \\ &+ \frac{1}{2}(1-x)\left(\operatorname{erf}\left(\frac{x}{\sqrt{16t}}\right) - \operatorname{erf}\left(\frac{x-1}{\sqrt{16}}\right)\right) \\ &+ \frac{2\sqrt{t}}{\sqrt{\pi}}\left(e^{-\frac{(x+1)^2}{16t}} - e^{-\frac{x^2}{16t}}\right) + \frac{2\sqrt{t}}{\sqrt{\pi}}\left(e^{-\frac{(x-1)^2}{16t}} - e^{-\frac{x^2}{16t}}\right). \end{aligned} \tag{9.19}$$

□

We end this section by looking into the solution of the non-homogeneous heat problem with initial condition,

$$\begin{cases} u_t - ku_{xx} = f(x,t), & -\infty < x < \infty,\ t > 0 \\ u(x,0) = \phi(x), & -\infty < x < \infty \end{cases} \tag{9.20}$$

for given function f, and ϕ. The derivation of the solution of (9.20) depends on *Dumamel's principle* and we ask the reader to consult Section 10.5 or with [59], [106], [136]. The next theorem is stated without proof.

Theorem 9.4 *The heat equation given by* (9.20) *has the solution*

$$u(x,t) = \int_{-\infty}^{\infty} S(x-y,t)\phi(y)dy + \int_0^t \int_{-\infty}^{\infty} S(x-y,t-\tau)f(y,\tau)dyd\tau, \quad for\, t > 0 \tag{9.21}$$

where $S(x,t)$ *is given by* (9.17).

9.3 Heat Equation on Semi-Infinite Domain—Dirichlet Condition

In the previous section, we considered the heat problem on $\mathbb{R}$. Presently, we want to make use of our previous results and obtain an explicit solution for the heat equation on the unbounded semi-infinite domain $[0,\infty]$. Now it makes sense to add a boundary condition at $x = 0$, that is, $u(0,t) = 0$. Thus, we consider the heat problem with Dirichlet boundary condition at the end point $x = 0$,

$$\begin{cases} v_t - kv_{xx} = 0, & 0 < x < \infty,\ t > 0 \\ v(x,0) = \phi(x), & x > 0 \\ v(0,t) = 0, & t > 0 \end{cases} \tag{9.22}$$

Our aim is to find a solution of (9.22) as we did for the heat problem over the entire real line. There will be no need to start from scratch, but instead we will reintroduce the problem over the entire real line by extending the initial data to the whole line. Whatever method we use to extend to the negative half-line, we should make sure that the boundary condition is automatically satisfied by the solution of the problem on the whole line that arises from the extended data. For heat problems with Dirichlet condition, one would choose the *odd extension* of the initial data $\phi(x)$. If $\psi(x)$ is odd, then $\psi(x) = -\psi(-x)$, from which we get $2\psi(0) = 0$, or $\psi(0) = 0$. This is true for any odd function. We make it formal in the next lemma.

Lemma 26 *Let $f : (-\infty, \infty) \to \mathbb{R}$ be an odd function $(f(x) = -f(-x))$, that is continuous at $x = 0$ then $f(0) = 0$.*

Proof *Since f is continuous at $x = 0$ and odd, we have*

$$f(0) = \lim_{x\to 0^+} f(x) = -\lim_{x\to 0^+} f(x) = \lim_{x\to 0^-} f(x) = -\lim_{x\to 0^-} f(x) = -f(0).$$

this gives $2f(0) = 0$, or $f(0) = 0$. This completes the proof.

The next lemma assures us that if the initial data is odd, then the solution of the heat equation over the real line is also odd.

Lemma 27 *Let $u(x,t)$ be the solution of the heat equation on $-\infty < x < \infty$. If the initial data $\phi(x) = u(x,0)$ is odd, then for all $t \geq 0$, $u(x,t)$ is an odd function of x.*

Proof *By* (9.18) *the solution is given by*

$$\begin{aligned} u(-x,t) &= \frac{1}{\sqrt{4\pi kt}} \int_{-\infty}^{\infty} e^{-\frac{(-x-y)^2}{4kt}} \phi(y)dy = \frac{1}{\sqrt{4\pi kt}} \int_{\infty}^{-\infty} e^{-\frac{(-x+yy)^2}{4kt}} \phi(-y)(-dy), \ (y \mapsto -y) \\ &= \frac{1}{\sqrt{4\pi kt}} \int_{-\infty}^{\infty} e^{-\frac{(-x+y)^2}{4kt}} \phi(-y)dy = -\frac{1}{\sqrt{4\pi kt}} \int_{-\infty}^{\infty} e^{-\frac{(x-y)^2}{4kt}} \phi(y)dy \\ &= -u(x,t). \end{aligned}$$

This completes the proof.

Thus, by Lemma 26 and Lemma 27, we see that if the initial data $\phi(x)$ is odd, then $u(x,t)$ is odd. Since

$$u(x,t) + u(-x,t)$$

solves the heat problem, we have

$$2u(0,t) = 0,$$

and hence $u(0,t) = 0$ for any $t > 0$, which is exactly the boundary condition in (9.22) for v.

In summary, if one extends the initial data to an odd function on the whole real line, then the solution with the extended initial data automatically satisfies the Dirichlet boundary condition of (9.22).

We now remind you of the following definition.

Definition 9.1 *The odd extension of a function $f(x)$ denoted by $f_o(x)$ is defined as*

$$f_o(x) = \begin{cases} f(x), & x > 0 \\ -f(-x), & x < 0 \\ 0, & x = 0 \end{cases} \tag{9.23}$$

The odd extension f_o is defined for negative x by reflecting the $f(x)$ with respect to the vertical axis, and then with respect to the horizontal axis. This procedure produces a function whose graph is symmetric with respect to the origin, and thus it is odd.

For example. if $f(x) = x$, $x > 0$, then $f_o(x) = x$ for $-\infty < x < \infty$. In light of the above discussion we recast the heat problem in (9.22) with extended data as

$$\begin{cases} u_t - ku_{xx} = 0, & -\infty < x < \infty,\ t > 0 \\ u(x,0) = \phi_o(x) \end{cases} \tag{9.24}$$

We already know from (9.16) that the solution of (9.24) is given by

$$u(x,t) = \int_{-\infty}^{\infty} S(x-y,t)\phi_o(y)dy, \ \text{for } t > 0. \tag{9.25}$$

Since v is a solution for $x \geq 0$, we have $v(x,t) = u(x,t)$ for $x \geq 0$. Notice that

$$v(x,0) = u(x,0)\Big|_{x>0} = \phi_o(x)\Big|_{x>0} = \phi(x),$$

and $v(0,t) = u(0,t) = 0$, since $u(x,t)$ is an odd function of x. Thus, $v(x,t)$ satisfies the boundary condition in (9.24). Substituting $\phi_o(x)$ into the solution given by (9.25), we obtain by splitting the integral over two regions

$$\begin{aligned} u(x,t) &= \int_0^{\infty} S(x-y,t)\phi_o(y)dy + \int_{-\infty}^{0} S(x-y,t)\phi_o(y)dy \\ &= \int_0^{\infty} S(x-y,t)\phi(y)dy - \int_{-\infty}^{0} S(x-y,t)\phi(-y)dy. \end{aligned}$$

Substituting y for $-y$ in the second integral leads to

$$u(x,t) = \int_0^{\infty} S(x-y,t)\phi(y)dy - \int_0^{\infty} S(x+y,t)\phi(y)dy.$$

Using (9.17)

$$S(x,t) = \frac{1}{\sqrt{4\pi kt}} e^{-\frac{x^2}{4kt}},$$

we may write the solution formula for (9.22) as follows:

$$v(x,t) = \frac{1}{\sqrt{4\pi kt}} \int_0^{\infty} \left[e^{-\frac{(x-y)^2}{4kt}} - e^{-\frac{(x+y)^2}{4kt}} \right] \phi(y)dy, \ \text{for } t > 0. \tag{9.26}$$

Example 9.2 *Consider the heat equation in (9.22) with $\phi(x) = u_0$ for constant u_0. Substituting $\phi(y) = u_0$ into the solution $v(x,t)$ in (9.26), we obtain*

$$v(x,t) = \frac{u_0}{\sqrt{4\pi kt}} \int_0^\infty \left[e^{-\frac{(x-y)^2}{4kt}} - e^{-\frac{(x+y)^2}{4kt}} \right] dy, \; for \, t > 0.$$

Making the change of variable $s = \frac{x-y}{\sqrt{4kt}}$, we arrive at

$$\frac{u_0}{\sqrt{4\pi kt}} \int_0^\infty e^{-\frac{(x-y)^2}{4kt}} ds = -\frac{u_0}{\sqrt{\pi}} \int_{\frac{x}{\sqrt{4kt}}}^{-\infty} e^{-s^2} ds = \frac{u_0}{\sqrt{\pi}} \int_{-\infty}^{\frac{x}{\sqrt{4kt}}} e^{-s^2} ds.$$

Similarly, by letting $s = \frac{x+y}{\sqrt{4kt}}$, we arrive at

$$\frac{u_0}{\sqrt{4\pi kt}} \int_0^\infty e^{-\frac{(x+y)^2}{4kt}} ds = \frac{u_0}{\sqrt{\pi}} \int_{\frac{x}{\sqrt{4kt}}}^{\infty} e^{-s^2} ds.$$

Thus, using the fact that $\int e^{-s^2} ds$ is even, we have that

$$\begin{aligned} v(x,t) &= \frac{u_0}{\sqrt{\pi}} \left[\int_{-\infty}^{\frac{x}{\sqrt{4kt}}} e^{-s^2} ds - \int_{\frac{x}{\sqrt{4kt}}}^{\infty} e^{-s^2} ds \right] \\ &= \frac{u_0}{\sqrt{\pi}} \left[\int_{-\infty}^{0} e^{-s^2} ds + \int_0^{\frac{x}{\sqrt{4kt}}} e^{-s^2} ds - \left(\int_0^\infty e^{-s^2} ds - \int_0^{\frac{x}{\sqrt{4kt}}} e^{-s^2} ds \right) \right] \\ &= u_0 \frac{2}{\sqrt{\pi}} \int_0^{\frac{x}{\sqrt{4kt}}} e^{-s^2} ds = u_0 \, \text{erf}\left(\frac{x}{\sqrt{4kt}} \right). \end{aligned}$$

□

In the heat problem (9.22) we considered the Dirichlet boundary condition $v(0,t) = 0$, and derived the solution given by (9.26). Currently, we are interested in finding the solution for (9.22), but with a nonzero boundary condition. That is, we consider the heat problem with a non-homogeneous boundary condition

$$\begin{cases} v_t - kv_{xx} = 0, & 0 < x < \infty, \, t > 0 \\ v(x,0) = 0, & x > 0 \\ v(0,t) = p(t), & t > 0 \end{cases} \tag{9.27}$$

where the function p is differentiable. Our approach is to make a change of variables and reduce (9.27) to a problem with homogeneous boundary condition. To do so we let

$$u(x,t) = v(x,t) - p(t).$$

Then (9.27) is transformed into the heat problem

$$\begin{cases} u_t - ku_{xx} = -p'(t), & 0 < x < \infty, \, t > 0 \\ u(x,0) = -p(0), & x > 0 \\ u(0,t) = 0, & t > 0 \end{cases} \tag{9.28}$$

This is a non-homogeneous heat problem with Dirichlet boundary condition. Therefore, the problem can be solved by recasting it over the real line by the concept of odd extensions and then use (9.21) to obtain its solution. To stay compatible with previous notation we let

$$\phi(x) = -p(0), \quad \text{and} \quad f(x,t) = -p'(t).$$

Then

$$\phi_o(x) = \begin{cases} -p(0), & x > 0 \\ p(0), & x < 0 \\ 0, & x = 0 \end{cases} \quad \text{and } f_o(x,t) = \begin{cases} -p'(t), & x > 0 \\ p'(t), & x < 0 \\ 0, & x = 0 \end{cases}$$

Then, by (9.21) the solution may be written in terms of the heat kernel

$$u(x,t) = \int_{-\infty}^{\infty} S(x-y,t)\phi_0(y)dy + \int_0^t \int_{-\infty}^{\infty} S(x-y,t-\tau)f_o(y,\tau)dyd\tau, \text{ for } t > 0. \tag{9.29}$$

By substituting ϕ_o and f_o into (9.29), we obtain the explicit solution

$$\begin{aligned} u(x,t) &= -\frac{1}{\sqrt{4\pi kt}} \int_0^{\infty} \left(e^{-\frac{(x-y)^2}{4kt}} - e^{-\frac{(x+y)^2}{4kt}} \right) p(0)dy \\ &- \int_0^t \int_0^{\infty} \frac{1}{\sqrt{4\pi k(t-\tau)}} \left(e^{-\frac{(x-y)^2}{4k(t-\tau)}} - e^{-\frac{(x+y)^2}{4k(t-\tau)}} \right) p'(\tau)dyd\tau. \end{aligned} \tag{9.30}$$

Finally, the solution of our original heat problem (9.27) is

$$v(x,t) = u(x,t) + p(t),$$

where $u(x,t)$ is given by (9.30).

9.4 Heat Equation on Semi-Infinite Domain—Neumann Condition

Now we turn our attention to addressing the one-dimensional heat equation on half-line with the Neumann condition, $u_x(0,t) = 0$. We employ a similar procedure as in the previous subsection by extending the initial data to the negative half-line in such a fashion that the boundary condition is automatically satisfied. In the case of the Dirichlet condition, we used the odd extension to satisfy the initial data, whereas here we use the notion of *even extension*. We begin by considering the heat problem under Neumann condition at the end point $x = 0$,

$$\begin{cases} v_t - kv_{xx} = 0, & 0 < x < \infty, \ t > 0 \\ v(x,0) = \phi(x), & x > 0 \\ v_x(0,t) = 0, & t > 0 \end{cases} \tag{9.31}$$

Our aim is to find a solution to (9.31) as we did for the heat problem over the entire real line. There will be no need to start from scratch, but instead we will try to recast the problem over the entire real line by extending the initial data to the whole line. For heat problems with Neumann conditions, one would choose the *even extension* of the initial data $\phi(x)$. Note that if $\psi(x)$ is even, then $\psi(x) = \psi(-x)$, from which we get $\psi'(x) = -\psi'(-x)$, and hence $2\psi'(0) = 0$, or $\psi'(0) = 0$. This is true for any even function. We make it formal in the next lemma.

Lemma 28 *Let $f : (-\infty, \infty) \to \mathbb{R}$ be an even function $(f(x) = f(-x))$, that is differentiable at $x = 0$ then $f'(0) = 0$.*

Proof *Using the definition of the derivative of any function g, we have*

$$g'(x) = \lim_{h \to 0} \frac{g(x+h) - g(x)}{h}.$$

Since f is differentiable and satisfies $f(h) = f(-h)$, it follows that

$$\begin{aligned} f'(0) &= \lim_{h \to 0} \frac{f(h) - f(0)}{h} = \lim_{h \to 0} \frac{f(-h) - f(0)}{h} \\ &= -\lim_{h \to 0} \frac{f(0) - f(-h)}{h} = -f'(0). \end{aligned}$$

This implies $2f'(0) = 0$, or $f'(0) = 0$. This completes the proof.

We now present a lemma similar to Lemma 27, specifically for even initial data.

Lemma 29 *Let $u(x,t)$ be the solution of the heat equation on $-\infty < x < \infty$. If the initial data $\phi(x) = u(x,0)$ is even, then for all $t \geq 0$, $u(x,t)$ is an even function of x.*

As before, by Lemma 29 we have that if the initial data $\phi(x)$ is even, then $u(x,t)$ is even. Thus, by Lemma 28, $u_x(0,t) = 0$ for any $t > 0$. Hence u automatically satisfies the Neumann boundary condition at $x = 0$. We have the following definition.

Definition 9.2 *The even extension of a function $f(x)$ denoted by $f_e(x)$ is defined as*

$$f_e(x) = \begin{cases} f(x), & x > 0 \\ f(-x), & x < 0 \\ 0, & x = 0 \end{cases} \tag{9.32}$$

In light of the above discussion we recast the heat problem in (9.31) with extended data as

$$\begin{cases} u_t - ku_{xx} = 0, & -\infty < x < \infty, \ t > 0 \\ u(x,0) = \phi_e(x) & \\ u_x(0,t) = 0, & t > 0 \end{cases} \tag{9.33}$$

We already know from (9.16) that the solution of (9.33) is given by

$$u(x,t) = \int_{-\infty}^{\infty} S(x-y,t)\phi_e(y)dy, \quad \text{for } t > 0. \tag{9.34}$$

Since v is a solution for $x \geq 0$, we have $v(x,t) = u(x,t)$ for $x \geq 0$. Notice that

$$v(x,0) = u(x,0)\Big|_{x>0} = \phi_e(x)\Big|_{x>0} = \phi(x),$$

and

$$v_x(0,t) = u_x(0,t) = 0,$$

since $u(x,t)$ is an even function of x. Thus, $v(x,t)$ satisfies the Neumann condition in (9.33).

Substituting $\phi_e(x)$ into the solution given by (9.34), we obtain by splitting the integral over two regions,

$$\begin{aligned} u(x,t) &= \int_0^{\infty} S(x-y,t)\phi_e(y)dy + \int_{-\infty}^{0} S(x-y,t)\phi_e(y)dy \\ &= \int_0^{\infty} S(x-y,t)\phi(y)dy + \int_{-\infty}^{0} S(x-y,t)\phi(-y)dy. \end{aligned}$$

Making the change of variables $y \mapsto -y$ in the second integral leads to

$$u(x,t) = \int_0^{\infty} S(x-y,t)\phi(y)dy - \int_{\infty}^{0} S(x+y,t)\phi(y)dy.$$

Using (9.17)

$$S(x,t) = \frac{1}{\sqrt{4\pi kt}} e^{-\frac{x^2}{4kt}},$$

and $v(x,t)$ in the above expression, we may write the solution formula for (9.31) as follows:

$$v(x,t) = \frac{1}{\sqrt{4\pi kt}} \int_0^{\infty} \left[e^{-\frac{(x-y)^2}{4kt}} + e^{-\frac{(x+y)^2}{4kt}} \right] \phi(y)dy, \quad t > 0. \tag{9.35}$$

We have the following example.

Example 9.3 Consider the heat equation given by (9.31) with $\phi(x) = u_0$ for constant u_0. Substituting $\phi(y) = u_0$ into the solution $v(x,t)$ in (9.35), we obtain

$$v(x,t) = \frac{u_0}{\sqrt{4\pi kt}} \int_0^{\infty} \left[e^{-\frac{(x-y)^2}{4kt}} + e^{-\frac{(x+y)^2}{4kt}} \right] dy, \quad \text{for } t > 0.$$

Making the change of variables $s = \frac{x-y}{\sqrt{4kt}}$ in the first integral and $s = \frac{x+y}{\sqrt{4kt}}$ in the second integral, we obtain

$$v(x,t) = \frac{u_0}{\sqrt{\pi}} \left[\int_{-\infty}^{\frac{x}{\sqrt{4kt}}} e^{-s^2} ds + \int_{\frac{x}{\sqrt{4kt}}}^{\infty} e^{-s^2} ds \right]$$

$$= \frac{u_0}{\sqrt{\pi}}\left[\int_{-\infty}^{0} e^{-s^2}ds + \int_0^{\frac{x}{\sqrt{4kt}}} e^{-s^2}ds + \left(\int_0^{\infty} e^{-s^2}ds - \int_0^{\frac{x}{\sqrt{4kt}}} e^{-s^2}ds\right)\right]$$

$$= u_0\frac{2}{\sqrt{\pi}}\int_0^{\infty} e^{-s^2}ds = u_0\frac{2}{\sqrt{\pi}}\left(\frac{\sqrt{\pi}}{2}\right) = u_0 \text{ (by Lemma 22).}$$

□

9.4.1 Exercises

Exercise 9.1 *Prove parts (a)–(c) of Lemma 25.*

Exercise 9.2 *For constant u_0, write the solution in terms of the error function for the heat problem*

$$\begin{cases} u_t - ku_{xx} = 0, & -\infty < x < \infty,\ t > 0 \\ u(x,0) = \phi(x), & -\infty < x < \infty \end{cases}$$

where

$$\phi(x) = \begin{cases} u_0, & |x| < \ell \\ 0, & |x| > \ell \end{cases}$$

Exercise 9.3 *Consider the heat problem*

$$\begin{cases} v_t - kv_{xx} = 0, & 0 < x < \infty,\ t > 0 \\ v(x,0) = \phi(x), & x > 0 \\ v(0,t) = 0, & t > 0 \end{cases}$$

with initial data

$$\phi(x) = \begin{cases} 0, & 0 < x < \ell \\ 1, & x > \ell \end{cases}$$

Show its solution $u(x,t)$ can be written as

$$v(x,t) = \frac{1}{2}\operatorname{erf}\left(\frac{\ell + x}{\sqrt{4kt}}\right) - \frac{1}{2}\operatorname{erf}\left(\frac{\ell - x}{\sqrt{4kt}}\right).$$

Exercise 9.4 *Solve the heat problem*

$$\begin{cases} u_t - ku_{xx} = 0, & -\infty < x < \infty,\ t > 0 \\ u(x,0) = e^{-2|x|}, & -\infty < x < \infty \end{cases}$$

Exercise 9.5 *Solve*

$$\begin{cases} u_t - ku_{xx} = 0, & 0 < x < \infty,\ t > 0 \\ u(x,0) = \phi(x), & x > 0 \\ u(0,t) = 0, & t > 0 \end{cases}$$

(a)

$$\phi(x) = \begin{cases} 1 - x, & 0 < x < 1 \\ 0, & x \geq 1 \end{cases}$$

(b)

$$\phi(x) = \begin{cases} xe^{-x^2}, & 0 < x < 1 \\ 0, & x \geq 1 \end{cases}$$

Exercise 9.6 *Provide all the details in obtaining* (9.30).

Exercise 9.7 *Consider the nonlinear heat equation*

$$u_t - ku_{xx} + bu_x^2 = 0, \quad \infty < x < \infty, \quad t > 0,$$

subject to an initial condition $u(x,0) = f(x)$. *This form of PDE makes its presence in stochastic optimal control theory. In this question you will derive a representation formula for the solution* $u(x,t)$.
(a) Define the Cole-Hopf transformation $w(x,t) = e^{-\frac{b}{k}u(x,t)}$. *Show that* w *is a solution of the linear heat equation* $w_t - kw_{xx} = 0$.
(b) Use the fundamental solution of the heat equation to solve for $w(x,t)$.
(c) Invert the Cole-Hopf transformation to find a formula for u.

Exercise 9.8 *Solve the heat equation* $u_t - u_{xx} = 0$ *on the entire real line* $-\infty < x < \infty$ *with initial condition* $u(x,0) = \cos(x)$ *without using the fundamental solution of the heat equation. Use your answer to deduce the value of the integral*

$$\int_{-\infty}^{\infty} \cos(x) e^{-\frac{x^2}{4t}} \, dx.$$

Hint: Search for a solution of the form $u(x,t) = h(t)\cos(x)$.

Exercise 9.9 *Consider a very long beam with a quadratic cross-section given by*

$$0 \le x \le L, \quad 0 \le y \le L$$

which lies on a plane where the temperature is maintained at 0 *degrees. The other surfaces of the beam are exposed to ambient air at a uniform temperature of* 10 *degrees. The beam extends infinitely along the z-axis, i.e.,*

$$-\infty \le z \le \infty.$$

Find the stationary temperature distribution $T(x,y)$ *in the beam.*

Exercise 9.10 *Solve*

$$\begin{cases} u_t - ku_{xx} = 0, & 0 < x < \infty,\ t > 0 \\ u(x,0) = 0, & x > 0 \\ u(0,t) = p(t), & t > 0 \end{cases}$$

(a)

$$p(t) = \begin{cases} 1, & 0 < t < 1 \\ 0, & t \ge 1 \end{cases}$$

(b)

$$p(t) = 1, \quad t > 0.$$

Exercise 9.11 *Provide all the details in obtaining* (9.19).

Exercise 9.12 *For positive constant c, consider the heat equation with convection term*

$$u_t + cu_x - ku_{xx} = 0, \tag{9.36}$$

(a) Determine the values of α *and* β *so that the transformation*

$$u(x,t) = v(x,t)e^{\alpha x + \beta t}$$

transforms (9.36) *to the heat equation* $v_t - kv_{xx} = 0$.
(b) Find the solution of (9.36) *on* $-\infty < x < \infty$, *with initial condition* $u(x,0) = \phi(x)$.
(c) Find the solution of (9.36) *on* $x > 0$, *with initial and boundary conditions* $u(x,0) = \phi(x)$, *and* $u(x,0) = 0,\ t > 0$.

Exercise 9.13 *Let* $f : (-\infty,\infty) \to \mathbb{R}$ *be an even function* $(f(z) = f(-z))$, *that is differentiable. Use the definition of the derivative and show that* $f'(-x) + f'(x) = 0$.

Exercise 9.14 *Solve*

$$\begin{cases} u_t - ku_{xx} = 0, & 0 < x < \infty,\ t > 0 \\ u(x,0) = \phi(x), & x > 0 \\ u_x(0,t) = 0, & t > 0 \end{cases}$$

(a)

$$\phi(x) = \begin{cases} 1, & 0 < x < 1 \\ 0, & x \geq 1. \end{cases}$$

(b)

$$\phi(x) = xe^{-ax}.$$

(c)

$$\phi(x) = \begin{cases} 1 - x^2, & 0 < x < 1 \\ 0, & x \geq 1 \end{cases}$$

Exercise 9.15 *Consider the heat equation over* $\mathbb{R}$ *along with its solution given by* (9.18). *Show that if the initial function* $\phi(x)$ *is uniformly bounded on* $\mathbb{R}$, *then the solution* $u(x,t)$ *satisfies*

$$|u(x,t)| \leq \max_{-\infty < x < \infty} |\phi(x)|.$$

Hint: Make use of the substitution $s = \frac{x-y}{4kt}$.

10

Wave Equation on Unbounded Domains

In this chapter, we examine both homogeneous and non-homogeneous wave equations defined on unbounded domains. Our focus is on the existence of solutions, which we approach using the method of characteristics. This involves analyzing D'Alembert's solution, which relies on specifying initial conditions, including the initial displacement and velocity. The solutions to the wave equation are governed by the characteristics, or specific paths or curves along which information about the wave propagates. These characteristics are straight lines in space-time, moving at the wave speed c, and they play a crucial role in determining the evolution of the wave.

10.1 Wave equation and D'Alembert's Solution

In Section 5.1, we derived the wave equation. Here, we will develop D'Alembert's solution, examine the domain of dependence of its solutions, and analyze the non-homogeneous wave equation.

We begin by considering the hyperbolic PDE with Cauchy conditions in the following form

$$u_{tt} - c^2 u_{xx} = 0, \tag{10.1}$$

$$u(x,0) = f(x), \tag{10.2}$$

$$u_t(x,0) = g(x), \tag{10.3}$$

where the function f is twice continuously differentiable, and g is continuously differentiable. The function f represents the initial displacement, while g is the initial velocity. We employ the method of characteristics, particularly using formula (8.60), where we substitute y for t. Setting $A = -c^2$, $B = 0$, and $C = 1$, we obtain the differential equations

$$\frac{dt}{dx} = \frac{\pm c}{-c^2} = \pm\frac{1}{c},$$

DOI: 10.1201/9781003652540-10

whose solutions are the characteristic lines. Thus, the corresponding characteristic lines (solutions) are

$$c_1 = x - ct, \quad c_2 = x + ct.$$

Letting

$$\xi = x + ct, \quad \eta = x - ct,$$

we obtain

$$u_{xx} = u_{\xi\xi} + 2u_{\xi\eta} + u_{\eta\eta},$$

and

$$u_{tt} = c^2(u_{\xi\xi} - 2u_{\xi\eta} + u_{\eta\eta}).$$

Substituting into (10.1) we arrive at the canonical form $-4c^2 u_{\xi\eta} = 0$. Since $c \neq 0$, we must have

$$u_{\xi\eta} = 0,$$

whose general solution is given by

$$u(\xi, \eta) = F(\xi) + G(\eta),$$

where the functions F and G are arbitrary and required to be twice differentiable. In terms of t and x, the solution takes the form

$$u(x,t) = F(x + ct) + G(x - ct). \tag{10.4}$$

To determine the arbitrary functions F and G we apply the initial conditions or Cauchy conditions (10.2) and (10.3) and obtain

$$f(x) = F(x) + G(x), \tag{10.5}$$

and

$$g(x) = cF'(x) - cG'(x). \tag{10.6}$$

Integrating (10.6) from x_0 to x we arrive at

$$F(x) - G(x) = \frac{1}{c}\int_{x_0}^{x} g(s)ds + K, \tag{10.7}$$

where $x_0 \in \mathbb{R}$ and K are constants. Solving for F and G from (10.5) and (10.7) yields

$$F(x) = \frac{1}{2}\left[f(x) + \frac{1}{c}\int_{x_0}^{x} g(s)ds + K\right],$$

and

$$G(x) = \frac{1}{2}\left[f(x) - \frac{1}{c}\int_{x_0}^{x} g(s)ds - K\right].$$

Then by (10.4) the general solution takes the form

$$u(x,t) \;=\; \frac{1}{2}\left[f(x+ct) + \frac{1}{c}\int_{x_0}^{x+ct} g(s)ds + K\right]$$

$$+ \quad \frac{1}{2}\left[f(x-ct) - \frac{1}{c}\int_{x_0}^{x-ct} g(s)ds - K\right]$$
$$= \quad \frac{1}{2}[f(x+ct)+f(x-ct)]$$
$$+ \quad \frac{1}{2c}\left[\int_{x_0}^{x+ct} g(s)ds - \int_{x_0}^{x-ct} g(s)ds\right].$$

Combining the two integrals we arrive at the *D'Alembert solution* or the *classical solution*

$$u(x,t) = \frac{1}{2}[f(x+ct)+f(x-ct)] + \frac{1}{2c}\int_{x-ct}^{x+ct} g(s)ds. \tag{10.8}$$

This proves the existence of the solution of the one dimensional wave equation. It is simple to verify $u(x,t)$ given by (10.8) is a solution of the wave equation (10.1). Moreover, by a direct substitution into the solution (10.8), it is evident that the initial conditions uniquely determine (10.8). According to (10.8) the value $u(x_0,t_0)$ depends on the initial data f and g in the interval $[x_0 - ct_0,\ x_0 + ct_0]$ which is cut out of the initial line by the two characteristics lines with slopes $\pm\frac{1}{c}$ passing through the point (x_0,t_0). The interval $[x_0 - ct_0,\ x_0 + ct_0]$ on the line $t = 0$ is called the *domain of dependence*, as indicated in Fig. 10.1.

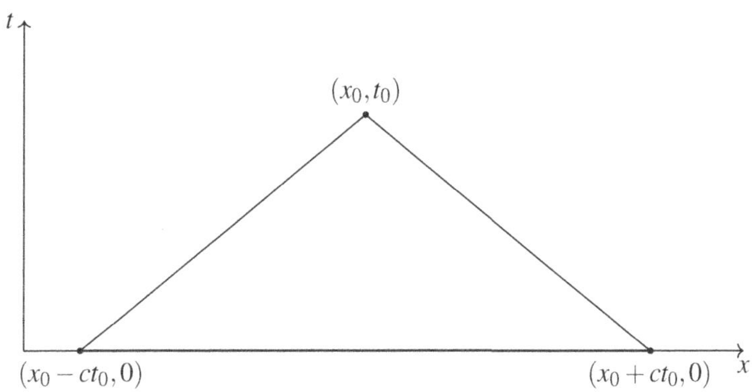

FIGURE 10.1: Domain of dependence.

Letting G denote the antiderivative of g vanishing at $x = 0$, we may write (10.8) in the form

$$u(x,t) = \frac{1}{2}\left[\left(f + \frac{1}{c}G\right)(x+ct) + \left(f - \frac{1}{c}G\right)(x-ct)\right],$$

where $G' = g$ and $G(0) = 0$. This exhibits u as the sum of a right-moving wave and a left-moving one, both with speed c.

Suppose that the initial data f and g are both bounded functions on $\mathbb{R}$. Let $||f||$ and $||g||$ denote the norms of f and g, respectively. Then, for all time $t > 0$, the solution u remains bounded and we have the estimate

$$||u(\cdot,t)|| \leq ||f|| + t||g||. \tag{10.9}$$

This shows that if the initial velocity is non-zero, the range of motion (at each point) may grow up to linearly in time. For instance, it is easy to check that for $f(x) = 0$, $g(x) = 1$, d'Alembert's formula gives

$$u(x,t) = t, \quad x \in \mathbb{R}, \quad t > 0.$$

In other words, the infinite string just moves upwards as a single straight line, with constant speed one. The following theorem addresses stability, stating that a small change in the initial data results in only a small change in the solution.

Theorem 10.1 (Stability) *Let $u^*(x,t)$ be another solution of (10.1)–(10.3) with initial data f^* and g^*. Define $|h| = \max_{-\infty<x<\infty} |h(x)|$, for $h : \mathbb{R} \to \mathbb{R}$ is continuous. Similarly, for $u = u(x,t)$, we define*

$$|u|_T = \max_{-\infty<x<\infty;|t|\leq T} |u(x,t)|.$$

Assume there is a small change in the initial data over a finite time T. That is, for small and positive ε we see that

$$|f - f^*| < \frac{\varepsilon}{2}, \quad |g - g^*| < \frac{\varepsilon}{2T}.$$

Then,

$$|u(x,t) - u^*(x,t)| < \varepsilon.$$

Proof *Using* (10.8) *we have*

$$\begin{aligned}
|u(x,t) - u^*(x,t)| &= \left|\frac{1}{2}(f(x+ct) - f^*(x+ct) + f(x-ct) - f^*(x-ct))\right. \\
&+ \left.\frac{1}{2c}\int_{x-ct}^{x+ct} (g(s) - g^*(s))ds\right| \\
&\leq \frac{1}{2}\left|f(x+ct) - f^*(x+ct)\right| + \frac{1}{2}\left|f(x-ct) - f^*(x-ct)\right| \\
&+ \frac{1}{2c}\int_{x-ct}^{x+ct} |g(s) - g^*(s)|ds \\
&\leq \frac{1}{2}(\varepsilon/2 + \varepsilon/2) + \frac{1}{2c}\int_{x-ct}^{x+ct} \frac{\varepsilon}{2T}ds \\
&\leq \varepsilon/2 + \frac{1}{2c}(2cT\frac{\varepsilon}{2T}) = \varepsilon.
\end{aligned}$$

Thus, for $|t| \leq T$, we have shown that

$$|u - u^*|_T < \varepsilon.$$

□

The D'Alembert solution given by (10.8) indicates that if an initial velocity or initial displacement is in the neighborhood of (x_0, t_0), it can only influence the region where $t > t_0$, bounded by the characteristic lines with slope $\pm\frac{1}{c}$ passing through the point (x_0, t_0), as shown in Fig. 10.2 with the initial time $t_0 = 0$.

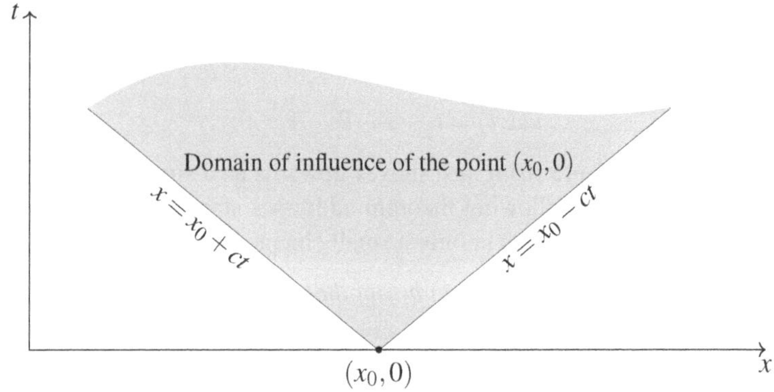

FIGURE 10.2: Domain of influence of the point $(x_0,0)$.

Example 10.1 Consider

$$4u_{tt} - 9u_{xx} = 0$$

subject to $u(x,0) = x^2$, $u_t(x,0) = \sin x$. Then $u(x,t)$ is given by the D'Alambert's solution (10.8) with $f = x^2$, $g = \sin x$, and $c^2 = \frac{9}{4}$. Thus,

$$\begin{aligned} u(x,t) &= \frac{1}{2}\left[(x+\frac{3}{2}t)^2 + (x-\frac{3}{2}t)^2\right] + \frac{1}{3}\int_{x-\frac{3}{2}t}^{x+\frac{3}{2}t} \sin s\,ds \\ &= x^2 + \frac{9}{4}t^2 + \frac{2}{3}\sin x \sin\frac{3}{2}t. \end{aligned}$$

□

Now that we displayed an example, let us take a closer look at the geometrical interpretation of (10.4). The term on the right-hand side of (10.4) is called the *progressive wave.* If we let $x^* = ct$, then the transformation $\xi = x + ct = x + x^*$ is a translation of the coordinate system to the left by x^*. Thus, $F(x+ct)$ is a wave that moves in the negative x direction with speed c without change in its shape. For example, $u(x,t) = \cos(x+ct)$ represents a cosine wave which moves in the negative x-direction with speed c without changing its shape. Similarly, $F(x-ct)$ is a wave which moves in the positive x-direction with speed c without change in its shape. Consequently, the solution

$$u(x,t) = F(x+ct) + G(x-ct),$$

is the sum of two waves traveling in opposite directions, and the shape of $u(x,t)$ will change with time.

Example 10.2 Consider the wave problem with zero initial velocity

$$u_{tt} - c^2 u_{xx} = 0,$$

subject to

$$u_t(x,0) = 0,$$

$$u(x,0) = \begin{cases} h, & |x| \leq a \\ 0, & |x| > a \end{cases}$$

This initial data corresponds to an initial disturbance of the string centered at $x = 0$ of height h. The solution is given by (10.8) with $g(x) = 0$. In other words,

$$u(x,t) = \frac{1}{2}[f(x+ct) + f(x-ct)].$$

We need to piece together the solution. Notice that

$$f(x+ct) = \begin{cases} h, & |x+ct| \leq a \\ 0, & |x+ct| > a \end{cases}$$

and

$$f(x-ct) = \begin{cases} h, & |x-ct| \leq a \\ 0, & |x-ct| > a \end{cases}$$

As a consequence the solution is defined piecewise over four different regions. We will only consider all regions for $t \geq 0$. It is clear from the definitions of $f(x+ct)$ and $f(x-ct)$ that the four regions are:

$$I = \{|x+ct| \leq a,\ |x-ct| \leq a\},$$

$$II = \{|x+ct| \leq a,\ |x-ct| > a\},$$

$$III = \{|x+ct| > a,\ |x-ct| \leq a\},$$

$$IV = \{|x+ct| > a,\ |x-ct| > a\},$$

with

$$u_I(x,t) = h, \quad u_{II}(x,t) = \frac{h}{2}, \quad u_{III}(x,t) = \frac{h}{2}, \quad u_{IV}(x,t) = 0.$$

The notation $u_I(x,t)$ stands for the value of u in region I, and so on. See Fig. 10.3. □

Example 10.3 Consider the wave problem with zero initial displacement

$$u_{tt} - c^2 u_{xx} = 0,$$

subject to

$$u(x,0) = 0,$$

$$u_t(x,0) = \begin{cases} g_0, & |x| \leq a \\ 0, & |x| > a \end{cases}$$

This is similar to the previous example but we will have to adjust the interval of integrations. However, here we have six different regions that we list.

$$I = \{x - ct < x + ct < -a < a\},$$

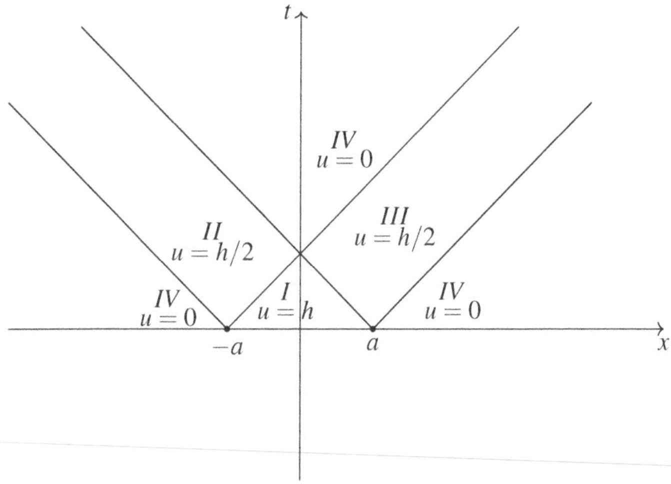

FIGURE 10.3: Different regions with different values of u.

$$\begin{aligned} II &= \{x-ct < -a < x+ct < a\}, \\ III &= \{x-ct < -a < a < x+ct\}, \\ IV &= \{-a < x-ct < x+ct < a\}, \\ V &= \{-a < x-ct < a < x+ct\}, \\ VI &= \{-a < a < x-ct < x+ct\}, \end{aligned}$$

with

$$u(x,t) = \begin{cases} 0 & \text{in } I \\ \frac{1}{2c}\int_{-a}^{x+ct} g_0 dx = \frac{g_0}{2c}(x+ct+a), & \text{in } II \\ \frac{1}{2c}\int_{-a}^{a} g_0 dx = \frac{g_0 a}{c}, & \text{in } III \\ \frac{1}{2c}\int_{x-ct}^{x+ct} g_0 dx = g_0 t, & \text{in } IV \\ \frac{1}{2c}\int_{x-ct}^{a} g_0 dx = \frac{g_0}{2c}(-x+ct+a), & \text{in } V \\ 0, & \text{in } VI \end{cases}$$

□

Next, we take a brief look at the effect of discontinuous initial data on the solution. Let $u(x,t)$ be given by (10.8). If $f(x)$ has a discontinuity at x_0, the function $u(x,t)$ will be discontinuous along the characteristic lines

$$x + ct = x_0 \quad \text{or} \quad x - ct = x_0.$$

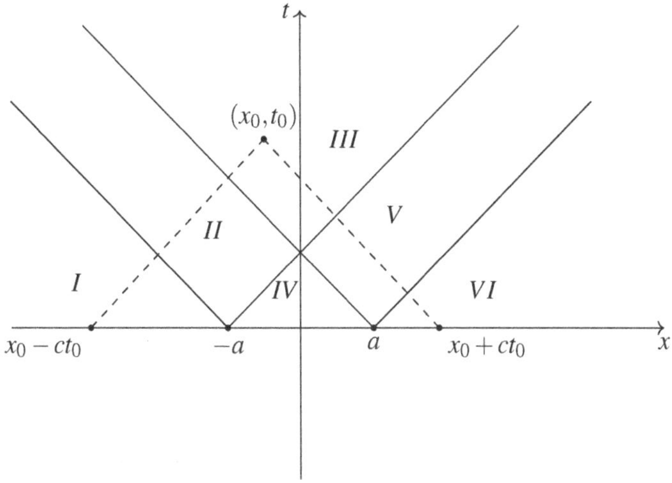

FIGURE 10.4: Different regions for the values of u when initial velocity is present.

This means that discontinuities in the initial condition travel along these characteristics. If $f(x)$ is continuous but $f'(x)$ or $f''(x)$ has a discontinuity at x_0, the first- or second-order partial derivatives of $u(x,t)$ will be discontinuous along characteristic lines. If $g(x)$ has a discontinuity at x_0, then either the first- or second-order derivatives of $u(x,t)$ will be discontinuous along the characteristic lines through $(x_0, 0)$. When $u(x,t)$ has discontinuities in its derivatives due to the discontinuities in $f(x)$ or $g(x)$, it can no longer be considered a classical solution. Instead, it is referred to as a **generalized solution**. A generalized solution allows for functions that might not be differentiable everywhere but still satisfy the equation in a weaker sense. If the initial data $f(x)$ and $g(x)$ have compact support, meaning they are nonzero only within a finite interval $[a,b]$, then the solution will also have compact support within a growing region defined by

$$x \in [a + ct, b - ct].$$

Outside this region, the solution remains identically zero.

Definition 10.1 (Support of a function) *The support of a function f, denoted by supp(f), is the set of points where the function is non-zero. More formally, it is the closure of the set where the function is not zero:*

$$supp(f) = \{x \in \mathbb{R} : f(x) \neq 0\}$$

The support can be thought of as the region where the function "lives" or is active.

Definition 10.2 (Compact support of a function) *A function f is said to have compact support if its support is contained within a compact set, meaning it is non-zero only on a closed and bounded interval. In simpler terms, the function is zero everywhere outside of some finite region.*

For example, the function $f(x) = \frac{1}{x}$ is non-zero for every point in the domain $(0,\infty)$ the support of this function is the entire domain itself. Therefore, the support is $(0,\infty)$. However, the support $(0,\infty)$ is not a compact set in the usual topology of the real line. This is because it does not include its "endpoint" at 0, meaning the support of the function is not bounded and therefore not compact. On the other hand, the function

$$f(x) = \begin{cases} 1-|x| & \text{if } |x| \leq 1 \\ 0 & \text{if } |x| \geq 1 \end{cases}$$

has compact support because it is non-zero only in the interval $[-1,1]$ and zero elsewhere.

The structure of discontinuities in the solution of the wave equation is entirely determined by the discontinuities in the initial data. If the initial data have compact support, then the influence of the initial conditions remains within a finite region expanding at speed c, ensuring that discontinuities appear only along the characteristic lines emanating from the edges of the compact support.

10.1.1 Exercises

Exercise 10.1 *Utilize the solution provided in* (10.8) *to demonstrate the estimate outlined in* (10.9).

Exercise 10.2 *Solve*

$$u_{tt} - 4u_{xx} = 0, \quad u(x,0) = \sin(x), \quad u_t(x,0) = \cos(x).$$

Exercise 10.3 *At points in space where no sources are present the spherical wave equation satisfies*

$$u_{tt} = c^2\left(u_{rr} + \frac{2}{r}u_r\right). \tag{10.10}$$

Equation (10.10) *is obtained by writing the homogeneous wave equation in spherical coordinates r,θ,ϕ and neglecting the angular dependence. Assume the initial functions*

$$u(r,0) = f(r), \quad u_t(r,0) = g(r).$$

Make the change of variables $v = ru$ to transform (10.10) *into the equation in v: $v_{tt} = c^2 v_{rr}$.*

Solve for v and then find the general solution of (10.10) *subject to the initial data.*

Exercise 10.4 *[Semi-infinite string with a fixed end] Consider a semi-infinite vibrating string with a fixed end, that is*

$$u_{tt} - c^2 u_{xx} = 0, \quad x > 0, \quad t > 0,$$

subject to

$$u(x,0) = f(x), \quad u_t(x,0) = g(x), \quad x \geq 0,$$

$$u(0,t) = 0, \quad t \geq 0.$$

Show that its solution is given by

$$u(x,t) = \frac{1}{2}[f(x+ct)+f(x-ct)] + \frac{1}{2c}\int_{x-ct}^{x+ct} g(s)ds, \quad \text{for} \quad x > ct,$$

$$u(x,t) = \frac{1}{2}[f(x+ct)-f(ct-x)] + \frac{1}{2c}\int_{ct-x}^{x+ct} g(s)ds, \quad \text{for} \quad x < ct.$$

Exercise 10.5 *Find the solution for*

$$u_{tt} - 4u_{xx} = 0, \quad x > 0, \quad t > 0,$$

subject to

$$u(x,0) = |\sin x|, \quad u_t(x,0) = 0, \quad x \geq 0,,$$
$$u(0,t) = 0, \quad t \geq 0.$$

Exercise 10.6 *Construct the solution for the wave problem with zero initial velocity.*

$$u_{tt} - u_{xx} = 0, \quad u_t(x,0) = 0,$$

$$u(x,0) = \begin{cases} 1, & |x| \leq 1 \\ 0, & |x| > 1 \end{cases}$$

for

(a) $0 < t < 1$, *(b)* $t > 1$, *(c)* $t = \frac{1}{2}$.

Exercise 10.7 *Construct the solution for the wave problem with zero initial displacement.*

$$u_{tt} - u_{xx} = 0, \quad u(x,0) = 0,$$

$$u_t(x,0) = \begin{cases} 1-x^2, & |x| < 1 \\ 0, & |x \geq 1 \end{cases}$$

In addition, find $u(x, \frac{1}{2})$.

Exercise 10.8 *Construct the solution for the wave problem with zero initial velocity.*

$$u_{tt} - u_{xx} = 0, \quad u_t(x,0) = 0,$$

$$u(x,0) = \begin{cases} 1-|x|, & |x| < 1 \\ 0, & |x \geq 1. \end{cases}$$

Exercise 10.9 *Assume that* $u(x,t)$

$$u_{tt} - u_{xx} = 0, \quad 0 < x < l, \quad t \geq 0,$$

with Robin boundary conditions

$$u(0,t) - u_x(0,t) = 0, \quad u(l,t) + u_x(l,t) = 0.$$

Show that the energy function

$$E(t) = \frac{1}{2}\int_0^l \left(u_t^2(x,t) + u_x^2(x,t)\right)dx + \frac{1}{2}u^2(0,t) + \frac{1}{2}u^2(l,t)$$

is conserved. That is $\frac{d}{dt}E(t) = 0$.

Exercise 10.10 *Suppose that the initial data of the wave equation have a compact support on* $[-R,R]$. *Use* (10.8) *to show the solution* $u(x,t)$ *has a compact support on* $[-R-ct, R+ct]$. *In other words, the size of the support grows linearly with* t.

Exercise 10.11 *Consider the bump function*

$$f(x) = \begin{cases} e^{-\frac{1}{1-4x^2}} & \text{for } |x| < \frac{1}{2} \\ 0 & \text{for } |x| \geq 1 \end{cases}$$

which is smooth, even, reaches a maximum of 1 *at* $x = 0$, *and with compact support* $[-\frac{1}{2}, \frac{1}{2}]$. *Consider the wave problem* (10.1)–(10.3) *with* $u(x,0) = f(x)$, *and* $u_t(x,0) = 0$.

(a) *Find the solution* $u(x,t)$ *when* $c = 1$.

(b) *What is the earliest time the graph of* $u(x,t)$ *(as a function of* x*) shows two separate 'bump functions' identical to the original one, moving to the right and to the left? Sketch the graph of* $u(x,2)$.

(c) *Now fix* $x_0 = 3$, *and sketch the graph of* $u(3,t)$ *as a function of* $t > 0$.

Exercise 10.12 *Let* u *be a solution to the wave equation*

$$u_{tt} - u_{xx} = 0 \quad in \quad \mathbb{R} \times [0,\infty).$$

Assume that both $u(x,0)$ *and* $u_t(x,0)$ *are smooth and compactly supported. Define the kinetic and potential energy as*

$$E_1(t) = \int_{\mathbb{R}} u_t^2\,dx, \quad E_2(t) = \int_{\mathbb{R}} u_x^2\,dx,$$

respectively. Show that there exists T *such that for all* $t \geq T$, $E_1(t) \leq E_2(t)$.

10.2 Non-Homogeneous Wave Equation

In this section we look at the Non-Homogeneous wave equation

$$u_{tt} - c^2 u_{xx} = h(x,t) \tag{10.11}$$

subject to

$$u(x,0) = f(x), \quad u_t(x,0) = g(x), \tag{10.12}$$

where the function h is assumed to be continuous with respect to both arguments. We show that the solution of (10.11) along with the initial data (10.12) is given by

$$\begin{aligned} u(x,t) &= \frac{1}{2}[f(x+ct)+f(x-ct)] + \frac{1}{2c}\int_{x-ct}^{x+ct} g(s)ds \\ &+ \frac{1}{2c}\int\int_{\triangle(x,t)} h(\tau,s)d\tau ds, \end{aligned} \quad (10.13)$$

where $\triangle(x,t)$ is shown in Fig. 10.4. We will do this by piecing together the solution

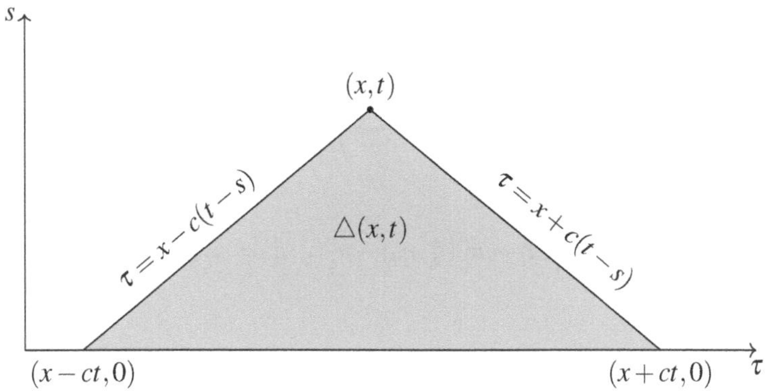

FIGURE 10.5: Non-homogeneous wave equation.

of the homogeneous problem (10.1)–(10.3), which has the solution given by (10.8), and the solution u^p of

$$u^p_{tt} - c^2 u^p_{xx} = h(x,t), \quad (10.14)$$

subject to

$$u^p(x,0) = 0, \quad u^p_t(x,0) = 0. \quad (10.15)$$

Solving for x and t in the transformation

$$\xi = x + ct, \quad \eta = x - ct,$$

we obtain

$$x = \frac{\xi+\eta}{2}, \quad t = \frac{\xi-\eta}{2c}. \quad (10.16)$$

Under the same transformation, we saw the left side, or homogeneous part of (10.14) reduces to

$$-4c^2 u_{\xi\eta} = 0.$$

Therefore, (10.14) takes the form

$$u^p_{\xi\eta} = -\frac{1}{4c^2}h(\xi,\eta). \quad (10.17)$$

Setting $t = 0$ in (10.16) we immediately have $\xi = \eta$. Thus, the first initial condition of (10.15) reduces to

$$u^p(\xi,\xi) = 0.$$

Using $u_x^p = u_\xi^p \xi_x + u_\eta^p \eta_x = u_\xi^p + u_\eta^p$ we have $u_x^p(x,0) = 0$, implies that

$$u_\xi^p(\xi,\xi) + u_\eta^p(\xi,\xi) = 0.$$

Similarly, $u_t^p = u_\xi^p \xi_t + u_\eta^p \eta_t = cu_\xi^p - cu_\eta^p$. Thus, the second boundary condition of (10.15) reduces to

$$cu_\xi^p(\xi,\xi) - cu_\eta^p(\xi,\xi) = 0.$$

From the last two equations above, it is immediate that $u_\xi^p(\xi,\xi) = u_\eta^p(\xi,\xi) = 0$. Fix a point (x_0,t_0). Then the corresponding point in the characteristic variables is (ξ_0,η_0). In order to find the value of the solution at this point we begin by integrating (10.17) in term of η from ξ to η_0 and obtain

$$\int_\xi^{\eta_0} u_{\xi\eta}^p d\eta = -\frac{1}{4c^2}\int_\xi^{\eta_0} h(\xi,\eta)d\eta.$$

However,

$$\int_\xi^{\eta_0} u_{\xi\eta}^p d\eta = u_\xi^p(\xi,\eta_0) - u_\xi^p(\xi,\xi) = u_\xi^p(\xi,\eta_0).$$

As a result, we have

$$u_\xi^p(\xi,\eta_0) = -\frac{1}{4c^2}\int_\xi^{\eta_0} h(\xi,\eta)d\eta = \frac{1}{4c^2}\int_{\eta_0}^{\xi} h(\xi,\eta)d\eta. \tag{10.18}$$

Similar to above, the integral

$$\int_{\eta_0}^{\xi_0} u_\xi^p(\xi,\eta)d\xi = u^p(\xi_0,\eta_0) - u^p(\xi_0,\xi_0) = u^p(\xi_0,\eta_0). \tag{10.19}$$

Integrating (10.18) with respect to ξ from η_0 to ξ_0 and then using (10.19) we arrive at

$$\begin{aligned} u^p(\xi_0,\eta_0) &= \frac{1}{4c^2}\int_{\eta_0}^{\xi_0}\int_{\eta_0}^{\xi} h(\xi,\eta)d\eta d\xi \\ &= \frac{1}{4c^2}\int\int_\triangle h(\xi,\eta)d\xi d\eta, \end{aligned} \tag{10.20}$$

where the double integral is taken over the triangle of dependence of the point (x_0,t_0), as shown in Fig. 10.5. Left to transform the double integral in (10.20) to a double integral in terms of the variables (x,t). For $\xi = x+ct$, $\eta = x-ct$, we have

$$J = \begin{vmatrix} \xi_x & \xi_y \\ \eta_x & \eta_y \end{vmatrix} = \begin{vmatrix} 1 & c \\ 1 & -c \end{vmatrix} = -2c \neq 0.$$

Thus,

$$u^p(\xi,\eta) = \frac{1}{4c^2}\int\int_{\triangle(x,t)} h(\tau,s)|J|d\tau ds = \frac{1}{2c}\int\int_{\triangle(x,t)} h(\tau,s)d\tau ds, \tag{10.21}$$

where $\triangle(x,t)$ is shown in Fig. 10.5. Finally, adding (10.21) to (10.8), we obtain (10.13). Note that the function

$$u_1(x,t) = \frac{1}{2c}\int\int_{\triangle(x,t)} h(\tau,s)d\tau ds,$$

defines a function of class C^2 in (x,t), assuming only h is continuous in (x,t).

For illustrational purpose, we provide the following two examples.

Example 10.4 Consider

$$4u_{tt} - 9u_{xx} = 4xt$$

subject to $u(x,0) = x^2$, $u_t(x,0) = \sin x$. Here $u(x,t)$ is given by the D'Alambert's solution (10.13) with $f = x^2$, $g = \sin x$, $h(x,t) = xt$ and $c^2 = \frac{9}{4}$. With the aid of Example 10.1 we have

$$\begin{aligned} u(x,t) &= x^2 + \frac{9}{4}t^2 + \frac{2}{3}\sin x \sin\frac{3}{2}t + \int_0^t \int_{x-\frac{3}{2}(t-s)}^{x+\frac{3}{2}(t-s)} \tau s d\tau ds \\ &= x^2 + \frac{9}{4}t^2 + \frac{2}{3}\sin x \sin\frac{3}{2}t + \frac{1}{3}\int_0^t \frac{s}{2}[x^2 - c^2(t-s)^2]ds \\ &= x^2 + \frac{9}{4}t^2 + \frac{2}{3}\sin x \sin\frac{3}{2}t + \frac{t^2x^2}{12} - \frac{3t^4}{96}. \text{ (See Fig. 10.6).} \end{aligned}$$

□

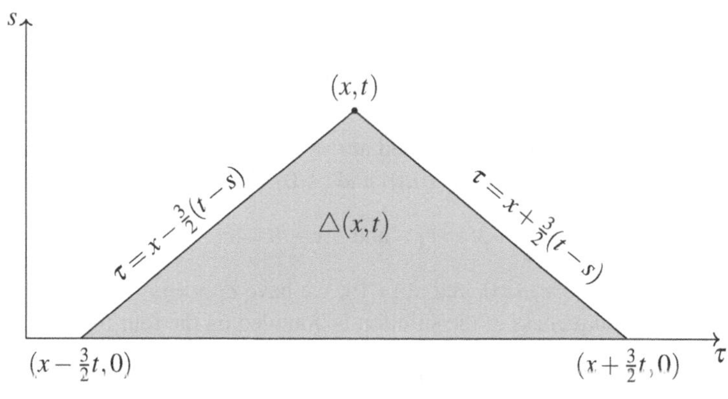

FIGURE 10.6: $\triangle(x,t)$.

Example 10.5 Consider

$$u_{tt} - 9u_{xx} = \frac{12}{t^2+1}$$

subject to $u(x,0) = x$, $u_t(x,0) = e^{-x}$.
(a) Find the solution.

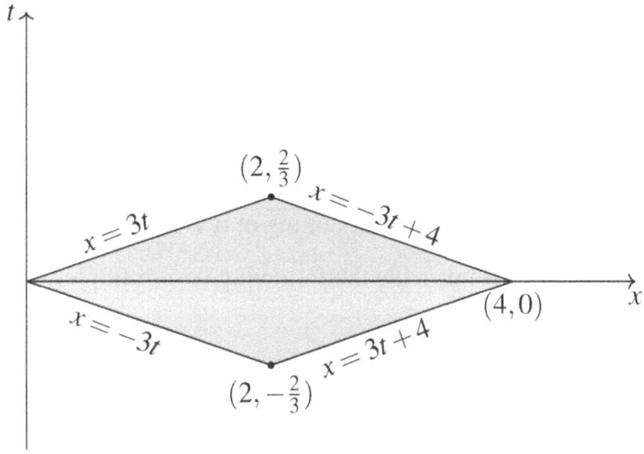

FIGURE 10.7: Region for existence and uniqueness.

(b) Determine the region where the solution is uniquely defined when $0 \leq x \leq 4$.
For (a) the solution is given by

$$\begin{aligned}
u(x,t) &= \frac{1}{2}[(x+3t)+(x-3t)] + \frac{1}{6}\int_{x-3t}^{x+3t} e^{-s} ds \\
&+ 2\int_0^t \int_{x-3(t-s)}^{x+3(t-s)} \frac{1}{s^2+1} d\tau ds \\
&= x + \frac{1}{6}\left(e^{-(x-3t)} - e^{-(x+3t)}\right) + 12\int_0^t \frac{t-s}{s^2+1} ds \\
&= x + \frac{1}{6}\left(e^{-(x-3t)} - e^{-(x+3t)}\right) + 12t\tan^{-1}(t) - 6\ln(t^2+1).
\end{aligned}$$

For part (b), the initial data is prescribed at $t = 0$ over the interval $0 \leq x \leq 4$. Thus, we need to consider the two points $(0,0)$ and $(4,0)$. The characteristic lines are

$$x - 3t = c_1 \quad \text{and} \quad x + 3t = c_2.$$

At $(0,0)$, we have $c_1 = c_2 = 0$, and at $(4,0)$, we have $c_1 = c_2 = 4$. Thus, the region of existence and uniqueness of the solution is bounded by the four lines.

$$x - 3t = 0, \quad x + 3t = 0, \quad x - 3t = 4, \quad \text{and} \quad x + 3t = 4,$$

as shown in Fig. 10.7. □

10.2.1 Exercises

Exercise 10.13 *By a direct substitution verify that*

$$u(x,t) = \int_0^t \int_{x+c(t-s)}^{x-c(t-s)} h(\tau,s)\, d\tau ds,$$

solves

$$u_{tt} - c^2 u_{xx} = h(x,t).$$

Exercise 10.14 *Solve*

$$u_{tt} - 4u_{xx} = \cos x \sin t, \quad u(x,0) = x, \quad u_t(x,0) = \sin(x).$$

Exercise 10.15 *Consider*

$$2u_{tt} - 18u_{xx} = \frac{36}{t^2+1}, \quad u(x,0) = x^2, \quad u_t(x,0) = e^{-2x}.$$

(a) Find the solution.
(b) Determine and sketch the region where the solution is uniquely defined when $0 \le x \le 6$.

Exercise 10.16 *Solve*

$$u_{tt} - u_{xx} = 1, \quad u(x,0) = \sin x, \quad u_t(x,0) = x.$$

Exercise 10.17 *Suppose that all* f, g *and* h *in* (10.13) *are bounded. Obtain the estimate on the solution* u,

$$\|u(\cdot,t)\| \le \|f\| + t\|g\| + \frac{t^2}{2}\|h\|.$$

Exercise 10.18 (KdV) *Consider the Korteweg-de Vries (KdV) equation, which describes traveling waves and is given by*

$$u_t + 6uu_x + u_{xxx} = 0, \quad -\infty < x < \infty, \quad t > 0.$$

We seek a traveling wave solution of the form

$$u(x,t) = f(\xi), \quad \textit{where} \quad \xi = x - ct$$

under the conditions that guarantee solitary wave,

$$f(x),\ f'(x),\ f''(x) \to 0 \quad \textit{as} \quad x \to \pm\infty.$$

(a) Substituting into the KdV equation, and obtain the nonlinear third-order ordinary differential equation

$$f''' - cf' + 6ff' = 0.$$

(b) Taking into consideration the conditions that guarantee a solitary wave, show that

$$\frac{df}{dx} = \pm f\sqrt{x - 2f},$$

where c is a constant, and then obtain the solution

$$f(x) = \frac{c}{2}\operatorname{sech}^2\left(\frac{\sqrt{c}}{2}(x-k)\right)$$

where k is a constant of integration.

10.3 Vibrating String with Fixed Ends

Obtaining the solution of a vibrating string with fixed ends is more complicated than solving a wave equation or an infinite string due to the repeated reflection of waves from boundaries. We want to find the solution of

$$\begin{cases} u_{tt} - c^2 u_{xx} = 0, & 0 < x < l,\ t > 0 \\ u(x,0) = f(x), & 0 \le x \le l \\ u_t(x,0) = g(x), & 0 \le x \le l \\ u(0,t) = 0 = u(l,t), & t \ge 0 \end{cases} \tag{10.22}$$

In what to follow, we establish the existence of the solution. For its uniqueness, we refer to Exercise 10.19. As before, the characteristic lines are given by

$$\xi = x + ct, \quad \eta = x - ct.$$

From (10.4) we have

$$u(x,t) = F(x+ct) + G(x-ct), \tag{10.23}$$

where the functions F and G are arbitrary and differentiable. Equation (10.23) is valid and over the domain

$$0 \le x + ct \le l, \quad \text{and} \quad 0 \le x - ct \le l.$$

Moreover, the solution is uniquely determined by the initial data in the region

$$t \le \frac{x}{c}, \quad t \le \frac{l-x}{c}, \quad t \ge 0.$$

For the fixed end $u(0,t) = 0$, $t \ge 0$, we have

$$0 = u(0,t) = F(ct) + G(-ct). \tag{10.24}$$

Letting $\zeta = -ct$, we obtain from (10.24) that

$$G(\zeta) = -F(-\zeta), \quad \zeta \le 0. \tag{10.25}$$

Equation (10.25) extends the range of G to negative values and can similarly be used to extend the range of F. Applying the initial data to (10.23), we obtain from Section 10.1 that

$$F(\xi) = \frac{1}{2}\left[f(\xi) + \frac{1}{c}\int_0^{\xi} g(s)ds + K\right], \quad 0 \le \xi = x + ct \le l, \tag{10.26}$$

$$G(\eta) = \frac{1}{2}\left[f(\eta) - \frac{1}{c}\int_0^{\eta} g(s)ds - K\right], \quad 0 \le \eta = x - ct \le l. \tag{10.27}$$

Using (10.25) in combination with (10.26) and (10.27), we arrive at the result by setting $G(\zeta) = -F(-\zeta)$ that

$$\begin{aligned} \frac{1}{2}\left[f(\zeta) - \frac{1}{c}\int_0^{\zeta} g(s)ds\right] &= -\frac{1}{2}\left[f(-\zeta) + \frac{1}{c}\int_0^{-\zeta} g(s)ds\right] \\ &= -\frac{1}{2}\left[f(-\zeta) - \frac{1}{c}\int_0^{\zeta} g(-s)ds\right]. \end{aligned}$$

By comparing both sides of the above expression, we immediately see that (10.25) is satisfied when

$$f(\zeta) = -f(-\zeta), \quad \text{and} \quad g(\zeta) = -g(-\zeta).$$

In other words, we must extend the functions f and g to be odd functions with respect to $x = 0$. Now, we turn our attention to the boundary condition $0 = u(l,t)$. As before, we have

$$0 = u(l,t) = F(l+ct) + G(l-ct).$$

Letting $\zeta = l + ct$ in the above equation, we get

$$F(\zeta) = -G(2l-\zeta), \quad \zeta \geq l. \tag{10.28}$$

This equation extends the range of F to positive values $l \leq \zeta \leq 2l$. As before, setting $F(\zeta) = -G(2l-\zeta)$, we arrive at

$$\begin{aligned} \frac{1}{2}\left[f(\zeta) + \frac{1}{c}\int_0^{\zeta} g(s)ds\right] &= -\frac{1}{2}\left[f(2l-\zeta) - \frac{1}{c}\int_0^{2l-\zeta} g(s)ds\right] \\ &= -\frac{1}{2}\left[f(2l-\zeta) + \frac{1}{c}\int_{2l}^{\zeta} g(2l-\tau)d\tau\right]. \end{aligned}$$

By comparing both sides of the above expression, we immediately arrive at

$$f(2l-\zeta) = -f(\zeta), \quad \text{and} \quad \int_0^{\zeta} g(s)ds = -\int_{2l}^{\zeta} g(2l-\tau)d\tau.$$

Differentiating the second expression with respect to ζ, we obtain

$$g(2l-\zeta) = -g(\zeta).$$

These conditions on f and g means that we can extend these functions to $l \leq \zeta \leq 2l$ by performing an odd extension about $x = l$. In summary, the conditions on f and g are

$$\begin{cases} f(x) = -f(-x), & -l \leq x \leq 0 \\ f(2l-x) = -f(x), & l < x < 2l \end{cases} \tag{10.29}$$

and

$$\begin{cases} g(x) = -g(-x), & -l \le x \le 0 \\ g(2l-x) = -g(x), & l \le x \le 2l \end{cases} \tag{10.30}$$

To make some sense out of (10.29) and (10.30), we notice that

$$f(x+2l) = -f(2l-(x+2l)) = -f(-x) = f(x).$$

A similar situation occurs for the function g. Thus, f and g need to be periodic odd extensions of the original functions with period $2l$. Now we try to piece the solution together. Replacing ζ by η in (10.25) and then using (10.26) we arrive at an expression for $G(\eta)$. That is for $0 \le -\eta \le l$, or $-l \le \eta \le 0$, we have

$$G(\eta) = -F(-\eta) = -\frac{1}{2}\left[f(-\eta) + \frac{1}{c}\int_0^{-\eta} g(s)ds + K\right]. \tag{10.31}$$

Similarly, replacing ζ by ξ in (10.28) and then using (10.27) we arrive at an expression for $F(\eta)$. In other words, for $0 \le 2l - \xi \le l$, or $l \le \xi \le 2l$, we have that

$$F(\xi) = -G(2l-\xi) = -\frac{1}{2}[f(2l-\xi) - \frac{1}{c}\int_0^{2l-\xi} g(s)ds - K]. \tag{10.32}$$

Equations (10.31) and (10.32) show that we can obtain $F(\xi)$ for every $\xi \ge 0$ and $G(\eta)$ for every $\eta \le l$. Hence the solution is determined for all $0 \le x \le l$ and $t \ge 0$. Let f_p and g_p denote the odd extensions of $2l$-periodic of f and g, respectively. Then,

$$f_p(x) = \begin{cases} f(x), & 0 < x < l \\ -f(-x), & -l < x < 0 \end{cases}, \quad g_p(x) = \begin{cases} g(x), & 0 < x < l \\ -g(-x), & -l < x < 0 \end{cases}$$

Consider the wave problem on the whole real line with the extended initial data

$$\begin{cases} v_{tt} - c^2 u_{xx} = 0, & -\infty < x < \infty, \ t > 0 \\ v(x,0) = f_p(x), & 0 \le x \le l \\ v_t(x,0) = g_p(x), & 0 \le x \le l \end{cases}$$

With this set up we automatically have $v(0,t) = v(l,t) = 0$, and the restriction

$$u(x,t) = v(x,t)\big|_{0 \le x \le l}$$

will solve (10.22). Finally, using the D'Alembert solution we see that

$$u(x,t) = \frac{1}{2}[f_p(x+ct) + f_p(x-ct)] + \frac{1}{2c}\int_{x-ct}^{x+ct} g_p(s)ds, \quad 0 < x < l. \tag{10.33}$$

Example 10.6 Let's obtain the odd extension of the function

$$f(x) = \begin{cases} x, & 0 \le x \le \frac{\ell}{2} \\ \ell - x, & \frac{\ell}{2} \le x \le \ell \end{cases}$$

The function defines the initial profile of a plucked string at $x = 0$ and at $x = \ell$. To construct the odd periodic extension of this initial function, we will need to extend the function beyond the interval $[0, \ell]$ such that the extended function is odd and satisfies the boundary conditions. The odd periodic extension of $f(x)$ should satisfy the condition that $f(-x) = -f(x)$. For an odd extension of $f(x)$, the function values for $x > \ell$ and $x < 0$ will mirror the behavior within the interval $[0, \ell]$ with sign reversal. The odd extension over $[-\ell, \ell]$ is

$$f_o(x) = \begin{cases} x, & -\frac{\ell}{2} \le x \le 0 \\ -\ell - x, & -\ell \le x \le -\frac{\ell}{2} \end{cases}$$

The periodic odd extension of $f(x)$ over the whole real lines can be written in the general formula

$$f_{\text{ext}}(x) = \begin{cases} x - 2k\ell, & \text{if } x \in [2k\ell, 2k\ell + \frac{\ell}{2}] \\ \ell - (x - 2k\ell), & \text{if } x \in [2k\ell + \frac{\ell}{2}, 2k\ell + \ell] \\ x - 2k\ell, & \text{if } x \in [2k\ell - \frac{\ell}{2}, 2k\ell) \\ -[\ell + (x - 2k\ell)], & \text{if } x \in [2k\ell - \ell, 2k\ell - \frac{\ell}{2}) \end{cases}$$

for $k = 0, \pm 1, \pm 2, \ldots$ This defines the odd periodic extension of the given initial function, subject to fixed boundary conditions at $x = 0$ and $x = \ell$. Now, suppose that we aim to solve (10.22) with the given function $f(x)$ and $g(x) = 0$. Defining $z_1 = x + ct$ and $z_2 = x - ct$, and applying the formula (10.33), the solution is given by

$$u(x,t) = \frac{1}{2}\left[f_{\text{ext}}(z_1) + f_{\text{ext}}(z_2)\right].$$

Example 10.7 Consider system (10.22) with $f(x) = 0$, $l = 1$ and

$$g(x) = \begin{cases} 0, & \text{if } 0 < x < \frac{1}{4} \\ 1, & \text{if } \frac{1}{4} \le x \le \frac{3}{4} \\ 0, & \text{if } \frac{3}{4} < x < 1 \end{cases}$$

Then the odd 2-periodic extension of $g(x)$ over the interval $[0, 2]$, is

$$g_o(x) = \begin{cases} 0, & \text{if } 0 < x < \frac{1}{4} \\ 1, & \text{if } \frac{1}{4} < x < \frac{3}{4} \\ 0, & \text{if } \frac{3}{4} < x < \frac{5}{4} \\ -1, & \text{if } \frac{5}{4} < x < \frac{7}{4} \\ 0, & \text{if } \frac{7}{4} < x < 2 \end{cases}$$

Define the antiderivative function G of g_o by

$$G(x) = \int_0^x g_0(s)\, ds.$$

Then,

$$G(x)=\begin{cases}\int_0^x 0\,ds, & \text{if } 0<x<\frac{1}{4}\\ \int_0^{\frac{1}{4}} 0\,ds+\int_{1/4}^x 1\,ds, & \text{if } \frac{1}{4}\le x\le\frac{3}{4}\\ \int_0^{\frac{1}{4}} 0\,ds+\int_{1/4}^{3/4} 1\,ds+\int_{3/4}^x 0\,ds, & \text{if } \frac{3}{4}<x<\frac{5}{4}\\ \int_0^{\frac{1}{4}} 0\,ds+\int_{1/4}^{3/4} 1\,ds+\int_{3/4}^{5/4} 0\,ds+\int_{5/4}^x (-1)\,ds, & \text{if } \frac{5}{4}<x<\frac{7}{4}\\ \int_0^{\frac{1}{4}} 0\,ds+\int_{1/4}^{3/4} 1\,ds+\int_{3/4}^{5/4} 0\,ds+\int_{5/4}^{7/4} (-1)\,ds+\int_{7/4}^x 0\,ds, & \text{if } \frac{7}{4}<x<2\end{cases}$$

$$=\begin{cases}0, & \text{if } 0<x<\frac{1}{4}\\ x-\frac{1}{4}, & \text{if } \frac{1}{4}\le x\le\frac{3}{4}\\ \frac{1}{2}, & \text{if } \frac{3}{4}<x<\frac{5}{4}\\ -x+\frac{7}{4}, & \text{if } \frac{5}{4}<x<\frac{7}{4}\\ 0, & \text{if } \frac{7}{4}<x<2\end{cases}$$

Let $z_1=x+ct$ and $z_2=x-ct$. Using the solution given by (10.33), we have

$$u(x,t)=\frac{1}{2c}\left[G(z_1)+G(z_2)\right].$$

Now suppose that $c=1$ and we want to find $u(\frac{1}{2},\frac{1}{4})$. Then,

$$z_1=x+t=\frac{1}{2}+\frac{1}{4}=\frac{3}{4},\quad z_2=x-t=\frac{1}{2}-\frac{1}{4}=\frac{1}{4},$$

and,

$$G(z_1)=G(\frac{3}{4})=\frac{3}{4}-\frac{1}{4}=\frac{1}{2},\quad G(z_2)=G(\frac{1}{4})=\frac{1}{4}-\frac{1}{4}=0.$$

Consequently,

$$u(\frac{1}{2},\frac{1}{4})=\frac{1}{2}\left[\frac{1}{2}+0\right]=\frac{1}{4}.$$

What follows, we bring back Fourier series to obtain the solution. Suppose that we have a function f that is piecewise continuous on the interval $(0,l)$. We have already seen that the *Fourier sine series* of f is

$$f(x)=\sum_{n=1}^{\infty} b_n \sin\frac{n\pi x}{l},\quad 0<x<l, \tag{10.34}$$

where the coefficients b_n, $n=1,2,\ldots$, are given by

$$b_n=\frac{2}{l}\int_0^l f(x)\sin\frac{n\pi x}{l}\,dx, \tag{10.35}$$

where we have assumed that term-by-term integration is valid. We know that since

$$f(0)=0,\quad \text{and}\quad f(l)=0,$$

the series (10.34) converges uniformly to $f(x)$ for all $0<x<l$. It is clear that the series converges to zero when $x=0$ and when $x=l$. Moreover, the series in (10.34) converges to the *odd periodic extension,* with period $2l$, of f for all values of $x \in \mathbb{R}$. We have the following example.

Example 10.8 Let
$$f(x)=x, \quad 0<x<\pi.$$
Then,
$$b_n=\frac{2}{\pi}\int_0^{\pi} x\sin nx\,dx=\frac{2}{\pi}\Big[-\frac{x\cos nx}{n}\Big|_0^{\pi}+\frac{1}{n}\int_0^{\pi}\cos nx\,dx\Big]=2\frac{(-1)^{n+1}}{n},\ n=1,2,\ldots$$
Thus,
$$f_p=2\sum_{n=1}^{\infty}\frac{(-1)^{n+1}}{n}\sin nx, \quad 0<x<\pi. \tag{10.36}$$
It can be shown that this series converges to $f(x)=x$ when $0<x<\pi$. □

Example 10.9 Consider
$$\begin{cases} u_{tt}=u_{xx}, & 0<x<\pi,\ t>0\\ u(x,0)=x, & 0\le x\le \pi\\ u_t(x,0)=0, & 0\le x\le \pi\\ u(0,t)=0=u(\pi,t), & t\ge 0\end{cases}$$
Using (10.33) and (10.36) we arrive at
$$\begin{aligned} u(x,t) &= \frac{1}{2}[f_p(x+ct)+f_p(x-ct)]\\ &= \frac{1}{2}\sum_{n=1}^{\infty}2\frac{(-1)^{n+1}}{n}\big[\sin(n(x+t))+\sin(n(x-t))\big].\end{aligned}$$
Clearly, the above solution satisfies $u(0,t)=0,\ u(\pi,t)=0$. □

Example 10.10 Consider
$$\begin{cases} u_{tt}=u_{xx}, & 0<x<\pi,\ t>0\\ u(x,0)=0, & 0\le x\le \pi\\ u_t(x,0)=x, & 0\le x\le \pi\\ u(0,t)=0=u(\pi,t), & t\ge 0.\end{cases}$$
Using (10.33) and (10.36) we arrive at
$$\begin{aligned} u(x,t) &= \frac{1}{2}\int_{x-t}^{x+t} g_p(s)ds,\\ &= \sum_{n=1}^{\infty}\frac{(-1)^{n+1}}{n}\int_{x-t}^{x+t}\sin ns\,ds=\sum_{n=1}^{\infty}\frac{(-1)^{n+1}}{n}\Big[-\frac{\cos ns}{n}\Big|_{x-t}^{x+t}\Big]\\ &= \sum_{n=1}^{\infty}\frac{(-1)^{n+1}}{n^2}\Big[-\cos n(x+t)+\cos n(x-t)\Big].\end{aligned}$$
Clearly, the above solution satisfies $u(0,t)=0,\ u(\pi,t)=0$. □

10.3.1 Exercises

Exercise 10.19 *[Uniqueness] In this exercise, we demonstrate the uniqueness of the solution to* (10.22).

(a) *Consider the energy $E(t)$ defined by*

$$E_u(t) = \frac{1}{2}\int_0^L \left(u_t^2 + c^2 u_x^2\right) dx.$$

Show that $\frac{dE}{dt} = 0$ *along the solutions of* (10.22) *and consequently,* $E(t)$ *is constant over time.*

(b) *Suppose that $u_1(x,t)$ and $u_2(x,t)$ are two solutions of the wave equation with the same initial and boundary conditions. Define*

$$v(x,t) = u_1(x,t) - u_2(x,t).$$

Show that $E_v(0) = 0$, *where* $E_v(t) = \frac{1}{2}\int_0^L \left(v_t^2 + c^2 v_x^2\right) dx$.

(c) *Conclude* $u_1 = u_2$, *which proves the solution of* (10.22) *is unique.*

Exercise 10.20 *Consider*

$$\begin{cases} u_{tt} = u_{xx}, & 0 < x < \pi,\ t > 0 \\ u(x,0) = x^3, & 0 \le x \le \pi \\ u_t(x,0) = 0, & 0 \le x \le \pi \\ u(0,t) = 0 = u(\pi,t), & t \ge 0 \end{cases}$$

(a) *Show that*

$$f_p(x) = 2\sum_{n=1}^{\infty}(-1)^{n+1}\frac{(n\pi)^2 - 6}{n^3}\sin nx, \quad 0 < x < \pi.$$

(b) *Find the solution $u(x,t)$.*

Exercise 10.21 *Consider*

$$\begin{cases} u_{tt} = u_{xx}, & 0 < x < \pi,\ t > 0 \\ u(x,0) = 1, & 0 \le x \le \pi \\ u_t(x,0) = 0, & 0 \le x \le \pi \\ u(0,t) = 0 = u(\pi,t), & t \ge 0 \end{cases}$$

(a) *Show that*

$$f_p(x) = \frac{4}{\pi}\sum_{n=1}^{\infty}\frac{\sin(2n-1)x}{2n-1}, \quad 0 < x < \pi.$$

(b) *Find the solution $u(x,t)$.*

Exercise 10.22 *Consider*

$$\begin{cases} u_{tt} = u_{xx}, & 0 < x < \pi,\ t > 0 \\ u(x,0) = \pi - x, & 0 \leq x \leq \pi \\ u_t(x,0) = 0, & 0 \leq x \leq \pi \\ u(0,t) = 0 = u(\pi,t), & t \geq 0 \end{cases}$$

(a) Show that

$$f_p(x) = 2\sum_{n=1}^{\infty} \frac{\sin nx}{n}, \quad 0 < x < \pi.$$

(b) Find the solution $u(x,t)$.

Exercise 10.23 *Consider*

$$\begin{cases} u_{tt} = u_{xx}, & 0 < x < 1,\ t > 0 \\ u(x,0) = 1, & 0 \leq x \leq 1 \\ u_t(x,0) = 0, & 0 \leq x \leq 1 \\ u(0,t) = 0 = u(1,t), & t \geq 0 \end{cases}$$

(a) Show that

$$f_p(x) = \frac{2}{\pi}\sum_{n=1}^{\infty} \frac{(-1)^{n+1}}{n} \sin n\pi x, \quad 0 < x < 1.$$

(b) Find the solution $u(x,t)$.

Exercise 10.24 *Consider*

$$\begin{cases} u_{tt} = u_{xx}, & 0 < x < 1,\ t > 0 \\ u(x,0) = x(1-x^2), & 0 \leq x \leq 1 \\ u_t(x,0) = 0, & 0 \leq x \leq 1 \\ u(0,t) = 0 = u(1,t), & t \geq 0 \end{cases}$$

(a) Show that

$$f_p(x) = \frac{12}{\pi^3}\sum_{n=1}^{\infty} \frac{(-1)^{n+1}}{n^3} \sin n\pi x, \quad 0 < x < 1.$$

(b) Find the solution $u(x,t)$.

Exercise 10.25 *Solve*

$$\begin{cases} u_{tt} - u_{xx}, & 0 < x < 1,\ t > 0 \\ u(x,0) = 0, & 0 \leq x \leq 1 \\ u_t(x,0) = x(1-x^2), & 0 \leq x \leq 1 \\ u(0,t) = 0 = u(1,t), & t \geq 0 \end{cases}$$

Exercise 10.26 *Solve*

$$\begin{cases} u_{tt} = 4u_{xx}, & 0 < x < 1,\ t > 0 \\ u(x,0) = 0, & 0 \leq x \leq 1 \\ u_t(x,0) = x(1-x), & 0 \leq x \leq 1 \\ u(0,t) = 0 = u(1,t), & t \geq 0 \end{cases}$$

Exercise 10.27 *In Exercise 10.25 set $u_t(x,0) = g(x) = x(1-x^2)$, $0 \le x \le 1$. Find the periodic odd extension of g over the interval $[-2,2]$ and find the solution u.*

Exercise 10.28 *Consider system (10.22) with $l = 3$ and*

$$f(x) = g(x) = \begin{cases} 1, & 0 \le x < \frac{1}{2} \\ x, & \frac{1}{2} \le x < \frac{3}{2} \\ 2, & \frac{3}{2} \le x < 3 \end{cases}$$

Find the odd periodic extensions of f and g over the interval $[0,6]$ and then write down the solution $u(x,t)$.

Ans:

$$f_o(x) = g_o(x) = \begin{cases} 1, & 0 \le x < \frac{1}{2} \\ x, & \frac{1}{2} \le x < \frac{3}{2} \\ 2, & \frac{3}{2} \le x < 3 \\ -1, & 3 \le x < 3.5 \\ -(6-x), & 3.5 \le x < 4.5 \\ -2, & 4.5 \le x < 6 \end{cases}$$

Exercise 10.29 *Consider system (10.22) with $l = 1 = c$, $g(x) = 0$ and*

$$f(x) = \begin{cases} 0, & 0 \le x < 0.45 \\ 20(x-0.45), & 0.45 \le x < 0.5 \\ 20(0.55-x), & 0.5 \le x \le 0.55 \\ 0, & 0.55 \le x \le 1 \end{cases}$$

Compute (a) $u(0.1,0.6)$, (b) $u(0.45,0.1)$.
Ans: (a) -0.5, (b) 0.

10.4 Wave Equation on Semi-Infinite Domain

We take advantage of the development of solutions to the wave equation on both infinite and finite domains, with either Dirichlet or Neumann conditions, to establish the solution of the wave equation on a semi-unbounded domain with either Dirichlet or Neumann conditions.

We begin by considering the wave equation for a semi-infinite string with a fixed end, corresponding to the Dirichlet condition.

$$\begin{cases} u_{tt} - c^2 u_{xx} = 0, & 0 < x < \infty, \, t > 0 \\ u(x,0) = f(x), & x > 0 \\ u_t(x,0) = g(x), & x > 0 \\ u(0,t) = 0, & t > 0 \end{cases} \tag{10.37}$$

Using the same concept as in the wave equation with Dirichlet conditions, we extend $f(x)$ and $g(x)$ to their odd extensions, $f_o(x)$ and $g_o(x)$, respectively. We then solve the wave equation on the entire real line with the initial conditions $u(x,0) = f_o(x)$ and $u_t(x,0) = g_o(x)$.

In other words, we consider

$$\begin{cases} u_{tt} - c^2 u_{xx} = 0, & -\infty < x < \infty,\ t > 0 \\ u(x,0) = f_o(x), \quad u_t(x,0) = g_o(x), & x > 0 \end{cases} \tag{10.38}$$

D'Alembert formula gives

$$u(x,t) = \frac{1}{2}[f_o(x+ct) + f_o(x-ct)] + \frac{1}{2c}\int_{x-ct}^{x+ct} g_o(s)ds. \tag{10.39}$$

By Exercise 10.34, $u(x,t)$ given in (10.39) is odd and so $u(0,t) = 0$. Now we try to make some sense out of the solution in (10.39).
Remember that $x > 0$. We do this in two cases.
(a) First, suppose that that $x > ct$. Then $x + ct \geq 0$ and $x - ct \geq 0$, and so

$$f_o(x+ct) = f(x+ct) \quad \text{and} \quad f_o(x-ct) = f(x-ct).$$

In addition, on $[x-ct, x+ct]$ we have $g_o(s) = g(s)$. Thus if $x > ct$ we have from (10.39) that

$$u(x,t) = \frac{1}{2}[f(x+ct) + f(x-ct)] + \frac{1}{2c}\int_{x-ct}^{x+ct} g(s)ds.$$

(b) Second, if $0 < x < ct$. Then $x + ct \geq 0$ and $x - ct < 0$, and so

$$f_o(x+ct) = f(x+ct) \quad \text{and} \quad f_o(x-ct) = -f(ct-x).$$

Moreover, on $[x-ct, 0]$ we have $g_o(s) = -g(-s)$ and on $[0, x+ct]$ we have $g_o(s) = g(s)$. Therefore if $0 < x < ct$ we have from (10.39) that

$$\begin{aligned} u(x,t) &= \frac{1}{2}[f(x+ct) - f(ct-x)] + \frac{1}{2c}\int_{x-ct}^{0} (-g(-s))ds + \frac{1}{2c}\int_0^{x+ct} g(s)ds \\ &= \frac{1}{2}[f(x+ct) - f(ct-x)] - \frac{1}{2c}\int_0^{ct-x} g(s)ds + \frac{1}{2c}\int_0^{x+ct} g(s)ds \ (s \mapsto -s) \\ &= \frac{1}{2}[f(x+ct) - f(ct-x)] + \frac{1}{2c}\int_{ct-x}^{ct+x} g(s)ds. \end{aligned}$$

In summary, the solution of the half-line wave equation with Dirichlet boundary condition is

$$u(x,t) = \begin{cases} \frac{1}{2}[f(x+ct) + f(x-ct)] + \frac{1}{2c}\int_{x-ct}^{x+ct} g(s)ds, & x \geq ct \\ \frac{1}{2}[f(x+ct) - f(ct-x)] + \frac{1}{2c}\int_{ct-x}^{ct+x} g(s)ds, & 0 < x < ct \end{cases} \tag{10.40}$$

We note that for this solution to exist, f must be twice continuously differentiable and g must be continuously differentiable, and in addition,

$$f(0) = f''(0) = g(0) = 0.$$

Fig. 10.8 shows the two regions of the existence of the solution.

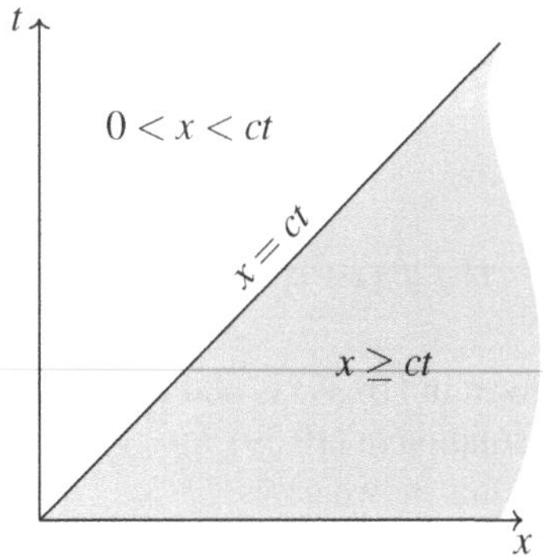

FIGURE 10.8: Regions of existence of the solution.

Now we turn our attention to establishing the solution of the wave equation on semi unbounded domain with Neumann condition. We begin by considering the wave equation of semi-infinite string with a Neumann condition

$$\begin{cases} u_{tt} - c^2 u_{xx} = 0, & 0 < x < \infty,\ t > 0 \\ u(x,0) = f(x), & x > 0 \\ u_t(x,0) = g(x), & x > 0 \\ u_x(0,t) = 0, & t > 0 \end{cases} \tag{10.41}$$

Using the same concept as we did for the wave equation with Dirichlet condition, we use the even extensions $f_e(x)$ and $g_e(x)$ of $f(x)$ and $g(x)$, respectively and solve the wave equation on the whole real line with initial conditions $u(x,0) = f_e(x)$ and $u_t(x,0) = g_e(x)$. In other words, we consider

$$\begin{cases} u_{tt} - c^2 u_{xx} = 0, & & -\infty < x < \infty,\ t > 0 \\ u(x,0) = f_e(x), & u_t(x,0) = g_e(x), & x > 0 \end{cases} \tag{10.42}$$

D'Alembert formula gives

$$u(x,t) = \frac{1}{2}[f_e(x+ct) + f_e(x-ct)] + \frac{1}{2c}\int_{x-ct}^{x+ct} g_e(s)ds. \tag{10.43}$$

By Exercise 10.33, $u(x,t)$ given by (10.43) is even and since the derivative of an even function is odd, and so u_x will be odd in x, and hence $u_x(0,t) = 0$. As before we can simplify (10.39). We do this in two cases and recall that $x > 0$.

(a) First, suppose that that $x > ct$. Then $x + ct \geq 0$ and $x - ct \geq 0$, and so

$$f_e(x+ct) = f(x+ct) \quad \text{and} \quad f_e(x-ct) = f(x-ct).$$

Moreover, on $[x-ct, x+ct]$ we have $g_e(s) = g(s)$. Thus if $x > ct$ we have from (10.43) that

$$u(x,t) = \frac{1}{2}[f(x+ct) + f(x-ct)] + \frac{1}{2c}\int_{x-ct}^{x+ct} g(s)ds.$$

(b) Second, if $0 < x < ct$. Then $x+ct \geq 0$ and $x-ct < 0$, and so

$$f_e(x+ct) = f(x+ct) \quad \text{and} \quad f_e(x-ct) = f(ct-x).$$

Moreover, on $[x-ct, 0]$ we have $g_e(s) = g(-s)$ and on $[0, x+ct]$ we have $g_e(s) = g(s)$. Therefore, if $0 < x < ct$ we have from (10.43) that

$$\begin{aligned} u(x,t) &= \frac{1}{2}[f(x+ct) + f(ct-x)] + \frac{1}{2c}\int_{x-ct}^{0} g(-s)ds + \frac{1}{2c}\int_0^{x+ct} g(s)ds \\ &= \frac{1}{2}[f(x+ct) + f(ct-x)] + \frac{1}{2c}\int_0^{ct-x} g(s)ds + \frac{1}{2c}\int_0^{x+ct} g(s)ds \ (s \mapsto -s). \end{aligned}$$

In summary, the solution of the half-line wave equation with Neumann boundary condition is

$$u(x,t) = \begin{cases} \frac{1}{2}[f(x+ct) + f(x-ct)] + \frac{1}{2c}\int_{x-ct}^{x+ct} g(s)ds, & x \geq ct \\ \frac{1}{2}[f(x+ct) + f(ct-x)] + \frac{1}{2c}\left[\int_0^{ct-x} g(s)ds + \int_0^{x+ct} g(s)ds\right], & 0 < x < ct. \end{cases} \tag{10.44}$$

We note that for this solution to exist, f must be twice continuously differentiable and g must be continuously differentiable, and in addition,

$$f'(0) = g'(0) = 0.$$

Example 10.11 Consider the wave equation (10.37) with $c = 1$ and initial data

$$f(x) = \sin x, \quad g(x) = 0,$$

along with the Dirichlet condition $u(0,t) = 0$. Using (10.40), for $x > t$, we obtain

$$u(x,t) = \frac{1}{2}[\sin(x+t) + \sin(x-t)] = \sin(x)\cos(t).$$

On the other hand, for $0 < x < t$, we have

$$u(x,t) = \frac{1}{2}[\sin(x+t) - \sin(t-x)] = \sin(x)\cos(t).$$

Thus, the solution is given by

$$u(x,t) = \sin x \cos t, \quad x > 0.$$

□

Example 10.12 Consider the wave equation (10.37) with $c = 1$ and initial data

$$f(x) = \cos(x), \quad g(x) = 0,$$

along with the Neumann condition $u_x(0,t) = 0$. Using (10.44), one can easily verify that

$$u(x,t) = \cos t \cos x, \quad x > 0.$$

□

10.4.1 Exercises

Exercise 10.30 *Solve*

$$\begin{cases} u_{tt} - c^2 u_{xx} = 0, & 0 < x < \infty,\ t > 0 \\ u(x,0) = 0, & x > 0 \\ u_t(x,0) = \cos(x), & x > 0 \\ u(0,t) = 0, & t > 0 \end{cases}$$

Exercise 10.31 *Solve*

$$\begin{cases} u_{tt} - c^2 u_{xx} = 0, & 0 < x < \infty,\ t > 0 \\ u(x,0) = 0, & x > 0 \\ u_t(x,0) = \sin(x), & x > 0 \\ u(0,t) = 0, & t > 0 \end{cases}$$

Exercise 10.32 *Solve*

$$\begin{cases} u_{tt} - c^2 u_{xx} = 0, & 0 < x < \infty,\ t > 0 \\ u(x,0) = 0, & x > 0 \\ u_t(x,0) = 1, & x > 0 \\ u(0,t) = 0, & t > 0 \end{cases}$$

Exercise 10.33 *Let $u(x,t)$ be the D'alembert's solution on $-\infty < x < \infty$ of the wave equation given by 10.8. If the initial data given by $f(x) = u(x,0)$ and $g(x) = u_x(x,0)$ is even, then for all $t \geq 0$, $u(x,t)$ is an even function of x.*

Exercise 10.34 *Let $u(x,t)$ be the D'alembert's solution on $-\infty < x < \infty$ of the wave equation given by 10.8. If the initial data given by $f(x) = u(x,0)$ and $g(x) = u_x(x,0)$ is odd, then for all $t \geq 0$, $u(x,t)$ is an odd function of x.*

Exercise 10.35 *Solve*

$$\begin{cases} u_{tt} - c^2 u_{xx} = 0 & 0 < x < \infty,\ t > 0 \\ u(x,0) = 0, & x > 0 \\ u_t(x,0) = \sin(x), & x > 0 \\ u_x(0,t) = 0, & > 0 \end{cases}$$

Exercise 10.36 *Solve*

$$\begin{cases} u_{tt} - c^2 u_{xx} = 0, & 0 < x < \infty,\ t > 0 \\ u(x,0) = 0, & x > 0 \\ u_t(x,0) = \cos(x), & > 0 \\ u_x(0,t) = 0, & t > 0 \end{cases}$$

Exercise 10.37 *Solve*

$$\begin{cases} u_{tt} - c^2 u_{xx} = 0, & 0 < x < \infty,\ t > 0 \\ u(x,0) = 1, & x > 0 \\ u_t(x,0) = 0, & x > 0 \\ u_x(0,t) = 0, & t > 0 \end{cases}$$

Exercise 10.38 *Solve*

$$\begin{cases} u_{tt} - c^2 u_{xx} = 0, & 0 < x < \infty,\ t > 0 \\ u(x,0) = 0, & x > 0 \\ u_t(x,0) = 1, & x > 0 \\ u_x(0,t) = 0, & t > 0 \end{cases}$$

Exercise 10.39 *Consider the wave problem with non-homogeneous Dirichlet boundary condition*

$$\begin{cases} v_{tt} - c^2 v_{xx} = 0, & 0 < x < \infty,\ t > 0 \\ v(x,0) = f(x), & x > 0 \\ v_t(x,0) = g(x), & x > 0 \\ v(0,t) = p(t), & t > 0 \end{cases}$$

where the function p is twice differentiable. Make the change of variables

$$u(x,t) = v(x,t) - p(t)$$

and reduce the wave problem to a problem with homogeneous boundary condition and then use the D'alembert solution given by (10.13) *to find the solution u and then find the solution v of the original wave problem.*

Exercise 10.40 *Find the solution of*

$$\begin{cases} u_{tt} - c^2 u_{xx} = 0, & & 0 < x,\ t > 0 \\ u(0,t) = h(t), & & 0 < t \\ u(x,0) = f(x), & u_t(x,0) = g(x), & 0 < x \end{cases}$$

with $f, g, h \in C^2$ *and satisfying*

$$h(0) = f(0), \quad h'(0) = g(0), \quad h''(0) = c^2 f''(0).$$

Show that the obtained solution u has continuous second partial derivatives even on the characteristic line $x = ct$.

10.5 Duhamel's Principle for the Wave Equation

Duhamel's principle is used to solve the non-homogeneous one-dimensional wave equation by converting it into an integral representation. It allows solving problems with time-dependent forcing functions by expressing the solution as a superposition of solutions to simpler problems. The principle states that the response of a system to a general forcing function can be obtained by integrating the response to an impulsive force over time. This method is particularly useful in handling inhomogeneous boundary conditions and external sources in wave equations.

We begin by considering the wave equation with external force and homogeneous initial conditions

$$\begin{aligned} &u_{tt} - c^2 u_{xx} = h(x,t), \quad -\infty < x < \infty,\ t > 0, \\ &u(x,0) = 0, \quad -\infty < x < \infty, \\ &u_t(x,0) = 0, \quad -\infty < x < \infty, \end{aligned} \tag{10.45}$$

where the function h is assumed to be continuous with respect to both arguments. The main idea of *Duhamel's principle* is to transfer the effect of the external force function to the initial velocity by introducing a new partial differential equation system. Specifically, we assume that $U(x,t,s)$ is a solution to the transformed system

$$\begin{aligned} &U_{tt} - c^2 U_{xx} = 0, \quad -\infty < x < \infty,\ t > 0, \\ &U(x,0,s) = 0, \quad -\infty < x < \infty, \\ &U_t(x,0,s) = h(x,s), \quad t > 0. \end{aligned} \tag{10.46}$$

Theorem 10.2 *If $U(x,t,s)$ is a solution of (10.46), then the solution $u(x,t)$ is given by*

$$u(x,t) = \int_0^t U(x,t-s,s)\,ds, \quad t \geq 0. \tag{10.47}$$

Proof *Let u be given by (10.47). Then, differentiating with respect to t, we obtain*

$$u_t(x,t) = \int_0^t U_t(x,t-s,s)\,ds + U(x,0,t) = \int_0^t U_t(x,t-s,s)\,ds.$$

Differentiating once more,

$$\begin{aligned} u_{tt}(x,t) &= \int_0^t U_{tt}(x,t-s,s)\,ds + U_t(x,0,t) - 0 = c^2 \int_0^t U_{xx}(x,t-s,s)\,ds + h(x,t) \\ &= c^2 \frac{\partial^2}{\partial x^2}\left(\int_0^t U(x,t-s,s)\,ds\right) + h(x,t) = c^2 u_{xx}(x,t) + h(x,t), \end{aligned}$$

which is exactly the PDE in (10.45). To check the initial conditions,

$$u(x,0) = \int_0^0 U(x,0-s,s)\,ds = 0,$$

and using the calculation of u_t performed above, we have

$$u_t(x,0) = \int_0^0 U_t(x,0-s,s)\,ds = 0.$$

This completes the proof. □

Now, we tackle the problem with forcing function and having non-homogeneous initial conditions. Thus, we consider

$$\begin{aligned} &u_{tt} - c^2 u_{xx} = h(x,t), \quad -\infty < x < \infty,\ t > 0, \\ &u(x,0) = f(x), \\ &u_t(x,0) = g(x). \end{aligned} \tag{10.48}$$

In order to be able to use Duhamel's principle to solve (10.48), we consider the two auxiliary systems

$$\begin{aligned} &v_{tt} - c^2 v_{xx} = 0, \quad -\infty < x < \infty,\ t > 0, \\ &v(x,0) = f(x), \\ &v_t(x,0) = g(x), \end{aligned} \tag{10.49}$$

and

$$\begin{aligned} &w_{tt} - c^2 w_{xx} = h(x,t), \quad -\infty < x < \infty,\ t > 0, \\ &w(x,0) = 0, \\ &w_t(x,0) = 0. \end{aligned} \tag{10.50}$$

The problem given by (10.49) is homogeneous and can be solved using the d'Alembert solution. On the other hand, the system given by (10.50) is non-homogeneous, so we apply Theorem 10.2 to determine its solution.

That is, to apply Duhamel's principle to (10.50), we consider, as before, the transformed system

$$\begin{aligned} &W_{tt} - c^2 W_{xx} = 0, \quad -\infty < x < \infty,\ t > 0, \\ &W(x,0,s) = 0, \\ &W_t(x,0,s) = h(x,s), \quad t > 0. \end{aligned} \tag{10.51}$$

Then, the solution $W(x,t,s)$ of (10.51) can be found by previous methods. We have the following theorem.

Theorem 10.3 *Let v and w be solutions of (10.49) and (10.50), respectively. Then the solution $u(x,t)$ of (10.48) is given by*

$$u(x,t) = v(x,t) + w(x,t),$$

where

$$w(x,t) = \int_0^t W(x,t-s,s)\,ds, \quad t \geq 0.$$

We have the following example.

Example 10.13 Consider the non-homogeneous wave problem

$$\begin{aligned}
&u_{tt} - u_{xx} = -1, \quad -\infty < x < \infty,\ t > 0,\\
&u(x,0) = \sin x,\\
&u_t(x,0) = x.
\end{aligned}$$

Based on (10.49) we have the homogeneous problem

$$\begin{aligned}
&v_{tt} - v_{xx} = 0, \quad -\infty < x < \infty,\ t > 0,\\
&v(x,0) = \sin x,\\
&v_t(x,0) = x.
\end{aligned}$$

Using the d'Alembert solution given by (10.8), we arrive at

$$\begin{aligned}
v(x,t) &= \frac{1}{2}[\sin(x+t) + \sin(x-t)] + \frac{1}{2}\int_{x-t}^{x+t} \tau\, d\tau\\
&= \frac{1}{2}[\sin(x+t) + \sin(x-t)] + xt.
\end{aligned}$$

Formulating system (10.50) yields

$$\begin{aligned}
&w_{tt} - w_{xx} = -1, \quad -\infty < x < \infty,\ \ t > 0,\\
&w(x,0) = 0,\\
&w_t(x,0) = 0.
\end{aligned}$$

To apply Theorem 10.2 to find $w(x,t)$, we use (10.51) to define the auxiliary system

$$\begin{aligned}
&W_{tt} - c^2 W_{xx} = 0, \quad -\infty < x < \infty,\ t > 0,\\
&W(x,0,s) = 0,\\
&W_t(x,0,s) = -1.
\end{aligned}$$

Its solution can be easily obtained using (10.13). In particular,

$$W(x,t,s) = \int_0^t \int_{x-(t-s)}^{x+(t-s)} d\tau\, ds = -t^2.$$

Consequently, by Theorem 10.2,

$$w(x,t) = \int_0^t W(x,t-s,s)\, ds = -\int_0^t (t-s)^2\, ds = -\frac{t^2}{2}.$$

Finally, the solution to our original problem is

$$u(x,t) = v(x,t) + w(x,t) = \frac{1}{2}[\sin(x+t) + \sin(x-t)] + xt - \frac{t^2}{2}.$$

□

Duhamel's principle easily accommodate vibration of a string with fixed ends. Suppose that there is a force $f(x,t)$ in the PDE for the wave equation

$$u_{tt} = c^2 u_{xx} + h(x,t), \quad 0 < x < L, \, t > 0,$$

with initial and boundary conditions

$$u(x,0) = 0, \quad u_t(x,0) = 0, \quad 0 < x < L,$$

$$u(0,t) = 0, \quad u(L,t) = 0, \quad t > 0.$$

First, move the force to the initial velocity. The new system is

$$U_{tt} = c^2 U_{xx}, \quad 0 < x < L, \, t > 0,$$

$$U(x,0,s) = 0, \quad U_t(x,0,s) = h(x,s), \quad 0 < x < L,$$

$$U(0,t,s) = 0, \quad U(L,t,s) = 0, \quad t > 0.$$

Then the solution to the original problem is

$$u(x,t) = \int_0^t U(x,t-s,s)\,ds, \quad t \geq 0.$$

Example 10.14 Consider the non-homogeneous wave equation with fixed ends

$$u_{tt} = c^2 u_{xx} + \sin 3x, \quad 0 < x < \pi, \, t > 0,$$

$$u(x,0) = 0 = u_t(x,0), \quad 0 < x < L,$$

$$u(0,t) = 0 = u(\pi,t) = 0, \quad t > 0.$$

Putting the source in the initial velocity, we arrive at the new system

$$U_{tt} = c^2 U_{xx}, \quad 0 < x < \pi, \, t > 0,$$

$$U(x,0,s) = 0, \quad U_t(x,0,s) = \sin 3x, \quad 0 < x < \pi,$$

$$U(0,t,s) = 0, \quad U(\pi,t,s) = 0, \quad t > 0.$$

Using separation of variables of the form $U(x,t,s) = X(x)T(t)$, we arrive at the eigenvalues and corresponding eigenfunctions

$$\lambda_n = n^2, \quad X_n(x) = \sin nx, \quad n = 1, 2, \ldots$$

Hence, its general solution can be written as

$$U(x,t,s) = \sum_{n=1}^{\infty} [a_n \cos cnt + b_n \sin cnt] \sin nx.$$

Applying both initial conditions, we obtain the solution

$$U(x,t,s) = \frac{1}{3c} \sin 3ct \sin 3x.$$

Finally,

$$\begin{aligned} u(x,t) &= \int_0^t U(x,t-s,s)\,ds = \frac{1}{3c}\int_0^t \sin 3c(t-s)\sin 3x\,ds \\ &= \frac{\sin 3x}{3c}\int_0^t \sin 3c(t-s)\,ds = \frac{\sin 3x}{3c}\left[\frac{-\cos 3c(t-s)}{3c}\right]_0^t \\ &= \frac{\sin 3x}{9c^2}[1-\cos 3ct]. \end{aligned}$$

10.5.1 Exercises

Exercise 10.41 *Prove Theorem 10.3.*

Exercise 10.42 *Redo Example 10.1 using Duhamel's principle.*

Exercise 10.43 *Redo Example 10.5 using Duhamel's principle.*

Exercise 10.44 *Redo Exercise 10.16 using Duhamel's principle.*

Exercise 10.45 *Consider the non-homogeneous wave equation with a point source whose amplitude varies with time*

$$u_{tt} - c^2 u_{xx} = f(t)\delta(x), \quad -\infty < x < \infty, \quad t > 0,$$

with zero initial conditions

$$u(x,0) = 0, \quad u_t(x,0) = 0.$$

(a) Use Duhamel's principle to show that the solution can be written as

$$u(x,t) = \frac{1}{2c}\int_0^t f(\tau)H(c(t-\tau)-|x|)\,d\tau,$$

where H is the Heaviside step function.

(b) Compute the explicit form of $u(x,t)$ when $f(t) = t^2$.

Exercise 10.46 *Consider the non-homogeneous wave equation with a spatially distributed source that activates at $t = 0$*

$$u_{tt} - c^2 u_{xx} = g(x)H(t), \quad -\infty < x < \infty, \quad t > 0,$$

with zero initial conditions

$$u(x,0) = 0, \quad u_t(x,0) = 0.$$

(a) Use Duhamel's principle to express the solution $u(x,t)$ as an integral involving $g(x)$.

(b) Specialize your result to the case $g(x) = \delta(x-a)$, where $a > 0$ is a constant, and determine the explicit solution $u(x,t)$.

Ans:

(a) $u(x,t) = \frac{1}{2c}\int_0^t \int_{x-c(t-\tau)}^{x+c(t-\tau)} g(y)\,dy\,d\tau.$

(b) $u(x,t) = \begin{cases} \frac{t-|x-a|/c}{2c}, & \text{if } |x-a| \le ct, \\ 0, & \text{otherwise}. \end{cases}$

Exercise 10.47 *Consider the non-homogeneous wave equation with a moving point source:*

$$u_{tt} - c^2 u_{xx} = \delta(x - ct), \quad -\infty < x < \infty, \quad t > 0,$$

with zero initial conditions

$$u(x,0) = 0, \quad u_t(x,0) = 0.$$

(a) *Use Duhamel's principle to find the solution* $u(x,t)$.

(b) *Analyze the behavior of the solution as* t *increases. Does the solution remain bounded? Explain the physical reason for your observation.*

Exercise 10.48 *Use Duhamel's principle to solve the initial value problem*

$$u_{tt} - u_{xx} = t\sin x, \quad -\infty < x < \infty, \quad t > 0,$$

with initial conditions

$$u(x,0) = 0, \quad u_t(x,0) = 0.$$

Ans: $u(x,t) = \sin x\,(t - \sin t)$.

Exercise 10.49 *Use Duhamel's principle to solve the initial value problem*

$$u_{tt} - c^2 u_{xx} = xe^{-t}, \quad -\infty < x < \infty, \quad t > 0,$$

with initial conditions

$$u(x,0) = x^2, \quad u_t(x,0) = 0.$$

Ans: $u(x,t) = x^2 + c^2t^2 + x(t + e^{-t} - 1)$.

Exercise 10.50 *Prove Theorem 9.4.*

Bibliography

[1] Adams, R.A. and Fournier, J.J.F., *Sobolev Spaces,* Academic Press, 2003.

[2] Alinhac, S. and Gérard, P., *Pseudo-Differential Operators and the Nash-Moser Theorem,* American Mathematical Society, 2007.

[3] Arnold, V.I., *Lectures on Partial Differential Equations,* Springer, 2004.

[4] Ambrosio, L., Colombo, M., and Figalli, A.A., *Transport Approach to Nonlinear Partial Differential Equations,* American Mathematical Society, 2021.

[5] Bellman, R., *Stability Theory of Differential Equations,* McGraw-Hill Book Company, New York, London, 1953.

[6] Berezansky, L. and Braverman, E., *Exponential stability of difference equations with several delays: Recursive approach,* Adv. Difference. Equ. Vol. 2009, Article ID 104310, 13 pages.

[7] Brown, J.W. and Churchill, R.V., *Fourier Series and Boundary Value Problems,* 8th edition, McGrawHill, 2012.

[8] Burgers, J.M., *A mathematical model illustrating the theory of turbulence,* Adv. Appl. Mech. 1 (1948) 171–191.

[9] Byerly, W.E., *Fourier Series,* Dover Publications, New York, 1959.

[10] Carleson, L., *On the convergence and growth of partial sums of Fourier Series,* Acta Mathematica, 116 (1966) 135–157.

[11] Carslaw, H.S., *Introduction to the Theory of Fourier Series and Integrals,* Dover Publications, New York, 1950.

[12] Carslaw, H.S. and Jaeger, J.C., *Operational Methods in Applied Mathematics,* Oxford University Press, London, 1949.

[13] Carslaw, H.S. and Jaeger, J.C., *Conduction of Heat in Solids (Second Edition),* Oxford University Press, Oxford, 1986.

[14] Churchill, R.V., *Operational Mathematics (Third Edition),* McGraw-Hill, New York, 1972.

[15] Coddington, E.A. and Levinson, N., *Theory of Ordinary Differential Equations,* McGraw-Hill, New York, 1955.

[16] Cole, R., *Theory of Ordinary Differential Equation,* Appleton-Century-Crofts, 1968.

[17] Constanda, C., *Integral Methods in Science and Engineering,* CRC Press, 2000.

[18] Copson, E.T., *Asymptotic Expansions,* Cambridge University Press, Cambridge, 1965.

[19] Coulson, C.A. and Jeffrey, A., *Waves: A Mathematical Approach to Common Types of Wave Motion (Second Edition),* Longman, London, 1977.

[20] Courant, R., *Variational methods for the solutions of problems of equilibrium and vibrations,* Bull. Amer. Math. Soc. 49 (1943) 10–30.

[21] Courant, R. and Hilbert, D., *Methods of Mathematical Physics, Volume 1 (1953), Volume 2 (1962),* Interscience, New York.

[22] Courant, R., Friedrichs, K., and Lewy, H., *Über die partiellen differenzengleichungen der mathematischen Physik,* Mathematische Annalen 100 (1928) 32–74.

[23] Crank, J., *Mathematics of Diffusion (Second Edition),* Clarendon Press, Oxford, 1975.

[24] Crank, J. and Nicholson, P., *A practical method for numerical evaluation of solutions of partial differential equations of the heat conduction type,* Proc. Camb. Phil. Soc. 43 (1947) 50–67.

[25] Davis, H.F., *Fourier Series and Orthogonal Functions,* Allyn and Bacon, New York, 1963.

[26] Debnath, L., *On asymptotic treatment of the transient development of surface waves,* Appl. Sci. Res. 21 (1969) 24–36.

[27] Debnath, L., *Unsteady axisymmetric capillary-gravity waves in a viscous liquid,* Indian J. Pure and Appl. Math. 14 (1983) 540–553.

[28] Debnath, L., *Nonlinear Waves,* Cambridge University Press, Cambridge, England, 1983.

[29] Debnath, L., *Nonlinear Water Waves,* Academic Press, Boston, 1994.

[30] Debnath, L., *Integral Transforms and Their Applications,* CRC Press, Boca Raton, Florida, 1995.

[31] Debnath, L., *Nonlinear dynamics of water waves and breaking phenomena,* Proc. Indian Nat. Sci. Acad. Special Volume (2002) 683–753.

[32] Debnath, L., *Nonlinear Partial Differential Equations for Scientists and Engineers (Second Edition),* Birkhauser Verlag, Boston, 2005.

[33] Debnath, L. and Bhatta, D., *Integral Transforms and Their Applications (Second Edition),* CRC Press, Boca Raton, Florida, 2006.

[34] Debnath, L. and Mikusinski, P., *Introduction to Hilbert Spaces with Applications (Third Edition),* Elsevier Academic Press, London, 2005.

[35] Debnath, L. and Rollins, D., *The Cauchy-Poisson waves in an inviscid rotating stratified liquid,* Internat. J. Nonlinear Mech. 27 (1992) 405–412.

[36] Debnath, L. and Shivamoggi, B.K., *Three-dimensional nonlinear Schrödinger equation for finite-amplitude gravity waves in a fluid,* Nuovo Cimento 94B (1986) 140–148; 99B (1987) 247.

[37] Dennemeyer, R., *Partial Differential Equations and Boundary Value Problems,* McGraw-Hill, New York, 1968.

[38] Ditkin, V.A. and Prudnikov, A.P., *Integral transforms and Operational Calculus,* Pergamon Press, Oxford, 1965.

[39] Doetsch, G., *Introduction to the Theory and Applications of the Laplace Transformation (Translated by Walter Nader),* Springer Verlag, New York, 1970.

[40] Driver, R.D., *Introduction to Ordinary Differential Equations,* Harper & Row Publishers, New York, 1978.

[41] Duff, G.F.D., *Partial Differential Equations,* University of Toronto, Toronto, 1956.

[42] Duff, G.F.D. and Naylor, D., *Differential Equations of Applied Mathematics,* John Wiley, New York, 1966.

[43] Dutta, M. and Debnath, L., *Elements of the Theory of Elliptic and Associated Functions with Applications,* World Press, Calcutta, 1965.

[44] Epstein, B., *Partial Differential Equations,* McGraw-Hill, New York, 1962.

[45] Erdélyi, A., Magnus, W., Oberhettinger, F., and Tricomi, F., *Tables of Integral Transforms (Vols. I and II),* McGraw-Hill, New York, 1954.

[46] Folland, G.B., *Fourier Analysis and Its Applications,* Brooks/Cole Publishing Company, Pacific Grove, California, 1992.

[47] Forsythe, G.E. and Wasow, W.R., *Finite-Difference Methods for Partial Differential Equations,* John Wiley, New York, 1967.

[48] Fourier, J.B.J., *The Analytical Theory of Heat (Translated by A. Freeman),* Dover Publications, New York, 1955.

[49] Fox, C., *An Introduction to the Calculus of Variations,* Oxford University Press, Oxford, 1963.

[50] Garabedian, P.R., *Partial Differential Equations,* John Wiley, New York, 1964.

[51] Gelfand, I.M. and Fomin, S.V., *Calculus of Variations (Translated by R.A. Silverman),* Prentice-Hall, Englewood Cliffs, New Jersey, 1963.

[52] Ghosh, K.K. and Debnath, L., *Some exact solutions of nonlinear shallow water equations,* Internal. J. Nonlinear Mech. 31 (1997) 104–108.

[53] Greenberg, M., *Applications of Green's Functions in Science and Engineering,* Prentice-Hall, Englewood Cliffs, New Jersey, 1971.

[54] Grunberg, G., *A new method of solution of certain boundary problems for equations of mathematical physics permitting of a separation of variables,* J. Phys. 10 (1946) 301–320.

[55] Haberman, R., *Mathematical Models: Mechanical Vibrations, Population Dynamics and Traffic Flow,* Prentice Hall, Englewood Cliffs, New Jersey, 1977.

[56] Hadamard, J., *Lectures on Cauchy's Problem in Linear Partial Differential Equations,* Dover Publications, New York, 1952.

[57] Haight, F.A., *Mathematical Theories of Traffic Flow,* Academic Press, New York, 1963.

[58] Hartman, P., *Ordinary Differential Equations,* John Wiley & Sons, Inc., New York, 1964.

[59] Hellwig, G., *Partial Differential Equations,* Blaisdell, Waltham, Massachusetts, 1964.

[60] Hildebrand, F., *Methods of Applied Mathematics,* Prentice-Hall, Englewood Cliffs, New Jersey, 1965.

[61] Huebuer, K.H., Dewhirst, D.L., Smith, D.L., and Byrom, T.G., *The Finite Element Method for Engineers (Fourth Edition),* John Wiley & Sons, New York, 2001.

[62] Irving, J. and Mullineux, N., *Mathematics in Physics and Engineering,* Academic, New York, 1959.

[63] Jackson, D., *Fourier Series and Orthogonal Polynomials,* Mathematical Association of America Monograph, 1941.

[64] Jaswon, M.A., *Integral equation method in potential theory. I,* Proc. R. Soc. London, A 275 (1963) 23–32.

[65] Jeffrey, A., *Applied Partial Differential Equations: An Introduction,* Academic Press, 2003.

[66] Jeffrey, A., *Advanced Engineering Mathematics,* Academic Press, Boston, 2002.

[67] Jeffrey, A. and Engelbrecht, J., *Nonlinear Waves in Solids,* Springer-Verlag, New York, 1994.

[68] Jeffrey, A. and Taniuti, T., *Nonlinear Wave Propagation,* Academic Press, New York, 1964.

[69] Jeffreys, H., *Operational Methods in Mathematical Physics,* Cambridge University Press, New York, 1931.

[70] Jeffreys, J. and Jeffreys, B.S., *Methods of Mathematical Physics (Third Edition),* Cambridge University Press, Cambridge, 1956.

[71] John, F., *Partial Differential Equations (Fourth Edition),* Springer-Verlag, New York, 1982.

[72] Johnson, R.S., *The KdV equation and related problems in water wave theory, in Nonlinear Waves (ed. L. Debnath),* 25–43, Cambridge University Press, Cambridge, 1983.

[73] Johnson, R.S., *A Modern Introduction to the Mathematical Theory of Water Waves,* Cambridge University Press, Cambridge, 1997.

[74] Jones, D.S., *Generalized Functions,* Academic Press, New York, 1966.

[75] Keener, J.P., *Principles of Applied Mathematics: Transformation and Approximation,* Addison-Wesley, Reading, Massachusetts, 1988.

[76] Kelley, W. and Peterson, A., *The Theory of Differential Equations, Classical and Qualitative,* Pearson Prentice Hall, 2004.

[77] Knobel, R., *An Introduction to the Mathematical Theory of Waves,* American Mathematical Society, Providence, 1999.

[78] Korteweg, D.J. and de Vries, G., *On the change of form of long waves advancing in a rectangular canal, and on a new type of long stationary waves,* Phil. Mag. 39 (1895) 422–443.

[79] Koshlayokov, N.S., Smirnov, M.M., and Gliner, E.B., *Differential Equations of Mathematical Physics,* Holden-Day, New York, 1964.

[80] Kreider, D., Kuller, R., Ostberg, D., and Perkins, F., *An Introduction to Linear Analysis,* Addison-Wesley, Reading, Massachusetts, 1966.

[81] Kreyszig, E., *Advanced Engineering Mathematics (Seventh Edition),* John Wiley, New York, 1993.

[82] Lamb, H., *Hydrodynamics (Sixth Edition),* Cambridge University Press, Cambridge, England, 1932.

[83] Lax, P.D., *Weak solutions of nonlinear hyperbolic equations and their numerical computation,* Comm. Pure Appl. Math. 7 (1954) 159–193.

[84] Lax, P.D., *The initial-value problem for nonlinear hyperbolic equations in two independent variables,* Ann. Math. Stud. (Princeton) 33 (1954) 211–299.

[85] Lax, P.D., *Hyperbolic systems of conservation law, II,* Comm. Pure Appl. Math. 10 (1957) 537–566.

[86] Lax, P.D., *Partial Differential Equations, Lectures on Hyperbolic Equations,* Stanford University Press, 1963.

[87] Lax, P.D., *Development of singularities of solutions of nonlinear hyperbolic partial differential equations,* J. Math. Phys. 5 (1964) 611–613.

[88] Lax, P.D., *Integrals of nonlinear equations of evolution and solitary waves,* Comm. Pure Appl. Math. 21 (1968) 467–490.

[89] Lax, P.D., *The formation and decay of shock waves,* Amer. Math. Monthly 79 (1972) 227–241.

[90] Lax, P.D., *Hyperbolic Systems of Conservation Lays and the Mathematical Theory of Shock Waves,* Society for Industrial and Applied Mathematics, Philadelphia, 1973.

[91] Lax, P.D., *Periodic solutions of the Korteweg-de Vries equation,* Comm. Pure Appl. Math. 28 (1975) 141–188.

[92] Lax, P.D. and Wendroff, B., *Systems of conservation laws,* Comm. Pure Appl. Math. 13 (1960) 217–237.

[93] Leibovich, S. and Seebass, A.R., *Nonlinear Waves,* Cornell University Press, Ithaca and London, 1972.

[94] Lighthill, M.J., *Introduction to Fourier Analysis and Generalized Functions,* Cambridge University Press, Cambridge, 1964.

[95] Lighthill, M.J., *Waves in Fluids,* Cambridge University Press, Cambridge, 1980.

[96] Logan, D.J., *Applied Partial Differential Equations,* Springer, 1998.

[97] Logan, J.D., *Applied Mathematics (Second Edition),* John Wiley, New York, 1997.

[98] Love, A.E.H., *A Treatise on the Mathematical Theory of Elasticity (Fourth Edition),* New York, Dover, 1944.

[99] Mainardi, F., *On the initial-value problem for the fractional diffusion-wave equation, in Waves and Stability in Continuous Media (ed. S. Rionero and T. Ruggeri),* World Scientific, Singapore, 1994, 246–251.

[100] Mainardi, F., *The time fractional diffusion-wave equation,* Radiofisika 38 (1995) 20–36.

[101] Mainardi, F., *Fractional relaxation-oscillation and fractional diffusion-wave phenomena,* Chaos, Solitons, Fractals 7 (1996) 1461–1477.

[102] McLachlan, N.W., *Complex Variable and Operational Calculus with Technical Applications,* Cambridge University Press, London, 1942.

[103] McLachlan, N.W., *Modern Operational Calculus,* Macmillan, London, 1948.

[104] Mikhlin, S.G., *Variational Methods in Mathematical Physics,* Pergamon Press, Oxford, 1964.

[105] Mikhlin, S.G., *Linear Equations of Mathematical Physics,* Holt, Rinehart and Winston, New York, 1967.

[106] Miller, K.S., *Partial Differential Equations in Engineering Problems,* Prentice Hall, Englewood Cliffs, New Jersey, 1953.

[107] Mitchell, A.R. and Griffiths, D.F., *Finite Difference Method in Partial Differential Equations,* John Wiley, New York, 1980.

[108] Morse, P.M. and Freshbach, H., *Methods of Theoretical Physics,* Volume 1 and 2, McGraw-Hill, New York, 1953.

[109] Myint-U, T., *Ordinary Differential Equations,* Elsevier North Holland, Inc., New York, 1978.

[110] Nigmatullin, R.R., *The realization of the generalized transfer equation in a medium with fractal geometry,* Phys. Sta. Sol. (b) 133 (1986) 425–430.

[111] Oberhettinger, F., *Tables of Bessel Transforms,* Springer-Verlag, New York, 1972.

[112] Petrovsky, I., *Lectures on Partial Differential Equations,* Interscience, New York, 1954.

[113] Picard, E., *Traité d'Analyse,* Gauthier-Villars, Paris, 1896.

[114] Pinkus, A. and Zafrany, S., *Fourier Series and Integral Transforms,* Cambridge University Press, Cambridge, 1997.

[115] Pipes, L.A., *Applied Mathematics for Engineers and Physicists,* McGraw-Hill, New York, 1958.

[116] Qiao, Z., *Generation of the hierarchies of solitons and generalized structure of the commutator representation,* Acta. Appl. Math. Sinica 18 (1995) 287–301.

[117] Raffoul, Y.N., *Advanced Differential Equations,* Academic Press/Elsevier, April, 2022.

[118] Raffoul, Y.N., *Applied Mathematics for Scientists and Engineers,* Taylor & Francis/ Chapman and Hill/CRC September, 2023.

[119] Raven, R.H., *Mathematics of Engineering Systems,* McGraw-Hill, New York, 1966.

[120] Reif, F., *Fundamentals of Statistical and Thermal Physics,* McGraw-Hill, New York, 1965.

[121] Roach, G.F., *Green's Functions (Second Edition),* Cambridge University Press, Cambridge, 1982.

[122] Rogosinski, W.W., *Fourier Series,* Chelsea, London, 1950.

[123] Sagan, H., *Boundary and Eigenvalue Problems in Mathematical Physics,* John Wiley, New York, 1966.

[124] Sansone, G., *Orthogonal Functions,* Interscience, New York, 1959.

[125] Schwartz, L., *Théorie des distributions,* Volume I (1950) and Volume II (1951), Herman and Cie, Paris.

[126] Scott, E.J., *Transform Calculus with an Introduction to Complex Variables,* Harper, New York, 1955.

[127] Seeley, R., *An Introduction to Fourier Series and Integrals,* W.A. Benjamin, New York, 1966.

[128] Smirnov, V.I., *Integral Equations and Partial Differential Equations,* Addison-Wesley, Reading, Massachusetts, 1964.

[129] Smith, M.G., *Introduction to the Theory of Partial Differential Equations,* Van Nostrand, Princeton, New Jersey, 1967.

[130] Smoller, J., *Shock Waves and Reaction-Diffusion Equations (Second Edition),* Springer-Verlag, New York, 1995.

[131] Sneddon, I.N., *Fourier Transforms,* McGraw-Hill, New York, 1951.

[132] Sneddon, I.N., *Elements of Partial Differential Equations,* McGraw-Hill, New York, 1957.

[133] Sneddon, I.N., *The Use of Integral Transforms,* McGraw-Hill, New York, 1972.

[134] Sobolev, S.L., *Partial Differential Equations of Mathematical Physics,* Addison-Wesley, Reading, Massachusetts, 1964.

[135] Sokolnikoff, I.S., *Mathematical Theory of Elasticity,* McGraw-Hill, New York, 1956.

[136] Stakgold, I., *Green's Functions and Boundary Value Problems,* Wiley-Interscience, New York, 1979.

[137] Strauss, W.A., *Partial Differential Equations: An Introduction (Second Edition),* Wiley, 2007.

[138] Tikhonov, A.N. and Samarskii, A.A., *Equations of Mathematical Physics,* Macmillan, New York, 1963.

[139] Timoshenko., P. and Goodier, J.N., *Theory of Elasticity (Third Edition),* McGraw-Hill, 1970.

[140] Virkar, A.N., Jackson, T.B., and Cutler, R.B., *Thermodynamic and kinetic effects of oxygen removal on the thermal conductivity of aluminum nitride,* Journal of the American Ceramic Society, 72 (1989): 2031–2042.

[141] Whitham, G.B., *Linear and Nonlinear Waves,* John Wiley, New York, 1976.

[142] Whittaker, E.T. and Watson, G.N., *A Course of Modern Analysis (Fourth Edition),* Cambridge University Press, Cambridge, England, 1996.

Index

L^2 norm, 382
L^2-convergence, 38

Absolute convergence, 61, 65, 69, 123
Absolutely integrable, 70
Air pressure, 165
Air resistance, 159, 174
Analytic, 240
Annulus, 142, 154
Arbitrary intervals, 19, 52
Arbitrary periods, 19
Auto-correlation function, 321
Average value, 7

Beam, 185
Beam vibration, 181
Bessel's equation, 186, 216
Bessel's inequality, 35, 37
Best approximation in the mean, 87
Binomial theorem, 243
Blow-up, 359
Boundary value problems, 93, 158
Buckling, 226
Bump function, 408
Burger's equation, 94, 356

Canonical form, 368
Cauchy conditions, 366
Cauchy problem, 347, 375
Cauchy's inequality, 61, 62
Cesàro summation, 71
Characteristic curves, 346
Characteristic lines, 341
Characteristics, 367
Classical solution, 400, 405
Compact support, 406, 408
Complementary error function, 290, 296, 333
Complex form of Fourier series, 29, 33, 373
Complex Fourier integral representation, 285
Conservation law, 362
Constant coefficients, 337
Convection, 397
Convergence convergence, 61
Cylindrical coordinates, 139

D'Alembert's solution, 398, 423, 424
Damped wave equation, 165
Density, 159
Derivation of the wave equation, 158, 398
Derivatives in Fourier transform, 309
Differential operator, 94
Dilation, 383
Dirichlet boundary conditions, 114, 123, 133, 154, 230, 235, 237, 256, 295, 329, 388, 422, 423, 427
Dirichlet kernel, 45, 69
Discontinuity, 3
Discriminant, 62
Domain of dependence, 400
Double Fourier series, 174, 236
Duhamel's principle, 427, 432

Eigenfunctions, 101, 197, 248
Eigenfunctions expansion, 127, 251

Eigenvalues, 101, 197, 248
Elastic beam, 184
Electromagnetic waves, 158
Elliptic, 367
Energy conservation, 407
Energy estimate, 373, 375
Energy function, 379, 382, 407
Equilibrium temperature, 107, 110, 114, 116, 120, 180
Error function, 290, 333
Estimates, 400, 413
Euler formula, 30
Euler's dilogarithm function, 65
Even extension, 393, 424
Even function, 13, 287, 303
Even Legendre series, 253
Expansion in series, 192
Exponential decay, 386
Extreme value theorem, 203

Fejér kernel, 71
Fixed ends, 160, 414
Fluid dynamics, 133
Forced vibrating string, 167
Fourier Integral Theorem, 279
Fourier Convergence Theorem, 45
Fourier cosine series, 16, 22
Fourier cosine transform, 303
Fourier integrals, 271
Fourier pointwise convergence, 47, 52
Fourier series, 1, 19, 61
Fourier sine series, 16, 22, 418
Fourier sine transform, 303
Fourier transform, 286, 301
Frequency, 33, 171, 178
Frequency domain, 73, 301, 308, 312, 320
Fubini's theorem, 320
Fundamental solution, 328, 387

Gaussian function, 322
General Dirichlet problem, 138
Generalized Fourier series, 81, 87, 129, 193
Generalized series, 252
Generalized solution, 405
Gibb's phenomenon, 55, 58, 60, 61, 70
Gram-Schmidt process, 91
Gravity, 159
Green's formula, 127
Green's function, 328, 387

Harmonic, 139, 151
Heat equation, 93, 94, 97, 98, 101
Heat flow, 97, 123, 325
Heat flow on a disc, 142
Heat flow on an annulus, 142
Heat flow with a source, 128
Heat flux, 98
Heat kernel, 387, 392
Heat source, 98, 120, 126
Heat source time and space dependent, 123
Heaviside unit step function, 314, 383
Hilbert spaces, 91
Homogeneous, 96
Homogeneous boundaries, 102, 106
Homogeneous solution, 96
Hooke's law, 158
Hyperbolic, 367

Improper integral, 290
Infinite rod, 326
Infinite string, 297
Initial curve, 347, 367
Insulated, 98
Insulated boundary conditions, 101
Insulated homogeneous heat problem, 102, 106
Integral curve, 346, 352
Integral surface, 345, 347
Integrations of Fourier series, 73
Invariance properties of the heat equations, 383

Invariant, 383
Inverse Fourier transform, 301, 305
Irregular singular point, 216

Jacobian, 368
Jump discontinuity, 4, 8

KdV, 413
Kinematic viscosity, 296
Kinetic energy, 408

L'Hôpital's, 58
Lagrange's identity, 128
Laplace equation, 94, 139, 329
Laplace integrals, 276
Laplacian, 139, 213, 255
Laplacian in circular domain, 144
Laplacian in cylindrical, 140
Laplacian in polar, 140
Laplacian in spherical, 140
Laplacian on a disk, 146
Laplacian on an annulus, 144
Laplacian-spherical coordinates, 255
Left-hand derivative, 3
Left-hand limit, 3
Legendre equation, 240
Legendre function of the second kind, 243
Legendre polynomials, 240
Legendre series, 253
Legendre series extension, 252
Linear, 94
Linear combinations, 383

Mass conservation, 165
Maximum principle, 119
Mean, 7
Mean value, 151
Mean-square deviation, 88
Mean-square error, 38
Modulation, 308
Moment of inertia, 185
Momentum equation, 165

Navier–Stokes equation, 296
Neumann boundary conditions, 101, 116, 132, 165, 231, 237, 256, 296, 334, 392, 422, 424, 425
Newton second law, 159, 165
Non-characteristic, 347, 367
Non-homogeneous, 96, 97
Non-homogeneous boundary conditions, 116, 120, 167, 296, 391, 427
Non-homogeneous heat equation, 379
Non-homogeneous heat flow, 115, 116
Non-homogeneous heat problem, 388
Non-homogeneous Sturm–Liouville problem, 265
Non-homogeneous wave equation, 408
Non-insulated homogeneous heat problem, 110
Nonuniform String, 201
Norm, 82
Normalized auto-correlation function, 321, 322
Normed vector space, 82

Odd extension, 389, 390, 422
Odd function, 13, 287, 303
Odd Legendre series, 253
One-way wave equation, 356
Operations, 61
Order, 94
Orthogonal, 5, 81, 189
Orthogonality, 33, 35
Orthogonality in Bessel's functions, 227
Orthogonality in Legendre polynomials, 248
Orthonormal functions, 83

Parabolic, 367
Parametric Bessel's function, 227

Parametric Legendre equation, 261
Parseval's equation, 39
Parseval's formula, 34, 319
Partial differential equations, 93
Particular solution, 96, 384
Periodic boundary conditions, 15, 235, 257, 288
Periodic even extension, 14, 16, 22
Periodic odd extension, 16, 22, 416
Piecewise continuous, 3, 58
Piecewise smooth, 3, 66
Pointwise convergence, 38, 45, 62
Poisson integral, 149
Polar coordinates, 139, 291
Potential energy, 408
Principle of superposition, 95
Progressive wave, 402
Propagator, 387
Properties of Fourier transform, 308

Quantum mechanics, 255
Quasi-Linear equations, 345

Rademacher system, 87
Ratio test, 123
Rayleigh quotient, 196, 197
Regular point, 216
Regular singular point, 216
Regular Sturm–Liouville problem, 187
Resonance, 171
Riemann sum, 384
Riemann–Lebesgue Lemma, 279
Right-hand derivative, 3
Right-hand limit, 3
Robin boundary condition, 237
Robin boundary conditions, 205, 213, 231, 329, 407
Rodrigues' Formula, 243

Scaling, 383
Schrödinger's equation, 328
Schwartz inequality, 86
Second-order PDEs, 366
Self-adjoint, 227, 240
Semi-infinite rod, 294, 325
Semi-infinite string, 406
Separation of variables, 101, 431
Shock path, 362
Signal compression, 319
Signal processing, 1, 73, 81, 279, 286, 301, 319
Singular point, 216
Singular Sturm–Liouville problem, 192, 240, 255
Solitary wave, 413
Sound waves, 158
Source function, 387
Span, 83
Spatial translation, 383
Spherical coordinates, 139
Square error, 382
Square-integrable, 35
Stability, 381, 382, 401
stability, 401
Standing wave, 160, 171
Steady-state solution, 107, 110, 116, 118, 123, 143
Struck string, 164
Sturm–Liouville operator, 188
Sturm–Liouville problem, 101, 134, 161, 177, 186, 197

Tension force, 159
Theory of Fourier series, 35
Thermal conductivity, 98
Thermal diffusivity, 98
Thermal energy, 97
Thin pipe, 165
Time-dependent boundary conditions, 123
Traffic flow, 365
Transformation, 338
Transport equation, 94
Transverse vibration, 181
Traveling wave, 413

Uniform convergence, 38, 58, 61, 62, 65, 69
Uniformly bounded, 37
Uniformly bounded solution, 397
Unique solution for heat equation, 379
Unsteady diffusion equation, 296

Variable coefficients, 342
Vector space, 82
Vertical column, 226
Vibrating string, 160, 297, 414
Vibration of a beam, 158

Wave equation, 94, 158, 398
Wave equation on semi-infinite domain, 422
Wave motion, 297
Weierstrass M-test, 61, 63
Weight function, 5, 189